DISCARD

Myasthenia Gravis and Related Disorders

CURRENT CLINICAL NEUROLOGY

Daniel Tarsy, MD, Series Editor

Myasthenia Gravis and Related Disorders

Second Edition

Edited by

Henry J. Kaminski, MD

Department of Neurology & Psychiatry, Saint Louis University,
St. Louis, MO

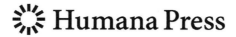 Humana Press

Editor
Henry J. Kaminski
Department of Neurology & Psychiatry
Saint Louis University
St. Louis, MO

ISBN: 978-1-58829-852-2   e-ISBN: 978-1-59745-156-7
DOI 10.1007/978-1-59745-156-7

Library of Congress Control Number: 2008934047

Printed on acid-free paper

springer.com

Preface

The objective of the second edition of *Myasthenia Gravis and Related Disorders* is identical to the first, to provide the clinician and the scientist a common source for understanding the complex disorder, myasthenia gravis (MG). Although the first edition appeared in 2003, there have been surprising advances in the field that recommend a new publication. The basic sciences continue to refine acetylcholine receptor structure and the dysregulation of the immune system that modulates MG; identification of the initial triggers still appears far away. MG remains a challenging disorder to recognize and treat, with patients frequently complaining of delays in diagnosis, complications of treatment, and poor response to therapies. Muscle-specific kinase antibody-associated MG has been clinically defined and the pathogenic nature of the antibodies has gained considerable support. New mutations as causes of congenital myasthenias have been discovered. Some incremental, therapeutic advancement has occurred and these are discussed. Since the first edition, the controversy of whether thymectomy benefits in the light of modern immunosuppressive therapies has motivated the implementation of an NIH-funded trial. Robert Paul and colleagues have expanded their original discussion of the most challenging aspects of MG, its impact on patients' psychological makeup. This subject is commonly neglected in MG texts. The new edition has been supplemented with chapters which discuss rigorous clinical assessments of patients for research trials and the epidemiology and genetics of MG.

Related to MG by clinical presentation or pathophysiology are Lambert-Eaton syndrome, congenital myasthenias, and toxic neuromuscular junction disorders. A chapter discusses neuromyotonia because of its similarity in autoimmune pathology to MG, and the occasional overlap of neuromyotonia with MG. These diseases are increasingly being defined at a molecular level. We do not use "myasthenic" when referring to the Lambert-Eaton syndrome. This is done in deference to Vanda Lennon and the late Edward Lambert, who prefer this terminology.

The second edition retains the "personal approach" of the authors regarding treatment. A challenge for MG investigation is its rarity, variability in clinical course, and heterogeneous nature, i.e., thymoma-associated, age-related differences, and clinical phenotypes (ocular vs. bulbar vs. generalized). Therefore, clinical trials are few, and robust evidence base is lacking and expert opinion often the bulwark of treatment recommendations. The individual opinion is particularly evident in the thymectomy, ocular myasthenia, and the treatment chapters.

I thank all the contributing authors, and am indebted, once again, to Humana Press for making the book a reality. Appreciation is extended to the National Eye Institute, the National Institute of Neurological Disorders and Stroke, and the Myasthenia Gravis Foundation of America that have supported my research as well as the work of many of the contributors. I also thank the Muscular Dystrophy Association for their support of neuromuscular research. To my patients who have given me more than I could ever return. Thank you.

I dedicate the work to a wonderful young boy, Adam, and his mother, Linda Kusner, from whom all good things flow.

Henry J. Kaminski, MD

Contents

Contributors

MARK A. AGIUS, MD • Department of Neurology, University of California, Davis, CA

RICHARD J. BAROHN, MD • Department of Neurology, University of Kansas Medical Center, Kansas City, KS

DAVID BEESON, PhD • Neurosciences Group, Weatherall Institute of Molecular Medicine, The John Racliffe and Oxford University, Oxford, UK

MICHAEL BENATAR, MBChB, DPhil • Department of Neurology, Emory University, Atlanta, GA

WEN-YU CHUANG, MD • Department of Pathology, Chang Gung University, Taoyuan, Taiwan

BIANCA M. CONTI-FINE, MD • Department of Biochemistry, Molecular Biology and Biophysics, University of Minnesota, St. Paul, MN

ROBERT B. DAROFF, MD • Case Western Reserve University and University Hospitals of Cleveland, Cleveland, OH

BRENDA DIETHELM-OKITA • Department of Biochemistry, Molecular Biology and Biophysics, University of Minnesota, St. Paul, MN

J. AMERICO FERNANDES FILHO, MD • Department of Neurological Sciences, University of Nebraska Medical Center, Omaha, NE

JAMES M. GILCHRIST, MD • Department of Clinical Neuroscience, Brown Medical School, Providence, RI

CHARLES M. HARPER, Jr., MD • Department of Neurology, Mayo Clinic Foundation, Rochester, MN

IAN HART, PhD, FRCP • Department of Neurological Science, Walton Centre for Neurology and Neurosurgery, Liverpool, UK

JAMES F. HOWARD, MD • Department of Neurology, The University of North Carolina, Chapel Hill, NC

ALFRED JARETZKI, III, MD • Department of Surgery, Columbia Presbyterian Medical Center and Columbia University, New York, NY

HENRY J. KAMINSKI, MD • Department of Neurology & Psychiatry, Saint Louis University, St. Louis, MO

BASHAR KATIRJI, MD, FACP • Department of Neurology, Case Western Reserve University and University Hospitals of Cleveland, Cleveland, OH

JAN B.M. KUKS, MD • Department of Neurology, University Hospital Groningen, Groningen, The Netherlands

JEREMIAH W. LANFORD, MD • Department of Neurology, University of Virginia, Charlottesville, VA

VANDA A. LENNON, MD, PhD • Departments of Neurology and Immunology, Mayo Clinic Foundation, Rochester, MN

JON M. LINDSTROM, PhD • Department of Neuroscience, Medical School of the University of Pennsylvania, Philadelphia, PA

ALEXANDER MARX, MD • Institute of Pathology, University Medical Center Mannheim and University of Heidelberg, Mannheim, Germany

MONICA MILANI • Department of Biochemistry, Molecular Biology and Biophysics, University of Minnesota, St. Paul, MN

LARRY L. MULLINS, PhD • Department of Pediatrics, University of Oklahoma Health Sciences Center, Oklahoma City, OK

NORMA OSTLIE, MS • Department of Biochemistry, Molecular Biology and Biophysics, University of Minnesota, St. Paul, MN

ROBERT H. PAUL, PhD • Department of Psychology, University of Missouri, St. Louis, MO

LAWRENCE H. PHILLIPS, II MD • Department of Neurology, University of Virginia, Charlottesville, VA

DAVID P. RICHMAN, MD • Department of Neurology, University of California, Davis, CA

ROBERT L. RUFF, MD, PhD • Neurology Service, Louis Stokes Veterans Affairs Medical Center; Departments of Neurology and Neurosciences, Case Western Reserve University Cleveland, OH

JOSHUA R. SONETT, MD • Department of Surgery, Columbia Presbyterian Medical Center and Columbia University, New York, NY

PHILIPP STRÖBEL, MD • Institute of Pathology, University Medical Center Mannheim and University of Heidelberg, Mannheim, Germany

JOSE I. SUAREZ, MD • Departments of Neurology and Neurosurgery, Baylor College of Medicine, Houston, TX

ANGELA VINCENT, MB, FRCPath • Institute of Molecular Medicine, John Radcliffe Hospital and Oxford University, Oxford, UK

EROBOGHENE E. UBOGU, MD • Department of Neurology, Baylor College of Medicine, Houston, TX

WEI WANG, MD, PhD • Department of Biochemistry, Molecular Biology and Biophysics, University of Minnesota, St. Paul, MN

GIL I. WOLFE, MD • Department of Neurology, University of Texas Southwestern Medical School, Dallas, TX

Series Editor Introduction

This second edition of *Myasthenia Gravis and Related Disorders* edited by Henry Kaminski provides a very useful new and updated source of information concerning this fascinating and complicated disease. Similar to the first volume published in 2003, basic information concerning structure and function of the neuromuscular junction and the interesting immunologic pathophysiology of the disease are followed by chapters concerning its varied clinical presentations; clinical and laboratory diagnosis; pharmacologic, immunologic, and surgical treatments; discussion of its impact on mood and quality of life; and the various myasthenic syndromes. Several new authors have been included, recent information concerning the epidemiology, genetics, and immunology of myasthenia gravis has been added, and most chapters have been expanded with new references.

The history of our understanding of myasthenia gravis is an interesting one. Early on, recognition of its abnormal neurophysiology led to a remarkably rational basis for treating patients with anticholinesterase drugs. Later on, treatment outcomes greatly improved due to increased attention to aggressive neurocritical care. Although thymectomy and corticosteroid treatment were first introduced in the 1940 s, the serendipitous discovery in the early 1970 s of the role of acetylcholine receptor autoantibodies in myasthenia gravis led to an increased and more rational emphasis on immunological treatments. Ironically, and perhaps due to obvious successes, the various available treatments have rarely been subjected to controlled clinical trials. The resulting differences in approaches to treatment which continue to the present time are thoughtfully and usefully reviewed in the clinical and treatment chapters of this volume. This book successfully reviews all aspects of myasthenia gravis and should provide impetus to continued research in order to develop new understandings in the years ahead.

Daniel Tarsy MD
Beth Israel Deaconess Medical Center
Harvard Medical School

Boston, Massachusetts

Color Plates

Color plates follow p. xxx

Neuromuscular Junction Physiology and Pathophysiology

Eroboghene E. Ubogu and Robert L. Ruff

1. INTRODUCTION

The aim of this chapter is to increase the reader's understanding of five factors: (1) the structure of the neuromuscular junction, (2) the mechanisms triggering release of vesicles of acetylcholine from the nerve terminal, (3) the two ion channels on the postsynaptic membrane (Na^+ channels and acetylcholine receptors) that are critical for converting the chemical signal for the nerve terminal into a propagated action potential on the muscle fiber, (4) the safety factor for neuromuscular transmission and characteristics of the neuromuscular junction contribute to the safety factor and (5) the mechanisms by which disease compromises the safety factor, leading to neuromuscular transmission failure.

2. MOTOR NERVE PROPERTIES

Skeletal muscle fibers are innervated by large motor neurons of the anterior horn of the spinal cord [1]. Each anterior horn cell gives rise to a single large myelinated motor nerve fiber or axon. Action potentials travel along motor axons by saltatory conduction, a process by which action potentials jump from node of Ranvier to node of Ranvier, with little current leaving the axon in the internodal region. The structure of the axons optimizes saltatory conduction in two ways:

(1) The internodal regions of the axons are covered by insulating layers of myelin produced by Schwann cells. The myelin reduces the current loss across the internodal region by increasing the effective transmembrane resistance and decreasing the capacitance between the axon and the extracellular space [2, 3].
(2) The nodes of Ranvier have high concentrations of sodium (Na^+) channels, which produce the depolarizing current of the action potential. Vertebrate nodes of Ranvier contain about 2000 channels/μm^2 [2, 4]. The high density of Na^+ channels reduces the threshold for generating an action potential [4, 5]. In addition, nodes of Ranvier have few potassium (K^+) channels [5]. Since outward K^+ currents counter the depolarizing action of Na^+ currents, the paucity of K^+ channels at the nodes of Ranvier minimizes inhibition of the action potential and enables the action potential to rapidly propagate along the axon at rates >50 m/s.

2.1. Distal Motor Nerve Properties

The distal portion of each motor nerve fiber branches into 20–100 thinner fibers. In mature mammalian muscle, each distal motor nerve fiber innervates a single muscle fiber with one nerve terminal (known as *en plaque* morphology). However, amphibian, reptile and certain mammalian muscles (extraocular muscles, tensor tympani, stapedius, some laryngeal muscles and the some muscles in the tongue) [6, 7] also contain fibers with multiple synaptic contacts, with the neuromuscular junction having en grappe synapses (each en grappe endplate has a diameter of 10–16 μm, compared to about 50 μm with en plaque endplates) [3, 8, 9, 10]. The muscle fibers

From: *Current Clinical Neurology*
Myasthenia Gravis and Related Disorders
Edited by: H.J. Kaminski, DOI 10.1007/978-1-59745-156-7_1, © Humana Press, New York, NY

innervated by a single motor nerve axon are called the motor unit. The terminal motor nerve branches are up to 100 μm long and are unmyelinated [3].

The unmyelinated terminal motor nerve branches contain delayed rectifier and inward rectifier K^+ channels as well as Na^+ channels [5, 11]. Therefore, the amplitude and duration of the action potential in the terminal nerve fibers are controlled by K^+ channels as well as by Na^+ channels. The nerve terminal has few if any Na^+ channels; hence the action potential passively propagates into the nerve terminal. The lack of nerve terminal Na^+ channels and the presence of K^+ channels prevent action potentials from reverberating among the distal nerve branches. Neuromyotonia, also called Isaac's syndrome, is a disorder of neuromuscular transmission in which a single action potential propagating down a motor nerve produces repeated action potentials in the nerve terminals [12]. In many instances, neuromyotonia results from autoantibodies that interfere with the functioning of nerve terminal-delayed rectifier K^+ channels [12]. Neuromyotonia demonstrates that the nerve terminal-delayed rectifier K^+ channels regulate nerve terminal membrane excitability [13].

2.2. The Nerve Terminal

Acetylcholine (ACh) is stored in vesicles within the nerve terminal (Figs. 1 and 2) [3, 14]. The ACh-containing vesicles are aligned near release sites, called active zones, in the nerve terminal, where the vesicles will fuse with the presynaptic nerve terminal membrane [3]. The release sites are positioned over the clefts between the tops of the secondary synaptic folds of the postsynaptic muscle membrane [3, 15, 16]. Transmitter release requires Ca^{+2} influx. The Ca^{+2} channels responsible for ACh release are P/Q type [17]. However, N-type calcium channels are probably also present on mammalian motor nerve terminals [18, 19, 20]. The Ca^{2+} channels which trigger ACh release are distributed as two parallel double rows with approximately five channels per row, a spacing of about 20 nm between rows and a spacing of about 60 nm between the double rows of channels at the active zone [21, 22].

2.2.1. Role of Ca^{2+} Channels in Transmitter Release

High concentration of Ca^{2+} channels at the active zones enables the Ca^{2+} concentration to reach 100–1000 μM quickly in the nerve terminal regions where vesicle fusion occurs [15, 16]. A normal nerve terminal action potential does not fully activate the nerve terminal Ca^{2+} channels

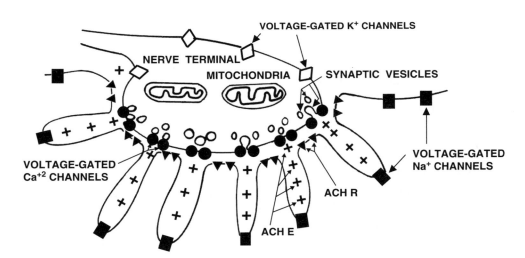

Fig. 1. Schematic representation of the neuromuscular junction. Details are discussed throughout the chapter.

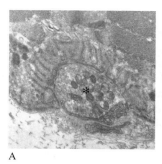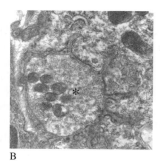

A B

Fig. 2. Electron micrograph of (**A**) normal mouse diaphragm neuromuscular junction and (**B**) mouse diaphragm from a mouse with experimentally induced myasthenia gravis. The asterisk marks the nerve terminal. Note the small round structures in the nerve terminal, which are the synaptic vesicles. In B, there is a loss of synaptic folds and within the synaptic cleft one can appreciate globular material, which likely represents degenerated junctional folds (Courtesy of Dr. Henry Kaminski.)

because the duration of the action potential is <1 ms, and the nerve terminal Ca^{2+} channels are activated with a time constant of >1.3 ms [19]. Increasing the duration of the nerve terminal action potential by blocking delayed rectifier potassium channels with tetraethylammonium (TEA) or 3,4-diaminopyridine (3,4-DAP) will increase calcium entry and increase ACh release [19, 23, 24]. In Lambert–Eaton syndrome (LES), antibodies directed against nerve terminal Ca^{2+} channels reduce the Ca^{2+} channel complement [25, 26, 27, 28, 29, 30, 31]. Neuromuscular transmission is impaired owing to a deficiency of vesicles released by the nerve terminal in response to an action potential. Treatment with 3,4-DAP improves neuromuscular transmission in LES by increasing the time that Ca^{2+} channels are stimulated resulting in increased Ca^{2+} entry which partially compensates for the deficiency of Ca^{2+} channels [32].

The approximation of synaptic vesicles and the presynaptic nerve terminal membrane may be opposed by electrostatic forces owing to the similar polarity of surface charges on the nerve terminal and vesicle membranes. Ca^{2+} may bind to the membrane surfaces and neutralize the negative surface charges, thereby removing a restraint to membrane fusion [33]. Ca^{2+} may also open Ca^{2+}-activated cationic channels, and the entry of cations may also reduce the negative surface charges on the synaptic vesicle and nerve terminal membranes [34]. Ca^{2+} also triggers conformational changes in large molecules that allow synaptic vesicles to detach from the cytoskeleton and that actively trigger membrane fusion [35]. Ca^{2+} entry may trigger phosphorylation of proteins including synaptotagmin [36, 37]. Synexin and members of the SNAP family of proteins are also modified by Ca^{2+} [38].

2.2.2. Synaptic Vesicle Fusion

Synaptic vesicle fusion is a complex process that involves conformational changes in multiple proteins on the vesicle and nerve terminal plasma membranes. The precise sequence of events leading to release of synaptic vesicle contents has not been completely elucidated. However, over the last decade, some of the molecular mechanisms involved in the process have been characterized and the events leading to synaptic vesicle release continue to be defined [39, 40, 41]. Current conceptualization suggests that vesicles need to go through a process called docking prior to fusion, a process that brings synaptic vesicles in proximity with the nerve terminal membrane. Next, the vesicles undergo priming that makes them competent to respond to the Ca^{2+} signal.

Before docking occurs, syntaxin-1 binds to munc18-1 and synaptobrevin binds to a synaptic vesicle protein, synaptophysin. The interaction of these proteins inhibits the formation of the docking complex. During docking, munc18-1 dissociates from the syntaxin-1 and synaptophysin from synaptobrevin allowing the synaptic core complex to form. Three proteins, two on the plasma membrane (syntaxin-1 and synaptic vesicle associate protein 25 or SNAP 25) and one on the synaptic vesicle membrane (synaptobrevin), are thought to form a docking complex. These three proteins are SNARE proteins, characterized by the 70-residue SNARE motif [10]. *N*-ethylmaleimide sensitive factor (NSF) and α-soluble NSF attachment protein associate to form a fusion complex with the docking proteins. NSF, which is an ATPase, crosslinks multiple core complexes into a network and ATP hydrolysis leads to hemifusion of the vesicle and presynaptic membranes. ATP hydrolysis appears to occur before Ca^{2+} influx. Synaptotagmin, located on the synaptic vesicle membrane, probably acts as the Ca^{2+} sensor. Although the Ca^{2+}-binding capacity of synaptotagmin is low, it increases significantly when they bind to phospholipid membranes [10].

How synaptotagmin serves to trigger the rapid discharge of synaptic vesicle contents is not known. The cytoplasmic portion of synaptotagmin includes two regions with high homology to the Ca^{2+}- and phospholipid-binding domains of a protein kinase. Synaptotagmin probably binds phospholipid of the plasma membranes and syntaxin. Binding of Ca^{2+} to synaptotagmin may alter interactions of synaptotagmin with membrane lipids and syntaxin transforming allowing the membranes to fully fuse. Also, Ca^{2+} may bind to the membrane surfaces and neutralize the negative surface charges, thereby removing a restraint to membrane fusion [33]. Exocytosis is somewhat inefficient in that only one in every 3–10 pulses of calcium leads to exocytosis and only one of many docked vesicles fuse.

After releasing their contents to the synaptic cleft, synaptic vesicle membrane appears to recycle by three processes [42, 43]. One involves the synaptic vesicle membrane fusing completely with the plasma membrane, followed by recycling of membrane components via clathrin-dependent mechanisms. After vesicular reuptake, the clathrin-coated vesicles shed their coats and translocate into the nerve terminal interior. The synaptic vesicle membrane fuses with endosomes and new vesicles bud from the endosomes. The new vesicles accumulate ACh and other substances into the vesicles by active transport and translocate back to the active zone either by diffusion or by a cytoskeletal transport process. Several of the synaptic vesicle-associated proteins are targets for proteolytic cleavage by botulinum toxins.

Synaptic vesicle membrane recycling may also occur by more rapid methods. The "kiss-and-run" mechanism involved endocytosis of fused vesicular membrane, and the vesicles are recycled independent of endosomes. The "kiss-and-stay" process involves the transient opening and closing of a fusion pore and recycling in proximity to the active zone. These fast recycling mechanisms have been investigated in highly active central nervous system nerve terminals and may not be as prominent at most neuromuscular junctions. Ocular motor neurons fire at high rates and the "kiss-and-run" mechanism may be of particular importance for these junctions to function [10].

The abundant mitochondria present in the nerve terminal play an important role in buffering intracellular Ca^{2+} in addition to energy production required to facilitate synaptic release, neurotransmitter synthesis and transporters of ions and ACh to load synaptic vesicles [10, 40, 44]. During repetitive discharges of action potentials, there is a rapid increase in cytosolic Ca^{2+} concentration, followed by a slower increase. Blockade of mitochondrial Ca^{2+} uptake results in rapid increases in cytosolic Ca^{2+} throughout the duration of the stimulus [45, 46, 47]. Post-tetanic facilitation, the enhancement of synaptic transmission following high-frequency stimulation, is also mediated by the slow Ca^{2+} release over minutes [48].

3. THE SYNAPTIC CLEFT

ACh diffuses across the synaptic cleft to activate ACh receptors (AChR). Each synaptic vesicle fusion releases about 10,000 ACh molecules into the synaptic cleft [49]. ATP is also released by synaptic vesicle fusion and the released ATP may modulate transmitter release of postsynaptic transmitter sensitivity [50]. An action potential propagating into the nerve terminal stimulates the fusion of between 50 and 300 synaptic vesicles (i.e., the normal quantal content is between 50 and

300) [51]. Diffusion of ACh across the synaptic cleft is very rapid owing to the small distance to be traversed and the relatively high diffusion constant for Ach [52].

3.1. Synatpic Cleft Acetylcholine Esterase

Acetylcholine esterase (AChE) in the basal lamina of the postsynaptic membrane accelerates the decline in concentration of ACh in the synaptic cleft as does diffusion of ACh out of the cleft [53]. Inactivation of AChE prolongs the duration of action of ACh on the postsynaptic membrane and slows the decay of the ACh-induced endplate current [54]. The concentration of AChE is approximately 3000 molecules/μm^2 of postsynaptic membrane [53], which is about five- to eightfold lower than the concentration of AChRs [55]. The concentration of AChE in the secondary synaptic folds is sufficiently high so that most of the ACh entering a synaptic cleft is hydrolyzed. Consequently, the secondary synaptic folds act like sinks that terminate the action of ACh and prevent AChRs from being activated more than once in response to released ACh [56]. The space between the nerve terminal and the postsynaptic membrane is about 50 nm^3.

3.2. Alteration of AChE in Neuromuscular Diseases

In mammalian nerve and muscle, abnormal cholinergic transmission, as seen in myasthenia gravis (MG) and experimental autoimmune myasthenia gravis (EAMG) in rats, has been shown to induce enhanced transcription and shifted alternative splicing of AChE pre-mRNA with resultant production of the normally rare, "readthrough" AChE variant (AChE-R) [57, 58]. The synaptic AChE described previously forms multimers that are attached to the postsynaptic membrane through the proline-rich PRiMA anchor, while AChE-R appears as soluble monomers that lack the carboxyl-terminal cysteine that is essential for linkage to PRiMA [59, 60]. AChE-R is believed to be involved in nonsynaptic hydrolysis of ACh and synaptic morphogenesis [58].

In the immediate response to acute stress or acetylcholine esterase exposure, elevated AChE-R attenuates the initial hyperexcitation [61]. However, continued accumulation of AChE-R may be detrimental, as it prolongs the state of cholinergic impairment [62, 63], increases adhesive and noncatalytic activities of AChE [64, 65] and is associated with muscle pathologies [66], thereby predicting additional long-term effects under AChE-R overproduction. In moribund EAMG rats, 4 week daily oral administration of specific antisense oligonucleotide sequences that selectively lower AChE-R in blood and muscle without affecting AChE improved survival, neuromuscular strength and clinical status [58], further highlighting the pathologic relevance of AChE-R and the potential for mRNA-targeted therapy for disorders of chronic cholinergic dysfunction.

3.3. Extracellular Matrix in the Synaptic Cleft

The extracellular matrix in the neuromuscular junction is a complex collection of proteins that regulate the synthesis of postsynaptic proteins as well as the concentration of ACh. The endplate basement membrane is enriched with collagen IV (α2-, α4- and α5 chains) and contains several forms of laminin (laminin-4, laminin-9 and laminin-11), all of which bind to α-dystroglycan in the endplate membrane (Fig. 3 and Color Plate 1, following p. x). Laminin-4 also binds to integrin. The laminin family of proteins forms a complex network in the synaptic space that anchors other extracellular matrix proteins including agrin, perlecan and entactin. The collagen-tailed form of AChE in the synapse binds to perlecan, which in turn can bind to α-dystroglycan.

In addition to binding laminin and perlecan, α-dystroglycan also binds agrin, integrin and the MASC (myotube-associated specificity component)/MuSK (muscle-specific kinase) complex. Agrin, MASC and MuSK are associated with the formation and maintenance of AChR clustering [35, 67, 68]. Rapsyn is the molecule that specifically links AChRs [69]. The high synthesis of AChR component subunits at the neuromuscular junction is due in part to ARIA (AChR-inducing activity), a molecule that is released by the nerve terminal. ARIA activates ErbB receptor tyrosine

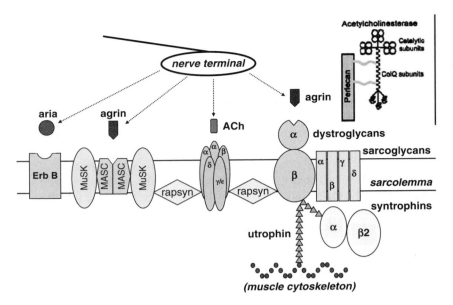

Fig. 3. Schematic representation of postsynaptic region of the neuromuscular junction. Ach, acetylcholine; ARIA, acetylcholine receptor-inducing activity; MASC, myotube-associated specificity component; MuSK, muscle-specific kinase. (*see* Color Plate 1, following p. 311)

kinases in the postsynaptic membrane. The ErbB receptors regulate the expression of AChR subunits by the subsynaptic nuclei.

3.4. ACh-Binding Protein

Three to five Schwann cells form a cap in close apposition to the nerve terminal and their processes extend into the synaptic cleft, coming with a few micrometers of the active zones [70]. Schwann cells play critical roles in several aspects of neuromuscular junction function and formation, including modulation of synaptic transmission, nerve terminal growth and maintenance, axonal sprouting and nerve regeneration [71, 72]. Recent experiments utilizing co-cultures of *Lymnaea* cholinergic neurons and specific subsets of oligodendroglial and Schwann cells demonstrated that the glial cells altered the ability of cholinergic neurons to generate excitatory postsynaptic potentials [73]. As a consequence, a 210 residue protein, with some sequence similarity with subunits of the Cys-loop family of ligand-gated ion channels, called ACh-binding protein (AChBP) was identified. Under appropriate conditions, AChBP causes suppression of cholinergic synaptic transmission [73].

Under conditions of active presynaptic transmitter release, AChBP may diminish or terminate the continuing acetylcholine response or raise basal AChBP concentration to the point that subsequent responses to acetylcholine are decreased. This process may occur by acetylcholine activation of both postsynaptic AChRs and AChRs located on synaptic glial cells, augmenting AChBP release with subsequent increase in its synaptic cleft concentration with reduction in acetylcholine available to bind to postsynaptic AChRs [10].

4. POSTSYNAPTIC MEMBRANE SPECIALIZATION

The postsynaptic membrane area is increased by folding into secondary synaptic clefts or folds. AChRs are concentrated at the tops of the secondary synaptic folds and are firmly anchored to the dystrophin-related protein complex through rapsyn (Fig. 3) [69, 74]. Rapsyn is important in clustering AChRs at the endplate during synaptogenesis and rapsyn-deficient transgenic mice do not cluster AChRs, utrophin and other dystrophin-related complex proteins [74, 75, 76, 77]. Clusters of AChRs are connected to the cytoskeleton via associations with dystroglycan and sarcoglycan protein complexes [35, 67]. The dystroglycan and sarcoglycan complexes connect to utrophin which in turn connects to cytoskeleton by binding to actin (Fig. 3).

Utrophin and dystrobrevin also connect to β1-syntrophin and β2-syntrophin, which in turn associates with nitric oxide synthase (NOS) [78, 79]. NOS produces the free radical gas nitric oxide (NO) that participates in signaling of many cellular processes. The presence of NOS at the neuromuscular junction suggests that NO could diffuse from its site of synthesis to affect target proteins in the nerve and muscle. Na^+ channels are concentrated in the depths of the secondary synaptic clefts (Fig. 1) [74, 80, 81, 82, 83, 84, 85]. Both the Na^+ channels and the AChRs are rigidly located in the endplate membrane. Na^+ channels are locked in the membrane by their associations with ankyrin, the sarcoglycan complex, the dystroglycan complex, dystrobrevins and dystrophin/utrophin.

4.1. How the NMJ Accommodates to Muscle Contraction

Given the complex molecular composition of the nerve terminal, the extracellular matrix and the endplate membrane and the precise alignment of the active zones on the nerve terminal with the clefts between the endplate synaptic folds, how does the neuromuscular junction accommodate to muscle fiber stretching and contracting? Surprisingly, the high efficiency of neuromuscular transmission is maintained during muscle stretch and contraction. The safety factor for neuromuscular transmission does not change for muscles at different lengths from 80 to 125% of the resting length [86]. The constancy of neuromuscular transmission is accomplished by the neuromuscular junction remaining rigid so that the endplate membrane does not deform when the muscle fiber changes length. The change in muscle fiber length is accomplished by folding and unfolding of the extrajunctional membrane while the endplate region remains rigid. The extreme stiffness of the endplate membrane adds mechanical and electrical stability to the neuromuscular junction, which enables the active zones on the nerve terminal to remain precisely aligned with the clefts of the synaptic folds and enables the safety factor for neuromuscular transmission to remain constant during muscle activity.

4.2. Postsynaptic Na^+ and AChR Channels

The density of AChRs at the endplate is about 15,000–20,000 receptors/μm^2 [2, 55]. Away from the endplate, the concentration of AChRs is about 1000-fold lower with a slight increase in AChR density at the tendon ends of the muscle fibers [87]. The relatively high concentration of AChRs at the endplate in part results because the muscle nuclei near the endplate preferentially express genes of the AChR subunits [75, 77]. AChRs continually turn over, with old receptors internalized and degraded. The removed receptors are replaced with new receptors. The AChRs are not recycled. Early in development of skeletal muscle the half-life of AChRs is short, about 13–24 h [88]. At a mature endplate, the half-life of AChRs is about 8–11 days. Crosslinking of receptors by AChR autoantibodies dramatically shortens AChR half-life by accelerating internalization of the receptors [89, 90].

The AChRs are firmly anchored to the cytoskeleton [74, 75, 76, 77]. Na^+ channels are also concentrated in the endplate region [74, 80, 81, 82, 83, 84, 85]. The increased density of Na^+ channels raises the safety factor for neuromuscular transmission [82, 83, 85, 91, 92, 93]. In addition, a higher concentration of Na^+ channels may be needed because at the endplate, two action potentials must be generated, one traveling toward each tendon end of the muscle fiber. The density of Na^+ channels varies with fiber type [83, 84, 85]. Fast-twitch fibers have about 500–550 Na^+ channels/μm^2 on the endplate membrane and slow-twitch fibers have about

100–150 Na$^+$ channels/μm^2. In acquired autoimmune MG, antibodies directed at AChRs attack the endplate membrane. The antibodies fix complement resulting in loss of endplate membrane. Na$^+$ channels as well as AChRs are lost from the endplate in MG, and the loss of endplate Na$^+$ channels reduces the safety factor for neuromuscular transmission [93].

The size of the depolarization of the endplate membrane produced by a single motor nerve action potential (endplate potential) is determined by the product of the number of vesicles (or quanta) of ACh released (quantal content) and size of the endplate depolarization to a single vesicle (quantal size), with an adjustment for the AChR reversal potential [2]. Diseases that interfere with ACh release (such as LES) reduce the quantal content. Conditions that reduce the sensitivity of the postsynaptic membrane to ACh (such as MG) reduce quantal size.

5. SAFETY FACTOR FOR NEUROMUSCULAR TRANSMISSION

The safety factor (SF) for neuromuscular transmission can be defined as

$$SF = (EPP)/(E_{AP} - E_M)$$

where EPP is the endplate potential amplitude, E_M is the membrane potential and E_{AP} is the threshold potential for initiating an action potential [94]. Several factors contribute to increase the safety factor for neuromuscular transmission of fast- compared to slow-twitch mammalian skeletal muscle fibers [95]. The nerve terminal morphology and transmitter release properties of axons innervating fast- and slow-twitch fibers are different. The quantal contents of synapses on rodent fast-twitch fibers are larger than for slow-twitch fibers [95, 96]. The postsynaptic sensitivities of fast-twitch fibers are also greater than slow-twitch fibers. Endplates on fast-twitch fibers depolarize more in response to quantitative iontophoresis of ACh [97]. Fast-twitch endplates have a higher concentration of Na$^+$ channels compared to slow-twitch endplates [82, 83, 84, 85, 92, 93]. The increased Na$^+$ current on fast-twitch fibers may be also needed because fast-twitch fibers require larger depolarizations to initiate contraction compared with slow-twitch fibers [98].

The differences in synaptic transmission for fast- and slow-twitch fibers may be in response to the different functional properties of fast- and slow-twitch motor units. Slow-twitch motor units in mammals *in vivo* are tonically active at slow rates whereas fast-twitch motor units are phasically active at high rates [99]. Under these conditions, transmitter depletion and other factors may not appreciably compromise neuromuscular transmission for slow-twitch fibers while fast-twitch fibers may suffer from reduction in the endplate potential amplitude [95, 100, 101, 102]. The effective safety factor for slow-twitch motor units from rat soleus muscle at a steady-state firing rate of 10 Hz is about 1.8 [95]. The safety factor for neuromuscular transmission for fibers from the fast-twitch rat extensor digitorum longus muscle stimulated at 40 Hz drops from 3.7 for the first stimulation to 2.0 after 200 stimuli [95]. Differences in safety factor among muscles may contribute to the varied clinical manifestations observed among patients with neuromuscular transmission disorders. The compromise of safety factor underlies all neuromuscular transmission disorders, which are detailed in this text.

REFERENCES

1. Burke RE. The structure and function of motor units. In: Karpati G, Hilton-Jones D, Griggs RC, eds. Disorders of voluntary muscle. 7th ed. Cambridge: Cambridge University Press; 2001:3–25.
2. Ruff RL. Ionic Channels II. Voltage- and agonist-gated and agonist-modified channel properties and structure. Muscle Nerve 1986;9:767–86.
3. Salpeter MM. The vertebrate neuromuscular junction. New York: Alan R. Liss, Inc.; 1987.
4. Ruff RL. Ionic Channels: I. The biophysical basis for ion passage and channel gating. Muscle Nerve 1986;9:675–99.
5. Black JA, Kocsis JD, Waxman SG. Ion channel organization of the myelinated fiber. Trends Neurosci 1990;13:48–54.

6. Han Y, Wang J, Fischman DA, Biller HF, Sanders I. Slow tonic muscle fibers in the thyroarytenoid muscles of human vocal folds; a possible specialization for speech. Anat Rec 1999;256(2):146–57.

7. Kaminski HJ, Li Z, Richmonds C, Ruff RL, Kusner L. Susceptibility of ocular tissues to autoimmune diseases. Ann NY Acad Sci 2003;998:362–74.

8. Morgan DL, Proske U. Vertebrate slow muscle: its structure, pattern of innervation, and mechanical properties. Physiol Rev 1984;64(1):103–69.

9. Ruff RL, Kaminski HJ, Maas E, Spiegel P. Ocular muscles: physiology and structure-function correlations. Bull Soc Belg Ophthalmol 1989;237:321–52.

10. Hughes BW, Kusner LL, Kaminski HJ. Molecular architecture of the neuromuscular junction. Muscle Nerve 2006;33(4):445–61.

11. Reid G, Scholz A, Bostock H, Vogel W. Human axons contain at least five types of voltage-dependent potassium channel. J Physiol 1999;518:681–96.

12. Shillito P, Molenaar PC, Vincent A, et al. Acquired neuromyotonia: evidence for autoantibodies directed against K + channels of peripheral nerves. Ann Neurol 1995;38:714–22.

13. Nagado T, Arimura K, Sonoda Y, et al. Potassium current suppression in patients with peripheral nerve hyperexcitability. Brain 1999;122:2057–66.

14. Katz B. Nerve muscle and synapse. New York: McGraw-Hill Co.; 1966.

15. Augustine GJ, Adler EM, Charlton MP. The calcium signal for transmitter secretion from presynaptic nerve terminals. Ann NY Acad Sci 1991;635:365–81.

16. Smith SJ, Augustine GJ. Calcium ions, active zones and synaptic transmitter release. Trends Neurosci 1988;10:458–64.

17. Protti DA, Sanchez VA, Cherksey BD, Sugimori M, Llinas RR, Uchitel OD. Mammalian neuromuscular transmission blocked by funnel web toxin. Ann NY Acad Sci 1993;681:405–7.

18. Catterall WA, De Jongh K, Rotman E, et al. Molecular properties of calcium channels in skeletal muscle and neurons. Ann NY Acad Sci 1993;681:342–55.

19. Stanley EF. Presynaptic calcium channels and the transmitter release mechanism. Ann NY Acad Sci 1993; 681:368–72.

20. Wray D, Porter V. Calcium channel types at the neuromuscular junction. Ann NY Acad Sci 1993;681:356–67.

21. Engel AG. Review of evidence for loss of motor nerve terminal calcium channels in Lambert-Eaton Myasthenic Syndrome. Ann NY Acad Sci 1991;635:246–58.

22. Pumplin DW, Reese TS, Llinas R. Are the presynaptic membrane particles calcium channels? Proc Natl Acad Sci USA 1981;78:7210–3.

23. Lindgren CA, Moore JW. Calcium current in motor nerve endings of the lizard. Ann NY Acad Sci 1991;635:58–69.

24. Stanley EF, Cox C. Calcium channels in the presynaptic nerve terminal of the chick ciliary ganglion giant synapse. Ann NY Acad Sci 1991;635:70–9.

25. Leys K, Lang B, Johnston I, Newsom-Davis J. Calcium channel autoantibodies in the Lambert-Eaton myasthenic syndrome. Ann Neurol 1991;29:307–414.

26. Lambert EH, Eaton LM, Rooke ED. Defect of neuromuscular transmission associated with malignant neoplasm. Am J Physiol 1956;187:612–3.

27. Motomura M, Johnston I, Lang B, Vincent A, Newsom-Davis J. An improved diagnostic assay for Lambert-Eaton myasthenic syndrome. J Neurol Neurosurg Psychiatr 1995;58:85–7.

28. Nagel A, Engel AG, Lang B, Newsom-Davis J, Fukuoka T. Lambert-Eaton syndrome IgG depletes presynaptic membrane active zone particles by antigenic modulation. Ann Neurol 1988;24:552–8.

29. Lang B, Johnston I, Leys K, et al. Autoantibody specificities in Lambert-Eaton myasthenic syndrome. Ann NY Acad Sci 1993;681:382–91.

30. Viglione MP, Blandino JKW, Kim YI. Effects of Lambert-Eaton syndrome serum and IgG on calcium and sodium currents in small-cell lung cancer cells. Ann NY Acad Sci 1993;681:418–21.

31. Rosenfeld MR, Wong E, Dalmau J, et al. Sera from patients with Lambert-Eaton Myasthenic Syndrome recognize the b-subunit of Ca^{2+} channel complexes. Ann NY Acad Sci 1993;681:408–11.

32. McEvoy KM, Windebank AJ, Daube JR, Low PA. 3,4-Diaminopyridine in the treatment of Lambert-Eaton myasthenic syndrome. N Engl J Med 1989;321:1567–71.

33. Niles WD, Cohen FS. Video-microscopy studies of vesicle-planar membrane adhesion and fusion. Ann NY Acad Sci 1991;635:273–306.

34. Ehrenstein G, Stanley EF, Pocotte SL, Jia M, Iwasa KH, Krebs KE. Evidence of a model of exocytosis that involves calcium-activated channels. Ann NY Acad Sci 1991;635:297–306.

35. Vincent A. The neuromuscular junction and neuromuscular transmission. In: Karpati G, Hilton-Jones D, Griggs RC, eds. Disorders of voluntary muscle. 7th ed. Cambridge: Cambridge University Press; 2001:142–67.

36. Perin MS, Fried VA, Mignery G, Jahn R, Südhof TC. Phospholipid binding by a synaptic vesicle protein homologous to the regulatory region of protein kinase C. Nature 1990;345:260–3.

37. Perin MS, Brose N, Jahn R, Südhof TC. Domain structure of synaptotagmin (p65). J Biol Chem 1991;266:623–9.

38. Zimmerberg J, Curran M, Cohen FS. A lipid/protein complex hypothesis for exocytotic fusion pore formation. Ann NY Acad Sci 1991;681:307–17.
39. Boonyapisit K, Kaminski HJ, Ruff RL. Disorders of neuromuscular junction ion channels. Am J Med 1999;106(1):97–113.
40. Poage RE, Meriney SD. Presynaptic calcium influx, neurotransmitter release, and neuromuscular disease. Physiol Behav 2002;77(4–5):507–12.
41. Sudhof TC. The synaptic vesicle cycle. Annu Rev Neurosci 2004;27:509–47.
42. Galli T, Haucke V. Cycling of synaptic vesicles: how far? How fast! Sci STKE 2004;2004(264):re19.
43. Gandhi SP, Stevens CF. Three modes of synaptic vesicular recycling revealed by single-vesicle imaging. Nature 2003;423(6940):607–13.
44. Erecinska M, Silver IA. Ions and energy in mammalian brain. Prog Neurobiol 1994;43(1):37–71.
45. David G, Barrett EF. Mitochondrial Ca2+ uptake prevents desynchronization of quantal release and minimizes depletion during repetitive stimulation of mouse motor nerve terminals. J Physiol 2003;548(Pt 2):425–38.
46. David G, Talbot J, Barrett EF. Quantitative estimate of mitochondrial [Ca2+] in stimulated motor nerve terminals. Cell Calcium 2003;33(3):197–206.
47. Talbot JD, David G, Barrett EF. Inhibition of mitochondrial Ca2+ uptake affects phasic release from motor terminals differently depending on external [Ca2+]. J Neurophysiol 2003;90(1):491–502.
48. Tang Y, Zucker RS. Mitochondrial involvement in post-tetanic potentiation of synaptic transmission. Neuron 1997;18(3):483–91.
49. Miledi R, Molenaar PC, Polak RL. Electrophysiological and chemical determination of acetylcholine release at the frog neuromuscular junction. J Physiol (Lond) 1983;334:245–54.
50. Etcheberrigaray R, Fielder JL, Pollard HB, Rojas E. Endoplasmic reticulum as a source of Ca^{2+} in neurotransmitter secretion. Ann NY Acad Sci 1991;635:90–9.
51. Katz B, Miledi R. Estimates of quantal content during chemical potentiation of transmitter release. Proc R Soc Lond 1979;205:369–78.
52. Land BR, Harris WV, Salpeter EE, Salpeter MM. Diffusion and binding constants for acetylcholine derived from the falling phase of miniature endplate currents. Proc Natl Acad Sci USA 1984;81:1594–8.
53. McMahan UJ, Sanes JR, Marshall LM. Cholinesterase is associated with the basal lamina at the neuromuscular junction. Nature 1978;271:172–4.
54. Katz B, Miledi R. The binding of acetylcholine to receptors and its removal from the synaptic cleft. J Physiol (Lond) 1973;231:549–74.
55. Land BR, Salpeter EE, Salpeter MM. Kinetic parameters for acetylcholine interaction in intact neuromuscular junction. Proc Natl Acad Sci USA 1981;78:7200–4.
56. Colquhoun D, Sakmann B. Fast events in single-channel currents activated by acetylcholine and its analogues at the frog muscle end-plate. J Physiol (Lond) 1985;369:501–57.
57. Soreq H, Seidman S. Acetylcholinesterase – new roles for an old actor. Nat Rev Neurosci 2001;2(4):294–302.
58. Brenner T, Hamra-Amitay Y, Evron T, Boneva N, Seidman S, Soreq H. The role of readthrough acetylcholinesterase in the pathophysiology of myasthenia gravis. Faseb J 2003;17(2):214–22.
59. Perrier AL, Massoulie J, Krejci E. PRiMA: the membrane anchor of acetylcholinesterase in the brain. Neuron 2002;33(2):275–85.
60. Seidman S, Sternfeld M, Ben Aziz Aloya R, Timberg R, Kaufer Nachum D, Soreq H. Synaptic and epidermal accumulations of human acetylcholinesterase are encoded by alternative 3'-terminal exons. Mol Cell Biol 1995;15:2993–3002.
61. Kaufer D, Friedman A, Seidman S, Soreq H. Acute stress facilitates long-lasting changes in cholinergic gene expression. Nature 1998;393(6683):373–7.
62. Shapira M, Tur-Kaspa I, Bosgraaf L, et al. A transcription-activating polymorphism in the ACHE promoter associated with acute sensitivity to anti-acetylcholinesterases. Hum Mol Genet 2000;9(9):1273–81.
63. Meshorer E, Erb C, Gazit R, et al. Alternative splicing and neuritic mRNA translocation under long-term neuronal hypersensitivity. Science 2002;295(5554):508–12.
64. Darboux I, Barthalay Y, Piovant M, Hipeau-Jacquotte R. The structure-function relationships in Drosophila neurotactin show that cholinesterasic domains may have adhesive properties. Embo J 1996;15(18):4835–43.
65. Sternfeld M, Ming G, Song H, et al. Acetylcholinesterase enhances neurite growth and synapse development through alternative contributions of its hydrolytic capacity, core protein, and variable C termini. J Neurosci 1998;18(4):1240–9.
66. Lev-Lehman E, Evron T, Broide RS, et al. Synaptogenesis and myopathy under acetylcholinesterase overexpression. J Mol Neurosci 2000;14(1–2):93–105.
67. Holland PC, Carbonetto S. The extracellular matrix of skeletal muscle. In: Karpati G, Hilton-Jones D, Griggs RC, eds. Disorders of voluntary muscle. 7th ed. Cambridge: Cambridge University Press; 2001:103–21.
68. Gautam M, Noakes PG, Moscoso L. Defective neuromuscular synaptogenesis in agrin-deficient mutant mice. Cell 1996;85:525–35.

69. Gautam M, Noakes PG, Mudd J. Failure of postsynaptic specialization to at neuromuscular junctions of rapsyn-deficient mice. Nature 1995;377:232–6.

70. Auld DS, Robitaille R. Perisynaptic Schwann cells at the neuromuscular junction: nerve- and activity-dependent contributions to synaptic efficacy, plasticity, and reinnervation. Neuroscientist 2003;9(2):144–57.

71. Langley JN. On the reaction of cells and of nerve endings to certain poisons, chiefly regards the reaction of striated muscle to nicotine and to curare. J Physiol 1905;33:374–413.

72. Reddy LV, Koirala S, Sugiura Y, Herrera AA, Ko CP. Glial cells maintain synaptic structure and function and promote development of the neuromuscular junction in vivo. Neuron 2003;40(3):563–80.

73. Smit AB, Syed NI, Schaap D, et al. A glia-derived acetylcholine-binding protein that modulates synaptic transmission. Nature 2001;411:261–7.

74. Flucher BE, Daniels MP. Distribution of Na + channels and ankyrin in neuromuscular junctions is complementary to that of acetylcholine receptors and the 43 kD protein. Neuron 1989;3:163–75.

75. Sanes JR, Johnson YR, Kotzbauer PT, et al. Selective expression of an acetylcholine receptor-lacZ transgene in synaptic nuclei of adult muscle fibers. Development 1991;113:1181–91.

76. Martinou JC, Falls DI, Fischback GD, Merlie JP. Acetylcholine receptor-inducing activity stimulates expression of the epsilon-subunit gene of the muscle acetylcholine receptor. Proc Natl Acad Sci USA 1991;88:7669–73.

77. Merlie JP, Sanes JR. Concentration of acetylcholine receptor mRNA in synaptic regions of adult muscle fibers. Nature 1985;317:66–8.

78. Brenman JE, Chao DS, Gee SH, et al. Interaction of nitric oxide synthase with the postsynaptic protein PSD-95 and a1-synthophin mediated by PDZ domains. Cell 1996;84:757–67.

79. Kusner LL, Kaminski HJ. Nitric oxide synthase is concentrated at the skeletal muscle endplate. Brain Res 1996;730:238–42.

80. Haimovich B, Schotland DL, Fieles WE, Barchi RL. Localization of sodium channel subtypes in rat skeletal muscle using channel-specific monoclonal antibodies. J Neurosci 1987;7:2957–66.

81. Roberts WM. Sodium channels near end-plates and nuclei of snake skeletal muscle. J Physiol (Lond) 1987;388:213–32.

82. Ruff RL. Na current density at and away from end plates on rat fast- and slow-twitch skeletal muscle fibers. Am J Physiol 1992;262 (Cell Physiol. 31):C229–C34.

83. Ruff RL, Whittlesey D. Na$^+$ current densities and voltage dependence in human intercostal muscle fibres. J Physiol (Lond) 1992;458:85–97.

84. Ruff RL, Whittlesey D. Na$^+$ currents near and away from endplates on human fast and slow twitch muscle fibers. Muscle Nerve 1993;16:922–9.

85. Ruff RL, Whittlesey D. Comparison of Na$^+$ currents from type IIa and IIb human intercostal muscle fibers. Am J Physiol 1993;265 (Cell Physiol. 34):C171–C7.

86. Ruff RL. Effects of length changes on Na$^+$ current amplitude and excitability near and far from the end-plate. Muscle Nerve 1996;19:1084–92.

87. Kuffler SW, Yoshikami D. The distribution of acetylcholine sensitivity at the post-synaptic membrane of vertebrate skeletal twitch muscles: Iontophoretic mapping in the micron range. J Physiol (Lond) 1975;244:703–30.

88. Salpeter MM, Loring RH. Nicotinic acetylcholine receptors in vertebrate muscle: Properties, distribution and neural control. Prog Neurobiol 1985;25:297–325.

89. Kao I, Drachman D. Myasthenic immunoglobulin accelerates acetylcholine receptor degradation. Science 1977;196:526–8.

90. Merlie JP, Heinemann S, Lindstrom JM. Acetylcholine receptor degradation in adult rat diaphragms in organ culture and the effect of anti-acetylcholine receptor antibodies. J Biol Chem 1979;254:6320–7.

91. Caldwell JH, Campbell DT, Beam KG. Sodium channel distribution in vertebrate skeletal muscle. J Gen Physiol 1986;87:907–32.

92. Milton RL, Lupa MT, Caldwell JH. Fast and slow twitch skeletal muscle fibres differ in their distributions of Na channels near the endplate. Neurosci Lett 1992;135:41–4.

93. Ruff RL, Lennon VA. Endplate voltage-gated sodium channels are lost in clinical and experimental myasthenia gravis. Ann Neurol 1998;43:370–9.

94. Banker BQ, Kelly SS, Robbins N. Neuromuscular transmission and correlative morphology in young and old mice. J Physiol (London) 1983;339:355–75.

95. Gertler RA, Robbins N. Differences in neuromuscular transmission in red and white muscles. Brain Res 1978;142:255–84.

96. Tonge DA. Chronic effects of botulinum toxin on neuromuscular transmission and sensitivity to acetylcholine in slow and fast skeletal muscle of the mouse. J Physiol (London) 1974;241:127–39.

97. Sterz R, Pagala M, Peper K. Postjunctional characteristics of the endplates in mammalian fast and slow muscles. Pflügers Arch 1983;398:48–54.

98. Laszewski B, Ruff RL. The effects of glucocorticoid treatment on excitation-contraction coupling. Am J Physiol 1985;248 (Endocrinol. Metab. 11):E363–E9.

99. Hennig R, Lømo T. Firing patterns of motor units in normal rats. Nature 1985;314:164–6.
100. Kelly SS, Robbins N. Sustained transmitter output by increased transmitter turnover in limb muscles of old mice. J Neurosci 1986;6:2900–7.
101. Lev-Tov A. Junctional transmission in fast- and slow-twitch mammalian muscle units. J Neurophysiol 1987;57:660–71.
102. Lev-Tov A, Fishman R. The modulation of transmitter release in motor nerve endings varies with the type of muscle fiber innervated. Brain Res 1986;363:379–82.

2

Acetylcholine Receptor Structure

Jon M. Lindstrom

1. INTRODUCTION

Nicotinic acetylcholine receptors (AChRs) are acetylcholine-gated cation channels *(1)*. They play a critical postsynaptic role in transmission between motor nerves and skeletal muscles and in autonomic ganglia *(2, 3)*. In the central nervous system, they also act presynaptically and extrasynaptically to modulate transmission by facilitating the release of many transmitters *(4, 5)*. In the skin *(6)*, bronchial and vascular epithelia *(7, 8)*, and other non-neuronal tissues *(9)* they also mediate intercellular communication.

Abnormalities of AChRs are responsible for several human diseases. Mutations in AChRs are known to cause congenital myasthenic syndromes *(10, 11)* and the rare autosomal-dominant nocturnal frontal lobe form of epilepsy (ADNFLE) *(12)*. Autoimmune responses to AChRs are known to cause myasthenia gravis (MG) *(13)*, certain dysautonomias *(14, 15)*, and some forms of pemphigus *(16)*.

Nicotine acting on AChRs in the brain causes addiction to tobacco *(17)*. This is by far the largest medical problem in which AChRs play a direct role, and the largest preventable cause of disease, accounting for 430,000 premature deaths annually in the United States *(18)*.

Nicotine acting through AChRs has many physiological effects, including beneficial ones such as inducing vascularization, neuroprotection, cognitive enhancement, anxiolysis, and anti-nociception. Thus, nicotinic agents are leading compounds for the development of drugs to treat many diseases including Alzheimer s disease, Parkinson's disease, chronic pain, schizophrenia, and Tourette's syndrome *(19, 20, 21, 22)* as well as for smoking cessation *(23)*.

There are many known and potential subtypes of AChRs, each defined by the subunits which compose them *(1, 2)*. All AChRs are formed by five homologous subunits organized around a central cation channel. There are 17 known AChR subunits: $\alpha 1$–10, $\beta 1$–4, γ, δ, and ϵ. By contrast with the many subtypes of neuronal AChRs, there are only two subtypes of muscle AChRs. These are a fetal subtype with an $(\alpha 1)_2 \beta 1 \gamma \delta$ stoichiometry and an adult subtype with an $(\alpha 1)_2 \beta 1 \epsilon \delta$ stoichiometry.

AChRs are part of a gene superfamily which includes the genes for subunits of ionotropic receptors for glycine, gamma-amino butyric acid (GABA), and serotonin *(1, 24)*. The structural homologies of all these receptors, and the sorts of evolutionary steps which produced this diversity of receptors, have been elegantly illustrated by experiments. One showed that changing only three amino acids in the channel-lining part of an AChR subunit to amino acids found in receptors for GABA or glycine receptors resulted in AChRs with anion-selective channels like those of GABA or glycine receptors *(25)*. Another experiment showed that a chimera of the extracellular domain of an AChR subunit and the remainder of a serotonin receptor subunit produced an ACh-gated cation channel with the conductance properties of a serotonin receptor *(26)*.

Muscle AChRs are the best characterized of the AChRs *(1)*. The presence of a single type of synapse in skeletal muscle (with the exception of extraocular muscle, Chapter 5) facilitated studies of AChR synthesis, developmental plasticity, and electrophysiological function *(27, 28, 29)*. The presence of large amounts of muscle-like AChR in the electric organs of Torpedo species

From: *Current Clinical Neurology*
Myasthenia Gravis and Related Disorders
Edited by: H.J. Kaminski, DOI 10.1007/978-1-59745-156-7_2, © Humana Press, New York, NY

permitted the purification and characterization of AChRs, partial sequencing of their subunit proteins, cloning of the subunit cDNAs, and low-resolution electron crystallographic determination of their three-dimensional structure *(24, 27, 30, 31)*. Low stringency hybridization, starting with cDNAs for muscle AChR subunits, led to the cloning of subunits for neuronal AChRs *(27)*. Immunization with purified electric organ AChRs led to the discovery of experimental autoimmune myasthenia gravis (EAMG), the autoimmune nature of MG, and an immunodiagnostic assay for MG *(13, 32)*. Monoclonal antibodies initially developed as model autoantibodies led to not only the discovery of the main immunogenic region (MIR) on $\alpha 1$ subunits and the molecular basis of the autoimmune impairment of neuromuscular transmission in MG *(13, 33, 34)*, but also the immunoaffinity purification of neuronal nicotinic AChRs. mAbs have continued to provide useful tools for characterizing AChRs *(2)*.

This chapter reviews the basic structures of muscle and neuronal AChRs. It will describe the antigenic structure of muscle AChRs and consider how this accounts for the pathological mechanisms by which neuromuscular transmission is impaired in MG. This will be briefly contrasted with the antigenic structure of a neuronal AChR involved in autoimmune dysautonomia. This chapter also considers the optimized functional structure of muscle AChRs and how mutations impair AChR function in congenital myasthenic syndromes. The many AChR mutations identified in all of the muscle AChR subunits in myasthenic syndromes is contrasted with the few disease-causing mutations discovered, thus far, in neuronal AChR subunits.

2. SIZE AND SHAPE OF AChRS

Electron crystallography of two-dimensional helical crystalline arrays of AChRs in fragments of Torpedo electric organ membranes have revealed the basic size and shape of this muscle-type AChR to a resolution of 4 Å, as shown in Fig. 1 *(31, 35)*. Viewed from the side, a Torpedo AChR is roughly cylindrical, about 140 Å long and 80 Å wide. About 65 Å extend on the extracellular surface, 40 Å cross the lipid bilayer, and 35 Å extend below. Viewed from the top, the extracellular vestibule is a pentagonal tube with walls about 25 Å thick and a central pore about 20 Å in diameter. The channel across the membrane narrows to a close. Other evidence suggests that the open lumen of the channel becomes narrow (perhaps 7 Å across), sufficient only for rapid flow of hydrated cations like Na^+ or K^+. The five subunits are rod-like, oriented like barrel staves at a 10° angle around the central channel.

X-ray crystallography of a molluscan glial ACh-binding protein has revealed the structure of the extracellular domain of an AChR-like protein at atomic resolution *(37, 38, 39, 40)* (Fig. 2). Snail glia were found to release a water-soluble protein which modulated transmission by binding ACh. The cloned protein showed 24% sequence identity with the extracellular domain of human $\alpha 7$. $\alpha 7$ AChR subunits form homomeric AChRs. The extracellular domain cleaved by protein engineering from $\alpha 7$ assembles into water-soluble pentamers with the ligand-binding properties of native $\alpha 7$ AChRs, but it does so very inefficiently *(43)*. Thus, the molluscan ACh-binding protein probably contains sequence adaptations for efficient assembly and secretion as a water-soluble protein. Nonetheless, it provides a very good model for the basic structure of the extracellular domains of AChRs and other receptors in their superfamily. This is proven by the demonstration that the AChBP, with slight modification to match their interface, can form a chimera with the transmembrane portion of the $5HT_3$ receptor to form an ACh-gated cation channel *(44)*. Figure 2 shows that five ACh-binding protein subunits assemble as the extracellular domains of AChR subunits would around the vestibule of the channel. A buffer component was initially found to occupy what was expected to be the ACh-binding site, which is formed at the interface between subunits halfway up the side of the assembled protein. All of the contact amino acids for this site corresponded to ones in AChRs which had been identified by affinity labeling or mutagenesis studies *(24, 30, 45)*. Subsequently AChBP was crystallized in combinations with both agonists

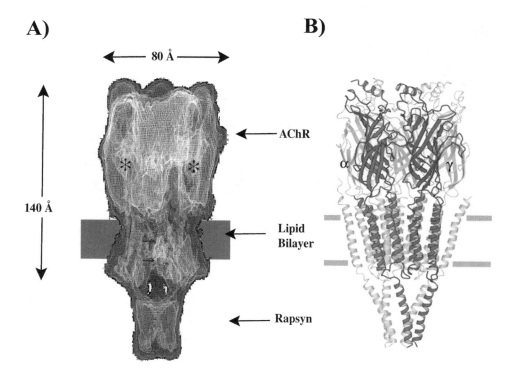

Fig. 1. Torpedo electric organ AChR structure determined by electron crystallography. (**A**) The large extracellular domain contrasts with the smaller domain on the cytoplasmic surface. Rapsyn is a 43,000 Da peripheral membrane protein through which muscle AChRs are linked to actin in the cytoskeleton to concentrate them at the tips of folds in the postsynaptic membrane adjacent to active zones in the presynaptic membrane at which ACh is released *(36)*. Modified from Unwin *(35)* with permission. (**B**) Here the polypeptide chains are shown as ribbon structures, highlighting the α1 and δ subunits, at whose interface one of the two ACh-binding sites in this (α1)2β1γδ AChR is formed. This figure is from Unwin *(31)*.

such as nicotine *(39)* and antagonists such as snake venom toxins *(40)*. This and other features of the structure will be discussed in more detail in subsequent sections. Basically, most recognizable features were found about where they were expected to be from studies of native AChRs, providing confidence that the structure of the ACh-binding protein is relevant to that of AChRs.

3. STRUCTURES OF AChR SUBUNITS

All AChR subunits share several features. Figure 3 diagrammatically shows the transmembrane orientation of the mature polypeptide chain of a generic AChR subunit. To produce the mature polypeptide sequence, a signal sequence of about 20 amino acids is removed from the N-terminus of each subunit during translation as the N-terminal domain crosses the membrane into the lumen of the endoplasmic reticulum.

The large N-terminal extracellular domain of each subunit, which consists of about 210 amino acids, contains a disulfide-linked loop (the cys loop) which is characteristic of all receptors in this superfamily. In α1 subunits, it extends from cysteine 128 to cysteine 142. This sequence is among the most conserved of all AChR subunit sequences. In the ACh-binding protein (Fig. 2), this loop is located near what would be the lipid bilayer or extracellular surface of the transmembrane

A)

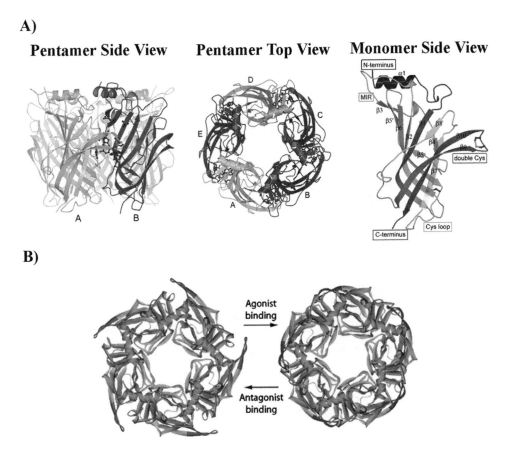

B)

Fig. 2. Mollusc ACh-binding protein structure determined by X-ray crystallography. (**A**) Atomic resolution structure of the ACh-binding protein reveals the basic structure of AChR extracellular domains. It is a 62 Å high cylinder 80 Å in diameter with an 18 Å diameter central hole. The structure is a homopentamer like α7 AChRs. There are five ACh-binding sites at subunit interfaces, rather than the two ACh-binding sites expected in a muscle AChR heteropentamer at interfaces between α1 and δ, γ, or ε subunits. The ACh-binding site is occupied by the buffer component *N*-2-hydroxyethylpiperazine-*N′*-2 ethanesulfonic acid (HEPES). The adjacent disulfide-linked cysteine pair corresponding to α1 192–193, which is characteristic of all AChR α subunits, is on a projection of what can be defined as the + side of the subunit. It intercalates with the − side of the adjacent subunit to form the ACh-binding site. As expected from studies of AChRs *(13, 41)*, the sequence corresponding to the MIR of α1 subunits is located at the extracellular tip and oriented out away from the central axis of the subunit. The Cys loop linked by a disulfide bond corresponding to that between cysteines 128 and 142 of α1 subunits, the signature loop characteristic of all subunits of the gene superfamily of which AChRs are a member, is located in the ACh-binding protein at the base near what would be the transmembrane portion of an AChR or the lipid bilayer. This loop sequence shows little homology with that of AChRs, and is more hydrophilic than the sequences characteristic of AChRs. This loop contributes to the water solubility of AChBP, but is at the interface between the extracellular and transmembrane domains of AChRs. Modified from Brejc et al. *(38)* with permission. (**B**) Shows a top view of an AChBP in its resting and active conformations. The C loop is open in the resting state and closed in the active and desensitized states. This part of the figure is from Hansen et al. *(40)*. Large movement of the C loop at the ACh-binding sites is propagated through the AChR to produce movement of the M2 transmembrane domain and ultimately opening of the channel gate located at the middle or bottom of the transmembrane domain *(42)*.

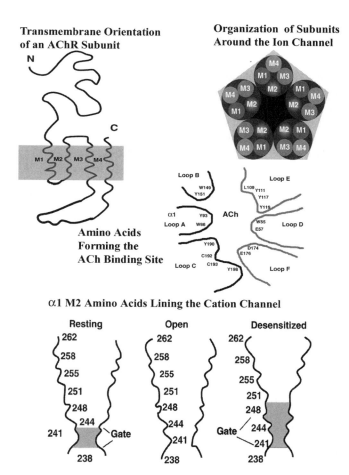

Fig. 3. Aspects of AChR structure. The transmembrane orientation of a generic AChR subunit is depicted in this diagram. The actual structure of the AChR shown in Fig. 1B and the large extracellular domain of the ACh-binding protein shown in Fig. 2 provide more details. The transmembrane domains M1–4 are depicted as largely α helical. The overall shape of the subunit is depicted as rod-like. Five of these rods assemble in a pentagonal array to form the AChR shown in Fig. 1. The subunits are organized around the ion channel so that the amphipathic M2 transmembrane domain from each subunit contributes to the lining of the channel. In muscle AChRs (e.g., with the subunit arrangement α1γα1δε), and other heteromeric AChRs (e.g., with the subunit arrangement α4β2α4β2β2), there are only two ACh-binding sites at interfaces between the + side of α subunits and complementary subunits, but small concerted conformation changes of all subunits are involved in activation and desensitization *(35, 44, 47)*. Thus, all subunits contribute to the conductance and gating of the channel, even if they are not part of an ACh-binding site. The amino acids lining the ACh-binding site have been identified by affinity labeling and mutagenesis studies *(24, 45)*, and have been found to correspond well to those identified in the crystal structure of the ACh-binding protein *(39, 40)*. Notice the predominance of aromatic amino acids in this region. As in ACh esterase *(62)*, the quaternary amine group of ACh is thought to be bound through interactions with π electrons of these aromatic amino acids rather than ionic interactions with acidic amino acids. Note also that the ACh-binding site is formed from amino acids from three different parts of the extracellular domain of the α subunit interacting with three parts of the complementary subunit, and that the interaction is at the interface between the + side and the α subunit and the − side of the complementary subunit. Thus, the site is ideally positioned to trigger small concerted conformation changes between subunits thereby permitting low-energy binding events in the extracellular domain to regulate opening, closing, and desensitization of the ion channel gate near the cytoplasmic vestibule of the channel *(31, 44)*. The amino acids lining the cation channel and accessible

regions *(38)*. It is hydrophobic in AChRs but hydrophilic in the binding protein. A proline in the loop, which is conserved in all AChR subunits, is missing in the binding protein. Mutating this proline to glycine disrupts assembly of AChR subunits and prevents transport to the surface of assembled AChRs *(46)*. The extracellular domains of AChR subunits contain one or more glycosylation sites, and in all but the α7–9 subunits, which can form homomeric AChRs, there is an N-glycosylation site at position 141 adjacent to the disulfide bond of the signature loop.

Three closely spaced, highly conserved, largely α helical transmembrane sequences (M1–M3) corresponding approximately to amino acids 220–310 extend between the large extracellular domain and the largest cytoplasmic domain. The N-terminal third of M1 and the hydrophilic side of M2 form each subunit's contribution to the lining of the channel *(1, 24, 47)*. This will be described in slightly more detail in a subsequent section on the channel and gate.

The large cytoplasmic domain between the transmembrane sequences M3 and M4 comprises 110–270 amino acids (α4 having by far the largest). This is the most variable region in sequence between subunits and between species. Consequently, many subunit-specific antibodies bind in this region *(48)*. The large cytoplasmic domain, by contrast with the extracellular domain is quite flexible in structure. Consequently many antibodies to this region bind to both native and denatured proteins, and visualization of this disordered region is difficult with crystallographic methods. The large cytoplasmic region of muscle AChRs interacts with rapsyn, a protein which links it to the cytoskeleton and thus helps to position it appropriately in the neuromuscular junction *(36)*. Proteins other than rapsyn are probably involved in interacting with the large cytoplasmic domain of neuronal AChRs to help transport them to and localize them at their various pre-, post-, and extrasynaptic sites of action, e.g., members of the PSD-95 family, but these proteins have not been characterized in detail *(49)*. The chaperone protein 14-3-3η binds to α4 at serine 441 in the large cytoplasmic domain, especially when it is phosphorylated, helping to increase conformational maturation or assembly *(50)*. Several chaperone proteins are known to participate in the conformational maturation and assembly of muscle AChR subunits *(51, 52, 53, 54)*. For example, COPI interacts with lysine 314 of α1 subunits. The transmembrane portion of M1 in α1 subunits contains an endoplasmic reticulum retention sequence that is obscured only after complete assembly of AChR pentamers *(54)*. The large cytoplasmic domain contains phosphorylation sites and perhaps other sequences thought to be involved in regulating the rate of desensitization *(55, 56)* and other properties such as intracellular transport *(57)*. Surprisingly, it even contains sequences that contribute to channel-gating kinetics *(58)*. This is because the rigid amphipathic α helices in the large cytoplasmic domain, which immediately precede M4, form intracellular portals that regulate access of ions to the central cation channel through the AChR *(31, 59)*.

A fourth transmembrane domain (M4), of about 20 amino acids, extends from the large cytoplasmic domain to the extracellular surface, leading to a 10–20 amino acid extracellular sequence. In the case of human α4 subunits, the C-terminal end of the α4 sequence has been found to form a site through which binding of estrogen enhances AChR function three- to sevenfold, while it inhibits the response of α3β2 AChRs *(60, 61)*. No functional role for the C-terminal sequence is yet known for other AChR subunits.

◀──

Fig. 3. *(Continued)* from either the extracellular or cytoplasmic surface have been determined largely by the substituted cysteine accessibility method (SCAM) *(47)*. The channel lining is thought to be formed by M2. The figure depicts the M1–M2 linker at the cytoplasmic end of M1 and M2. In the closed, resting state of the channel, only a short region is occluded and inaccessible to labeling. In the closed desensitized state, a larger region is occluded.

α subunits are defined by the presence of a disulfide-linked adjacent cysteine pair in the large extracellular domain homologous to α192 and 193 of α1 subunits. In all α subunits, other than α5 and perhaps α10, this is thought to contribute to the ACh-binding site. These cysteines, after reduction of the disulfide bond between them, were the targets of the first affinity labels developed for AChRs *(24)*. In the ACh-binding protein (Fig. 2), this cysteine pair is located at the tip of a protruding C loop on what can be defined as the + side of the subunit. The ACh-binding site is formed at the interface between the + side of an α subunit and the – side of an adjacent subunit, as will be discussed in somewhat more detail in a subsequent section.

4. ORGANIZATION OF SUBUNITS IN ACHR SUBTYPES

Homologous rod-like AChR subunits are organized in a pentagonal array around the central cation channel *(31, 35)*. In the muscle-type AChRs of Torpedo electric organ, the order of subunits around the channel is α1, γ, α1, δ, β1 *(24)*. In the adult muscle AChR subtype, ε substitutes for γ. Reflecting their similar functional roles and evolutionary origins *(63)* γ, δ, and ε have especially similar sequences, as do β1, β2, and β4. In cDNA expression systems, β2 and β4 can substitute for β1 in muscle AChR *(64)*. α7 can also substitute for α1 *(65)*, revealing basic homologies between AChR subunits rather than an AChR subtype which is thought to be important in vivo. The arrangement of subunits in major AChR subtypes is depicted in Fig. 4.

The primordial AChR was presumably homomeric, and heteromeric AChRs evolved subsequent to duplication of the primordial subunit *(64)*. AChR subunits are lettered in order of increasing molecular weight as they were discovered in the AChRs first purified from Torpedo electric organ, and they are numbered in the order in which their cDNAs were cloned *(27)*.

Homomeric AChRs can be formed by α7, α8, and α9 subunits *(1, 2)*. In mammals, α7 is the predominant homomeric AChR. Like muscle AChRs (though with much lower affinity) α7, α8, and α9 AChRs bind α bungarotoxin (αBgt) and related snake venom peptide toxins to their ACh-binding sites. Heteromeric neuronal AChRs containing α1–6 in combination with β2–4 subunits do not bind αBgt. α8 has been found only in chickens *(73)*. It can form heteromeric AChRs with α7. α9 can form heteromeric AChRs with α10 *(74, 75)*.

Heteromeric neuronal AChRs are formed from α subunits 2, 3, 4, or 6 in combination with β2 or β4 *(1, 2)*. α5 and β3 are present as additional subunits in AChRs in which ACh-binding sites are formed by other α and β subunits *(68, 69, 76, 77, 78, 79, 80)*. The AChR subunit pairs α2, α4; α3, α6; β2, β4; and α5, β3 are closely related in sequence, reflecting their origins by gene duplication and their similar functional roles *(63)*. The subunit compositions of these heteromeric neuronal AChRs can be as simple as α3β4α3β4β4 or α4β2α4β2β2, with two identical ACh-binding sites. They can be more complex, involving three (e.g., α3β4α3β4α5) or more types of AChR subunits (α4β2α6β2β3). These can have two identical ACh-binding sites (α4β2α4β2β2), two different ACh-binding sites with one kind of α subunit (α3β2α3β4α5), as in muscle AChRs (α1εα1δβ1), or two different ACh-binding sites with two different α subunits (α4β2α6β2β3). Many neurons express complex mixtures of ACh subunits *(81, 82, 83, 84)*. Ganglionic neurons usually express α3, α5, α7, β2, and β4 subunits, which could potentially form many subtypes, but the function of α3β4 tends to predominate postsynaptically *(85, 86)*. α7 assembles only as homomers in these neurons. Adjacent neurons, within the same nucleus, can express different complex combinations of AChR subunits *(82, 87)*. It remains to be determined which AChR subtypes are expressed from these subunit combinations, where they are located within the neurons, and what functional roles they play in postsynaptic transmission, presynaptic modulation and extrasynaptic plasticity. What is clear is that AChRs can play much more complex regulatory roles than their classic postsynaptic role in high-safety factor neuromuscular transmission might suggest.

In muscle AChRs, ACh-binding sites are formed at the interfaces of the + side of α1 with the – side of γ, δ, or ε subunits, and β1 subunits do not participate in forming ACh-binding sites *(24)*.

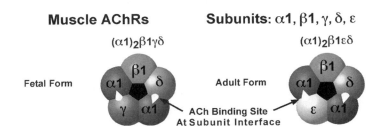

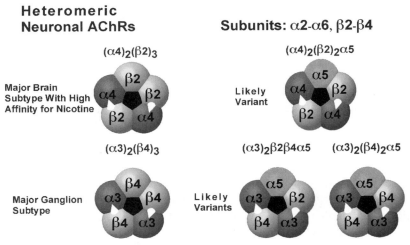

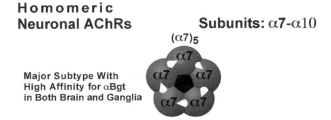

Fig. 4. Prominent AChR subtypes. AChR subtypes are defined by their subunit compositions. The arrangement of AChR subunits around the channel is known for Torpedo electric organ muscle-type AChRs *(24)* and inferred for other types based on this example. The subunit stoichiometry of some neuronal AChR subtypes is known *(66, 67)*, and it is known that ACh-binding sites can be formed at specific interfaces of α1–4 or α6 with β2 or β4 subunits but not with α5 or β3 *(61, 68, 69)*. Stoichiometries such as $(\alpha4)_3 (\beta2)_2$ have been found in cell lines and may exist in vivo *(70, 71, 72)*. All subunits are expected to participate in the structure of the channel and in conformation changes associated with activation and desensitization. Modified from Lindstrom *(13)*.

In heteromeric neuronal AChRs, α5 and β3 subunits are thought to occupy positions comparable to β1 *(76, 88)*. They contribute to both ion channel properties and agonist potency not only because they contribute to the lining of the channel but also because the ease with which the agonist-induced conformation changes of the whole AChR molecule take place in the course of activation or desensitization depends on the structures of all moving parts of the AChR *(89)*.

5. ACETYLCHOLINE-BINDING SITES

ACh-binding sites are formed at the interfaces between subunits. This is shown in high resolution in the case of the ACh-binding protein in Fig. 2 and diagrammatically in Fig. 3. Both affinity-labeling studies of native AChRs *(24, 45)* and the structure of the ACh-binding protein *(39, 40)* reveal that the ACh-binding site does not contain negatively charged amino acids to bind positively charged ACh. The ACh-binding site contains many aromatic amino acids, and it is thought that interactions between the π electrons of these amino acids and the quaternary amine play an important role in binding. The ACh-binding site of ACh esterase also works this way *(62)*.

Tetramethylammonium is the simplest form of an agonist. The small movements initiated by the low-affinity binding at just this portion of the site must be sufficient to trigger channel opening. The fundamental features of an AChR agonist are a quaternary or tertiary amine which binds to the site and permits closure of the C loop over the site *(40*, Fig. 2*)*.

By contrast with small agonists, antagonists tend to be larger and are often multivalent. These prevent closing of the C loop. In the extreme case of large snake venom toxins like α bungarotoxin, binding of the toxin leaves the C loop propped open even more than in the resting conformation *(40*, Fig. 2*)*. This also helps to explain the slow rate of binding of snake venom toxins. They can only fit into the site when it assumes a relatively rare especially wide open conformation.

Snake venom toxins such as α bungarotoxin and cobratoxin have been extremely important for identifying, localizing, quantifying, purifying, and characterizing AChRs *(27, 65)*. These are peptides of ~9000 mw with a flat three-fingered structure. Only the tip of the longest finger enters the ACh-binding site of muscle AChRs *(40, 90)*. Snake venom toxins evolved from families of proteins of similar structure, some of which have recently been found to interact competitively or allosterically with AChRs. Lynx1 is tethered to membranes by a glycophosphatidyl anchor, colocalizes with α4β2 AChRs in the brain, promotes expression, and enhances desensitization of these AChRs *(91)*. SLURP-1 is found in both anchored and soluble forms, is secreted by keratinocytes, and potentiates responses of α7 AChRs found in keratinocytes *(92)*. Mutations in SLURP-1 cause Mal de Meleda, an inflammatory skin disease characterized by hypoproliferation and release of TNFα from macrophages. SLURP-1 binds competitively to keratinocyte AChRs *(93)*. SLURP-2 is secreted from keratinocytes and acts as an agonist on keratinocyte α3 AChRs *(94)*. SLURP-2 delays keratinocyte differentiation and prevents apoptosis, whereas SLURP-1 facilitates keratinization and programmed cell death. This suggests that the cholinergic signaling through AChR subtypes was involved in intracellular communication before it evolved to specialized use in rapid signaling with the evolution of the nervous system. The regulation of gene expression evident in cholinergic effects on keratinocytes may well be reflected in trophic effects of AChRs within the nervous system. Within the primitive nervous system of *C. elegans*, for example, there are many more AChR-like subunits than are found in humans *(95)*. Even the puffer fish genome has more AChR subunits than humans, reflecting this evolutionary trend *(96)*. Simple bacterial homologues of voltage-gated ion channels have proven extremely useful for crystal structure determination of these ion channels *(97, 98)*. Precursors of AChRs have been recognized in bacteria, but AChBP secreted by mollusk glia has proven extremely useful for structural studies of the extracellular domain of AChRs and the structures of their ACh-binding sites *(37, 38, 39, 40)*.

The ACh-binding sites are located halfway up the extracellular domain *(39, 40)*, far removed from the gate at the middle or cytoplasmic end of the channel whose opening they regulate *(47)*. This regulation is thought to occur by means of small global conformation changes of the AChR protein which may involve slight changes in the angles of the rod-like subunits to produce an iris-like regulation of channel opening *(35, 42)*. Both the ACh-binding site *(99)* and the cation channel *(47)* change conformation between the resting, open, and desensitized states. The amino acids lining the ACh-binding site come from three parts of the + side of the α subunit and three parts of the − side of the complementary subunit *(24, 39, 40)*. It would seem that such an arrangement would be ideal for initiating and communicating small motions involved in ligand binding to motions along subunit interfaces throughout the molecule. The binding energies involved are small, as must be the differences between the resting and open conformations as the channel flickers open.

It is argued that the two ACh-binding sites in muscle AChRs should differ in affinity for ACh ($K_D = \mu M$ versus mM) in order to ensure the rapid opening and closing of the channel *(1, 28)*. The properties of the two ACh-binding sites in muscle AChRs differ because they are formed at the interface of α1 with γ, δ, or ε subunits *(100)*. The α1γ sites have higher affinity, producing longer channel opening at low ACh concentrations, as may be appropriate for fetal AChRs during synapse development (and perhaps useful when fetal AChRs help to compensate for AChR loss after autoimmune or genetic damage). The actual kinetics of gating, surprisingly, depends not just on properties of the ACh site or gate, but also on sequences in the large cytoplasmic domain *(58)*. During an endplate potential, ACh is present in the synaptic cleft at concentrations in the mM range for less than a millisecond *(101)*. Normally, this ensures a substantial safety factor for transmission because the ACh saturates the AChRs and the AChRs are in excess over what is needed to provide sufficient current to trigger an action potential (Chapter 1). When the number of AChRs is reduced in autoimmune or congenital myasthenia, the current may be insufficient, or become insufficient on successive stimuli, as fewer vesicles of ACh are released and AChRs accumulate in the desensitized state. Then some synthesis of the fetal form of AChR with higher affinity for ACh and consequently longer burst and channel open times may be critical for achieving sufficient current to more nearly ensure transmission.

Competitive antagonists ideally keep AChRs in the resting conformation. They are usually multivalent quaternary or tertiary amines thought to interact both at the ACh site and peripheral sites to stabilize the resting state. Some, like curare, may actually be very low-efficacy agonists *(102)*. αBgt and related snake venom peptides of 8000–9000 Da are large flat molecules that might cover 800–1200 Å of AChR surface *(103)*, corresponding perhaps to a 30×30 Å square centered on the ACh-binding site. Only the tip of one finger of this structure enters the actual ACh-binding site *(40, 103)*, but this occurs with high affinity and stabilizes the AChR in a resting state.

Exposure to ACh or other agonists for long periods causes desensitization, which is a conformation (or perhaps a set of conformations) characterized by a closed channel and higher affinity for agonists. In the normal course of neuromuscular transmission, desensitization is not a limiting factor. However, ACh esterase inhibitors given to treat MG (by increasing the concentration and duration of ACh to compensate for the loss of AChRs) can, at excessive doses, further impair transmission by causing accumulation of desensitized AChRs *(104)*. Nerve gases and insecticides that act as esterase inhibitors can have similar effects *(105)*. The depolarizing block surgical muscle relaxant succinylcholine should also have a similar component to its action in addition to producing a depolarizing blockade of action potential generation. Nicotine in tobacco users is present for many hours ($t_{1/2}$ for clearance = 2 hours) at an average concentration around 0.2 μM, and rises to nearly 1μM briefly after inhalation of smoke *(106)*. The low affinity of muscle AChRs for nicotine normally prevents much effect on neuromuscular transmission. Nicotine, acting on several neuronal AChR subtypes with a wide range of affinities for nicotine, produces many behavioral effects (addiction, tolerance, anxiolysis, cognitive enhancement, antinociception) resulting from a complex mixture of activation, desensitization, and upregulation effects on the AChRs *(2, 107, 108)*. AChRs containing α4 and β2 subunits, which account for most high-affinity nicotine binding in the brain, can account for the most prominent effects of nicotine: reward, tolerance, and sensitization *(17)*.

Binding of agonists at two ACh-binding sites is required for efficient activation of AChRs *(100)*. Antagonist blockage of any one site is usually sufficient to prevent activation *(109)*. Homomeric AChRs, such as α7 AChRs, have five ACh-binding sites. Desensitization at any one of these may be sufficient to prevent activation, probably accounting for the very rapid desensitization characteristic of such AChRs *(110)*.

6. CATION CHANNEL AND ITS GATE

The cation channel is lined with amino acids from the M2 transmembrane domains, as depicted in Fig. 3. Much of the identification of amino acids lining the channel pore has been achieved using the substituted cysteine accessibility method (SCAM) *(24, 47)*. In this method, successive amino acids along a putative transmembrane domain such as M1 or M2 are replaced by cysteine. This introduces a free thiol group, usually without greatly altering AChR function. Then a thiol-alkylating agent which contains a positively charged amino group (e.g., aminoethyl methanethiosulfonate (MTSEA)) is applied from outside or inside the cell in which the mutated AChR is expressed. If MTSEA covalently reacts and blocks the channel, it is assumed that the substituted cysteine was exposed on the interior of the channel. From the periodicity at which cysteines are exposed, it can be inferred whether the domain has α helical or another secondary conformation, and accessibility of the cysteine in the resting, open, or desensitized conformation determines the outer limits of the channel gate. Most of M1 and M2 appear to be in an α helical conformation. In the resting state, the open lumen of the channel extends to nearly the cytoplasmic surface with only a small occluded region between α1 G240 and T244. The highly conserved sequence α1 G240, E241, K242, in the M1–M2 linker immediately preceding the cytoplasmic end of M2, lines the narrowest part of the channel in which the gate is located. Small motions in this region could open and close the channel. In the desensitized state, nearly half of the channel is occluded over a region extending from G240 to L251. Other mutagenesis studies have also helped to define the structure of the channel *(111)*. Several polar or charged rings of amino acids lining the channel formed by homologous amino acids from each subunit form the selectivity filter. Electron crystallography images the channel as lined with a cylinder of α helical M2 sequences, and this structure is surrounded by an arrangement of the other three transmembrane domains *(112)*. Both the extracellular and intracellular vestibules are strongly electronegative, which would provide a stable environment for cations passing through the channel *(31)*. Access of cations to the inner vestibule is through narrow lateral portals formed by cytoplasmic α helices immediately preceding M4 *(31)*. Electrophysiological studies suggest that three or four polar amino acids lining these portals are critical for allowing passage of ions into the inner lumen of the channel in all members of the Cys loop family of receptors *(59)*.

7. ANTIGENIC STRUCTURE AND THE MAIN IMMUNOGENIC REGION (MIR)

Immunization with native α1 AChRs provokes antibodies directed primarily at the extracellular surface *(48, 113)*. This is because more than half of the antibodies to AChR in an immunized animal or an MG patient are directed at the MIR *(33, 34, 114)*. This is a highly conformation-dependent epitope, thus immunization with denatured AChRs or subunits, by default, provokes antibodies to the cytoplasmic surface, whose epitopes are much less conformation dependent

(48). Synthetic peptides can be used to induce antibodies to many parts of the AChR sequence, but many of the antibodies will not bind to AChRs in their native conformation *(48)*.

Monoclonal antibodies (mAbs) from rats and mice immunized with AChRs have provided excellent model autoantibodies and structural probes for AChRs *(2, 34, 114)*. Half or more of the mAbs to native α1 AChRs are directed at the MIR *(33)*. Such antibodies compete with each other for binding and prevent the binding of more than half of the autoantibodies from MG patients *(34, 114)*. Thus, the antibodies bind to overlapping regions, but not necessarily to identical epitopes. The crystal structures of the binding sites of two mAbs to the MIR have been determined *(115, 116)*. The two antibody-binding sites differ because they recognize different aspects of the MIR and do not present a clear negative image of the whole structure. One is a crescent-shaped crevice and the other is relatively planar with a surface area of 2865 Å.

The structure of the MIR is being defined with increasing precision. Both absolutely conformation-dependent mAbs to the MIR and others which also bind to denatured α1 subunits with lower affinity depend critically on the same α1 amino acids 68 and 71, as shown by mutagenesis studies *(117)*. The dependence on the native conformation of the AChR for binding of rat mAbs to the MIR and MG patient autoantibodies to the MIR suggests that several sequences which are adjacent only in the native conformation contribute to the formation of the cluster of MIR epitopes recognized by rat mAbs and MG patient autoantibodies. Several lines of evidence showed that the MIR is oriented so that an antibody bound to one α1 subunit cannot also bind the other α1 subunit in the same AChR, but can efficiently crosslink adjacent AChRs *(118)*. Cryoelectronmicroscopy confirmed and extended these conclusions, showing that the MIR was located at the extracellular tip of α1 subunits and angled away from the central axis of the AChR *(41)*. In the ACh-binding protein, the sequence which would form the MIR in an α1 subunit is located at the extracellular tip of the subunit near the N-terminus in a loop directed away from axis of the subunit, as shown in Fig. 2 *(38)*. Because the sequence of this region in the ACh-binding protein is not closely homologous to that of α1 subunits, the precise conformation of the MIR in α1 subunits must differ from that of this sequence of the binding protein. An antibody bound to the MIR would cover an even larger area of the α1 surface than is covered by a bound αBgt molecule. The separation between the MIR at the tip of the subunit and the ACh-binding site halfway down the side of the subunit reveals why both antibodies to the MIR and αBgt can bind simultaneously to α1 AChRs. Few autoantibodies in MG are directed at the ACh-binding site *(118)*. These facts form the structural basis of the basic immunodiagnostic assay for MG in which ^{125}I-labeled αBgt is used to specifically label detergent-solubilized human α1 AChRs to permit quantitation of MG patient autoantibodies in an immune precipitation assay *(119)*.

The fundamental pathological significance of the MIR derives from its basic structure. This is represented diagramatically in Fig. 5. The MIR is highly immunogenic. This may be because it has a novel shape, is easily accessible on the protein surface, and is present in two copies per AChR to permit multivalent binding and crosslinking of receptor immunoglobulins to efficiently stimulate B cells. Because the MIR is easily accessible on the extracellular surface, antibodies can easily bind to it and fix complement. α1 AChRs are densely packed in a semicrystalline array at the tips of folds in the postsynaptic membrane, promoting high-avidity binding of antibodies and focal lysis of the postsynaptic membrane, but allowing the postsynaptic membrane to reseal after the membrane fragments containing antibodies, AChR, and complement are shed *(120)*. This process contributes to impairment of transmission both by destroying AChRs and by disrupting the morphology of the synapse in which presynaptic active zones of ACh release are normally located near postsynaptic patches of AChRs. The MIR is oriented at an angle which promotes crosslinking of AChR by antibodies. This triggers antigenic modulation, which is a crosslinking-induced increase in the rate of AChR internalization and lysosomal destruction *(121, 122, 123)*. mAbs to the MIR and most serum antibodies do not impair AChR function *(124)*. This is probably because the antibodies do not interfere with ACh binding and because antibodies bound

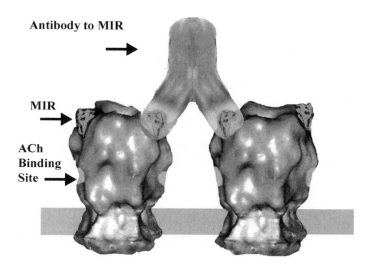

Antibody to MIR

MIR

ACh Binding Site

Fig. 5. Structural basis of the pathological significance of the MIR. It is important to remember that the MIR is a region of closely spaced epitopes, not a single epitope. The MIR is easily accessible on the extracellular surface. This permits binding of autoantibodies and complement. The MIR has a highly immunogenic conformation. There are two copies of the MIR. This contributes to its immunogenicity by permitting aggregation of receptor immunoglobulins on B cells. It contributes to synaptic pathology by permitting crosslinking of AChRs by antibodies to cause antigenic modulation. The MIR is oriented to facilitate crosslinking of adjacent AChRs by antibody while not permitting crosslinking of $\alpha 1$ subunits within an AChR. The MIR is away from the ACh-binding site and subunit interfaces. Thus, antibodies to the MIR do not inhibit AChR function competitively or allosterically. Also, this permits binding of both antibodies and αBgt simultaneously, so that immunodiagnostic assays can be done by indirect immunoprecipitation of detergent-solubilized AChRs labeled with ^{125}IαBgt. AChRs are densely packed in a semicrystalline array at the tips of folds in the postsynaptic membrane. Thus, bound antibody and complement can cause focal lysis and shedding of AChR-rich membrane fragments, while the postsynaptic membrane can reseal. The result is loss of AChR and disrupted synaptic morphology, but not lethal lysis of the muscle fiber.

to the MIR are far from the subunit interfaces where the subtle movements take place mediating activation and desensitization.

Some rat mAbs to the MIR on $\alpha 1$ bind very well to human $\alpha 1$, as well as $\alpha 3$, and $\alpha 5$ subunits (which have similar MIR sequences) *(2, 76)*. These mAbs do not recognize denatured $\alpha 3$ even though they do recognize denatured $\alpha 1$. MG patient antisera do not bind significantly to human $\alpha 3$ AChRs, and human autoantibodies to $\alpha 3$ do not bind to $\alpha 1$ *(14)*. Thus, the MIR epitopes recognized by rats and humans are not identical, even though they are close enough that antibodies to them compete for binding.

Prominent antibody epitopes on the cytoplasmic surfaces of $\alpha 1$ subunits and other subunits have been determined using antisera and antibodies *(34, 48)*. These epitopes are pathologically irrelevant because autoantibodies cannot bind to them in vivo.

The spectrum of autoantibodies to AChRs produced in MG patients is dominated by the MIR, but includes other parts of the AChR and closely resembles the spectrum seen in animals with EAMG or in dogs with idiopathic MG *(113, 125)*. Many MG sera react better with extrajunctional AChRs *(126)*. This may reflect that the endogenous immunogen in MG has γ subunits, and these may be more immunogenic because they are not normally expressed in adults. One unusual MG patient serum was found to be specific for γ and occluded the $\alpha 1\gamma$ ACh-binding site, thereby inhibiting function *(127)*. Rare mothers have been found who make autoantibodies only to γ

subunits *(128)*. These have no effect on the mother (whose AChRs have only ε subunits), but they paralyze or kill fetuses, causing arthrogryposis multiplex congenita.

α1 subunits also predominate in the T-cell response to AChRs, but epitopes have been found on all subunits *(129, 130)*. Several α1 T-cell epitopes seem to predominate in MG, but there is disagreement between laboratories on which sequences comprise these epitopes. In Lewis rats, α1 100–116 is a dominant T-cell epitope, but in brown Norway rats α172–205 predominates, as does α152–70 in buffalo rats *(131)*. There is probably a similar diversity in human groups. Pathologically significant T-cell epitopes could derive from any part of an AChR subunit because the T cells do not bind to native AChRs in vivo, but to peptide fragments digested by professional antigen-presenting cells like dendritic cells, or interested amateurs like AChR-reactive B cells.

Before an AChR epitope can be recognized by the antigen receptor of a T-helper cell that will collaborate with an AChR-stimulated B lymphocyte to produce plasma cells-secreting autoantibodies to AChRs, the AChR peptide must first be bound by an MHC class II antigen-presenting protein *(132)*. The proteolytic processing mechanisms of the antigen-presenting cell and the binding properties of the various MHC class II proteins restrict the peptides which can be recognized by T cells *(133)*. This is probably reflected in part in the higher incidence of HLA-A1, B8, and DR3 class II MHC determinants in young onset Caucasian MG patients and different MHC determinants in other groups *(134)*. Inbred mice become resistant to EAMG subsequent to a single amino acid change in the I-A$_B$ protein *(135, 136)*. The human α1 sequence 144–156 is recognized only when it is presented by HLA-DR4 class II protein variants with glycine at position 86 and not by a variant with valine at this position *(134)*. If humans were as inbred as these mouse strains, and if AChRs had only a few T-cell epitopes, MG would be much more genetically constrained than is actually the case.

8. INDUCTION OF THE AUTOIMMUNE RESPONSE TO AChRS IN MG

The mechanisms which induce and sustain the autoimmune response to α1 AChRs in MG are not known. These mechanisms may differ in various forms of human MG, and mechanisms may differ between species. Human MG is a remitting and exacerbating disease in which the autoimmune response to α1 AChRs persists for years *(13, 123)*. Canine MG is an acute disease in which the autoimmune response to AChRs and muscle weakness usually remit within 6 months, except when the MG is associated with a neoplastic growth *(137)*. This suggests that in humans a source of immunogen persists which does not usually happen in dogs. For example, a chronic occult infection in humans might involve a tissue which expresses α1 AChR in an unusual amount, place, or state of post-translational modification, causing it to be immunogenic. In dogs, this infection might not persist. In either case, if chronic infection provided the immunogen, then appropriate antibiotic therapy might provide the cure.

MG in humans can be initiated, then terminated when the immunogen is removed. Rheumatoid arthritis patients treated with penicillamine sometimes develop autoimmune MG *(138, 139)*. It is thought that, in these patients with a predisposition to autoimmune responses, covalent reaction of penicillamine with thiol groups on α1 AChRs haptenizes them to produce new antigenic sites and provoke an autoimmune response *(140)*. Ending the penicillamine treatment ends the MG within a couple of months. Neonatal MG occurs in babies of mothers with MG as a result of passive transfer of autoantibodies from the mother *(141, 142)*. This too spontaneously remits within a couple of months. If passive or active EAMG is not sufficiently severe to be lethal, the animals recover as the response to the immunogen diminishes *(13)*. All four examples (penicillamine-induced MG, neonatal MG, passive and acute EAMG) illustrate that an autoimmune assault on neuromuscular junction AChRs sufficient to cause muscle weakness is not sufficient to present enough AChR to the immune system to sustain the autoimmune response.

Something special must happen in human MG to chronically provide sufficient immunogen to sustain the autoimmune response.

A paraneoplastic autoimmune response may be present in some MG patients. Twelve percent of MG patients have thymoma, and 35% of thymoma patients have MG *(143)*. These patients characteristically make high levels of autoantibodies to AChRs and to several structural proteins found within muscle cells, e.g., titin *(144, 145, 146)*. Thymic tumor cells do not usually express α1 AChRs *(130)*, but traces of AChR have been detected in thymus myoid cells, thymic epithelia, and dendritic cells *(147, 148)*. The autoantibodies to AChRs might result from a bystander adjuvant effect of the disruption caused by the tumor to an immune regulatory organ which contains traces of AChRs. Alternatively, viruses or other factors which cause the tumor might also cause disruption of AChR-expressing cells. In Lambert–Eaton syndrome (LES), tumor immunogens play a much more obvious role in provoking the autoimmune response. Sixty percent of these patients have small cell lung carcinomas *(149)* (as a result of addiction to tobacco through neuronal AChRs) *(2, 109)*. These carcinomas express voltage-sensitive calcium channels, which are the target of the autoimmune response in LES. The muscular weakness which these patients exhibit due to autoantibody impairment of ACh release *(149)*, often for years before the tumor is discovered, is the trade-off for development of a successful immune response to the tumor voltage-gated calcium channel which holds the otherwise rapidly fatal tumor in check.

In most MG patients, the immunogen is likely to be native muscle AChR or a closely related molecule, because the spectrum of autoantibody specificities in MG and EAMG is very similar *(114, 125)*. It may often be a fetal muscle AChR, given the selective reactivity for such AChRs by many MG patients *(126, 150)*. Denatured AChR subunits or synthetic fragments of AChRs are extremely inefficient at inducing EAMG because they lack the MIR *(151, 152, 153)*. Thus, molecular mimicry of the AChR by a bacteria or virus expressing a protein with a short stretch of sequence similar to the AChR is unlikely to induce MG. However, bacterial DNA can be a potent adjuvant, and microbial-associated autoimmune responses have been found in connection with both relatively obvious *(154, 155)* and more cryptic bacterial infections *(156)*.

9. AUTOIMMUNE MECHANISMS WHICH IMPAIR NEUROMUSCULAR TRANSMISSION IN MG AND EAMG

EAMG has been induced in many species by immunization with purified Torpedo electric organ AChR *(118)*. Native AChR is quite immunogenic, and EAMG has been induced with syngeneic AChR *(157)*, even in the absence of adjuvants *(158)*. The most detailed studies have been in Lewis rats *(118, 120, 159)*. Some other rat strains produce high levels of antibodies to AChRs, but are resistant to EAMG *(160)*. Some mouse strains are more resistant than others, but all are more resistant than Lewis rats *(136)*.

Among MG patients, the absolute concentration of antibodies to AChR does not correlate closely with severity *(161, 162)*, but, in general, patients with only ocular signs have lower concentrations than do patients with generalized MG, and changes in clinical state are usually paralleled by changes in antibody concentrations *(163)*.

The basic mechanisms by which autoantibodies to AChRs impair neuromuscular transmission appear to be very similar in MG and chronic EAMG *(118, 120)*.

There are forms of both MG and EAMG which are passively transferred by autoantibodies to AChRs. Passive transfer of EAMG is associated with an antibody and complement-dependent, transient phagocyte-mediated attack on the postsynaptic membrane which greatly amplifies the pathological potency of the bound antibodies *(164, 165)*. In neither chronic EAMG nor MG is antibody-dependent phagocytic attack or cytotoxic T-lymphocyte attack on the postsynaptic membrane observed *(120)*. Repeated injection of mice with large amounts of MG patient IgG causes a mild form of muscular weakness *(166, 167)*. Mothers with MG can passively transfer MG to their babies *(141, 142)*. Usually this is mild and transient, ending as maternal IgG is cleared from the baby.

Active immunization of Lewis rats with AChR in adjuvant results in chronic EAMG after about 30 days *(157)*. If sufficient AChR and adjuvants are used, this can be preceded by an acute phase of EAMG 8–11 days after immunization, which involves antibody and complement-dependent phagocytes, as in passive EAMG *(157, 168, 169, 170)*. In both chronic EAMG and MG, ACh sensitivity is decreased and decrementing electromyogram responses are observed as transmission becomes more likely to fail on successive stimulae *(171, 172)*. This is because decreased postsynaptic sensitivity to ACh reduces the safety factor for transmission, so failure becomes more likely as fewer vesicles of ACh are released on successive stimulae due to depletion of docked vesicles. Serum autoantibody concentration is high, there is loss of more than half of muscle AChRs, and many of those which remain have antibodies bound *(157, 173)*. The structure of the postsynaptic membrane is disrupted: the folded structure is simplified, AChR content is reduced, antibodies and complement are bound, and focal lysis occurs with shedding of membrane fragments into the synaptic cleft *(168, 174, 175)*. In addition to AChR loss through complement-mediated lysis, loss occurs due to antigenic modulation following crosslinking of AChRs by antibody *(121, 122, 176)*. Only a small fraction of antibodies to AChR are capable of inhibiting AChR function *(177, 178, 179, 180, 181, 182)*. The lack of direct effect on AChR function by most antibodies to AChR is illustrated by the observation that if rats are depleted by the C3 component of complement to prevent both lysis and phagocytic invasion, then passive transfer of IgG from rats with chronic EAMG can label at least 67% of the AChRs without causing weakness *(183)*.

10. EFFECTS OF AChR MUTATIONS IN CONGENITAL MYASTHENIC SYNDROMES

The structure of muscle AChRs is optimized for this functional role *(28)*. Studies of congenital myasthenic syndromes due to mutations in AChR subunits *(1, 10, 11, 120)* have clearly demonstrated that it is detrimental to either increase *(184)* or decrease *(185)* affinity for ACh, or increase *(186)* or decrease *(187)* the channel-opening time. Such changes often result in large pleotropic changes in the number of AChRs and the morphology of the neuromuscular junction, further contributing to impairment of neuromuscular transmission. Below, some examples of the many congenital myasthenic syndromes elegantly characterized by Andrew Engel *(10, 11)* and his coworkers will be described briefly with the intention of relating defects in aspects of the AChR structures previously described to specific functional defects and the resulting medical effects.

Synthesis of AChR subunits can be impaired or prevented by mutations in the promotor region or frameshift mutations that result in truncation of the subunit *(188, 189, 190, 191)*. These sorts of mutations are more frequent in ε subunits than others because, in humans, expression of γ is induced which partially compensates for the loss of ε. Expansion of the endplate area also helps to compensate for loss of AChRs. In mice, knockout of the ε subunit is perinatal lethal *(192)*. Deletion mutants of α1 subunits are not seen, presumably because this would be lethal.

Mutations both within and outside the ACh-binding region alter agonist potency and efficacy. Reciprocally, these mutations alter channel-opening kinetics. A mutation which replaces α1 glycine 153 with a serine (in loop B of the ACh-binding site) decreases the rate of ACh dissociation, resulting in repeated channel opening during prolonged ACh binding and increased desensitization *(184)*. The resulting 100-fold increase in affinity which prevents the normal rapid termination of transmission results in excitotoxic damage, including reduced AChR amount, increased desensitization, and morphological alterations of the postsynaptic region due to overloading with cations. ACh esterase inhibitors would not be beneficial therapy. This mutation was less disabling to the patient than were some mutations in the M2 channel-lining region which, by increasing the stability of the open state of the channel, prolonged channel opening fivefold longer than did this increase in affinity for ACh *(186)*. A similar increase in agonist-binding affinity, activation, and desensitization was also caused by a mutation in the $\alpha 1$ M2 domain V249F at a position that is not on the hydrophilic lining of the channel *(193)*. An ε P121L mutation near the E loop of the ACh-binding site reduced the affinity for ACh in the open and desensitized states, resulting in infrequent and brief episodes of channel opening and, consequently, impaired transmission causing weakness *(185)*. The amount of AChR and synaptic morphology were normal due to the absence of excitotoxicity in this mutation. An $\alpha 1$ N217K mutation in the M1 region was found to cause a 20-fold increase in the affinity of ACh for the resting state of the AChR *(194)*. This N is a conserved amino acid in the synaptic third of the M1 transmembrane domain, i.e., not a contact amino acid in the ACh-binding site but part of the lumen of the channel, between the ACh-binding site and the conserved P121 C122 in the middle of M1, a region which is known to move during the process of AChR activation *(195)*. Efficacy and potency of agonists is effected by mutations in the M2 channel-forming part of the AChR because some of the mutations so destabilize the resting conformation that even binding of choline (usually not an effective agonist) is sufficient to activate the AChR *(196)*. Thus, the levels of choline in serum are sufficient to cause spontaneous activation of AChRs and contribute to excitotoxic damage in these long-channel congenital myasthenic syndromes. Experimental mutations in similar regions in M2 of $\alpha 7$ AChRs cause some antagonists to behave as agonists *(197)* by facilitating the transition from the resting to the open state. Choline is selective as an agonist, though a weak one, for $\alpha 7$ AChRs *(198)*. This may be because the five ACh-binding sites on $\alpha 7$ homomers permit the low-affinity binding of a sufficient number of choline molecules to trigger activation by this very weak agonist.

Mutations in the M2 channel-forming part of the AChR frequently affect the duration of channel opening, usually causing prolonged channel opening resulting in excitotoxicity. For example, the T264P mutation in the εM2 hydrophilic surface causes prolonged channel openings *(186)*. The α V249F mutation on the hydrophobic surface of M2 also causes an autosomal-dominant slow-channel myasthenic syndrome *(193)*. Agonist potency is increased and desensitization is enhanced sufficiently to impair function at physiological rates of stimulation. Excitotoxic effects include loss of AChR, degeneration of junctional folds, cluttering of an enlarged synaptic cleft with bits of membranous debris, degeneration of synaptic mitochondria and other organelles, and apoptosis of some junctional nuclei. Mutations in the M2 regions of several AChR subunits cause slow-channel syndromes *(191)*. Other mutations can cause excessively rapid closure of the channel, but also excessive conductance, so that the net effect is also excitotoxic *(187)*.

Myasthenic syndromes have also been identified in other parts of AChR subunits. One was found to involve two recessive mutations in ε *(199)*. The C128S mutation at one end of the disulfide-linked signature loop in the extracellular domain inhibits AChR assembly. Then the ε 1245 ins 18 mutation, which duplicates six amino acids in the C-terminal half of the large cytoplasmic domain, surprisingly, forms AChRs with altered opening kinetics and reduced net openings. That mutations in and out of the ACh-binding site effect agonist potency, and mutations in and out of the channel effect gating, is all evidence that activation and desensitization of AChRs involve both local and global conformational changes; and mutations in distant parts of the AChR protein can effect the stability of various conformation states and the kinetics of transition between them. Clinically, it is interesting that a patient and her similarly affected siblings with these two mutations went through life exhibiting muscle weakness and fatigue, but the crisis which provoked detailed study was when the 36-year-old patient underwent surgery and was given a curariform muscle relaxant which resulted in paralysis lasting several hours due to her low safety factor for transmission.

Another striking feature of mutations causing congenital myasthenic syndromes is that there are a lot of these mutations. Often a recessive mutation does not produce a phenotype until two occur simultaneously. Thus far, more than 60 such mutations have been found *(10, 11)*. It seems likely that similar varieties of mutations will be found in other receptors and channels.

11. NEURONAL AChR SUBTYPES AND FUNCTIONAL ROLES

There are many potential and a few known prominent neuronal AChR subtypes, as indicated in Fig. 4 *(2)*. α4β2 AChRs account for most of the high-affinity nicotine binding in brain. α2 AChRs in primate brain may assume some of the roles of α4 AChRs in rodent brains *(200)*. Some of these also have α5 or perhaps β3 subunits replacing the β2 subunit equivalent to β1 in muscle AChRs *(201, 202)*. Autonomic ganglia postsynaptic AChRs are predominantly α3β4 AChRs, but α3β4α5 AChRs and subtypes containing β2 are also present. In brain aminergic neurons, α6 in combination often with β3 and β2 or β4 and sometimes α3 or α4 can form a variety of complex subtypes *(68, 82, 83, 84, 203, 204)*. α7 homomers are found both in autonomic ganglia and in brain. Neurons often contain complex mixtures of AChR subtypes *(81, 82)*. In autonomic ganglia, α3 AChRs perform a postsynaptic role in transmission similar to that of muscle AChRs, but many brain AChRs are involved in pre- and extrasynaptic modulation of the release of many transmitters *(205)* and some extrasynaptic α7 AChRs may play roles in trophic regulation *(85)*.

A good way to get an overview of the functional roles of neuronal AChRs is to examine the features of mice in which a particular subunit has been knocked out or replaced with a hyperactive mutant form.

Knock out of α3 subunits is neonatal lethal *(206)*. They lack autonomic transmission and die as a result of a distended and infected bladder.

Knock out of α4 subunits causes loss of most high-affinity nicotine binding in brain, loss of nicotine-induced anti-nociception, and increased anxiety *(207, 208)*. Replacement of α4 with an excitotoxic M2 mutation was neonatal lethal, but a heterozygote survived which had reduced expression of the mutation due to some expression of the neo cassette involved in making the mutant mouse *(209)*. These mice lost dopaminergic neurons in the substantia nigra, exhibited altered motor behavior and learning, and increased anxiety. Knock in of hyperactive α4 subunits was used to show that AChRs containing α4 subunits can account for the rewarding, tolerance, and sensitizing effects of nicotine *(17)*. These hypersensitive AChRs also cause changes in the sleep–wake cycle and increase sensitivity to nicotine-induced seizures *(210)*.

Knock out of α6 subunits does not reduce the total number of AChRs in dopaminergic nerve endings in the striatum because of a compensatory increase in the α4 subunits which are expressed in these neurons to compensate for the loss of α6 in α6β2β3 and α6α4β2β3 AChRs *(211, 212)*. α6-containing AChRs are selectively expressed in these nerve endings while α4 AChRs are selectively expressed in the cell bodies of these neurons in the ventral tegmental area. α6 AChRs are selectively lost in MPTP-induced animal models of Parkinson s disease *(213)* and may be target for drug therapy in this disease *(21)*.

Knock out of α7, surprisingly, produced few evident problems *(214, 215)*. The most evident phenotype is reduced fertility. Knock in of an excitotoxic M2 mutant α7 was neonatally lethal *(216)*. A problem in interpretation of knock out mice is that developmental compensation may minimize the apparent functional importance of the knock out subunit. The problem with knocking in a lethally excitotoxic mutant subunit is that any neuron which at any time in its development expresses some of the subunit risks dying, thus exaggerating the apparent functional importance of the subunit. Multinucleate muscle cells are large and rather resilient to the lethal excitotoxic effects of mutant AChRs, but smaller neurons may not be, especially early in development when their ability to buffer excess calcium ions entering through hyperactive AChRs is low *(217)*.

Knock out of α9 prevented cochlear efferent stimulation *(218)*, as expected from its post-synaptic role there. α10 subunits are functionally equivalent to β subunits, and α9α10 AChRs usually have the stoichiometry $(α9)_2(α10)_3$ *(219)*. These AChRs are unusual in their physiological role in that entry of Ca^{++} through these AChRs triggers Ca^{++}-sensitive K^+ channels resulting in a net inhibitory postsynaptic response *(220)*.

Knock out of β2 altered some learning behaviors, caused increased neuronal death with aging, prevented nicotine-induced anti-nociception, and prevented nicotine reinforcement (i.e., addiction) *(221)*. Selective expression of β2 only in the ventral tegmental area of β2 knock out mice results in nicotine-induced dopamine release and nicotine self-administration *(221)*. Nicotine induces wakefulness in wild-type but not β2 knock out mice, and knock out mice exhibit disrupted sleep patterns *(222)*. Wild-type mouse pups exposed to nicotine exhibit reduced growth, unstable breathing, as well as impaired arousal and catecholamine synthesis *(223)*. β2 knock out pups exhibit the same symptoms. This suggests that these effects of nicotine result from desensitization of AChRs containing β2 subunits. Deletion of the β2 subunit altered development of tolerance to nicotine and eliminated nicotine-induced upregulation of the amount of brain AChR *(225)*. Nicotine induces the amount of AChRs *(226, 227, 228)*. This is much more prominent with β2-containing than β4-containing AChRs *(228, 229)*. Upregulation results primarily from nicotine binding to assembly intermediates, thereby causing a conformation change which promotes their assembly *(72, 230)*. Nicotine also slows the rate of AChR destruction *(72, 226, 228)*. High concentrations of nicotine can increase the expression of human muscle AChRs in cell lines *(231)*, but in vivo smoking-associated concentrations of nicotine only upregulate brain *(232, 233)* not muscle AChRs.

Knock out of β3 subunits *(224)* selectively reduces, but does not eliminate, expression of the α6-containing nigrostriatal AChRs *(203)* with which they are associated. This alters striatal dopamine release, locomotor activity, and acoustic startle behaviors.

Knock out of β4 produced viable mice, indicating that β2 expression must have compensated for loss of β4 in α3 AChRs in autonomic neurons *(234)*. When both β2 and β4 were knocked out, the effects were even more lethally severe than the knock out of the α3 subunit.

Neuronal nicotinic AChRs are directly responsible for addiction to the nicotine in tobacco *(2, 107, 108)*, and are more peripherally involved in many other neurological problems *(2, 19)*.

12. AUTOIMMUNE IMPAIRMENT OF NEURONAL AChRS

Low levels of autoantibodies to α3 AChRs were found in 41% of patients with idiopathic or paraneoplastic dysautonomias, and 9% of patients with postural tachycardia syndrome, idiopathic gastrointestinal dysmotility, or diabetic autonomic neuropathy *(14)*. Antibody levels were higher in the more severely affected and decreased with improvement. Low levels of autoantibodies to α3 or α7 AChRs have also been found in a few patients with MG, LES, Guillain–Barre Syndrome, or chronic inflammatory demyelinating polyneuropathy who also exhibited symptoms of autonomic dysfunction *(235)*. Unlike human MG, but like canine MG *(137)*, these autoimmune dysautonomias are often monophasic illnesses *(14)*. Details of the mechanisms by which the autoantibodies impair autonomic transmission are unknown, but they probably include some or all of the mechanisms observed in MG and EAMG. As with MG, plasmapheresis of autoimmune autonomic neuropathy produces dramatic improvement *(236)*. Experimental autoimmune autonomic neuropathy (EAAN) can be induced by immunization of rabbits with the bacterially expressed extracellular domain of the α3 subunit *(237, 238)*, and serum from these rabbits can passively transfer EAAN to mice *(239)*.

It is interesting that rat mAbs to the MIR react well with human α3 AChRs *(76)*, yet MG patient antisera in general do not react with human α3AChRs *(14)*, AAN patient autoantibodies do not react with muscle AChRs *(14)*, and rats with EAMG do not exhibit obvious autonomic symptoms *(13)*. Thus, despite the similarities in primary sequences in the 66–76 regions of α1, α3, α5, and β3 subunits, the conformation dependence of MIR antibody binding and the large areas of a protein occluded by a bound antibody can account for the apparent differences in MIR antibody specificity between rats and humans, despite mutual competition for binding to α1 subunits.

Low levels of AChR α3, α5, α7, α9 and β2 and β4 AChR subunits have been reported in human keratinocytes, and ACh has been reported to modulate keratinocyte adhesion, proliferation, migration, and differentiation *(7)*. Autoantibodies in pemphigus patients have been found directed not only at desmoglien, but also at α9 AChRs, and a novel keratinocyte ACh-binding protein termed pemphaxin *(16)*. Pemphaxin is not the equivalent of the snail ACh-binding protein *(38, 240)*. Autoimmune responses to neuronal AChRs found in other non-neuronal tissues *(7, 8, 9)* have not been reported yet.

A priori, antibody-mediated autoimmune responses to receptors in brain would seem unlikely due to the presence of the blood/brain barrier. However, autoantibodies to glutamate receptors have been found in some Rasmussen's encephalitis patients and certain forms of cerebellar degeneration *(241, 242, 243)*. Immunization of rabbits with bacterially expressed glutamate receptor protein causes lethal Rasmussen's-like seizures resulting from autoantibodies with agonist activity which cause excitotoxic damage *(241)*. Thus, there is precedent for antibody-mediated autoimmune responses to brain neuronal AChRs, and it is possible that such responses could involve neuronal AChRs.

13. EFFECTS OF HUMAN NEURONAL AChR MUTATIONS

Autosomal-dominant nocturnal frontal lobe epilepsy (ADNFLE) is a rare form of epilepsy (resembling night terrors) which is caused by mutations in either α4 or β2 subunits in the M2 region *(12, 244, 245, 246)*. There are mutations known in both α4 and in β2. A conundrum is that the α4 mutations involve loss of function, whereas the β2 mutations involve gain of function. For example, replacement of α4 serine 247 in the M2 channel lining with a phenylalanine reduced channel function as a result of use-dependent functional upregulation, faster desensitization, slower recovery from desensitization, less inward rectification, and virtual elimination of Ca^{++} permeability *(247)*. Although many functions of α4β2 AChRs in brain are unknown, α4β2 AChRs have been shown to promote the release of the inhibitory transmitter GABA *(248)*. Thus, reduced function of α4β2 AChRs might produce the excessive activation characteristic of epilepsy; if under the right circumstances in the wake/sleep cycle, reduced AChR function caused reduced inhibition. However, ADNFLE mutations in β2 subunits cause excessive function when expressed with α4 *(245, 246)*. As an example, a V287M mutation in a conserved valine near the C-terminal end of M2 causes a 10-fold increase in ACh potency and reduced desensitization *(245)*. Since β2 can also function in combinations with α2, α3, and α6, it is possible that one of these combinations or α4β2 AChRs in a different circuit could account for ADNFLE. It has been proposed that a common feature shared by all ADNFLE mutations is reduced Ca^{++} potentiation of the response to ACh *(250)*. Clearly, it is much more difficult to study neuronal AChR mutations than muscle AChRs because the many complex functions of neuronal AChRs in many overlapping circuits are not well known and because biopsy material is not usually available or easy to study.

The autonomic dysfunction and neonatal lethality of α3 or β2 plus β4 knock outs in mice *(187, 234)* resemble the human disease megacystic-microcolon-hypoperistalsis syndrome. No AChR mutations have yet been found to account for this disease *(249)*, although there is some evidence for loss of α3 AChRs in these patients *(251)*.

The large number of mutations found in muscle AChRs *(1, 10)* makes it seem likely that there are many neuronal AChR mutation syndromes to be discovered. Instead of affecting a single subtype of AChR, and producing symptoms of one basic type, varying the degrees or features of myasthenia, congenital neuronal AChR mutation syndromes will probably be discovered which affect a wide range of AChR subtypes (252, 253, 254). These mutations might produce hyper- and hypo- function, and result in a wide range of pleotropic phenotypes affecting both the central and peripheral nervous systems as well as the non-neuronal tissues in which these AChRs are expressed.

Acknowledgments The Lindstrom laboratory is supported by grants from the NIH.

REFERENCES

1. Sine S and Engel A. Recent advances in Cys-loop receptor structure and function. Nature 2006;440:448–454
2. Lindstrom J: The structures of neuronal nicotinic receptors. In Clementi F, Gotti C, Fornasari D eds. Neuronal Nicotinic Receptors Handbook Exp Pharmacol, New York: Springer, 2000:144:101–162.
3. Berg D, Shoop R, Chang K, et al. Nicotinic acetylcholine receptors in ganglionic transmission. In Clementi F, Gotti C, Fornasari D eds. Neuronal Nicotinic Receptors Handbook Exp Pharmacol, New York: Springer, 2000:144:247–270.
4. Zoli M. Distribution of cholinergic neurons in the mammalian brain with special reference to their relationship with neuronal nicotinic receptors. In Clementi F, Gotti C, Fornasari D eds. Neuronal Nicotinic Receptors Handbook Exp Pharmacol, New York: Springer, 2000:144:13–30.
5. Kaiser S, Soliokov L, Wonnacott S. Presynaptic neuronal nicotinic receptors: pharmacology, heterogeneity, and cellular mechanisms. In Clementi F, Gotti C, Fornasari D eds. Neuronal Nicotinic Receptors Handbook Exp Pharmacol, New York: Springer, 2000:144:193–212.
6. Grando S. Cholinergic control of epidermal cohesion. Experimental Dermatology. 2006;15:265–282.
7. Maus A, Pereira E, Karachunski P, et al. Human and rodent bronchial epithelial cells express functional nicotinic acetylcholine receptors. Mol Pharmacol 1998;54:779–788.
8. Macklin K, Maus A, Pereira E, et al. Human vascular endothelial cells express functional nicotinic acetylcholine receptors. J Pharmacol Exp Ther 1998;287:435–439.
9. Sekhon HS, Jia YB, Raab R, et al. Prenatal nicotine increases pulmonary alpha 7 nicotinic receptor expression and alters fetal lung development in monkeys. J Clin Invest 1999;103: 637–647.
10. Engel A, Ohno K, and Sine S. Sleuthing molecular targets for neurological diseases at the neuromuscular junction. Nat Rev Neurosci 2003;4:399–352.
11. Steinlein OK. Neuronal nicotinic receptors in human epilepsy. Eur J Pharmacol 2000;393:243–247.
12. Combi R, Dalpra L, Tenchini ML, Ferini-Strambi L. Autosomal dominant nocturnal frontal lobe epilepsy – A critical overview. J Neurol 2004;251:923–934.
13. Lindstrom J. Acetylcholine receptors and myasthenia. Muscle Nerve 2000;23:453–477.
14. Vernino S, Low PA, Fealey RD, et al. Autoantibodies to ganglionic acetylcholine receptors in autoimmune autonomic neuropathies. N Eng J Med 2000;343:847–855.
15. Vernino S, Ermilov L, Sha L, Szurszewski J, Law P, and Lennon V. Passive transfer of autoimmune autonomic neuropathy to mice. J Neurosci. 2004;24:7037–7042.
16. Grando SA. Autoimmunity to keratinocyte acetylcholine receptors in pemphigus. Dermatology 2000;201:290–295.
17. Tapper A, McKinney S, Nashmi R, Schwarz J, Deshpande P, Labarca C, Whiteaker P, Marks M, Collins A, and Lester H. Nicotine activation of alpha4* receptors: sufficient for reward, tolerance, and sensitization. Science. 2004;306:1029–1032.
18. Peto R, Lopez AD, Boreham J, et al. Mortality from tobacco in developed-countries – indirect estimation from national vital-statistics. Lancet 1992;339:1268–1278.
19. Lloyd GK, Williams M. Neuronal nicotinic acetylcholine receptors as novel drug targets. J Pharmacol Exp Ther 2000;292:461–467.
20. Heeschen C, Jang J, Weis M, et al. Nicotine stimulates angiogenesis and promotes tumor growth and atherosclerosis. Nat Med 2001;7:833–839.
21. Quik M, McIntosh JM. Striatal alpha6* nicotinic acetylcholine receptors: potential targets for Parkinson s disease therapy. J Pharm Exp. 2006;316:481–489.
22. Hurst RS, Hajos M, Raggenbass M, et al. A novel positive allosteric modulator of the alpha7 neuronal nicotinic acetylcholine receptor: in vitro and in vivo characterization. J Neurosci 2005;25:4396–4405.
23. Coe JW, Brooks PR, Wirtz MC, et al. 3,5-Bicyclic aryl piperidines: a novel class of alpha4beta2 neuronal nicotinic receptor partial agonists for smoking cessation. Bioorg Med Chem Lett 2005;15:4889–97.
24. Karlin A, Akabas MH. Toward a structural basis for the function of nicotinic acetylcholine receptors and their cousins. Neuron 1995;15:1231–1244.
25. Galzi J-L, Devillers-Thiery A, Hussy N, Bertrand S, Changeux J-P and Bertrand D. Mutations in the channel domain of a neuronal nicotinic receptor convert ion selectivity from cationic to anionic. Nature 1992;359:500–505.

26. Eisele JL, Bertrand S, Galzi J-L, Devillers-Thiery A, Changeux J-P and Bertrand D. Chimeric nicotinic serotonergic receptor combines distinct ligand-binding and channel specificities. Nature 1993;366:479–483.

27. Lindstrom J. Purification and cloning of nicotinic acetylcholine receptors. In Arneric S, Brioni D, eds. Neuronal nicotinic receptors: pharmacology and therapeutic opportunities. New York, John Wiley and Sons, Inc, 1999;3–23.

28. Jackson MB. Perfection of a synaptic receptor – kinetics and energetics of the acetylcholine-receptor. Proc Nat Acad Sci USA 1989;86:2199–2203.

29. Kummer TT, Misgeld T, Sanes JR. Assembly of the postsynaptic membrane at the neuromuscular junction: paradigm lost. Curr Opin Neurobiol 2006;16:74–82.

30. Changeux J-P, Edelstein SJ. Nicotinic acetylcholine receptors: from molecular biology to cognition. Odile Jacob/ Johns Hopkins University Press, New York: 2005.

31. Unwin N. Refined structure of the nicotinic acetylcholine receptor at 4 Å resolution. J Mol Biol. 2005;346:967–989.

32. Patrick J, Lindstrom J. Autoimmune response to acetylcholine receptor. Science 1973;180:871–2.

33. Tzartos SJ. Lindstrom JM. Monoclonal antibodies used to probe acetylcholine receptor structure: localization of the main immunogenic region and detection of similarities between subunits. Proc Nat Acad Sci USA 1980;77:755–9.

34. Tzartos SJ, Barkas T, Cung MT, et al. Anatomy of the antigenic structure of a large membrane autoantigen, the muscle-type nicotinic acetylcholine receptor. Immunol Rev 1998;163:89–120.

35. Unwin N. Nicotinic acetylcholine receptor and the structural basis of fast synaptic transmission. Phil Tran Roy Soc London B 2000;1404:1813–1829.

36. Maimone MM, Merlie JP. Interaction of the 43 kd postsynaptic protein with all subunits of the muscle nicotinic acetylcholine-receptor. Neuron 1993;11:53–66.

37. Smit AB, Syed NI, Schaap D, et al. A glia-derived acetylcholine-binding protein that modulates synaptic transmission. Nature 2001;411:261–268.

38. Brejc K, van Dijk WJ, Klaassen, et al. Crystal structure of an ACh-binding protein reveals the ligand-binding domain of nicotinic receptors. Nature 2001;411:269–276.

39. Celie PH, van Rossum-Fikkert SE, van Dijk WJ, et al. Nicotine and carbamylcholine binding to nicotinic acetylcho-line receptors as studied in AChBP crystal structures. Neuron 2004;41:907–914.

40. Hansen SB, Sulzenbacher G, Huxford T, et al. Structures of Aplysia AChBP complexes with nicotinic agonists and antagonists reveal distinctive binding interfaces and conformations. EMBO J 2005;24:3635–46.

41. Beroukhim R, Unwin N. 3-Dimensional location of the main immunogenic region of the acetylcholine-receptor. Neuron 1995;15:323–331.

42. Lee WY, Sine SM. Principal pathway coupling agonist binding to channel gating in nicotinic receptors. Nature 2005;438:243–7.

43. Wells GB, Anand R, Wang F, Lindstrom J. Water-soluble nicotinic acetylcholine receptor formed by α7 subunit extracellular domains. J Biol Chem 1998;273:964–973.

44. Bouzat C, Gumilar F, Spitzmaul G, et al. Coupling of agonist binding to channel gating in an ACh-binding protein linked to an ion channel. Nature 2004;430:896–900.

45. Grutter T, Changeux J-P. Nicotinic receptors in wonderland. Trends Biochem Sci 2001;26:459–463.

46. Fu DX, Sine SM. Asymmetric contribution of the conserved disulfide loop to subunit oligomerization and assembly of the nicotinic acetylcholine receptor. J Biol Chem 1996;271:31479–31484.

47. Wilson GG, Karlin A. Acetylcholine receptor channel structure in the resting, open, and desensitized states probed with the substituted-cysteine-accessibility method. Proc Nat Acad Sci USA 2001;98:1241–1248.

48. Das MK, Lindstrom J: Epitope mapping of antibodies to acetylcholine receptor-alpha subunits using peptides synthesized on polypropylene pegs. Biochemistry 1991;30:2470–2477.

49. Conroy WG, Liu Z, Nai Q, et al. PDZ-containing proteins provide a functional postsynaptic scaffold for nicotinic receptors in neurons. Neuron 2003;38:759–71.

50. Jeanclos E, Lin L, Trevel M, Rao J, DeCoster M, Anand R. The chaperone protein 14-3-3 η interacts with the nicotinic receptor α4 subunit. J Biol Chem 2001; 276:28281–28290.

51. Keller S, Lindstrom J, Taylor P. Involvement of the chaperone protein calnexin and the acetylcholine receptor β subunit in the assembly and cell surface expression of the receptor. J Biol Chem 1996;271:22871–22877.

52. Keller D and Taylor P. Determinants responsible for assembly of the nicotinic acetylcholine receptor. J Gen Physiol 1999;113:171–176.

53. Keller SH, Lindstrom J, and Taylor P. Adjacent basic amino acid residues recognized by the COPI complex and ubiquitination govern endoplasmic reticulum to cell surface trafficking of the nicotinic acetylcholine receptor alpha-subunit. J Biol Chem 2001;276:18384–18391.

54. Wang JM, Zhang L, Yao Y, et al. A transmembrane motif governs the surface trafficking of nicotinic acetylcholine receptors. Nat Neurosci 2002;5:963–70.

55. Miles K, Huganir RL. Regulation of nicotinic acetylcholine-receptors by protein-phosphorylation. Mol Neurobiol 1988;2:91–124.

56. Fenster CP, Beckman ML, Parker JC, et al. Regulation of alpha 4 beta 2 nicotinic receptor desensitization by calcium and protein kinase C. Mol Pharmacol 1999;55:432–443.

57. Williams BM, Temburni MK, Levy MS, Bertrand S, Bertrand D, Jacob M. The long internal loop of the α3 subunit targets nAChRs to subdomains within individual synapses on neurons in vivo. Nat Neurosci 1998;1:557–562.
58. Bouzat C, Bren N, Sine SM. Structural basis of the different gating kinetics of fetal and adult acetylcholine receptors. Neuron 1994;13:1395–1402.
59. Hales T, Dunlop J, Deeb T, et al. Common determinants of single channel conductance within the large cytoplasmic loop of 5-hydroxytryptamine type 3 and alpha4beta2 nicotinic acetylcholine receptors. J Biol Chem 2006;281:8062–71.
60. Paradiso K, Zhang J, Steinbach J. The C terminus of the human nicotinic α4β2 receptor forms a binding site required for potentiation by an estrogenic steroid. J Neurosci 2001;21:6561–6568.
61. Curtis L, Buisson B, Bertrand S, Bertrand D. Potentiation of the α4β2 neuronal nicotinic acetylcholine receptor by estradiol. Mol Pharmacol 2002;61:127–135.
62. Sussman JL, Harel M, Frolow F, et al. Atomic-structure of acetylcholinesterase from *Torpedo-californica* – a prototypic acetylcholine-binding protein. Science 1991;253:872–879.
63. LeNovere N, Changeux JP. Molecular evolution of the nicotinic acetylcholine-receptor – an example of a multigene family in excitable cells. J Mol Evol 1995;40:155–172.
64. Duvoisin R, Deneris E, Patrick J, Heinemann S. The functional diversity of the neuronal nicotinic receptors is increased by a novel subunit: β4. Neuron 1989;3:487–496.
65. Couturier S, Bertrand D, Matter JM, Hernandez MC, Bertrand S, Millar N, Valera S, Barkas T, Ballivet M. A neuronal nicotinic acetylcholine receptor subunit (alpha 7) is developmentally regulated and forms a homo-oligomeric channel blocked by alpha-BTX. Neuron 1990;5:847–56.
66. Anand R, Conroy WG, Schoepfer R, Whiting P, Lindstrom J. Chicken neuronal nicotinic acetylcholine receptors expressed in Xenopus oocytes have a pentameric quaternary structure. J Biol Chem 1991;266:11192–11198.
67. Cooper E, Couturier S, Ballivet M. Pentameric structure and subunit stoichiometry of a neuronal nicotinic acetylcholine receptor. Nature 1991;350:235–238.
68. Kuryatov A, Olale F, Cooper J, Choi C, Lindstrom J. Human α6 AChR subtypes: subunit composition, assembly, and pharmacological responses. Neuropharmacology 2000;39:2570–2590.
69. Boorman J, Groot-Kormelink P, Silviotti L. Stoichiometry of human recombinant neuronal nicotinic receptors containing the β3 subunit expressed in Xenopus oocytes. J Physiol 2000;529:565–577.
70. Nelson ME, Kuryatov A, Choi CH, Zhou Y, Lindstrom J. Alternate stoichiometries of alpha4beta2 nicotinic acetylcholine receptors. Mol Pharmacol 2003;63:332–41.
71. Zhou Y, Nelson ME, Kuryatov A, Choi C, Cooper J, Lindstrom J. Human alpha4beta2 acetylcholine receptors formed from linked subunits. J Neurosci 2003;23:9004–15.
72. Kuryatov A, Luo J, Cooper J, Lindstrom J. Nicotine acts as a pharmacological chaperone to up-regulate human alpha4beta2 acetylcholine receptors. Mol Pharmacol 2005;68:1839–1851.
73. Schoepfer R, Conroy W, Whiting P, Gore M, Lindstrom J. Brain α-bungarotoxin binding-protein cDNAs and mAbs reveal subtypes of this branch of the ligand-gated ion channel gene superfamily. Neuron 1990;5:35–48.
74. Elgoyhen A, Vetter D, Katz E, Rothlin C, Heinemann S, Boulter J. Alpha 10: A determinant of nicotinic cholinergic receptor function in mammalian vestibular and cochlear mechanosensory hair cells. Proc Nat Acad Sci USA 2001;98:3501–3506.
75. Lustig LR, Peng H, Hiel H, Yamamoto T, Fuchs P. Molecular cloning and mapping of the human nicotinic acetylcholine receptor alpha 10 (CHRNA10). Genomics 2001;73: 272–283.
76. Wang F, Gerzanich V, Wells GB, et al. Assembly of human neuronal nicotinic receptor α5 subunits with α3, β2, and β4 subunits. J Biol Chem 1996;271:17656–17665.
77. Forsayeth JR, Kobrin E. Formation of oligomers containing the β3 and β4 subunits of the rat nicotinic receptor. J Neurosci 1997;17:1531–1538.
78. Fucile S, Barabino B, Palma E, et al. α5 subunit forms functional α3β4α5 nAChRs in transfected human cells. Neuro Rep 1997;8:2433–2436.
79. Vailati S, Hanke W, Bejan A, et al. Functional α6-containing nicotinic receptors are present in chick retina. Mol Pharmacol 1999;56:11–19.
80. Groot-Kormelink P, Luyten W, Colquhoun D, Silviotti L. A reporter mutation approach shows incorporation of the orphan subunit β3 into a functional nicotinic receptor. J Biol Chem 1998;273:15317–15320.
81. Conroy WG, Berg DK. Neurons can maintain multiple classes of nicotinic acetylcholine-receptors distinguished by different subunit compositions. J Biol Chem 1995;270:4424–4431.
82. Lena C, deKerchove d Exaerde AD, Cordero-Erausquin M, LeNovere N, Arroyo-Jimenez M, Changeux J-P. Diversity and distribution of nicotinic acetylcholine receptors in the locus ceruleus neurons. Proc Nat Acad Sci USA 1999;96:12126–12131.
83. Zoli M, Moretti M, Zanardi A, McIntosh JM, Clementi F, Gotti C. Identification of the nicotinic receptor subtypes expressed on dopaminergic terminals in the rat striatum. J Neurosci 2002;22(20):8785–8789.
84. Gotti C, Moretti M, Zanardi A, Gaimarri A, Champtiaux N, Changeux JP, Whiteaker P, Marks MJ, Clementi F, Zoli M. Heterogeneity and selective targeting of neuronal nicotinic acetylcholine receptor (nAChR) subtypes expressed on retinal afferents of the superior colliculus and lateral geniculate nucleus: identification of a new native nAChR subtype alpha3beta2(alpha5 or beta3) enriched in retinocollicular afferents. Mol Pharmacol 2005;68(4):1162–71.

85. Shoop RD, Chang KT, Ellisman MH, Berg D. Synaptically driven calcium transients via nicotinic receptors on somatic spines. J Neurosci 2001;21:771–781.

86. Nelson M, Wang F, Kuryatov A, Choi C, Gerzanich V, Lindstrom J. Functional properties of human nicotinic AChRs expressed in IMR-32 neuroblastoma cells resemble those of $\alpha 3 \beta 4$ AChRs expressed in permanently transfected HEK cells. J Gen Physiol 2001;118:563–582

87. Klink R, deKerchove d Exaerde A, Zoli M, Changeux J-P. Molecular and physiological diversity of nicotinic acetylcholine receptors in the midbrain dopaminergic nuclei. J Neurosci 2001;21:1452–1463.

88. Ramirez-Latorre J, Yu C, Qu X, Perin F, Karlin A, Role L. Functional contributions of $\alpha 5$ subunit to neuronal acetylcholine receptor channels. Nature 1996;380:347–351.

89. Gerzanich V, Wang F, et al. $\alpha 5$ subunit alters desensitization, pharmacology, Ca^{++} permeability and Ca^{++} modulation of human neuronal $\alpha 3$ nicotinic receptors. J Pharmacol Exp Ther 1998;286:311–320.

90. Tsetlin VI, Hucho F. Snake and snail toxins acting on nicotinic acetylcholine receptors: fundamental aspects and medical applications. FEBS Lett 2004;557:9–13.

91. Ibanez-Tallon I, Miwa JM, Wang HL, Adams NC, Crabtree GW, Sine SM, Heintz N. Novel modulation of neuronal nicotinic acetylcholine receptors by association with the endogenous prototoxin lynx1. Neuron 2002;33:893–903.

92. Chimienti F, Hogg RC, Plantard L, Lehmann C, Brakch N, Fischer J, Huber M, Bertrand D, Hohl D. Identification of SLURP-1 as an epidermal neuromodulator explains the clinical phenotype of Mal de Meleda. 1: Hum Mol Genet 2003;12:3017–24.

93. Arredondo J, Chernyavsky AI, Webber RJ, Grando SA. Biological effects of SLURP-1 on human keratinocytes. J Invest Dermatol 2005;125:1236–41.

94. Arredondo J, Chernyavsky AI, Jolkovsky DL, Webber RJ, Grando SA. SLURP-2: A novel cholinergic signaling peptide in human mucocutaneous epithelium. J Cell Physiol 2006;208:238–45.

95. Jones AK, Sattelle DB. Functional genomics of the nicotinic acetylcholine receptor gene family of the nematode, *Caenorhabditis elegans*. Bioessays 2004;26:39–49.

96. Jones AK, Elgar G, Sattelle DB. The nicotinic acetylcholine receptor gene family of the pufferfish, *Fugu rubripes*. Genomics 2003;82:441–51.

97. Jiang Y, Lee A, Chen J, Ruta V, Cadene M, Chait BT, MacKinnon R. X-ray structure of a voltage-dependent K + channel. Nature 2003;423:33–41.

98. Gouaux E, Mackinnon R. Principles of selective ion transport in channels and pumps. Science 2005;310:1461–5.

99. Bohler S, Gay S, Bertrand S, et al. Desensitization of neuronal nicotinic receptors conferred by N-terminal segments of the $\beta 2$ subunit. Biochemistry 2001;40:2066–2074.

100. Prince RJ, Sine SM. Acetylcholine and epibatidine binding to muscle acetylcholine receptors distinguish between concerted and uncoupled models. J Biol Chem 1999;274:19623–19629.

101. Kuffler S, Yoshikami D. The number of transmitter molecules in a quantum. J Physiol 1975;251:465–482.

102. Steinbach J, Chen Q. Antagonist and partial agonist actions of d-tubocurarine at mammalian muscle acetylcholine-receptors. J Neurosci 1995;15:230–240.

103. Malany S, Osaka H, Sine SM, Taylor P. Orientation of αneurotoxin at the subunit interfaces of the nicotinic acetylcholine receptor. Biochemistry 2000;39:15388–15398.

104. Magleby K. Neuromuscular transmission. In: Engel A, Franzini-Armstrong C, eds. Myology: Basic and Clinical, 2nd ed., vol. 1. ,New York: McGraw Hill Inc, 1994:442–463.

105. Gunderson CH, Lehmann CR, Sidell FR, Gabbari B. Nerve agents – a review. Neurology 1992;42:946–950.

106. Benowitz N. Pharmacology of nicotine: addiction and therapeutics. Ann Rev Pharmacol and Toxicol 1996;36:597–613.

107. Dani J, Ji D, Zhou F-M. Synaptic plasticity and nicotine addiction. Neuron 2001;31:349–352.

108. Picciotto M, Caldarone B, King S, Zachariou V. Nicotinic receptors in the brain: links between molecular biology and behavior. Neuropsychopharmacol 2000;22:451–465.

109. Kopta C, Steinbach J. Comparison of mammalian adult and fetal nicotinic acetylcholine receptors stably expressed in fibroblasts. J Neurosci 1994;14:3922–3933.

110. Papke RL, Meyer E, Nutter T, Uteshev V. $\alpha 7$ receptor-selective agonists and modes of $\alpha 7$ receptor activation. Euro J Pharmacol 2000;393:179–195.

111. Corringer PJ, LeNovere N, Changeux J-P. Nicotinic receptors at the amino acid level. Ann Rev Pharmacol Toxicol 2000;40:431–458.

112. Miyazawa A, Fujiyoshi Y, Unwin N. Structure and gating mechanism of the acetylcholine receptor pore. Nature 2003;423:949–55.

113. Froehner SC. Identification of exposed and buried determinants of the membrane-bound acetylcholine-receptor from *Torpedo-californica*. Biochemistry 1981;20:4905–4915.

114. Tzartos SJ, Seybold ME, Lindstrom JM. Specificities of antibodies to acetylcholine-receptors in sera from myasthenia-gravis patients measured by monoclonal-antibodies. Proc Nat Acad Sci USA 1982;79:188–192.

115. Kontou M, Leonidas D, Vatzaki E, et al. The crystal structure of the Fab fragment of a rat monoclonal antibody against the main immunogenic region of the human muscle acetylcholine receptor. Eur J Biochem 2000;267:2389–2396.

116. Poulas K, Eliopoulas E, Vatzoki E, et al. Crystal structure of Fab 198, an efficient protector of the acetylcholine receptor agonist myasthenogenic antibodies. Eur J Biochem 2001;268:3685–3693.

117. Saedi MS, Anand R, Conroy WG, Lindstrom J. Determination of amino-acids critical to the main immunogenic region of intact acetylcholine-receptors by in vitro mutagenesis. FEBS Lett 1990;267: 55–59.

118. Lindstrom J, Shelton GD, Fujii Y. Myasthenia gravis. Adv Immunol 1988;42:233–284.

119. Lindstrom J. An assay for antibodies to human acetylcholine receptor in serum from patients with myasthenia gravis. Clin Immunol Immunopath 1977;7:36–43.

120. Engel A, ed. Myasthenia Gravis and Myasthenic Disorders, Contemporary Neurology Series. New York, Oxford University Press, 1999.

121. Heinemann S, Bevan S, Kullberg R, Lindstrom J, Rice J. Modulation of the acetylcholine receptor by anti-receptor antibody. Proc Nat Acad Sci USA 1977;74:3090–3094.

122. Drachman DB, Angus CW, Adams RN, Michelson J, Hoffman G. Myasthenic antibodies cross-link acetylcholine receptors to accelerate degradation. New Eng J Med 1978;298:1116–1122.

123. Drachman D. The biology of myasthenia gravis. Ann Rev Neurosci 1981;4:195–225.

124. Blatt Y, Montal MS, Lindstrom J, Montal M. Monoclonal-antibodies specific to the β-subunit and γ-subunit of the torpedo acetylcholine-receptor inhibit single-channel activity. J Neurosci 1986;6:481–486.

125. Shelton GD, Cardinet GH, Lindstrom JM. Canine and human myasthenia-gravis autoantibodies recognize similar regions on the acetylcholine-receptor. Neurology 1988;38:1417–1423.

126. Weinberg CB, Hall ZW. Antibodies from patients with myasthenia-gravis recognize determinants unique to extra-junctional acetylcholine receptors. Proc Nat Acad Sci USA 1979;76: 504–508.

127. Burges J, Wray DW, Pizzighella S, Hall Z, Vincent A. A myasthenia-gravis plasma immunoglobulin reduces miniature endplate potentials at human endplates in vitro. Muscle Nerve 1990;13:407–413.

128. Vincent A, Newland C, Brueton L, et al. Arthrogryposis multiplex congenita with maternal autoantibodies specific for a fetal antigen. Lancet 1995;346:24–25.

129. Conti-Fine BM, Navaneetham D, Karachunski PI, et al. T cell recognition of the acetylcholine receptor in myasthenia gravis. Ann NY Acad Sci 1998;841:283–308.

130. Beeson D, Bond AP, Corlett L, et al. Thymus, thymoma, and specific T cells in myasthenia gravis. Ann NY Acad Sci 1998;841:371–387.

131. Fujii Y, Lindstrom J. Specificity of the t-cell immune-response to acetylcholine-receptor in experimental autoimmune myasthenia-gravis – response to subunits and synthetic peptides. J Immunol 1988;140:1830–1837.

132. Hohlfeld R, Wekerle H. The immunopathogenesis of myasthenia gravis. In Engel A, ed. Myasthenia Gravis and Myasthenic Disorders, Contemporary Neurology Series, New York, Oxford, 1999;56:87–110.

133. Raju R, Spack E, David C. Acetylcholine receptor peptide recognition in HLA DR3-transgenic mice: in vivo responses correlate with MHC-peptide binding. J Immunol 2001;167:1118–1124.

134. Ong B, Willcox N, Wordsworth P, et al. Critical role for the val/gly[86] HLA-DR-β dimorphism in autoantigen presentation to human T-cells. Proc Nat Acad Sci USA 1991;88:7343–7347.

135. Christadoss P, Lindstrom JM, Melvold RW, Talal N. IA subregion mutation prevents experimental autoimmune myasthenia-gravis. Immunogenetics 1985;21:33–38.

136. Christadoss P, Poussin M, Deng CS. Animal models of myasthenia gravis. Clin Immunol 2000;94:75–87.

137. Shelton D, Lindstrom J. Spontaneous remission in canine myasthenia gravis: implications for assessing human therapies. Neurology 2001;57:2139–2141.

138. Russell A, Lindstrom J. Penicillamine induced myasthenia gravis associated with antibodies to the acetylcholine receptor. Neurology 1989;28:847–849.

139. Vincent A, Newsom-Davis J, Martin V. Anti-acetylcholine receptor antibodies in D-penicillamine associated myasthenia gravis. Lancet 1978;I:1254.

140. Penn AS, Low BW, Jaffe IA, Luo L, Jacques J. Drug-induced autoimmune myasthenia gravis. Ann NY Acad Sci 1998;841:433–449.

141. Keesey J, Lindstrom J, Cokely A. Anti-acetylcholine receptor antibody in neonatal myasthenia gravis. N Eng J Med 1977;296:55.

142. Vernet der Garabedian B, Lacokova M, Eymard B, et al. Association of neonatal myasthenia gravis with antibodies against the fetal acetylcholine receptor. J Clin Invest 1994;94:555–559.

143. Willcox H. Thymic tumors with myasthenia gravis or bone marrow dyscrasias. In Peckham M, ed. Oxford Textbook of Oncology, New York, Oxford University Press, 1995;1562–1568.

144. Vincent A, Newsom-Davis J. Acetylcholine receptor antibody characteristics in myasthenia gravis. I: Patients with generalized myasthenia or disease restricted to ocular muscles. Clin Exp Immunol 1982;49:257–265.

145. Baggi F, Andreetta F, Antozzi C, et al. Anti-titin and antiryanodine receptor antibodies in myasthenia gravis patients with thymoma. Ann NY Acad Sci 1998;841:538–541.

146. Cikes N, Momoi M, Williams C, et al. Striational autoantibodies: quantitative detection by enzyme immunoassay in myasthenia gravis, thymoma and recipients of D-penicillamine or allogenic bone marrow. May Clin Proc 1998;63:474–481.

147. Wakkach A, Guyon T, Bruand C, Tzartos S, Cohen-Kaminsky S, Berrih-Aknin S. Expression of acetylcholine receptor genes in human thymic epithelial cells. Implications for myasthenia gravis. J Immunol 1996;157:3752–3760.

148. Zheng Y, Whatly L, Liu T, Levinson A. Acetylcholine receptor α subunit in mRNA expression in human thymus: augmented expression in myasthenia gravis and upregulation by interferon-δ. Clin Immunol 1999;91:170–177.

149. Vincent A, Lang B, Newsom-Davis. Autoimmunity to the voltage-gated calcium channel underlies the Lambert-Eaton myasthenic syndrome, a paraneoplastic disorder. TINS 1989;12:496–502.

150. Vincent A, Willcox N, Hill M, Curnow J, MacLennon C, Beeson D. Determinant spreading and immunoresponses to acetylcholine receptors in myasthenia gravis. Immunol Rev 1998;164:157–168.

151. Bartfeld D, Fuchs S. Specific immunosuppression of experimental autoimmune myasthenia gravis by denatured acetylcholine receptor. Proc Nat Acad Sci USA 1978:;75:4006–4010.

152. Lindstrom J, Einarson B, Merlie J. Immunization of rats with polypeptide chains from *Torpedo* acetylcholine receptor causes an autoimmune response to receptors in rat muscle. Proc Nat Acad Sci USA 1978;75:769–773.

153. Lindstrom J, Peng X, Kuryatov A, et al. Molecular and antigenic structure of nicotinic acetylcholine receptors. Ann NY Acad Sci 1998;841:71–86.

154. Bachmaier K, Nen N, de la Maza L, Pal S, Hessel A, Penninger J. *Chlamydia* infections and heart disease linked through antigenic mimicry. Science 1999;283:1335–1339.

155. Nachamkin I, Allos B, Ho T. *Campylobacter* species and Guillain Barré syndrome. Clin Microbio Rev 1998;11:555–567.

156. Faller G, Steininger H, Kranzlein J, et al. Antigastric autoantibodies in *Helicobacter pylori* infection: implications of histological and clinical parameters of gastritis. Gut 1997;41:619–623.

157. Lindstrom J, Einarson B, Lennon V, Seybold M. Pathological mechanisms in EAMG. I: Immunogenicity of syngeneic muscle acetylcholine receptor and quantitative extraction of receptor and antibody-receptor complexes from muscles of rats with experimental autoimmune myasthenia gravis. J Exp Med 1976;144:726–738.

158. Jermy A, Beeson D, Vincent A. Pathogenic autoimmunity to affinity-purified mouse acetylcholine receptor induced without adjuvant in BALB/c mice. Eur J Immunol 1993;23:973–976.

159. Lindstrom J. Experimental induction and treatment of myasthenia gravis. In: Engel A, ed. Myasthenia Gravis and Myasthenic Disorders. Contemporary Neurology Series, New York: Oxford University Press 1999;111–130.

160. Zoda T, Krolick K. Antigen presentation and T cell specificity repertoire in determining responsiveness to an epitope important in experimental autoimmune myasthenia gravis. J Neuroimmunol 1993;43:131–138.

161. Lindstrom J, Seybold M, Lennon V, Whittingham S, Duane D. Antibody to acetylcholine receptor in myasthenia gravis: prevalence, clinical correlates, and diagnostic value. Neurology 1976;26:1054–1059.

162. Vincent A, Newsom-Davis J. Acetylcholine receptor antibody as a diagnostic test for myasthenia gravis: results in 153 validated cases and 2,967 diagnostic assays. J Neurol Neurosurg Psych 1985;48:1246–1252.

163. Seybold M, Lindstrom J. Patterns of acetylcholine receptor antibody fluctuation in myasthenia gravis. Ann NY Acad Sci 1981;377:292–306.

164. Lindstrom J, Engel A, Seybold M, Lennon V, Lambert E. Pathological mechanisms in EAMG. II: Passive transfer of experimental autoimmune myasthenia gravis in rats with anti-acetylcholine receptor antibodies. J Exp Med 1976;144:739–753.

165. Tzartos S, Hochschwender S, Vasquez P, Lindstrom J. Passive transfer of experimental autoimmune myasthenia gravis by monoclonal antibodies to the main immunogenic region of the acetylcholine receptor. J Neuroimmunol 1987;15:185–194.

166. Toyka KV, Drachman DB, Pestronk A, Kao I. Myasthenia gravis: passive transfer from man to mouse. Science 1975;190:397–399.

167. Toyka KV, Birmberger KL, Anzil AP, Schlegel C, Besinger V, Struppler A. Myasthenia gravis: further electrophysiological and ultrastructural analysis of transmission failure in the mouse passive transfer model. J Neurol Neurosurg & Psych 1978;41:746–753.

168. Engel A, Tsujihata M, Lambert E, Lindstrom J, Lennon V. Experimental autoimmune myasthenia gravis: a sequential and quantitative study of the neuromuscular junction ultrastructure and electrophysiologic correlation. J Neuropath Exp Neurol 1976;35:569–587.

169. Engel A, Tsujihata M, Lindstrom J, Lennon V. End-plate fine structure in myasthenia gravis and in experimental autoimmune myasthenia gravis. Ann NY Acad Sci 1976;274:60–79.

170. Engel A, Sakakibara H, Sahashi K, Lindstrom J, Lambert E, Lennon V. Passively transferred experimental autoimmune myasthenia gravis. Neurology 1979;29:179–188.

171. Bevan S, Heinemann S, Lennon V, Lindstrom J. Reduced muscle acetylcholine sensitivity in rats immunized with acetylcholine receptor. Nature 1976;260:438–439.

172. Lambert E, Lindstrom J, Lennon V. End-plate potentials in experimental autoimmune myasthenia gravis in rats. Ann NY Acad Sci 1976;274:300–318.

173. Lindstrom J, Lambert E. Content of acetylcholine receptor and antibodies bound to receptor in myasthenia gravis, experimental autoimmune myasthenia gravis, and in Eaton-Lambert syndrome. Neurology 1978;28:130–138.

174. Sahashi K, Engel, Lindstrom, Lambert EH, Lennon V. Ultrastructural localization of immune complexes (IgG and C3) at the end-plate in experimental autoimmune myasthenia gravis. J Neuropath Exp Neurol 1978;37:212–223.

175. Sahashi K, Engel A, Lambert E, Howard F. Ultrastructural localization of the terminal and lytic ninth complement component (C9) at the motor end-plate in myasthenia gravis. J Neuropath Exp Neurol 1980;39:160–172.

176. Appel S, Anwyl R, McAdams M, Elias S. Accelerated degradation of acetylcholine receptor from cultured rat myotubes with myasthenia gravis sera and globulins. Proc Nat Acad Sci USA 1977;74:2130–2134.

177. Bufler J, Pitz R, Czep M, Wick M, Franke C. Purified IgG from seropositive and seronegative patients with myasthenia gravis reversibly blocks currents through nicotinic acetylcholine receptor channels. Ann Neurol 1998;43:458–464.

178. Burges J, Vincent A, Molenaar P, Newsom-Davis J, Peers C, Wray D. Passive transfer of seronegative myasthenia gravis to mice. Muscle Nerve 1994;17:1393–1400.

179. Donnelly D, Mihovilovic M, Gonzalez-Ros J, Ferragut J, Richman D, Martinez-Carrion M. A noncholinergic site-directed monoclonal antibody can impair agonist-induced ion flux in *Torpedo californica* acetylcholine receptor. Proc Nat Acad Sci USA 1984;81:7999–8003.

180. Fels G, Plumer-Wilk R, Schreiber M, Maelicke A. A monoclonal antibody interfering with binding and response of the acetylcholine receptor. J Biol Chem 1986;261:15746–15754.

181. Gomez G, Richman D. Anti-acetylcholine receptor antibodies directed against the α bungarotoxin binding site induce a unique form of experimental myasthenia. Proc Nat Acad Sci USA 1983;80:4089–4093.

182. Lang B, Richardson G, Rees J, Vincent A, Newsom-Davis J. Plasma from myasthenia gravis patients reduces acetylcholine receptor agonist-induced Na$^+$ flux into TE671 cell line. J Neuroimmunol 1988;19:141–148.

183. Lennon V, Seybold M, Lindstrom J, Cochrane C, Yulevitch R. Role of complement in pathogenesis of experimental autoimmune myasthenia gravis. J Exp Med 1978;147:973–983.

184. Sine SM, Ohno K, Bougat C, et al. Mutation of the acetylcholine-receptor alpha-subunit causes a slow-channel myasthenic syndrome by enhancing agonist binding-affinity. Neuron 1995;15:229–239.

185. Ohno K, Wang HL, Milone M, et al. Congenital myasthenic syndrome caused by decreased agonist binding affinity due to a mutation in the acetylcholine receptor epsilon subunit. Neuron 1996;17:157–170.

186. Ohno K, Hutchinson DO, Milone M, et al. Congenital myasthenic syndrome caused by prolonged acetylcholine-receptor channel openings due to a mutation in the M2 domain of the epsilon-subunit. Proc Nat Acad Sci USA 1995;92:758–762.

187. Engel A, Uchitel O, Walls T, Nagel A, Harper C, Bodensteiner J. Newly recognized congenital myasthenic syndrome associated with high conductance and fast closure of the acetylcholine receptor channel. Ann Neurol 1993;34:38–47.

188. Ohno K, Anlar B, Engel A. Congenital myasthenic syndrome caused by a mutation in the Ets-binding site of the promoter region of the acetylcholine receptor ε subunit gene. Neuromusc Dis 1999;9:131–135.

189. Ohno K, Anlar B, Ozdirim E, Grengman J, DeBleecker J, Engel A. Myasthenic syndromes in Turkish kinships due to mutations in the acetylcholine receptor. Ann Neurol 1998;44:234–241.

190. Ohno K, Quiram P, Milone M, et al. Congenital myasthenic syndromes due to heteroallelic nonsense/missense mutations in the acetylcholine receptor ε subunit gene: identification and functional characterization of six new mutations. Human Mol Gene 1997;6:753–766.

191. Engel A, Ohno K, Milone M, et al. New mutations in acetylcholine receptor subunit genes reveal heterogeneity in the slow channel congenital myasthenic syndrome. Hum Mol Genet 1996;5:1217–1227.

192. Missias A, Mudd J, Cunningham J, Steinbach J, Merlie J, Seines J. Deficient development and maintenance of postsynaptic specializations in mutant mice locking an adult acetylcholine receptor subunit. Development 1997;124:5075–5086.

193. Milone M, Wang H-L, Ohno, et al. Slow channel syndrome caused by enhanced activation, desensitization, and agonist binding affinity due to mutation in the M2 domain of the acetylcholine receptor α subunit. J Neurosci 1997;17:5651–5665.

194. Wang H-L, Auerbach A, Bren N, Ohno K, Engel A, Sine S. Mutation in the M1 domain of the acetylcholine receptor α subunit decreases the rate of agonist dissociation. J Gen Physiol 1997;109:757–766.

195. Lo D, Pinkham J, Stevens C. Role of a key cystein residue in the gating of the acetylcholine receptor. Neuron 1991;6:31–40.

196. Zhou M, Engel A, Auerbach A. Serum choline activates mutant acetylcholine receptors that cause slow channel congenital myasthenic syndromes. Proc Nat Acad Sci USA 1999;96:10466–10471.

197. Bertrand D, Devillers-Thiery A, Revah F, et al. Unconventional pharmacology of a neuronal nicotinic receptor mutated in the channel domain. Proc Nat Acad Sci USA 1992;89:1261–1265.

198. Alkondon M, Pereira E, Eisenberg H, Albuquerque E. Choline and selective antagonists identify two subtypes of nicotinic acetylcholine receptors that modulate GABA release from CAI interneurons in rat hippocampal slices. J Neurosci 1999;2693–2705.

199. Milone M, Wang H-L, Ohno K, et al. Mode switching kinetics produced by a naturally occurring mutation in the cytoplasmic loop of the human acetylcholine receptor ε subunit. Neuron 1998;20:575–588.

200. Han Z-Y, LeNovere N, Zoli M, Hill J, Champtiaux N, Changeux J-P. Localization of nAChR subunit mRNAs in the brain of *Macaca mulatta*. Eur J Neurosci 2000;12:3664–3674.

201. Vailati S, Moretti M, Balestra B, McIntosh M, Clementi F, Gotti C. β3 subunit is present in different nicotinic receptor subtypes in chick retina. Eur J Pharmacol 2000;393:23–30.

202. Gerzanich V, Wang F, Kuryatov A, Lindstrom J. alpha 5 Subunit alters desensitization, pharmacology, Ca + + permeability and Ca + + modulation of human neuronal alpha 3 nicotinic receptors. J Pharmacol Exp Ther 1998;286:311–20.

203. Vailati S, Moretti M, Longhi R, Rovati GE, Clementi F, Gotti C. Developmental expression of heteromeric nicotinic receptor subtypes in chick retina. Mol Pharmacol 2003;63(6):1329–37.

204. Gotti C, Moretti M, Clementi F, Riganti L, McIntosh JM, Collins AC, Marks MJ, Whiteaker P. Expression of nigrostriatal alpha 6-containing nicotinic acetylcholine receptors is selectively reduced, but not eliminated, by beta 3 subunit gene deletion. Mol Pharmacol 2005;67:2007–15.

205. Wonnacott S. Presynaptic nicotinic ACh receptors. Trends Neurosci 1997;20:92–98.

206. Xu W, Gelber S, Orr-Urtreger A, et al. Megacystis, mydriasis, and ion channel defect in mice lacking the α3 neuronal nicotinic receptor. Proc Nat Acad Sci USA 1999;96:5746–5751.

207. Marubio L, del Mar Arroyo-Jiminez M, Cordero-Erausquin M, et al. Reduced antinociception in mice lacking neuronal nicotinic receptor subunits. Nature 1999;398:805–810.

208. Ross S, Wong J, Clifford J, et al. Phenotypic characterization of an α4 neuronal nicotinic acetylcholine receptor subunit knockout mouse. J Neurosci 2000;20:6431–6441.

209. Labarca C, Schwarz J, Deshpande P, et al. Point mutant mice with hypersensitive α4 nicotinic receptors show dopaminergic defects and increased anxiety. Proc Nat Acad Sci USA 2001;98:2786–2791.

210. Fonck C, Nashmi R, Deshpande P, et al. Increased sensitivity to agonist-induced seizures, straub tail, and hippocampal theta rhythm in knock-in mice carrying hypersensitive alpha 4 nicotinic receptors. J Neurosci 2003;23:2582–2590.

211. Champtiaux N, Han Z-Y, Rossi F, et al. Distribution and pharmacology of alpha 6-containing nicotinic acetylcholine receptors analyzed with mutant mice. J Neurosci 2002;22:1208–17.

212. Champtiaux N, Gotti C, Cordero-Erausquin M, et al. Subunit composition of functional nicotinic receptors in dopaminergic neurons investigated with knock-out mice. J Neurosci 2003;23:7820–9.

213. Quik M, Vailati S, Bordia T, et al. Subunit composition of nicotinic receptors in monkey striatum: effect of treatments with 1-methyl-4-phenyl-1,2,3,6-tetrahydropyridine or L-DOPA. Mol Pharmacol 2005;67:32–41.

214. Orr-Urtreger A, Göldner F, Saeki M, et al. Mice deficient in the α7 neuronal nicotinic acetylcholine receptor lack α bungarotoxin binding sites and hippocampal fast nicotinic currents. J Neurosci 1997;17:9165 9171.

215. Franceschine D, Orr-Urtreger A, Yu W, et al. Altered baroreflex responses in α7 deficient mice. Behav Brain Res 2000;113:3–10.

216. Orr-Urtreger A, Broide R, Kasten M, et al. Mice homozygous for the L250T mutation in the α7 nicotinic acetylcholine receptor show increased neuronal apoptosis and die within 1 day of birth. J Neurochem 2000;74:2154–2166.

217. Berger F, Gage F, Vijayaraghavan S. Nicotinic receptor-induced apoptotic cell death of hippocampal progenitor cells. J Neurosci 1998;18:6871–6881.

218. Vetter D, Liberman M, Mann J, et al. Role of α9 nicotinic ACh receptor subunits in the development and function of cochlear efferent innervation. Neuron 1999;23:91–103.

219. Plazas PV, Katz E, Gomez-Casati ME, Bouzat C, Elgoyhen AB. Stoichiometry of the alpha9alpha10 nicotinic cholinergic receptor. J Neurosci 2005;25:10905–10912.

220. Fuchs PA. Synaptic transmission at vertebrate hair cells. Curr Opin Neurobiol 1996;6:514–9.

221. Maskos U, Molles BE, Pons S, et al. Nicotine reinforcement and cognition restored by targeted expression of nicotinic receptors. Nature 2005;436:103–107.

222. Lena C, Popa D, Grailhe R, Escourrou P, Changeux JP, Adrien J. Beta2-containing nicotinic receptors contribute to the organization of sleep and regulate putative micro-arousals in mice. J Neurosci 2004;24:5711–8.

223. Cohen G, Roux JC, Grailhe R, Malcolm G, Changeux JP, Lagercrantz H. Perinatal exposure to nicotine causes deficits associated with a loss of nicotinic receptor function. Proc Nat Acad Sci USA 2005;102:3817–3821.

224. Cui C, Booker TK, Allen RS, et al. The beta3 nicotinic receptor subunit: a component of alpha-conotoxin MII-binding nicotinic acetylcholine receptors that modulate dopamine release and related behaviors. J Neurosci 2003;23:11045–11053.

225. McCallum SE, Collins AC, Paylor R, Marks MJ. Deletion of the beta 2 nicotinic acetylcholine receptor subunit alters development of tolerance to nicotine and eliminates receptor upregulation. Psychopharmacology 2006;184:314–27.

226. Peng X, Gerzanich V, Anand R, Whiting PJ, Lindstrom J. Nicotine-induced increase in neuronal nicotinic receptors results from a decrease in the rate of receptor turnover. Mol Pharmacol 1994;46:523–530.

227. Peng X, Gerzanich V, Anand R, Wang F, Lindstrom J. Chronic nicotine treatment up-regulates alpha3 and alpha7 acetylcholine receptor subtypes expressed by the human neuroblastoma cell line SH–SY5Y. Mol Pharmacol 1997;51:776–84.

228. Wang F, Nelson ME, Kuryatov A, Olale F, Cooper J, Keyser K, Lindstrom J. Chronic nicotine treatment up-regulates human alpha3 beta2 but not alpha3 beta4 acetylcholine receptors stably transfected in human embryonic kidney cells. J Biol Chem 1998;273:28721–32.

229. Sallette J, Bohler S, Benoit P, et al. An extracellular protein microdomain controls up-regulation of neuronal nicotinic acetylcholine receptors by nicotine. 1: J Biol Chem 2004;279:18767–75.

230. Sallette J, Pons S, Devillers-Thiery A, Soudant M, et al. Nicotine upregulates its own receptors through enhanced intracellular maturation. Neuron 2005;46:595–607.

231. Luther MA, Schoepfer R, Whiting P, Casey B, Blatt Y, Montal MS, Montal M, Linstrom J. A muscle acetylcholine receptor is expressed in the human cerebellar medulloblastoma cell line TE671. J Neurosci 1989;9:1082–96.

232. Perry DC, Davila-Garcia MI, Stockmeier CA, Kellar KJ. Increased nicotinic receptors in brains from smokers: membrane binding and autoradiography studies. J Pharmacol Exp Ther 1999;289:1545–52.

233. Marks MJ, Pauly JR, Gross SD, Deneris ES, Hermans-Borgmeyer I, Heinemann SF, Collins AC. Nicotine binding and nicotinic receptor subunit RNA after chronic nicotine treatment. J Neurosci 1992;12:2765–84.

234. Xu W, Orr-Urtreger A, Nigro F, et al. Multiorgan autonomic dysfunction in mice lacking the β2 and β4 subunits of neuronal nicotinic acetylcholine receptors. J Neurosci 1999;19:9298–9305.

235. Balestra B, Moretti M, Longhi R, Mantegazza R, Clementi F, Gotti C. Antibodies against neuronal nicotinic receptor subtypes in neurological disorders. J Neuroimmunol 2000;102:89–97.

236. Schroeder C, Vernino S, Birkenfeld AL, et al. Plasma exchange for primary autoimmune autonomic failure. N Engl J Med 2005;353:1585–90.

237. Lennon VA, Ermilov LG, Szurszewski JH, Vernino S. Immunization with neuronal nicotinic acetylcholine receptor induces neurological autoimmune disease. J Clin Invest 2003;111:907–13.

238. Vernino S, Low PA, Lennon VA. Experimental autoimmune autonomic neuropathy. J Neurophysiol 2003;90:2053–9.

239. Vernino S, Ermilov LG, Sha L, Szurszewski JH, Low PA, Lennon VA. Passive transfer of autoimmune autonomic neuropathy to mice. J Neurosci 2004;24:7037–42.

240. Nguyen V, Ndoye A, Grando S. Pemphigus vulgaris antibody identifies pemphaxin – a novel keratinocyte annexin-like molecule binding acetylcholine. J Biol Chem 2000;275:29466–29476.

241. Rogers S, Andrews J, Gahring L, et al. Autoantibodies to glutamate receptor in GluR3 in Rasmussen s encephalitis. Science 1994;265:648–651.

242. Rogers S, Twyman R, Gahring L. The role of autoimmunity to glutamate receptors in neurological disease. Mod Med Today 1996;2:76–81.

243. Gahring L, Rogers S. Autoimmunity to glutamate receptors in Rasmussen s encephalitis: a rare finding or the tip of an iceberg? The Neuroscientist 1998;4:373–379.

244. Steinlein OK. Neuronal nicotinic receptors in human epilepsy. Eur J Pharmacol 2000;393:243–247.

245. Phillips HA, Favre I, Kirkpatrick M, et al. CHRNB2 is the second acetylcholine receptor subunit associated with autosomal dominant nocturnal frontal lobe epilepsy. Am J Hum Gen 2001;68:225–231.

246. De Fusco M, Becchetti A, Patrignani A, et al. The nicotinic receptor beta 2 subunit is mutant in nocturnal frontal lobe epilepsy. Nature Genet 2000;26:275–276.

247. Kuryatov A, Gerzanich V, Nelson M, Olale F, Lindstrom J. Mutation causing autosomal dominant nocturnal frontal lobe epilepsy alters Ca^{++} permeability, conductance, and gating of human α4β2 nicotinic acetylcholine receptors. J Neurosci 1997;17:9035–9047.

248. Alkondon M, Periera E, Eisenberg H, Albuquerque E. Nicotinic receptor activation in human cerebral cortical interneurons: a mechanism for inhibition and disinhibition of neuronal networks. J Neurosci 2000;20:66–75.

249. Lev-Lehman E, Bercovich D, Xu W, Stockton D, Beaudit A. Characterization of the human β4 nAChR gene and polymorphisms in CHRNA3 and CHRNB4. J Hum Genet 2001;46:362–366.

250. Rodrigues-Pinguet NO, Pinguet TJ, Figl A, Lester HA, Cohen BN. Mutations linked to autosomal dominant nocturnal frontal lobe epilepsy affect allosteric Ca2 + activation of the alpha 4 beta 2 nicotinic acetylcholine receptor. Mol Pharmacol 2005;68:487–501.

251. Richardson CE, Morgan JM, Jasani B, et al. Megacystis-microcolon-intestinal hypoperistalsis syndrome and the absence of the alpha3 nicotinic acetylcholine receptor subunit. Gastroenterology 2001;121:350–7.

252. Bocfuet, N., Prado de Carvalho, L., Cartaud, J., Neyton, J., Le Poupon, C, Taly, A., Grutter, T., Changrux, J-P., Corringer, P-J. A prokaryotic proton-gated ion channel from the nicotinic acetylcholine receptor family. Nature 2006; 445:116–119.

253. Hilf, R. and Dutzler, R. X-ray structure of a prokaryotic puntarueric ligand-gated ion channel. Nature 2008;452:375–379.

254. Lindstron, J., Luo, J. and Kuryaton, A Myasthenia groves and the tops and bottoms of AChRs: Antigenic structure of the MIR and specific immunosuppression of EAME using AChR cytoplasmic domains. Ann, NY Acad sci 2008;1132:29–41.

3

Immunopathogenesis of Myasthenia Gravis

Bianca M. Conti-Fine, Brenda Diethelm-Okita, Norma Ostlie, Wei Wang and Monica Milani

1. INTRODUCTION

Myasthenia gravis (MG) is a prototypic and perhaps the best characterized antibody (Ab)-mediated autoimmune disease: autoAb against the nicotinic acetylcholine receptor (AChR) at the neuromuscular junction (NMJ) cause the myasthenic symptoms *(1, 2, 3, 4)*. Yet also anti-AChR T cells have a crucial role in the pathogenesis of MG, because they permit and modulate the synthesis of the high-affinity Ab that cause AChR loss, damage of the NMJ, and failure of the neuromuscular transmission. Moreover, activation of potentially self-reactive CD4$^+$ T cells, commonly present in healthy people, may be the primary event in the pathogenesis of MG. This might occur because of cross-reactivity of self-reactive CD4$^+$ T cells with microbial antigens or because of the action of microbial superantigens (SAGs) *(5, 6)*.

Here, we will first give a short summary of the characteristics and pathogenic mechanisms of the anti-AChR Ab: more detailed information on these issues can be found elsewhere in this book. We will then review what is known about the characteristics of the CD4$^+$ T-cell responses in MG and describe how the anti-AChR CD4$^+$ T cells could be the target of specific immunosuppressive treatments. We will discuss also the role that the thymus might have in the development of the anti-AChR autoimmune response. Finally, we will present different models of the possible pathogenic mechanisms of MG.

2. ANTI-ACHR Ab IN MG AND IN EXPERIMENTAL AUTOIMMUNE MG (EAMG)

Most (90%) of MG patients have anti-AChR Ab in their serum and IgG at their NMJ *(1, 2, 3, 4)*. Different lines of evidence indicate that anti-muscle AChR Ab cause the failure of the neuromuscular transmission in MG:

- MG patients do not have detectable AChR-specific cytotoxic phenomena mediated by T cells (e.g., *7*);
- treatment of experimental animals with purified IgG from MG patients, or with anti-AChR Ab, induces the symptoms of MG *(8, 9, 10, 11)*;
- procedures that reduce the concentration of anti-AChR Ab, such as plasmapheresis, improve myasthenic weakness *(12)*.

About 10% of MG patients do not have detectable serum anti-AChR Ab *(13)*. Some "seronegative" MG patients may synthesize small amounts of highly pathogenic Ab, which quickly disappear from the serum because they bind to the NMJ *(14)*. Other seronegative MG patients make autoAb directed to muscle antigens other than the AChR, which can disrupt the neuromuscular transmission *(15, see below)*.

The AChR is a large protein which contains many potential Ab epitopes and can accommodate simultaneous binding of several different Ab *(16)*. In MG patients, the anti-AChR Ab recognize

From: *Current Clinical Neurology*
Myasthenia Gravis and Related Disorders
Edited by: H.J. Kaminski, DOI 10.1007/978-1-59745-156-7_3, © Humana Press, New York, NY

a complex epitope repertoire that is different in the individual subjects *(1, 2, 3, 4, 16)*. Even the anti-AChR Ab that recognize an individual epitope region are polyclonal and do not share a dominant idiotype *(17)*. The anti-AChR Ab of MG patients include different subclasses of the IgG heavy chain and different types of light chains *(1, 16)*. Thus, many B-cell clones contribute to the anti-AChR Ab response in MG.

The epitope recognized and the type of heavy chain used by the anti-AChR Ab influence their pathogenic potential. Pathogenic Ab need to recognize areas of the AChR that are accessible in the intact postsynaptic membrane of the NMJ. Furthermore, certain surface areas of the AChR, like the main immunogenic region (MIR), may be especially well situated to permit crosslinking of nearby AChR molecules *(16, 17, 18)*. Anti-AChR Ab against those epitopes may be especially pathogenic for two reasons. First, they can accelerate the AChR degradation *(19, 20, 21)* by a mechanism termed antigenic modulation *(22)*. Second, binding of several of those Ab to closely packed AChR molecules on the NMJ postsynaptic membrane would provide an ideal geometry for binding of complement and activation of the complement cascade *(16)*. The action of complement on the NMJ is an important pathogenic mechanism of anti-AChR Ab *(23, 24)*: anti-AChR Ab of subclasses that bind complement effectively likely have a high pathogenic potential *(25)*.

Immunization of experimental animals with AChR from different sources causes EAMG, an experimental disease which reproduces the clinical and electrophysiological features of MG *(1, 16)*. Because of the strong structural similarity of muscle-type AChR from even distant species *(26)*, EAMG can be induced by immunization with the easily purified AChR from the electric organ of *Torpedo* and *Electrophorus* fish. Fish electric tissue is embryologically a modified muscle, and it is exceedingly rich in AChR *(26)*. Also immunization with syngeneic AChR causes EAMG *(27, 28)*: this indicates that self-reactive T and B cells specific for muscle AChR survive clonal deletion and persist in healthy adult animals.

Patrick and Lindstrom first demonstrated the induction of EAMG in rabbits *(29)*. Later studies described the induction of EAMG in a variety of species (e.g., *30, 31, 32*). Particularly used and useful to study EAMG were the Lewis rats, which are very susceptible to EAMG (e.g., *31, 33, 34*), and different mouse strains which have individual and different susceptibility to EAMG (e.g., *32, 35, 36, 37*). Even the most susceptible strains are relatively resistant to EAMG, and their symptoms are usually subclinical *(1, 16)*. This is likely because mice have a lot of "spare" AChR at their NMJ, and they develop myasthenic weakness only when most of the AChR has been destroyed *(16)*. Yet mice are very useful to study EAMG mechanisms, because their immune system is well characterized and there are abundant reagents to identify their immune cells. Moreover, a variety of congenic mouse strains carry null mutations of genes for cytokines or other molecules important in the immune response; other strains carry transgenes encoding important immune receptors or mediators. C57Bl/6 mice are susceptible to EAMG, and they are the best-studied strain *(1, 3, 16)*. In EAMG, as in MG, high-affinity anti-AChR Ab, whose synthesis is modulated by anti-AChR $CD4^+$ T cells, cause the failure of neuromuscular transmission *(1, 3, 16)*.

The anti-AChR Ab cause the myasthenic symptoms, and in a given patient fluctuations in their serum concentration correlate loosely with the severity of the symptoms *(38)*. Yet there is little correlation between the serum anti-AChR Ab concentration of different patients and the severity of their symptoms (e.g., *39*). Some MG patients have high serum anti-AChR Ab and minimal weakness. In contrast, severely ill MG patients may have little serum anti-AChR Ab. Also in EAMG, the serum concentration of anti-AChR Ab correlates poorly with the severity of myasthenic symptoms (e.g., *25, 32, 35, 40, 41*). These findings suggest that only particular subpopulations of anti-AChR Ab are pathogenic, while other anti-AChR Ab are unable, or less able, to cause NMJ damage. The pathogenic Ab may be made in different amounts in individual MG patients and rodent strains.

3. EPITOPES RECOGNIZED BY ANTI-ACHR Ab IN MG AND EAMG

The epitope repertoire of anti-AChR Ab in MG patients and EAMG animals is heterogeneous and individual *(1)*. Yet most anti-AChR Ab in both MG and EAMG recognize epitopes formed by the α subunit *(1, 16)*. This might occur because the muscle AChR has two α subunits, and one copy of the other subunits *(26)*: the α subunit may be twice as effective as the other subunits at activating T and B cells.

3.1. The MIR

Many of the anti-AChR Ab recognize a set of largely overlapping epitopes on the AChR α subunit, termed main immunogenic region (MIR) *(16, 17)*. Anti-MIR Ab are highly pathogenic: they cause AChR loss in muscle cell cultures *(20)*, induce myasthenic weakness when injected into rodents *(42)*, and mediate the AChR loss in muscle cell cultures exposed to sera of MG patients *(21)*. Because of their abundance and pathogenic potential, anti-MIR Ab are especially important in MG development.

Each AChR molecule has two MIR located on well-defined bumps at the tip of each of the domains formed by an α subunit *(18)*. Residues within the sequence segment 67–76 of the α subunit contribute to formation of the MIR: Asp68, Pro69, Asp71, and Tyr72 are especially important for binding of anti-MIR Ab *(43, 44, 45)*.

Some structural features of the MIR may contribute to its immunodominance, as discussed elsewhere in this book. Also, most of the Ab to any protein antigen may recognize just a few epitopes: the dominance of the MIR may reflect an Ab response especially focused on one set of epitopes. Molecular mimicry between the MIR and epitopes on unrelated but common antigens might also relate to the immunodominance of the MIR. A sequence region of the AChR involved in forming the MIR is similar to a region of the pol-polyprotein of several human retroviruses (Fig. 1), and to a sequence region of the U1 small nuclear ribonucleoprotein—an autoantigen in systemic lupus erythematosus *(46)*. Cross-reactivity of the MIR with epitopes on microbial or autologous antigens might facilitate the Ab response to the MIR and contribute to the MIR immunodominance.

3.2. The Cholinergic Site

MG patients may have small amounts of Ab that recognize the AChR cholinergic binding site *(1, 16)*. Those Ab may be important in certain presentations of MG, because Ab against the cholinergic site cause an acute form of EAMG, without inflammation or necrosis at the NMJ

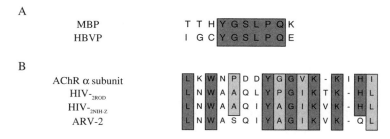

Fig. 1. **Panel A**: Homology between the sequence 66–75 of human MBP and the sequence 589–598 of the hepatitis B virus polymerase (HBVP). **Panel B**: Homology between the sequence segment α56–80 of human AChR and the polyprotein from the different human retroviruses (ARV-2 polyprotein segment 419–434, HIV-2 ROD pol plyprotein segment 448–463 and HIV-2NIH-Z pol-polyprotein segment 447–462).

(47). Ab that recognize the cholinergic site and block the AChR function might contribute to the myasthenic weakness in spite of their low concentration in the patients' sera. They might cause acute myasthenic crises and are likely responsible for the "Mary Walker phenomenon" *(48)*.

The Mary Walker phenomenon bears the name of the English neurologist who first realized that cholinesterase inhibitors could relieve the symptoms of MG *(49)*. This sobriquet refers to the increasing weakness that occurs in muscles which were not exercised, after vigorous exercise of other muscle groups. The Mary Walker phenomenon is consistent with the presence of soluble factors bound to muscle and released after exercise and able to block neuromuscular transmission. Anti-AChR Ab attached to the cholinergic binding sites at the NMJ may well be among those factors. The large concentrations of acetylcholine in the synaptic cleft of the NMJ after strenuous exercise, especially in the presence of cholinesterase inhibitors, might cause the detachment of anti-AChR Ab bound to the cholinergic site by competitive binding. The detached Ab could diffuse away, to reach and block the AChR of other muscles.

3.3. Non-AChR Antigens

The symptoms of most anti-AChR Ab negative MG patients improve after removal of the blood Ab by plasmapheresis *(12, 13)*. Moreover, their sera may cause AChR dysfunctions *in vitro* *(50)*. These findings suggested that Ab to the AChR, or to other proteins involved in the neuromuscular transmission, caused the myasthenic weakness in those patients. About 70% of the AChR seronegative patients have autoAb against the receptor tyrosine kinase MuSK which is a protein important in neuromuscular transmission *(15)*. Anti-MuSK Ab interfere with the agrin/MuSK/AChR clustering in myotubes and alter MuSK function at the adult NMJ *(15)*. The observation that rabbits immunized with the MuSK ectodomain manifested muscular weakness typical of MG and diminished AChR clustering at the NMJ provided a direct proof for the pathogenic role of anti-MuSK Ab *(51)*.

Other AChR-Ab "seronegative" patients may make small amounts of highly pathogenic Ab that bind the muscle AChR with high affinity *(14)* and are continuously removed from the bloodstream; consequently they may be undetectable in the serum. Yet other seronegative MG patients might have a plasma factor able to activate a second messenger pathway that leads to the phosphorylation and inactivation of the AChR, thus contributing to the impairment of the neuromuscular transmission *(52)*.

MG patients may have serum Ab against other muscle or synaptic proteins (such as a presynaptic membrane protein that binds β-bungarotoxin and the Ca^{++}-releasing sarcoplasmic membrane), whose block or destruction contributes to myasthenic symptoms *(53, 54)*.

MG patients without thymomas frequently have serum Ab to myofibrillar muscle proteins (myosin, tropomyosin, troponin, α-actinin, and actin) *(55)*, and the presence of anti-myosin Ab in the serum correlates with the severity of the disease *(56)*. Some Ab against myosin and fast troponin-1 cross-react with muscle AChR, and specifically with the MIR, suggesting that these contractile proteins bear epitopes structurally reminiscent of AChR epitopes *(57)*. MG patients with thymoma make Ab that recognize contractile proteins of striated muscle in amounts that correlate with that of the anti-AChR Ab *(58)*. Most of those Ab bind to titin *(59)*, a giant myofibrillar protein which may have a role in the elastic recoil of muscle fibers *(60)* or to the ryanodine receptor, the calcium channel of the sarcoplasmic reticulum *(61)*. In addition to Ab against muscle proteins, MG patients with thymoma may have also Ab directed against cytokines, such as IFN-α IL-12p40 and IL-12p70 *(62, 63, 64)*. The synthesis of these Ab is related to the presence of the tumor, because the anti-cytokine Ab titers decrease after thymectomy and increase again if the tumor recurs *(63, 64)*. It is not known whether the non-anti-AChR Ab of thymoma patients have a role in the pathogenesis of the myasthenic symptoms.

4. ANTI-ACHR CD4$^+$ T-HELPER CELLS IN MG

Most of the anti-AChR serum Ab of MG patients are high-affinity IgG, which are synthesized only after cytokines secreted by CD4$^+$ T-helper cells activate the B cells *(1, 2, 3, 4)*. Anti-AChR CD4$^+$ T cells are common in the blood and thymi of MG patients *(1, 3)*, and they have T-helper function *(65)*.

Several lines of evidence suggested that the CD4$^+$ T-helper cells have a critical role in MG pathogenesis. Treatment of MG patients with anti-CD4 Ab caused clinical and electrophysiological improvements, which correlated with a disappearance of the AChR-induced T-cell responses *in vitro (66)*. Also, a decrease in the anti-AChR activity of blood T cells is the only consistent and early effect of thymectomy *(67)*—a common and effective treatment in the management of MG *(68)*. Studies that used a cell transfer model *in vivo* demonstrated the pivotal role of AChR-specific CD4$^+$ T cells in rat EAMG: a mixture of B cells and CD4$^+$ T cells from rats immunized with AChR adoptively transferred into sublethally irradiated, thymectomized rats caused synthesis of anti-AChR Ab and EAMG symptoms, whereas transfer of B cells alone did not *(69)*. Other studies confirmed the essential role of CD4$^+$ T cells for rodent EAMG induction *(70, 71)*. The findings obtained in EAMG support an important role of anti-AChR CD4$^+$ T cells also in human MG.

Severe, combined immunodeficient (SCID) mice do not have functional B and T cells and tolerate xenografts *(72)*. When transplanted with fragments of thymus or with blood mononuclear cells (BMC) from MG patients, but not from healthy controls, SCID mice develop human anti-AChR Ab in the blood and at the NMJ and show myasthenic weakness *(13, 73, 74)*. The use of SCID mice engrafted with mixtures of immune cells from the blood of MG patients has proven that development of myasthenic weakness, and of human anti-AChR Ab in the serum and bound to the AChR at the NMJ, required the presence of CD4$^+$ T cells *(14)*. Furthermore, the inclusion of anti-AChR CD4$^+$ T cells as the only CD4$^+$ T cells induced synthesis of human anti-AChR Ab and myasthenic symptoms, whereas CD4$^+$-depleted BMC supplemented with CD4$^+$ T cells specific for irrelevant antigens did not *(14)*. These findings demonstrated directly that anti-AChR CD4$^+$ T cells are necessary for production of pathogenic anti-AChR Ab.

Healthy individuals usually have blood CD4$^+$ T cells specific for autoantigens, including the muscle AChR (e.g., *75, 76, 77, 78*). Yet the presence of potentially autoreactive CD4$^+$ T cells seldom leads to autoimmune diseases. Autoreactive T cells may survive clonal deletion during their maturation in the thymus because their T-cell receptors (TCRs) bind the self-epitope/MHC complex with low affinity (Fig. 2) *(79)*. T-cell activation requires effective binding of the TCR to the T epitope/MHC complex: thus, low-affinity autoreactive T cells may never be activated during the adult life. The common presence of autoreactive CD4$^+$ T cells in healthy people suggests that mechanisms of peripheral tolerance keep in check the activity of potentially self-reactive CD4$^+$ T cells. Failure of those mechanisms is a likely cause of autoimmune diseases, including MG. CD4$^+$ CD25$^+$ T cells that express the transcriptor factor Foxp-3 (commonly referred to as T-regulatory (Treg) cells) play a crucial role for maintenance of the peripheral tolerance toward self-antigens *(80)*: impairment of their function and/or their number has been correlated to the presence of many autoimmune disorders *(81, 80)*. MG patients may have a reduced function of the thymic pool of Treg cells compared to healthy individuals *(82)*. Moreover, the number of circulating Treg cells in MG patients increases after thymectomy and correlates with improvement of the symptoms *(83)*.

The epitopes recognized by the anti-AChR CD4$^+$ T cells of MG patients, the structure of their TCR, and the functional effects of the cytokines they secrete are important to understand the pathogenesis of MG. The first two characteristics may give clues about the mechanisms that trigger and maintain the anti-AChR response. The third may help understanding the mechanisms that cause the intermittent progress of MG and the different clinical presentations among MG patients.

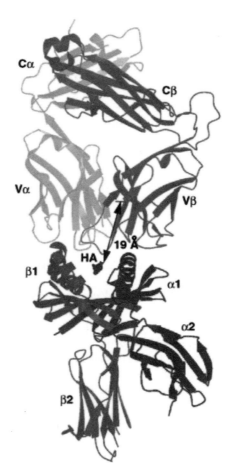

Fig. 2. Example of the complex TCR-peptide-MHC class II. The figure shows the interaction of an αβ TCR, the influenza HA peptide 306–318, and the HLA-DR1. The TCR is represented on the top of the figure; C and V refer, respectively, to the constant and variable domain of the α and β chains. The peptide is indicated as HA and the HLA-DR1 is at the bottom of the figure. The arrow indicates the distance between the C-terminus of the peptide and residue 3 of the TCR Vβ domain. Used with permission from Oxford University Press (EMBO Journal; 2000; Vol. 19, No. 21, page 5612).

4.1. Epitope Repertoire of Anti-AChR CD4$^+$ T Cells in MG

Several studies have characterized the epitope repertoire of the autoimmune CD4$^+$ T cells in MG patients extensively *(1)*. CD4$^+$ T cells recognize denatured sequence segments of the antigen: thus, biosynthetic and synthetic sequences of the human AChR could be used to identify sequence regions and even individual residues forming epitopes recognized by the CD4$^+$ T cells of MG patients *(1, 84, 85* and references therein).

Studies in experimental autoimmune diseases, including EAMG, have examined the characteristics and evolution in time of the epitope repertoire of self-reactive CD4$^+$ T cells. In rodent EAMG, most of the anti-AChR CD4$^+$ T cells recognize a limited number of epitopes, all formed by the AChR α subunit *(86, 87, 88)*. Prolonged immunization with *Torpedo* AChR (TAChR) of C57Bl/6 mice leads to a focusing of their CD4$^+$ repertoire onto a single short segment of the TAChR α subunit recognized by highly pathogenic CD4$^+$ T cells *(89* and references therein).

Also in rat EAMG most pathogenic anti-AChR CD4$^+$ T cells recognize one epitope sequence of the AChR α subunit *(86)*. After immunization with TAChR, C57Bl/6 mice genetically deficient in IL-4 (IL-4$^{-/-}$ mice) developed a more complex epitope repertoire than that of the wild-type strain *(90)*: their CD4$^+$ T cells respond intensely to a large number of sequence regions of the α subunit of both the TAChR used for the immunization and the autologous mouse AChR. The substantial autoimmune response and quick CD4$^+$ T-cell epitope spreading in IL-4$^{-/-}$ mice, together with the high susceptibility to EAMG and tendency to develop a chronic form of EAMG that is unique to this strain, are likely due to the lack in IL-4$^{-/-}$ mice of immunoregulatory Th3 cells, which need IL-4 as differentiation factor *(90)*. The characteristics of the epitope repertoire of anti-AChR CD4$^+$ T cells in rodents with an intact immune system are in contrast with what occurs in other experimental autoimmune diseases, where the recognition of the autoantigen by autoimmune CD4$^+$ T cells spreads in time to many epitopes, and even to other proteins of the target tissues *(91)*.

The finding that many anti-AChR Ab of MG patients recognize epitopes on the α subunit is consistent with the possibility that also in MG the anti-AChR CD4$^+$ T-cell response focuses on a few immunodominant epitopes. This may occur because of a preferential collaboration between CD4$^+$ T-helper and B cells that recognize epitopes within the same domain of the antigen *(1)*. Since in both MG and rodent EAMG the anti-AChR Ab response is focused on the α subunits and on the MIR, the anti-AChR CD4$^+$ T cells of MG patients might also recognize primarily one or more α subunit epitopes. Early studies found that CD4$^+$ T-cell lines propagated from the blood of MG patients using stimulation with *Torpedo* AChR recognized the α subunits almost exclusively *(92)*. Those results bolstered the hypothesis that the α subunit dominates the sensitization of CD4$^+$ T cells in MG. However, those early findings were likely due to the higher level of sequence identity among α subunits from different species than the other AChR subunits *(16, 26)*: the use of an AChR from a distant species like *Torpedo* to propagate anti-human AChR CD4$^+$ T cells likely favored propagation of CD4$^+$ T cells that recognized the highly conserved α subunit. MG patients have blood CD4$^+$ T cells that recognize each of the AChR subunits, including the adult AChR ε subunit and the embryonic AChR γ subunit *(93, 94, 95, 96, 97)*.[1] Also, CD4$^+$ T cells specific for any of the AChR subunits could be propagated *in vitro* from the blood of MG patients *(102, 103, 104, 105, 106, 107)*. Those anti-AChR CD4$^+$ T cells responded vigorously when challenged with purified mammalian muscle AChR: thus, they recognized epitopes that derived from processing of the AChR molecule.

MG patients who had generalized symptoms for several years usually had CD4$^+$ T cells that recognize all AChR subunits, whereas patients who had generalized symptoms for short periods of time usually recognized only a few subunits *(95)*. In generalized MG patients, the CD4$^+$ T-cell recognition of AChR subunits, once established, appeared to persist in time, and the intensity of the CD4$^+$ T-cell responses to the AChR subunits increased with duration of the disease *(95)*. These findings are consistent with spreading of the CD4$^+$ T-cell repertoire in MG, because they suggest that the number of CD4$^+$ T cells that become sensitized to the AChR, and possibly the number of AChR epitopes they recognize, increased with time. Longitudinal studies of the same MG patients verified that both the number of AChR-specific CD4$^+$ T cells and the number of epitopes they recognized increased as the disease progressed *(97)*. Comparison of the epitope repertoire of anti-AChR CD4$^+$ T-cell lines propagated from the same generalized MG patients, at time intervals of up to 10 years *(85*; Diethelm-Okita, Howard and Conti-Fine, unpublished), confirmed that the epitope repertoire of the anti-AChR CD4$^+$ T cells spreads to an increasing

---

[1]After innervation of the skeletal muscles, the γ subunit is substituted for by the ε subunit *(16, 26)*. However, the extrinsic ocular muscles (EOM) of adult mammals still express the embryonic γ subunit *(98, 99)* which is also expressed in the adult thymus *(100, 101)*. We will discuss later the implications of these findings for models of MG pathogenesis.

number of epitopes and demonstrated that the CD4$^+$ T cells sensitized to an epitope persisted for many years.

In summary, the results of several studies indicate that in generalized MG the CD4$^+$ T cells recognize complex repertoires of epitopes formed by all of the AChR subunits. The numerous epitopes recognized on the AChR by CD4$^+$ T cells from MG patients are consistent with spreading of the CD4$^+$ T-cell sensitization to increasingly larger parts of the AChR molecule as MG progresses. The different evolution of the epitope repertoire of the anti-AChR CD4$^+$ T cells in MG and EAMG underlines the different pathogenic mechanisms of spontaneous auto-immune diseases and their experimental models, induced by immunization with autoantigens in adjuvants.

Administration of synthetic peptide analogues of epitopes recognized by autoimmune CD4$^+$ T cells (altered peptide ligands, APL) may prevent and even cure certain experimental autoimmune diseases by shutting off an autoimmune response, at least temporarily *(108)*. The APL bind to the MHC class II molecules, and compete with the peptide epitopes derived from the autoantigen, without stimulating the epitope-specific CD4$^+$ T cells. By this mechanism, treatment with APL may anergize the pathogenic CD4$^+$ T cells or stimulate modulatory anti-inflammatory CD4$^+$ T cells, or both *(108, 109, 110)*. However, APL-based approaches might not be effective for MG management, because the anti-AChR CD4$^+$ T cells recognize many epitopes presented by different class II isotypes and alleles. Furthermore, the therapeutic effects of APL are ephemeral: they disappear when administration of the APL is discontinued.

4.2. CD4$^+$ T Cells of MG Patients Recognize "Universal", Immunodominant AChR Epitopes

A few short sequence regions of the AChR are recognized by the CD4$^+$ T cells of most or all MG patients regardless of their MHC type *(93, 94, 103, 104, 106, 107, 111)*. Moreover, MG patients have abundant CD4$^+$ T cells sensitized to epitopes within those AChR sequences *(95, 111)*. Thus, the AChR forms "universal", immunodominant epitopes for sensitization of self-reactive human CD4$^+$ T cells. Studies that used SCID mice engrafted with immune cells from MG patients demonstrated that CD4$^+$ T cells specific for those universal AChR epitopes induced synthesis of pathogenic anti-AChR Ab, and myasthenic symptoms *(14)*, suggesting that those CD4$^+$ T cells are important in MG pathogenesis.

Also exogenous protein antigens, such as tetanus and diphtheria toxoids, bear universal epitopes that sensitize CD4$^+$ T cells in most or all humans *(112)*. Universal epitope sequences are all flanked by relatively unstructured, highly mobile sequence loops exposed to the solvent: these structural features likely facilitate the release of the epitope sequence during antigen processing *(112 and references therein)*. Solvent-exposed loops would be easily accessible to the processing proteases. Human antigen presenting cells (APC) express a variety of MHC class II molecules, especially if they are heterozygous at the MHC locus *(113)*. Moreover, most human class II molecules do not have stringent sequence requirements for their peptide ligands *(114, 115, 116)*. Thus, most peptide sequences released upon antigen processing will likely encounter a human class II molecule able to bind and present them to CD4$^+$ T cells.

4.3. Unstable Recognition of AChR Epitopes by CD4$^+$ T Cells of Ocular MG Patients

The first symptoms of MG usually include, or are limited to, weakness of the extraocular muscles (EOM). In most MG patients the myasthenic weakness spreads over time to the skeletal muscles. However, in about 15% of them it remains limited to the EOM *(4, 117)*. The causes of the preferential involvement of the EOM in MG are unknown *(118)*. Physiological properties of the EOM synapses may have a role *(119)*: the motor neurons which innervate the EOM have high firing frequencies, which may make their synapses especially susceptible to myasthenic fatigue *(118, 120)*. Furthermore, the EOMs include tonic fibers, whose force of contraction is a function of the size of the depolarization at their endplates: any AChR loss, and the resulting reduction in endplate potential, would cause decreased strength of those tonic fibers. The EOMs are especially susceptible to present myasthenic weakness when a reduction of the AChR function occurs *(118, 120)*: thus, EOM weakness is the most common and earliest manifestation also in congenital MG, where genetic defects of the AChR expression or structure affect the AChR of all muscles to a similar extent *(121)*.

Patients with ocular MG have lower concentrations of serum anti-AChR Ab than generalized MG patients, and 50% of them do not have detectable Ab *(1, 22, 117)*. In ocular MG, the clinical symptoms may remain localized to the EOM because the modest anti-AChR Ab synthesis in these patients is insufficient to affect the function of muscles other than the EOM. Unique antigenic properties of the AChR expressed by the EOM might also contribute to the selective involvement of the EOM in ocular MG. The EOMs express both embryonic and adult AChR isotypes, whereas skeletal muscles express only adult AChR *(98, 122)*. Furthermore, some ocular MG patients have Ab that recognize specifically EOM synapses containing AChR with fetal properties *(120)*. Those findings suggested anti-AChR response of ocular MG patients may preferentially recognize the embryonic AChR and cause preferential weakness of the EOM *(95, 118)*. However, other studies contradicted that hypothesis *(97, 123, 124)*.

Several studies examined the T-cell recognition of the AChR in small groups of ocular MG patients. One study found that their CD4$^+$ T cells recognized consistently the embryonic γ subunit and had minimal and sporadic responses to the adult ε subunit *(95)*. In contrast, another study found that some ocular MG patients had CD4$^+$ T cells and Ab specific for adult AChR *(124)*. Those discrepancies might have been caused by a stable heterogeneity in the characteristics of the CD4$^+$ response in ocular MG patients, or by an unstable recognition of the AChR, with changes over time of the subunits recognized by the CD4$^+$ T cells. A later study *(97)* supported the latter possibility. That study found that CD4$^+$ T cells from ocular MG patients may recognize any AChR subunit: however, they seldom recognized all AChR subunits, even when the disease has lasted for many years *(97)*. Moreover, ocular MG patients had CD4$^+$ T-cell responses to the AChR subunits lower than those of generalized MG patients, and the repertoire of their anti-AChR CD4$^+$ T cells was unstable over time: CD4$^+$ T cells that recognized a given subunit appeared and disappeared over a few weeks or months *(97)*.

The low and unstable responses of the anti-AChR CD4$^+$ T cells agree with the sporadic and scarce serum anti-AChR Ab of ocular MG patients *(1, 22)*. In ocular MG, the symptoms might be limited to the EOM because of the modesty of the anti-AChR immune response and the high susceptibility of the EOM to myasthenic weakness. Ocular MG patients might still have regulatory immune mechanisms that counteract the activated pathogenic CD4$^+$ T cells and make them ineffective at driving a pathogenic Ab response. In generalized MG patients, the persistence and spreading of the anti-AChR CD4$^+$ T-cell response as the disease progresses suggest a more substantial failure of the normal mechanisms of down-regulation of self-reactive CD4$^+$ T cells.

5. TCR Vβ AND Vα USAGE BY ANTI-ACHR CD4$^+$ T CELLS OF MG PATIENTS

The TCR, which confers antigen specificity to the CD4$^+$ T cells, is a membrane protein that binds the complex between a class II MHC molecule and an epitope sequence of the antigen. The TCRs of anti-AChR CD4$^+$ T cells of MG patients are usually αβ heterodimers *(125, 126)*. The TCR β subunit is especially interesting because its variable region forms important structural elements of both the binding site for the epitope/MHC complex and the binding site for SAGs *(127, 128, 129)*.

The variable region genes of the β subunit result from somatic rearrangement of germline-encoded V (variable), D (diversity), and J (joining) segments. There are more than 100 different human TCR Vβ genes: they can be grouped into 24 families based on the extent of their sequence similarity *(130)*.

SAGs bind and link the Ag-binding domain of a TCR and a class II molecule on the surface of an APC *(131)*. Similar to what occurs after engagement of the TCR with the specific epitope/MHC molecule complex on the APC surface, crosslinking by a SAG permits costimulation and therefore activation of the CD4$^+$ T cell *(129, 131)*. Each SAG is specific for particular Vβ gene families *(129, 131)* and activates all CD4$^+$ T cells which express TCRs that utilize the preferred Vβ family or families. A SAG may initiate an autoimmune response by activating self-reactive T cells that express the preferred Vβ family *(132, 133)*.

The TCR antigen-binding site is formed by two sets of three sequence segments termed complementarity-determining regions (CDR) 1, 2, and 3: they are within the variable domains of each of the two TCR subunits *(127, 128, 129)*. CDR1 and CDR2 are formed by the V segment and CDR3 by the sequence at the joining of the V, D, and J segments of the β subunit, or the V and J segments of the α subunit, with the constant (C) region segment *(127, 128, 129)*. The V(D)J/C joining may be inaccurate, thus increasing the potential variability of CDR3. The CDR sequences, and especially CDR3, may give clues as to whether an antigen drove the selection of CDR displaying a similar sequence motif, compatible with epitope binding. On the other hand, clonal expansion in most patients of autoimmune CD4$^+$ T cells bearing TCR that use certain Vβ gene families, without any common CDR3 sequence motif, suggests the action of a SAG. Also the Vα gene family used by a TCR may influence the binding of certain SAGs *(134)*. Determination of the Vα gene family used by autoimmune TCR, and of the sequence of their CDR3 regions, may yield clues about the presence of antigen-driven clonal expansion, and perhaps also about the action of a SAG.

The search for clues supporting one or the other model is complicated by the evolution in time of the autoimmune CD4$^+$ T-cell responses, with spreading of the T-cell repertoire. Recruitment of CD4$^+$ T clones that recognize new epitopes and use different TCR genes may obscure the characteristics of the T cells whose activation triggered the autoimmune response.

Several studies have examined the TCR Vβ and Vα usage in MG patients. Some studies investigated the overall Vβ usage in the thymus and blood of MG patients; other studies, the Vβ and Vα usage by AChR-specific CD4$^+$ T-cell lines propagated from MG patients *(125, 126, 135, 136, 137, 138, 139)*.

The study of the Vβ usage in unselected T cells from the thymus or blood yielded conflicting results. Some studies *(135, 139)* did not find evidence of clonal expansion of T cells or of preferential Vβ usage in myasthenic thymi. In contrast, other studies *(136, 137, 138)* found evidence of clonal expansion and preferential Vβ usage in MG patients, although their results did not always agree on the Vβ families that seemed to be expanded. Those studies investigated a limited number of patients and used different assays of the TCR Vβ usage: a biased representation in small samples and the different shortcomings of the methods used may explain the discrepancies of those results.

Studies that examined AChR-specific CD4$^+$ T cells from MG patients *(125, 126, 138)* concluded that the Vα and Vβ repertoire of anti-AChR CD4$^+$ T cells was diverse and characteristic of the individual patient. In each patient the CD4$^+$ T cells specific for a given AChR epitope had relatively restricted Vα and Vβ usage, but the overall Vα and Vβ usage of the anti-AChR CD4$^+$ T cells was heterogeneous, because the anti-AChR CD4$^+$ T cells recognized many epitopes. CD4$^+$ T cells that recognized universal AChR epitopes used different Vβ families in different patients *(125)*.

In spite of the heterogeneity of their epitope repertoire and the Vβ and Vα usage, the anti-AChR CD4$^+$ T cells used the Vβ4 and Vβ6 families frequently, even when they recognized different AChR epitopes *(125)*. Even unselected blood CD4$^+$ T cells of MG patients used Vβ4 and Vβ6 to a significantly higher extent than the blood CD4$^+$ T cells of normal subjects *(139)*. These findings are consistent with the possibility that a SAG might have triggered the CD4$^+$ T-cell response to the AChR. Anti-AChR CD4$^+$ T cells of MG patients used also some Vα families with unexpectedly high frequency (Navaneetham et al., unpublished data), perhaps reflecting the action of a SAG in triggering the anti-AChR response *(134)*.

Experimental treatments to specifically suppress an undesirable immune response include attempts to interfere with the ability of a TCR to bind the specific peptide epitope/MHC molecule complex, by administration or induction of Ab that recognized the antigen-binding site of the TCR *(140, 141)*. This sort of treatment was successful in rodent experimental autoimmune encephalitis (EAE), where most of the pathogenic CD4$^+$ T cells initially recognize only one epitope of the myelin basic protein and use a single TCR Vβ family *(142)*. The diverse epitope repertoire and TCR Vβ and Vα usage of the anti-AChR CD4$^+$ T cells in MG make it unlikely that this approach will be useful for MG.

6. ROLES IN MG AND EAMG OF CYTOKINES SECRETED BY DIFFERENT CD4$^+$ SUBSETS

A network of cytokines modulates the development of normal and pathologic immune responses *(143, 144, 145)*. Cells that mediate acquired or innate immune responses and also non-immune cells can secrete cytokines and express cytokine receptors, which make them responsive to the presence of these chemical messengers. Identification of the cytokines produced by the anti-AChR T and B cells, and of the cytokines which modulate their activity, is a necessary step in the elucidation of the pathogenesis of MG. Also, future treatments of MG aimed at restoring tolerance to the AChR may include interventions in the cytokine network.

Some cytokines (e.g., IL-2) augment the activity of immune cells; other cytokines (e.g., the TGF-β family) down-regulate immune responses. Yet many cytokines may have complex and contrasting effects on the immune system because of their ability to influence the activity of a variety of cells: they might augment or decrease an immune response depending on the phase of the immune response when they exerted their action and the cells upon which they acted.

In Ab-mediated diseases, cytokines that induce proliferation and differentiation of B cells may cause an increase in the mature B cells that secrete high-affinity, pathogenic anti-AChR Ab. Other cytokines may increase the Ab synthesis indirectly by increasing the activity of CD4$^+$ T-helper cells and APC. Cytokines with anti-inflammatory activity, like IL-4, IL-10, and TGF-β, down-regulate activated APC and CD4$^+$ T cells and therefore may protect from autoimmune diseases. However, IL-4 and IL-10 are also growth and differentiation factors for B cells: thus, their effects on Ab-mediated autoimmune diseases may be complex.

Activated CD4$^+$ T cells differentiate into subtypes that secrete different cytokines *(143, 144, 145)*. The simplest classification of differentiated CD4$^+$ T cells discriminates between Th1, Th2, and Th3 cells. Th1 cells secrete proinflammatory cytokines, such as IFN-γ and IL-2. Th2 cells secrete anti-inflammatory cytokines, such as IL-4 and IL-10. IL-4 is a growth factor for Th3 cells which secrete TGF-β. Both Th1 and Th2 cells secrete cytokines that are proliferation and differentiation factors for B cells, and therefore induce Ab synthesis. In mice (but not in rats), Th1 cytokines induce synthesis of Ab that fix complement effectively, and therefore cause tissue damage, whereas Th2 cytokines induce synthesis of Ab that do not bind complement *(146)*. Also in humans the Th1-induced Ab (IgG1, IgG2, IgG3) are more efficient in binding and activating the complement cascade than the Th2-induced Ab (IgG4) *(147, 148)*.

Anti-AChR Th1 cells, and the cytokines they secrete, are likely important in the pathogenesis of MG. Usually, MG patients have abundant anti-AChR Th1 cells, which recognize many AChR epitopes, including several universal epitopes *(111)*. Experiments in SCID mice engrafted with immune cells from MG patients suggest that anti-AChR Th1 cells induce synthesis of pathogenic Ab: SCID mice grafted with B cells and macrophages from MG patients, supplemented with AChR-specific CD4$^+$ T-cell lines propagated with IL-2—a growth factor for Th1 cells—developed human anti-AChR Ab and myasthenic symptoms, whereas control mice that received the same blood B cells and macrophages without anti-AChR CD4$^+$ T cells did not develop anti-AChR Ab or myasthenic weakness *(14)*.

Studies on the presence of anti-AChR Th2 and Th3 cells in MG patients yielded inconclusive results. Some studies determined the cytokines secreted by blood lymphocytes of MG patients after challenge *in vitro* with AChR. They concluded that AChR challenge induced expression of a variety of cytokines, consistent with a response of both Th1 and Th2 cells, or of Th0 cells (the precursor to both Th1 and Th2 subsets, which can secrete small amounts of both Th1 and Th2 cytokines), or of all the three CD4$^+$ T-cell types *(149, 150, 151)*. Most anti-AChR CD4$^+$ T-cell lines propagated from the blood of MG patients secreted IL-2, but not IL-4, indicating that they comprised only Th1 cells *(107)*. This result was not surprising, since those CD4$^+$ T-cell lines had been propagated by cycles of stimulation with purified AChR and IL-2, which preferentially expands Th1 cells. Rarely, AChR-specific polyclonal CD4$^+$ T-cell lines from MG patients secreted both IL-2 and IL-4, suggesting that they may have comprised both Th1 and Th2 cells *(152)*.

Several studies have examined the role(s) of cytokines in rodent EAMG. Some studies determined the susceptibility to EAMG of mice strains either genetically deficient in a cytokine, a cytokine receptor, or an element involved in the cytokine signaling *(25, 41, 153, 154, 155, 156, 157, 158, 159, 160)* or hyper-producing a cytokine because of the presence of a transgene *(161)*. Other studies examined the effect on EAMG development of manipulating the concentration of different cytokines during the immunization with the AChR: administration of a cytokine increased its concentration, whereas administration of an anti-cytokine Ab neutralized the endogenous cytokine *(153, 162, 163)*. Other studies determined the pattern of secretion of different cytokines after treatments that induced tolerance to the AChR and protection from EAMG (e.g., *164, 165, 166, 167)*. Those studies consistently suggested that CD4$^+$ Th1 cells are involved in the synthesis of pathogenic anti-AChR Ab and development of EAMG, whereas IL-4 and TGF-β may have a protective role. Mice genetically deficient in IL-12, which is essential for development of Th1 cells, were resistant to EAMG induction: they synthesized Th2-induced anti-AChR Ab that bound to the AChR at the NMJ, but did not bind and activate complement, and therefore were ineffective at causing NMJ destruction *(25, 153)*. In contrast with the clear pathogenic role of IL-12 in EAMG, the role of IFN-γ—a key effector cytokine of Th1 cells—is still debated. Some studies found that mice genetically deficient in IFN-γ or in its receptor had increased resistance to EAMG as compared to the wild-type mice *(154, 155, 156)*. In contrast, another study showed that IFN-γ-deficient mice were as susceptible to EAMG as wild-type mice *(25)*. IFN-γ has complex effects: it is a potent proinflammatory stimulus *(146, 145)*, but it is also a differentiation factor for CD4$^+$CD25$^+$ Treg cells *(168, 169, 170)*. Thus, its absence might result in a functional deficit of Treg cells, and a propensity to develop autoimmune reactions (Wang, W., Milani, M. and Conti-Fine, B.M., data not published).

Among Th2 cytokines, IL-4 does not seem to have a pathogenic role in EAMG, because mice genetically deficient in IL-4 developed EAMG with even higher frequency and severity than wild-type mice *(41)*. Moreover, they developed a chronic form of EAMG after just one immunization with TAChR *(90)*, whereas induction of EAMG in the corresponding wild-type mice required several TAChR immunizations and the resulting myasthenic syndrome was generally short lived. Thus, IL-4 may have a protective role in EAMG either because of its anti-inflammatory action on Th1 cells and APC or because it is a growth factor for TGF-β-secreting modulatory CD4$^+$ T cells. Oral, nasal, or subcutaneous administration of AChR, AChR fragments, or AChR peptide sequences recognized by CD4$^+$ T cells induced AChR-specific CD4$^+$ T cells that secreted IL-4 and TGF-β and protected from EAMG induction and even reversed established myasthenic symptoms *(164, 166, 167, 171, 172)*. These treatments were ineffective in mice genetically deficient in IL-4, consistent with a protective role of this cytokine in EAMG *(41)*. Furthermore, protection from EAMG could be passively transferred into IL-4-deficient mice by the use of CD4$^+$ T cells from wild-type mice treated nasally with CD4$^+$ T-epitope peptides of the AChR *(173)*.

The Th2 effector cytokine, IL-10, reduces the expression of costimulatory molecules and cytokines by APC *(174, 175, 176, 177)* and inhibits the transcription of the IL-12 genes during the primary antigenic stimulus *(178)*, thereby reducing the activation of Th1 cells. Also, IL-10 affects CD4$^+$ T cells and resting T cells by inhibiting IL-2 production and T-cell growth *(179, 180)*. IL-10 can induce a long-term antigen-specific anergic state of CD4$^+$ T cells when it is present during antigenic challenge *(181, 182)*, and it increases activation-induced T-cell death mediated by Fas/Fas ligand *(183)*. Because of its ability to down-regulate Th1 cells and the synthesis of a variety of cytokines, IL-10 is considered as a possible therapy of undesirable immune responses *(184, 185)*.

One study examined the role of IL-10 in EAMG using C57Bl/6 (B6) mice that carried an IL-10 transgene controlled by the IL-2 promoter *(161)*. In those mice the T cells that synthesize transgenic IL-10 include activated Th1 cells, which produce IL-10 only at the time of their activation *(186)*. Because of the transient expression of the transgene, the immune system of these mice is not significantly different from the control littermates: the serum IgG levels, and the numbers and phenotype of T and B cells are normal *(186)*. Moreover, although IL-10 is a potent down-regulator of Th1 cells *(129, 130, 131)*, the IFN-γ synthesis in these transgenic mice is reduced but not abrogated *(186)*. The IL-10 transgenic mice were more susceptible to development of EAMG after immunization with TAChR than the wild-type B6 mice *(161)*. The IL-10 transgenic mice developed EAMG with greater frequency and more severe symptoms than wild-type B6 mice, and even after immunization with amounts of AChR too small to induce EAMG in wild-type mice. The IL-10 transgenic mice developed more anti-AChR Ab than non-transgenic animals. Thus, in this transgenic system, the action of IL-10 as proliferation and differentiation factor for B cells overshadows the anti-inflammatory and modulatory actions of IL-10. In support of a facilitatory role of IL-10 in the pathogenesis of EAMG, another study found that mutant B6 mice genetically deficient in IL-10, after immunization with TAChR, developed less severe myasthenic symptoms than wild-type mice, although they appeared to synthesize comparable amounts of anti-AChR Ab *(187)*.

The enhanced susceptibility to EAMG of IL-10 transgenic mice, the resistance to EAMG of mice genetically deficient in IL-10, and the pathogenic role of IL-10 in systemic lupus erythematosus, another Ab-mediated autoimmune disease *(188, 189)*, raise concerns about the suitability of IL-10 treatments for curbing Ab-mediated autoimmune diseases and suggest that IL-10 might be a target, rather than a tool, in the suppression of undesirable Ab responses. MG patients have increased blood levels of AChR-specific cells that secrete IL-10 *(190)*, a finding consistent with a pathogenic role of this cytokine in MG. Also the effects of IL-10 on T-cell-mediated experimental autoimmune responses are complex and conflicting (e.g., *191, 192, 193, 194*).

A recent study showed that another Th2 cytokine, IL-5, may also have a pathogenic role in EAMG *(149)*. Mice genetically deficient in IL-5 were resistant to EAMG induction: although they had a normal secondary immune response to the AChR, their primary response was impaired, and they had a reduced deposition of complement at the NMJ as compared to wild-type mice *(159)*.

In summary, the consistent presence in MG patients of anti-AChR Th1 cells that can induce synthesis of anti-AChR Ab and the findings that Th1 cytokines facilitate or even permit EAMG development in rodents suggest that Th1 cells have a preeminent pathogenic role in MG and EAMG. This is likely because of the ability of Th1 cells to drive the synthesis of anti-AChR Ab that bind and activate complement, and therefore can cause destruction of the NMJ effectively.

By analogy with the protective function in rodent EAMG of IL-4 and TGF-β, also in MG regulatory circuits that use these cytokines as mediators or effectors might modulate the anti-AChR response. If so, favoring the differentiation of anti-AChR CD4$^+$ T cells toward the Th2 and Th3 phenotypes may be a useful approach to turn off the anti-AChR immune response specifically. However, Th2 cytokines might also facilitate synthesis of anti-AChR Ab and development of EAMG because of their stimulating action on proliferation and differentiation of B cells. This caveat is supported by findings that the Th2 cytokines, IL-10 and IL-5, may facilitate synthesis of pathogenic anti-AChR Ab and development of EAMG. Thus, therapeutic approaches attempting to potentiate anti-AChR Th2 cells may be fraught with dangers. Much work still needs to be done to elucidate the intricacies of the cytokine networks in MG and EAMG and to make the intervention in such networks a viable option for MG treatment.

The Th1/Th2 balance can be influenced by factors outside the immune system. For instance, one study (185) suggested a link between sex hormones and Th1 cells that might explain why MG, like other autoimmune diseases, is more frequent in women than in men (1). That study found that estrogen treatments promoted EAMG in mice by augmenting the IL-12 secretion by T cells and by enhancing the production of the Th1-induced IgG2 subclasses (195).

7. CD8$^+$ CELLS IN MG AND EAMG

CD8$^+$ T cells, by virtue of their cytotoxic activity, may be effectors of immune and auto-immune responses. They have an important role in a variety of experimental autoimmune diseases that cannot be induced in animals depleted of CD8$^+$ T cells (e.g., 196, 197). Also a mouse Ab-mediated disease, a lupus-like disease induced by anti-DNA Ab, requires the presence of CD8$^+$ T cells: MHC class I-deficient mice, which do not have functional CD8$^+$ cells, are resistant to developing this syndrome (198). However, CD8$^+$ T cells may have also immunomodulatory functions and down-regulate immune responses. For example, CD8$^+$ T cells have a regulatory function in Theilers murine encephalomyelitis, a virus-induced demyelinating disease reminiscent of multiple sclerosis (199).

It is still unclear if CD8$^+$ T cells have a role in MG and EAMG. Different studies have examined this issue in mouse EAMG, with conflicting results. Mice deficient in β2 microglobulin do not express MHC class I molecules and do not have CD8$^+$ T cells, yet they developed EAMG after immunization with TAChR with higher incidence, earlier onset, and more severe symptoms than heterozygous littermates that had CD8$^+$ T cells (200). These results suggested that CD8$^+$ T cells might have regulatory rather than pathogenic functions in mouse EAMG. Consistent with those results, another study found that CD8$^+$ T-cell depletion augmented mouse susceptibility to EAMG (201). Yet other studies yielded results that suggested a pathogenic role for CD8$^+$ T cells in rodent EAMG. Lewis rats depleted in CD8$^+$ T cells developed EAMG with less severe symptoms, and lower serum concentration of anti-AChR Ab, than the control rats (202). Another study found that mice genetically deficient in CD8$^+$ T cells were resistant to EAMG induction. Thus, CD8$^+$ T cells are likely involved in the pathogenesis of EAMG, although it is not clear yet whether they have a pathogenic or a modulatory role, or both.

MG patients do not have detectable cytotoxic CD8$^+$ T-cell responses to the AChR or other components of the NMJ (7): thus, CD8$^+$ T cells are unlikely to have a significant pathogenic role in MG. However, they might have a modulatory role.

MG patients may have different numbers of blood CD8$^+$ T cells than normal subjects (203, 204). MG patients with modest symptoms or in clinical remission had numerous activated CD8$^+$ T cells in the blood, whereas severely affected patients had very few: an increase in activated CD8$^+$ T cells in the blood predicted an improvement in the symptoms (205). CD8$^+$ T cells inhibited the anti-AChR responses of CD4$^+$ T cells in vitro (93, 94, 205), and removal of CD8$^+$

T cells from cultures of BMC of MG patients caused an increase in Ab-producing B cells and anti-AChR Ab *(206, 207)*.

$CD8^+$ cells that recognize proteins expressed by activated anti-AChR $CD4^+$ T cells might mediate immunoregulatory circuits in MG patients *(208)*. Regulatory $CD8^+$ T cells may recognize epitopes derived from the TCR uniquely expressed by a $CD4^+$ T-cell clone (idiotypic epitopes) or from proteins expressed by all activated $CD4^+$ T cells (ergotypic epitopes). A precarious balance between the activated autoimmune $CD4^+$ T cells and $CD8^+$ T cells able to modulate the autoimmune $CD4^+$ T-cell reactivity might explain the relapsing character of MG and of many other autoimmune diseases.

8. THE THYMUS IN MG

The thymus has an important albeit still obscure role in the pathogenesis of MG. Usually, MG patients have an abnormal thymus, and thymectomy may improve the clinical course of their disease *(68)*. Most (70%) of the abnormal MG thymi have lymphoid follicular hyperplasia *(1, 2, 3, 4, 209, 210)*: the perivascular spaces are extended, disrupted, and filled with lymphoid tissue that resembles peripheral immune organs; and active germinal centers are present, similar to the secondary lymph follicles of peripheral lymphoid organs. Hyperplastic MG thymi have abnormally high numbers of mature T cells ($CD4^+$ $CD8^-$ or $CD4^-$ $CD8^+$), which usually are present in peripheral lymphoid organs only. Also, they contain anti-AChR T and B cells able to mount a productive anti-AChR Ab response: thus, transplantation of thymus fragments from MG patients into SCID mice induces appearance of human anti-AChR Ab, which binds the mouse NMJ and causes AChR loss *(73)*.

About 10% of MG patients have a thymoma *(1, 2, 3, 4, 209, 210)*. Thymomas are a heterogeneous group of tumors, and their histologic characteristics are not consistent with the presence of an ongoing immune response *(209)*. However, areas of lymphoid follicular hyperplasia surround the thymoma tissue of MG patients frequently.

8.1. AChR-Like Proteins Are Expressed in the Thymus

The anti-AChR response that occurs in MG thymi with lymphoid follicular hyperplasia may be directed to a protein or proteins synthesized within the thymus, which are similar to muscle AChR *(1, 2, 3, 4, 209, 210)*. Normal and MG thymi express protein(s) that cross-react with anti-muscle AChR Ab and bind α-bungarotoxin (α-BGT), a specific ligand of muscle AChR *(211, 212, 213, 214, 215, 216, 217)*. The thymus AChR or AChR-like protein might be the original auto-antigen that triggers the anti-AChR response in MG *(1, 209, 210)*.

It is not clear whether the thymus expresses true muscle-type AChR or AChR-like proteins with different antigenic structures and whether expression of AChR or AChR-like proteins is different in thymi from MG patients than from normal subjects. Human thymus may express mRNA transcripts for all the muscle AChR subunits. However, transcripts of the α and β subunits have been found consistently *(218, 219, 220, 221, 222, 223, 100, 224, 225, 226, 227)*, whereas those of the γ, δ, and ε subunits have not *(219, 221, 222, 223, 100, 227, 101, 228)*. Some studies found γ subunit, but not ε subunit transcript *(101, 100)*. In contrast, another study found expression of the ε subunit, but not of the γ subunit, mRNA in both normal and MG thymi *(223)*. Some studies found δ subunit protein or mRNA transcript in thymus tissue and in cultured myoid and epithelial cells from the thymus *(101, 219, 100, 227)*. Yet other studies found absence, or extreme reduction, or erratic expression of δ subunit transcripts in human thymi *(221, 222, 228)*. The results of studies that measure mRNA transcripts rather than the corresponding proteins need to be interpreted with caution, because the amounts of different transcripts in a tissue may not reflect the relative abundance of the encoded proteins. The conclusions of such studies will require verification by measurement of the corresponding AChR proteins.

Expression of even an incomplete set of muscle AChR subunits in the thymus must result in assembly of membrane proteins reminiscent of muscle AChR, because the thymus contains binding sites for α-BGT and AChR-specific Ab *(211, 212, 213, 214, 215, 216, 217)*. Transfection experiments have shown that an incomplete set of AChR subunits, which included only α, β, and γ subunits, assembled into an AChR-like protein that was inserted in the cell membrane and bound cholinergic ligands: an extra copy of the γ subunit substituted for the missing δ subunit *(229)*. The inconsistent presence of δ and ε subunit transcripts in the thymus led to the suggestion that the thymus expresses "triplet receptors" formed only by α, β, and γ subunits *(222)*. Such receptors, because of their different antigenic structure than muscle AChR, might trigger the autoimmune response in MG. However, the finding that thymus tissues from healthy subjects rarely express detectable δ subunit transcripts is inconsistent with that hypothesis *(221, 222, 228)*.

In summary, expression of AChR subunits in human thymus seems to be less tightly regulated than in muscle; the human thymus may express, at least intermittently, AChR-like proteins that have different subunit composition and antigenic structure than muscle AChR.

The thymus may express other AChR-like proteins. Thymomas of MG patients express an AChR-like protein which contains a single subunit about three times as large as a true AChR subunit *(209)*. This protein binds anti-AChR Ab that recognize a linear determinant of the α subunit, but it does not bind α-BGT or anti-MIR Ab: it might be involved in the anti-AChR response in MG associated with thymoma. MG patients with thymomas have several unique clinical and immunologic features, which may reflect a different initiation of their immune response *(1, 2, 3, 4, 209, 230)*. Thymomas also contain α-BGT binding proteins, which may be bona fide AChR *(209, 230)*.

8.2. Thymus Cells That Express AChR Proteins

Not all the cell types present in the thymus express AChR-like proteins. Also, different types of thymus cells may express different AChRs. Cultured human thymus epithelial cells and myoid cells expressed AChR-like proteins: however, epithelial cells expressed the ε subunit, not the γ subunit, whereas myoid cells expressed both ε and γ subunits *(219, 91)*. The different spectrum of AChR subunits expressed in various cell types, and the potential heterogeneity in cellular compositions of the samples of adult human thymi, may contribute to the inconsistent results of the studies on expression of AChR subunits in the thymus.

Several studies examined the expression of AChR subunits in normal and MG thymi: they could not identify any consistent characteristic of MG thymi that related to the anti-AChR response *(209)*. However, different studies found an increased synthesis of some AChR subunits in MG thymi, perhaps related to the hyperplastic changes common in MG patients *(218, 221)*.

Among the different types of cells present in the thymus, the myoid and epithelial cells have characteristics that make them attractive candidates as the source of AChR antigens that sensitize autoimmune CD4[+] T cells in MG. Myoid cells in both normal and MG thymi express AChR-like proteins, similar to both embryonic and adult types *(214, 217, 100, 224, 227)*. Myoid cells do not express MHC class II molecules. However, hyperplastic MG thymi are close to, and occasionally inside germinal centers, and in intimate contact with HLA-DR-positive reticulum cells *(209, 231)*, which might serve as APC of AChR epitopes.

Epithelial cells in normal and MG thymi contain AChR-like proteins that bind α-BGT *(212)*. Cultured epithelial cells express only AChR proteins that comprise the adult ε subunit and may not include all the subunits necessary for proper AChR function *(219)*. Their AChR-like proteins are non-functional, because they do not gate currents upon acetylcholine binding *(232)*. This is in contrast with the functional AChR expressed by cultured myoid cells, which have electrophysiological characteristics similar to muscle AChR and which include the δ subunit *(227)*. The inconsistent presence of δ subunit transcripts in human thymi, and especially in samples from

MG patients, is consistent with the possibility that the AChR-like proteins of epithelial cells—which are much more numerous than myoid cells—do not include the δ subunit.

The Th1 cytokines IFN-γ and TNF-α may upregulate the expression of AChR in both epithelial and myoid cells in the thymus *(233)*. In MG patients with hyperplastic thymi, the IFN-regulated genes are overexpressed as compared to thymi from control subjects: thus in MG patients, these Th1 cytokines could contribute to the onset of the disease by influencing the expression of AChR in the thymus *(233)*. Moreover, the thymus of MG patients overexpresses proteases involved in the MHC class II presentation pathway *(234)*. These findings support an important role of the thymus in the generation of autoreactive anti-AChR CD4[+] T cells.

9. PATHOGENIC MECHANISMS OF MG

The cause or causes of the breakdown of self-tolerance in autoimmune diseases are not known. Many potentially autoreactive CD4[+] T cells survive clonal deletion during the development of the immune system: they include CD4[+] T cells specific for self-antigens which, like muscle AChR, may be targets of autoimmune responses *(1)*. Yet in normal subjects their presence does not result in clinically significant autoimmune responses.

Clinical and epidemiological studies suggest that infections may facilitate the induction and relapses of autoimmune diseases *(132)*. Microbial infections might trigger autoimmune responses by activating potentially self-reactive CD4[+] T cells: this mechanism might apply to MG, because CD4[+] T cells play a crucial role in the synthesis of pathogenic anti-AChR Ab *(1, 2, 3, 4)*. Several models have attempted to explain how this might occur.

One model proposes that molecular mimicry between a microbial epitope recognized by CD4[+] T cells and a similar sequence of a self-antigen activates CD4[+] T cells cross-reactive with the self-antigen *(235)*. Many sequence segments of microbial antigens are similar to sequences of human proteins *(133)*. Moreover, amino acid identity may be needed only for a few amino acids within a CD4[+] epitope sequence *(236)*. Several sequence regions of human muscle AChR, including sequences forming epitopes recognized by CD4[+] T cells of MG patients, are similar to sequences of common microbes *(94, 106)*. This is not surprising, because certain short amino acid sequences—that may correspond to common tridimensional structural motifs—occur in the sequence of proteins much more frequently than expected statistically *(237)*. This may explain why fragments of alien proteins long enough to form CD4[+] T-cell epitopes frequently share sequence similarities with fragments of self-proteins.

Once tolerance to a self-epitope is broken, T cells that recognize that epitope and secrete proinflammatory cytokines can migrate into the tissue that contains the antigen and cause an inflammatory response and tissue destruction. Because of this, APC able to activate any potentially self-reactive CD4[+] T cells may present new epitopes derived from the same antigen, and even from other antigens released by the damaged tissue *(91, 142)*. Thus, inflammation and intermolecular and intramolecular epitope spreading may become self-maintaining processes *(238)*, which could cause chronic tissue destruction and further sensitization of CD4[+] T cells to an increasingly large repertoire of epitopes and antigens. This process has been termed "epitope spreading", and it results in sensitization of CD4[+] T cells to epitopes distinct from, and non-cross-reactive with, the inducing epitope *(91)*.

Microbial SAGs may also activate potentially autoimmune T cells. This mechanism might trigger insulin-dependent diabetes mellitus, rheumatoid arthritis, and perhaps multiple sclerosis *(239, 240, 241)*. The action of a SAG might explain the relapses common in autoimmune diseases: in mouse strains which develop EAE and whose pathogenic anti-MBP CD4[+] T cells use the Vβ8 gene family, the SAG staphylococcal enterotoxin B, which activates Vβ8[+] T cells, induced EAE relapses in mice that were in clinical remissions and triggered paralysis in mice with subclinical disease *(242)*.

The action of a SAG might trigger MG. In hyperplastic MG thymi, dendritic cells are unusually abundant, and the CD4$^+$ T cells cluster around them, forming rosette-like structures (209). This might be the place where a SAG released by virus-infected dendritic cells could activate the anti-AChR CD4$^+$ T cells and trigger MG. Anti-AChR CD4$^+$ T cells activated by a SAG in the thymus could migrate to the periphery and activate anti-AChR B cells in the lymph nodes. Encounter of the anti-AChR CD4$^+$ T cells with muscle AChR epitopes presented by APC at the inflamed NMJ, damaged by the binding of complement to complement-fixing anti-AChR Ab, could cause further activation of autoimmune CD4$^+$ T cells and release of proinflammatory cytokines. This would perpetuate this mechanism by causing further recruitment of professional APC and epitope spreading of the CD4$^+$ T-cell response.

A third model proposes that a viral infection may activate self-reactive immune cells and trigger a tissue-specific autoimmune response by "bystander activation" because of the inflammatory reaction to the virus-infected tissue (243). It has been proposed that the immune system responds to antigens according to not only whether they are perceived as self or non-self but also whether they are potentially dangerous (244). In this model, self-proteins may elicit an immune response if presented to CD4$^+$ T cells in situations of danger, that is, by professional APC in inflamed tissues. The inflammation caused by a viral infection would attract antiviral T cells that express proinflammatory cytokines, and professional APC able to present epitopes derived from tissue proteins. This would result in the activation of potentially self-reactive CD4$^+$ T cells recognizing those epitopes. Furthermore, proinflammatory cytokines, like IFN-γ, may stimulate expression of class II MHC molecules in cells that do not express them constitutively, including muscle cells (146, 145): this may facilitate presentation of self-epitopes and expansion of any primed self-reactive CD4$^+$ T cells.

Thus, an inflammatory response resulting from a viral infection in the thymus could cause professional APC to present epitopes derived from the thymus AChR proteins. This would activate potentially autoreactive anti-AChR CD4$^+$ T cells and initiate the autoimmune response in MG. This mechanism is especially attractive to explain the cases of MG that are associated with a thymoma. The immune response against the cancer cells expressing AChR-like antigens may trigger an autoimmune response that involves the muscle AChR and cause the myasthenic symptoms. This mechanism is similar to that proposed for a variety of paraneoplastic syndromes, including the Lambert–Eaton myasthenic syndrome (3).

Whatever the mechanism that triggers the anti-AChR response might be, the thymus, rather than the muscle, is a good candidate as the tissue where the immune response begins (1, 2, 3, 4, 209) because it expresses AChR-like proteins and contains cells that express MHC class II molecules (209, 231). Among these, the reticulum cells have an ideal location for processing and presentation of epitopes derived form the AChR expressed by the myoid cells, as discussed above. If the anti-AChR response begins in the thymus, the anti-AChR CD4$^+$ T cells would become sensitized first to epitopes on the AChR subunits that are expressed, or best expressed, in the thymus. When the muscle AChR becomes involved in the autoimmune response, the CD4$^+$ T-cell repertoire would spread to all muscle AChR subunits, even those that may not be expressed in the thymus. The finding that CD4$^+$ T cells from MG patients responded to the δ subunit significantly less than to other AChR subunits (95, 97) is consistent with the possibility that a thymus AChR-like protein lacking the δ subunit (221) triggered the sensitization of anti-AChR CD4$^+$ T cells. Furthermore, anti-AChR CD4$^+$ T cells of MG patients frequently recognize the γ subunit (106), which is usually expressed in the thymus (100) but not in adult muscle (26). This finding is consistent with the possibility that the CD4$^+$ T cells have been sensitized to an AChR outside the muscle, and possibly in the thymus.

The beneficial effects of thymectomy in MG (68) suggest that the thymus has a role in maintaining the anti-AChR response that causes MG symptoms. The therapeutic effects of thymectomy could be explained by removal of a source of antigenic stimulation. The symptom improvement usually appears several months after the surgery *(68)*, consistent with the long life span of activated antigen-specific T-helper cells *(245)*. Such cells must die out before the effects of thymectomy become evident.

The large number of epitopes recognized by any MG patient *(1)* indicates that when the anti-AChR response has caused clinical myasthenic weakness, the whole AChR is involved in the autoimmune response. Yet this does not exclude that the anti-AChR sensitization was initiated by molecular mimicry between a single viral or bacterial epitope and a small sequence region on the AChR, or by SAG activation of $CD4^+$ T cells expressing a particular Vβ gene family and recognizing just one, or a limited set, of AChR epitopes: activation of $CD4^+$ T cells against even one epitope may be followed by spreading of the $CD4^+$ response to the whole AChR.

The activated AChR-specific $CD4^+$ T-helper cells need to interact with anti-AChR B cells in order to produce a pathogenic response high-affinity anti-AChR Ab. B cells secreting Ab that bind the AChR with low affinity are common: about 10% of the monoclonal IgG in multiple myeloma patients bind muscle AChR *(246)*. Yet myelomas are seldom associated with MG: the low affinity for the AChR of the myeloma Ab may explain their inability to cause MG. The interaction of activated anti-AChR $CD4^+$ T-helper cells with B cells that make low-affinity anti-AChR Ab may trigger the mechanism of somatic mutation of the immunoglobulin genes and lead to the production of high-affinity, pathogenic anti-AChR Ab, and myasthenic symptoms.

Acknowledgments The studies carried out in the authors' laboratory and included in this review were supported by the NINDS grant NS23919 and by research grants from the Muscular Dystrophy Association of America (to BMC-F).

REFERENCES

1. Conti-Fine BM, Protti M, Bellone M, Howard JJF. Myasthenia Gravis: The Immunobiology of an Autoimmune Disease. Austin, TX: R.G. Landes Company; 1997.
2. Engel AG. The myasthenic syndromes. New York: Oxford University Press; 1999.
3. Richman D. Myasthenia Gravis and related diseases. Disorders of the neuromusculat junction. Ann N Y Acad Sci 1998; 841: 1–838.
4. Oosterhuis HJ. Myasthenia gravis. Groningen: Groningen Neurological Press; 1997.
5. Oldstone MB. Molecular mimicry and immune-mediated diseases. FASEB J 1998; 12: 1255–1265.
6. Brocke S, Hausmann S, Steinman L, Wucherpfennig KW. Microbial peptides and superantigens in the pathogenesis of autoimmune diseases of the central nervous system. Semin Immunol 1998; 10: 57–67.
7. Mokhtarian F, Pino M, Ofosu-Appiah W, Grob D. Phenotypic and functional characterization of T cells from patients with myasthenia gravis. J Clin Invest 1990; 86: 2099–2108.
8. Toyka KV, Drachman DB, Griffin DE, Pestronk A, Winkelstein JA, Fishbeck KH, Kao I. Myasthenia gravis. Study of humoral immune mechanisms by passive transfer to mice. N Engl J Med 1977; 296: 125–131.
9. Lindstrom JM, Engel AG, Seybold ME, Lennon VA, Lambert EH. Pathological mechanisms in experimental autoimmune myasthenia gravis. II. Passive transfer of experimental autoimmune myasthenia gravis in rats with anti-acetylcholine receptor antibodies. J Exp Med 1976; 144: 739–753.
10. Oda K, Korenaga S, Ito Y. Myasthenia gravis: passive transfer to mice of antibody to human and mouse acetylcholine receptor. Neurology 1981; 31: 282–287.
11. Lennon VA, Lambert EH. Myasthenia gravis induced by monoclonal antibodies to acetylcholine receptors. Nature 1980; 285: 238–240.
12. Cornelio F, Antozzi C, Confalonieri P, Baggi F, Mantegazza R. Plasma treatment in diseases of the neuromuscular junction. Ann N Y Acad Sci 1998; 841: 803–810.
13. Soliven BC, Lange DJ, Penn AS, Younger D, Jaretzki A, 3rd, Lovelace RE, Rowland LP. Seronegative myasthenia gravis. Neurology 1988; 38: 514–517.
14. Wang ZY. Myasthenia in SCID mice grafted with myasthenic patient lymphocytes: role of $CD4^+$ and $CD8^+$ cells. Neurology 1999; 52: 484–497.
15. Hoch W, McConville J, Helms S, Newsom-Davis J, Melms A, Vincent A. Auto-antibodies to the receptor tyrosine kinase MuSK in patients with myasthenia gravis without acetylcholine receptor antibodies. Nat Med 2001; 7: 365–368.
16. Lindstrom JM. Acetylcholine receptors and myasthenia. Muscle Nerve 2000; 23: 453–477.
17. Tzartos SJ, Cung MT, Demange P et al. The main immunogenic region (MIR) of the nicotinic acetylcholine receptor and the anti-MIR antibodies. Mol Neurobiol 1991; 5: 1–29.

18. Beroukhim R, Unwin N. Three-dimensional location of the main immunogenic region of the acetylcholine receptor. Neuron 1995; 15: 323–331.

19. Lindstrom J, Einarson B. Antigenic modulation and receptor loss in experimental autoimmune myasthenia gravis. Muscle Nerve 1979; 2: 173–179.

20. Conti-Tronconi B, Tzartos S, Lindstrom J. Monoclonal antibodies as probes of acetylcholine receptor structure. 2. Binding to native receptor. Biochemistry 1981; 20: 2181–2191.

21. Tzartos SJ, Sophianos D, Efthimiadis A. Role of the main immunogenic region of acetylcholine receptor in myasthenia gravis. An Fab monoclonal antibody protects against antigenic modulation by human sera. J Immunol 1985; 134: 2343–2349.

22. Drachman DB. Myasthenia gravis. N Engl J Med 1994; 330: 1797–1810.

23. Lennon VA, Seybold ME, Lindstrom JM, Cochrane C, Ulevitch R. Role of complement in the pathogenesis of experimental autoimmune myasthenia gravis. J Exp Med 1978; 147: 973–983.

24. Engel AG, Arahata K. The membrane attack complex of complement at the endplate in myasthenia gravis. Ann N Y Acad Sci 1987; 505: 326–332.

25. Karachunski PI, Ostlie NS, Monfardini C, Conti-Fine BM. Absence of IFN-gamma or IL-12 has different effects on experimental myasthenia gravis in C57BL/6 mice. J Immunol 2000; 164: 5236–5244.

26. Conti-Tronconi BM, McLane KE, Raftery MA, Grando SA, Protti MP. The nicotinic acetylcholine receptor: structure and autoimmune pathology. Crit Rev Biochem Mol Biol 1994; 29: 69–123.

27. Lindstrom JM, Einarson BL, Lennon VA, Seybold ME. Pathological mechanisms in experimental autoimmune myasthenia gravis. I. Immunogenicity of syngeneic muscle acetylcholine receptor and quantitative extraction of receptor and antibody-receptor complexes from muscles of rats with experimental autoimmune myasthenia gravis. J Exp Med 1976; 144: 726–738.

28. Granato DA, Fulpius BW, Moody JF. Experimental myasthenia in Balb/c mice immunized with rat acetylcholine receptor from rat denervated muscle. Proc Natl Acad Sci USA 1976; 73: 2872–2876.

29. Patrick J, Lindstrom J. Autoimmune response to acetylcholine receptor. Science 1973; 180: 871–872.

30. Tarrab-Hazdai R, Aharonov A, Abramsky O, Silman I, Fuchs S. Proceedings: Animal model for myasthenia gravis: acetylcholine receptor-induced myasthenia in rabbits, guinea pigs and monkeys. Isr J Med Sci 1975; 11: 1390.

31. Lennon VA, Lindstrom JM, Seybold ME. Experimental autoimmune myasthenia: A model of myasthenia gravis in rats and guinea pigs. J Exp Med 1975; 141: 1365–1375.

32. Berman PW, Patrick J. Experimental myasthenia gravis. A murine system. J Exp Med 1980; 151: 204–223.

33. Zoda T, Yeh TM, Krolick KA. Clonotypic analysis of anti-acetylcholine receptor antibodies from experimental autoimmune myasthenia gravis-sensitive Lewis rats and experimental autoimmune myasthenia gravis-resistant Wistar Furth rats. J Immunol 1991; 146: 663–670.

34. Biesecker G, Koffler D. Resistance to experimental autoimmune myasthenia gravis in genetically inbred rats. Association with decreased amounts of in situ acetylcholine receptor-antibody complexes. J Immunol 1988; 140: 3406–3410.

35. Christadoss P, Krco CJ, Lennon VA, David CS. Genetic control of experimental autoimmune myasthenia gravis in mice. II. Lymphocyte proliferative response to acetylcholine receptor is dependent on Lyt-1$^+$23- cells. J Immunol 1981; 126: 1646–1647.

36. Fuchs S, Nevo D, Tarrab-Hazdai R, Yaar I. Strain differences in the autoimmune response of mice to acetylcholine receptors. Nature 1976; 263: 329–330.

37. Christadoss P, Lindstrom JM, Melvold RW, Talal N. Mutation at I-A beta chain prevents experimental autoimmune myasthenia gravis. Immunogenetics 1985; 21: 33–38.

38. Lindstrom JM, Seybold ME, Lennon VA, Whittingham S, Duane DD. Antibody to acetylcholine receptor in myasthenia gravis. Prevalence, clinical correlates, and diagnostic value. Neurology 1976; 26: 1054–1059.

39. Roses AD, Olanow CW, McAdams MW, Lane RJ. No direct correlation between serum antiacetylcholine receptor antibody levels and clinical state of individual patients with myasthenia gravis. Neurology 1981; 31: 220–224.

40. Bellone M, Ostlie N, Lei SJ, Wu XD, Conti-Tronconi BM. The I-Abm12 mutation, which confers resistance to experimental myasthenia gravis, drastically affects the epitope repertoire of murine CD4$^+$ cells sensitized to nicotinic acetylcholine receptor. J Immunol 1991; 147: 1484–1491.

41. Karachunski PI, Ostlie NS, Okita DK, Conti-Fine BM. Interleukin-4 deficiency facilitates development of experimental myasthenia gravis and precludes its prevention by nasal administration of CD4$^+$ epitope sequences of the acetylcholine receptor. J Neuroimmunol 1999; 95: 73–84.

42. Tzartos S, Hochschwender S, Vasquez P, Lindstrom J. Passive transfer of experimental autoimmune myasthenia gravis by monoclonal antibodies to the main immunogenic region of the acetylcholine receptor. J Neuroimmunol 1987; 15: 185–194.

43. Tzartos SJ, Kokla A, Walgrave SL, Conti-Tronconi BM. Localization of the main immunogenic region of human muscle acetylcholine receptor to residues 67–76 of the alpha subunit. Proc Natl Acad Sci USA 1988; 85: 2899–2903.

44. Bellone M, Tang F, Milius R, Conti-Tronconi BM. The main immunogenic region of the nicotinic acetylcholine receptor. Identification of amino acid residues interacting with different antibodies. J Immunol 1989; 143: 3568–3579.

45. Barkas T, Gabriel JM, Mauron A et al. Monoclonal antibodies to the main immunogenic region of the nicotinic acetylcholine receptor bind to residues 61–76 of the alpha subunit. J Biol Chem 1988; 263: 5916–5920.

46. Manfredi AA, Bellone M, Protti MP, Conti-Tronconi BM. Molecular mimicry among human autoantigens. Immunol Today 1991; 12: 46–47.

47. Gomez CM, Richman DP. Anti-acetylcholine receptor antibodies directed against the alpha-bungarotoxin binding site induce a unique form of experimental myasthenia. Proc Natl Acad Sci USA 1983; 80: 4089–4093.

48. Conti-Fine BM, Kaminski HJ. Autoimmune neuromuscular transmission disorders: Myasthenia Gravis and Lambert-Eaton Myasthenic syndrome. Continuum 2001; 7: 56–95.

49. Walker MB. case showing the effect of Prostigmin on myasthenia gravis. Proc R Soc Med 1935; 28: 759–761.

50. Poea S, Guyon T, Bidault J, Bruand C, Mouly V, Berrih-Aknin S. Modulation of acetylcholine receptor expression in seronegative myasthenia gravis. Ann Neurol 2000; 48: 696–705.

51. Shigemoto K, Kubo S, Maruyama N et al. Induction of myasthenia gravis by immunization against muscle-specific kinase. J Clin Invest 2006; 116: 1016–1024.

52. Plested CP, Tang T, Spreadbury I, Littleton ET, Kishore U, Vincent A. AChR phosphorylation and indirect inhibition of AChR function in seronegative MG. Neurology 2002; 59: 1672–1673.

53. Aarli JA, Skeie GO, Mygland A, Gilhus NE. Muscle striation antibodies in myasthenia gravis. Diagnostic and functional significance. Ann N Y Acad Sci 1998; 841: 505–515.

54. Link H, Sun JB, Lu CZ, Xiao BG, Fredrikson S, Hojeberg B, Olsson T. Myasthenia gravis: T and B cell reactivities to the beta-bungarotoxin binding protein presynaptic membrane receptor. J Neurol Sci 1992; 109: 173–181.

55. Takaya M, Kawahara S, Namba T, Grob D. Antibodies against myofibrillar proteins in myasthenia gravis patients. Tokai J Exp Clin Med 1992; 17: 35–39.

56. Mohan S, Barohn RJ, Jackson CE, Krolick KA. Evaluation of myosin-reactive antibodies from a panel of myasthenia gravis patients. Clin Immunol Immunopathol 1994; 70: 266–273.

57. Mohan S, Barohn RJ, Krolick KA. Unexpected cross-reactivity between myosin and a main immunogenic region (MIR) of the acetylcholine receptor by antisera obtained from myasthenia gravis patients. Clin Immunol Immunopathol 1992; 64: 218–226.

58. Kuks JB, Limburg PC, Horst G, Dijksterhuis J, Oosterhuis HJ. Antibodies to skeletal muscle in myasthenia gravis. Part 1. Diagnostic value for the detection of thymoma. J Neurol Sci 1993; 119: 183–188.

59. Aarli JA. Titin, thymoma, and myasthenia gravis. Arch Neurol 2001; 58: 869–870.

60. Skeie GO. Skeletal muscle titin: physiology and pathophysiology. Cell Mol Life Sci 2000; 57: 1570–1576.

61. Baggi F, Andreetta F, Antozzi C. et al. Anti-titin and antiryanodine receptor antibodies in myasthenia gravis patients with thymoma. Ann N Y Acad Sci 1998; 841: 538–541.

62. Meager A, Wadhwa M, Dilger P, Bird C, Thorpe R, Newsom-Davis J, Willcox N. Anti-cytokine autoantibodies in autoimmunity: preponderance of neutralizing autoantibodies against interferon-alpha, interferon-omega and interleukin-12 in patients with thymoma and/or myasthenia gravis. Clin Exp Immunol 2003; 132: 128–136.

63. Shiono H, Wong YL, Matthews I et al. Spontaneous production of anti-IFN-alpha and anti-IL-12 autoantibodies by thymoma cells from myasthenia gravis patients suggests autoimmunization in the tumor. Int Immunol 2003; 15: 903–913.

64. Yoshikawa H, Sato K, Edahiro S et al. Elevation of IL-12p40 and its antibody in myasthenia gravis with thymoma. J Neuroimmunol 2006.

65. Hohlfeld R, Toyka KV, Miner LL, Walgrave SL, Conti-Tronconi BM. Amphipathic segment of the nicotinic receptor alpha subunit contains epitopes recognized by T lymphocytes in myasthenia gravis. J Clin Invest 1988; 81: 657–660.

66. Ahlberg R, Yi Q, Pirskanen R et al. Treatment of myasthenia gravis with anti-CD4 antibody: improvement correlates to decreased T-cell autoreactivity. Neurology 1994; 44: 1732–1737.

67. Morgutti M, Conti-Tronconi BM, Sghirlanzoni A, Clementi F. Cellular immune response to acetylcholine receptor in myasthenia gravis: II. Thymectomy and corticosteroids. Neurology 1979; 29: 734–738.

68. Gronseth GS, Barohn RJ. Practice parameter: thymectomy for autoimmune myasthenia gravis (an evidence-based review): report of the Quality Standards Subcommittee of the American Academy of Neurology. Neurology 1982; 55: 7–15.

69. Hohlfeld R, Kalies I, Ernst M, Ketelsen UP, Wekerle H. T-lymphocytes in experimental autoimmune myasthenia gravis. Isolation of T-helper cell lines. J Neurol Sci 1982; 57: 265–280.

70. Zhang GX, Xiao BG, Bakhiet M, van der Meide P, Wigzell H, Link H, Olsson T. Both CD4$^+$ and CD8$^+$ T cells are essential to induce experimental autoimmune myasthenia gravis. J Exp Med 1996; 184: 349–356.

71. Kaul R, Shenoy M, Goluszko E, Christadoss P. Major histocompatibility complex class II gene disruption prevents experimental autoimmune myasthenia gravis. J Immunol 1994; 152: 3152–3157.

72. Simpson E, Farrant J, Chandler P. Phenotypic and functional studies of human peripheral blood lymphocytes engrafted in SCID mice. Immunol Rev 1991; 124: 97–111.

73. Schonbeck S, Padberg F, Hohlfeld R, Wekerle H. Transplantation of thymic autoimmune microenvironment to severe combined immunodeficiency mice. A new model of myasthenia gravis. J Clin Invest 1992; 90: 245–250.

74. Martino G, DuPont BL, Wollmann RL et al. The human-severe combined immunodeficiency myasthenic mouse model: a new approach for the study of myasthenia gravis. Ann Neurol 1993; 34: 48–56.

75. Pette M, Fujita K, Kitze B, Whitaker JN, Albert E, Kappos L, Wekerle H. Myelin basic protein-specific T lymphocyte lines from MS patients and healthy individuals. Neurology 1990; 40: 1770–1776.

76. Martin R, Jaraquemada D, Flerlage M et al. Fine specificity and HLA restriction of myelin basic protein-specific cytotoxic T cell lines from multiple sclerosis patients and healthy individuals. J Immunol 1990; 145: 540–548.

77. Sommer N, Harcourt GC, Willcox N, Beeson D, Newsom-Davis J. Acetylcholine receptor-reactive T lymphocytes from healthy subjects and myasthenia gravis patients. Neurology 1991; 41: 1270–1276.

78. Kellermann SA, McCormick DJ, Freeman SL, Morris JC, Conti-Fine BM. TSH receptor sequences recognized by CD4$^+$ T cells in Graves' disease patients and healthy controls. J Autoimmun 1995; 8: 685–698.

79. Marrack P. T cell tolerance. Harvey Lect 1993; 89: 147–155.

80. Beissert S, Schwarz A, Schwarz T. Regulatory T cells. J Invest Dermatol 2006; 126: 15–24.

81. Sakaguchi S, Sakaguchi N, Asano M, Itoh M, Toda M. Immunologic self-tolerance maintained by activated T cells expressing IL-2 receptor alpha-chains (CD25). Breakdown of a single mechanism of self-tolerance causes various autoimmune diseases. J Immunol 1995; 155: 1151–1164.

82. Balandina A, Lecart S, Dartevelle P, Saoudi A, Berrih-Aknin S. Functional defect of regulatory CD4$^+$CD25$^+$ T cells in the thymus of patients with autoimmune myasthenia gravis. Blood 2005; 105: 735–741.

83. Sun Y, Qiao J, Lu CZ, Zhao CB, Zhu XM, Xiao BG. Increase of circulating CD4$^+$CD25$^+$ T cells in myasthenia gravis patients with stability and thymectomy. Clin Immunol 2004; 112: 284–289.

84. Conti-Fine BM, Navaneetham D, Karachunski PI et al. T cell recognition of the acetylcholine receptor in myasthenia gravis. Ann N Y Acad Sci 1998; 841: 283–308.

85. Diethelm-Okita B, Wells GB, Kuryatov A, Okita D, Howard J, Lindstrom JM, Conti-Fine BM. Response of CD4$^+$ T cells from myasthenic patients and healthy subjects of biosynthetic and synthetic sequences of the nicotinic acetylcholine receptor. J Autoimmun 1998; 11: 191–203.

86. Fujii Y, Lindstrom J. Specificity of the T cell immune response to acetylcholine receptor in experimental autoimmune myasthenia gravis. Response to subunits and synthetic peptides. J Immunol 1988; 140: 1830–1837.

87. Oshima M, Pachner AR, Atassi MZ. Profile of the regions of acetylcholine receptor alpha chain recognized by T-lymphocytes and by antibodies in EAMG-susceptible and non-susceptible mouse strains after different periods of immunization with the receptor. Mol Immunol 1994; 31: 833–843.

88. Bellone M, Ostlie N, Lei S, Conti-Tronconi BM. Experimental myasthenia gravis in congenic mice. Sequence mapping and H-2 restriction of T helper epitopes on the alpha subunits of Torpedo californica and murine acetylcholine receptors. Eur J Immunol 1991; 21: 2303–2310.

89. Bellone M, Ostlie N, Karachunski P, Manfredi AA, Conti-Tronconi BM. Cryptic epitopes on the nicotinic acetylcholine receptor are recognized by autoreactive CD4$^+$ cells. J Immunol 1993; 151: 1025–1038.

90. Ostlie N, Milani M, Wang W, Okita D, Conti-Fine BM. Absence of IL-4 facilitates the development of chronic autoimmune myasthenia gravis in C57BL/6 mice. J Immunol 2003; 170: 604–612.

91. Vanderlugt CL, Begolka WS, Neville KL et al. The functional significance of epitope spreading and its regulation by co-stimulatory molecules. Immunol Rev 1998; 164: 63–72.

92. Hohlfeld R, Toyka KV, Tzartos SJ, Carson W, Conti-Tronconi BM. Human T-helper lymphocytes in myasthenia gravis recognize the nicotinic receptor alpha subunit. Proc Natl Acad Sci USA 1987; 84: 5379–5383.

93. Manfredi AA, Protti MP, Wu XD, Howard JF, Jr., Conti-Tronconi BM. CD4$^+$ T-epitope repertoire on the human acetylcholine receptor alpha subunit in severe myasthenia gravis: a study with synthetic peptides. Neurology 1992; 42: 1092–1100.

94. Manfredi AA, Protti MP, Dalton MW, Howard JF, Jr., Conti-Tronconi BM. T helper cell recognition of muscle acetylcholine receptor in myasthenia gravis. Epitopes on the gamma and delta subunits. J Clin Invest 1993; 92: 1055–1067.

95. Wang ZY, Okita DK, Howard J, Jr., Conti-Fine BM. T-cell recognition of muscle acetylcholine receptor subunits in generalized and ocular myasthenia gravis. Neurology 1998; 50: 1045–1054.

96. Wang ZY, Okita DK, Howard JF, Jr., Conti-Fine BM. CD4$^+$ T cell repertoire on the epsilon subunit of muscle acetylcholine receptor in myasthenia gravis. J Neuroimmunol 1998; 91: 33–42.

97. Wang ZY, Diethelm-Okita B, Okita DK, Kaminski HJ, Howard JF, Conti-Fine BM. T cell recognition of muscle acetylcholine receptor in ocular myasthenia gravis. J Neuroimmunol 2000; 108: 29–39.

98. Horton RM, Manfredi AA, Conti-Tronconi B. The 'embryonic' gamma subunit of the nicotinic acetylcholine receptor is expressed in adult extraocular muscle. Neurology 1993; 43: 983–986.

99. Kaminski HJ, Fenstermaker R, Ruff RL. Adult extraocular and intercostal muscle express the gamma subunit of fetal AChR. Biophys J 1991; 59: 444a.

100. Geuder KI, Marx A, Witzemann V, Schalke B, Toyka K, Kirchner T, Muller-Hermelink HK. Pathogenetic significance of fetal-type acetylcholine receptors on thymic myoid cells in myasthenia gravis. Dev Immunol 1992; 2: 69–75.

101. Nelson S, Conti-Tronconi BM. Adult thymus expresses an embryonic nicotinic acetylcholine receptor-like protein. J Neuroimmunol 1990; 29: 81–92.

102. Protti MP, Manfredi AA, Straub C, Wu XD, Howard JF, Jr., Conti-Tronconi BM. Use of synthetic peptides to establish anti-human acetylcholine receptor CD4$^+$ cell lines from myasthenia gravis patients. J Immunol 1990; 144: 1711–1720.

103. Protti MP, Manfredi AA, Straub C, Howard JF, Jr., Conti-Tronconi BM. Immunodominant regions for T helper-cell sensitization on the human nicotinic receptor alpha subunit in myasthenia gravis. Proc Natl Acad Sci USA 1990; 87: 7792–7796.

104. Protti MP, Manfredi AA, Wu XD, Moiola L, Howard JF, Jr., Conti-Tronconi BM. Myasthenia gravis. T epitopes on the delta subunit of human muscle acetylcholine receptor. J Immunol 1991; 146: 2253–2261.

105. Protti MP, Manfredi AA, Howard JF, Jr., Conti-Tronconi BM. T cells in myasthenia gravis specific for embryonic acetylcholine receptor. Neurology 1991; 41: 1809–1814.

106. Protti MP, Manfredi AA, Wu XD, Moiola L, Dalton MW, Howard JF, Jr., Conti-Tronconi BM. Myasthenia gravis. CD4$^+$ T epitopes on the embryonic gamma subunit of human muscle acetylcholine receptor. J Clin Invest 1992; 90: 1558–1567.

107. Moiola L, Karachunski P, Protti MP, Howard JF, Jr., Conti-Tronconi BM. Epitopes on the beta subunit of human muscle acetylcholine receptor recognized by CD4$^+$ cells of myasthenia gravis patients and healthy subjects. J Clin Invest 1994; 93: 1020–1028.

108. Collins EJ, Frelinger JA. Altered peptide ligand design: altering immune responses to class I MHC/peptide complexes. Immunol Rev 1998; 163: 151–160.

109. Nicholson LB, Greer JM, Sobel RA, Lees MB, Kuchroo VK. An altered peptide ligand mediates immune deviation and prevents autoimmune encephalomyelitis. Immunity 1995; 3: 397–405.

110. Tsitoura DC, Holter W, Cerwenka A, Gelder CM, Lamb JR. Induction of anergy in human T helper 0 cells by stimulation with altered T cell antigen receptor ligands. J Immunol 1996; 156: 2801–2808.

111. Wang ZY, Okita DK, Howard J, Jr., Conti-Fine BM. Th1 epitope repertoire on the alpha subunit of human muscle acetylcholine receptor in myasthenia gravis. Neurology 1997; 48: 1643–1653.

112. Diethelm-Okita B, Okita DK, Kaminski HJ, Howard JF, Conti-Fine BM, Diethelm-Okita BM. Universal epitopes for human CD4$^+$ cells on tetanus and diphtheria toxins. J Neuroimmunol 2000; 108: 29–39.

113. Accolla RS, Auffray C, Singer DS, Guardiola J. The molecular biology of MHC genes. Immunol Today 1991; 12: 97–99.

114. Sinigaglia F, Hammer J. Rules for peptide binding to MHC class II molecules. APMIS 1994; 102: 241–248.

115. Diethelm-Okita BM, Raju R, Okita DK, Conti-Fine BM. Epitope repertoire of human CD4$^+$ T cells on tetanus toxin: identification of immunodominant sequence segments. J Infect Dis 1997; 175: 382–391.

116. Panina-Bordignon P, Demotz S, Corradin G, Lanzavecchia A. Study on the immunogenicity of human class-II-restricted T-cell epitopes: processing constraints, degenerate binding, and promiscuous recognition. Cold Spring Harb Symp Quant Biol 1989; 54 Pt 1: 445–451.

117. Sommer N, Melms A, Weller M, Dichgans J. Ocular myasthenia gravis. A critical review of clinical and pathophysiological aspects. Doc Ophthalmol 1993; 84: 309–333.

118. Kaminski HJ, Maas E, Spiegel P, Ruff RL. Why are eye muscles frequently involved in myasthenia gravis? Neurology 1990; 40: 1663–1669.

119. Ruff R, Kaminski H, Maas E, Spiegel P. Ocular muscles: physiology and structure-function correlations. Bull Soc Belge Ophthalmol 1989; 237: 321–352.

120. Kaminski HJ, Ruff RL. Ocular muscle involvement by myasthenia gravis. Ann Neurol 1997; 41: 419–420.

121. Engel AG, Ohno K, Sine SM. Congenital myasthenic syndromes: experiments of nature Myasthenic syndromes in Turkish kinships due to mutations in the acetylcholine receptor. J Physiol Paris 1998; 92: 113–117.

122. Kaminski HJ, Kusner LL, Block CH. Expression of acetylcholine receptor isoforms at extraocular muscle endplates. Invest Ophthalmol Vis Sci 1996; 37: 345–351.

123. Kaminski HJ, Kusner LL, Nash KV, Ruff RL. The gamma-subunit of the acetylcholine receptor is not expressed in the levator palpebrae superioris. Neurology 1995; 45: 516–518.

124. MacLennan C, Beeson D, Bujis AM, Vincent A, Newsom-Davis J. Acetylcholine receptor expression in human extraocular muscles and their susceptibility to myasthenia gravis. Ann Neurol 1997; 41: 423–431.

125. Raju R, Navaneetham D, Protti MP, Horton RM, Hoppe BL, Howard J, Jr., Conti-Fine BM. TCR V beta usage by acetylcholine receptor-specific CD4$^+$ T cells in myasthenia gravis. J Autoimmun 1997; 10: 203–217.

126. Melms A, Oksenberg JR, Malcherek G, Schoepfer R, Muller CA, Lindstrom J, Steinman L. T-cell receptor gene usage of acetylcholine receptor-specific T-helper cells. Ann N Y Acad Sci 1993; 681: 313–314.

127. Garcia KC, Teyton L, Wilson IA. Structural basis of T cell recognition. Ann Rev Immunol 1999; 17: 369–397.

128. Li H, Llera A, Malchiodi EL, Mariuzza RA. The structural basis of T cell activation by superantigens. Ann Rev Immunol 1999; 17: 435–466.

129. Hennecke J, Wiley DC. T cell receptor-MHC interactions up close. Cell 2001; 104: 1–4.

130. Wei S, Charmley P, Robinson MA, Concannon P. The extent of the human germline T-cell receptor V beta gene segment repertoire. Immunogenetics 1994; 40: 27–36.

131. Papageorgiou AC, Acharya KR. Microbial superantigens: from structure to function. Trends Microbiol 2000; 8: 369–375.

132. Rose NR. The role of infection in the pathogenesis of autoimmune disease. Semin Immunol 1998; 10: 5–13.

133. Todd JA, Steinman L. The environment strikes back. Curr Opin Immunol 1993; 5: 863–865.

134. Blackman MA, Woodland DL. Role of the T cell receptor alpha-chain in superantigen recognition. Immunol Res 1996; 15: 98–113.

135. Tesch H, Hohlfeld R, Toyka KV. Analysis of immunoglobulin and T cell receptor gene rearrangements in the thymus of myasthenia gravis patients. J Neuroimmunol 1989; 21: 169–176.

136. Grunewald J, Ahlberg R, Lefvert AK, DerSimonian H, Wigzell H, Janson CH. Abnormal T-cell expansion and V-gene usage in myasthenia gravis patients. Scand J Immunol 1991; 34: 161–168.

137. Truffault F, Cohen-Kaminsky S, Khalil I, Levasseur P, Berrih-Aknin S. Altered intrathymic T-cell repertoire in human myasthenia gravis. Ann Neurol 1997; 41: 731–741.

138. Xu BY, Giscombe R, Soderlund A, Troye-Blomberg M, Pirskanen R, Lefvert AK. Abnormal T cell receptor V gene usage in myasthenia gravis: prevalence and characterization of expanded T cell populations. Clin Exp Immunol 1998; 113: 456–464.

139. Navaneetham D, Penn AS, Howard JF, Jr., Conti-Fine BM. TCR-Vbeta usage in the thymus and blood of myasthenia gravis patients. J Autoimmun 1998; 11: 621–633.

140. Yang XD, Tisch R, McDevitt HO. Selective targets for immunotherapy in autoimmune disease. Chem Immunol 1995; 60: 20–31.

141. Vandenbark AA, Hashim GA, Offner H. T cell receptor peptides in treatment of autoimmune disease: rational and potential. J Neurosci Res 1996; 43: 391–402.

142. Miller SD, McRae BL, Vanderlugt CL, Nikcevich KM, Pope JG, Pope L, Karpus WJ. Evolution of the T-cell repertoire during the course of experimental immune-mediated demyelinating diseases. Immunol Rev 1995; 144: 225–244.

143. Dong C, Flavell RA. Th1 and Th2 cells. Curr Opin Hematol 2001; 8: 47–51.

144. Feldmann M, Brennan FM, Maini R. Cytokines in autoimmune disorders. Int Rev Immunol 1998; 17: 217–228.

145. O'Garra A. Cytokines induce the development of functionally heterogeneous T helper cell subsets. Immunity 1998; 8: 275–283.

146. Abbas AK, Murphy KM, Sher A. Functional diversity of helper T lymphocytes. Nature 1996; 383: 787–793.

147. Seder RA. Acquisition of lymphokine-producing phenotype by CD4$^+$ T cells. J Allergy Clin Immunol 1994; 94: 1195–1202.

148. Constant SL, Bottomly K. Induction of Th1 and Th2 CD4$^+$ T cell responses: the alternative approaches. Ann Rev Immunol 1997; 15: 297–322.

149. Yi Q, Ahlberg R, Pirskanen R, Lefvert AK. Acetylcholine receptor-reactive T cells in myasthenia gravis: evidence for the involvement of different subpopulations of T helper cells. J Neuroimmunol 1994; 50: 177–186.

150. Yi Q, Lefvert AK. Idiotype- and anti-idiotype-reactive T lymphocytes in myasthenia gravis. Evidence for the involvement of different subpopulations of T helper lymphocytes. J Immunol 1994; 153: 3353–3359.

151. Link J, Fredrikson S, Soderstrom M, Olsson T, Hojeberg B, Ljungdahl A, Link H. Organ-specific autoantigens induce transforming growth factor-beta mRNA expression in mononuclear cells in multiple sclerosis and myasthenia gravis. Ann Neurol 1994; 35: 197–203.

152. Moiola L, Protti MP, McCormick D, Howard JF, Conti-Tronconi BM. Myasthenia gravis. Residues of the alpha and gamma subunits of muscle acetylcholine receptor involved in formation of immunodominant CD4$^+$ epitopes. J Immunol 1994; 152: 4686–4698.

153. Moiola L, Galbiati F, Martino G et al. IL-12 is involved in the induction of experimental autoimmune myasthenia gravis, an antibody-mediated disease. Eur J Immunol 1998; 28: 2487–2497.

154. Balasa B, Deng C, Lee J, Bradley LM, Dalton DK, Christadoss P, Sarvetnick N. Interferon gamma (IFN-gamma) is necessary for the genesis of acetylcholine receptor-induced clinical experimental autoimmune myasthenia gravis in mice. J Exp Med 1997; 186: 385–391.

155. Balasa B, Deng C, Lee J, Christadoss P, Sarvetnick N. The Th2 cytokine IL-4 is not required for the progression of antibody-dependent autoimmune myasthenia gravis. J Immunol 1998; 161: 2856–2862.

156. Zhang GX, Xiao BG, Bai XF, van der Meide PH, Orn A, Link H. Mice with IFN-gamma receptor deficiency are less susceptible to experimental autoimmune myasthenia gravis. J Immunol 1999; 162: 3775–3781.

157. Deng C, Goluszko E, Tuzun E, Yang H, Christadoss P. Resistance to experimental autoimmune myasthenia gravis in IL-6-deficient mice is associated with reduced germinal center formation and C3 production. J Immunol 2002; 169: 1077–1083.

158. Goluszko E, Deng C, Poussin MA, Christadoss P. Tumor necrosis factor receptor p55 and p75 deficiency protects mice from developing experimental autoimmune myasthenia gravis. J Neuroimmunol 2002; 122: 85–93.

159. Poussin MA, Goluszko E, Franco JU, Christadoss P. Role of IL-5 during primary and secondary immune response to acetylcholine receptor. J Neuroimmunol 2002; 125: 51–58.

160. Wang W, Ostlie NS, Conti-Fine BM, Milani M. The susceptibility to experimental myasthenia gravis of STAT6–/– and STAT4–/– BALB/c mice suggests a pathogenic role of Th1 cells. J Immunol 2004; 172: 97–103.

161. Ostlie NS, Karachunski PI, Wang W, Monfardini C, Kronenberg M, Conti-Fine BM. Transgenic expression of IL-10 in T cells facilitates development of experimental myasthenia gravis. J Immunol 2001; 166: 4853–4862.

162. Sitaraman S, Metzger DW, Belloto RJ, Infante AJ, Wall KA. Interleukin-12 enhances clinical experimental auto-immune myasthenia gravis in susceptible but not resistant mice. J Neuroimmunol 2000; 107: 73–82.

163. Im SH, Barchan D, Maiti PK, Raveh L, Souroujon MC, Fuchs S. Suppression of experimental myasthenia gravis, a B cell-mediated autoimmune disease, by blockade of IL-18. FASEB J 2001; 15: 2140–2148.

164. Karachunski PI, Ostlie NS, Okita DK, Conti-Fine BM. Prevention of experimental myasthenia gravis by nasal administration of synthetic acetylcholine receptor T epitope sequences. J Clin Invest 1997; 100: 3027–3035.

165. Baggi F, Andreetta F, Caspani E et al. Oral administration of an immunodominant T-cell epitope downregulates Th1/Th2 cytokines and prevents experimental myasthenia gravis. J Clin Invest 1999; 104: 1287–1295.

166. Im SH, Barchan D, Fuchs S, Souroujon MC. Mechanism of nasal tolerance induced by a recombinant fragment of acetylcholine receptor for treatment of experimental myasthenia gravis. J Neuroimmunol 2000; 111: 161–168.

167. Wang ZY, Link H, Ljungdahl A et al. Induction of interferon-gamma, interleukin-4, and transforming growth factor-beta in rats orally tolerized against experimental autoimmune myasthenia gravis. Cell Immunol 1994; 157: 353–368.

168. Nishibori T, Tanabe Y, Su L, David M. Impaired development of CD4$^+$ CD25$^+$ regulatory T cells in the absence of STAT1: increased susceptibility to autoimmune disease. J Exp Med 2004; 199: 25–34.

169. Kelchtermans H, De Klerck B, Mitera T et al. Defective CD4$^+$CD25$^+$ regulatory T cell functioning in collagen-induced arthritis: an important factor in pathogenesis, counter-regulated by endogenous IFN-gamma. Arthritis Res Ther 2005; 7: R402–415.

170. Sawitzki B, Kingsley CI, Oliveira V, Karim M, Herber M, Wood KJ. IFN-gamma production by alloantigen-reactive regulatory T cells is important for their regulatory function in vivo. J Exp Med 2005; 201: 1925–1935.

171. Karachunski PI, Ostlie NS, Okita DK, Garman R, Conti-Fine BM. Subcutaneous administration of T-epitope sequences of the acetylcholine receptor prevents experimental myasthenia gravis. J Neuroimmunol 1999; 93: 108–121.

172. Im SH, Barchan D, Fuchs S, Souroujon MC. Suppression of ongoing experimental myasthenia gravis by oral treatment with an acetylcholine receptor recombinant fragment. J Clin Invest 1999; 104: 1723–1730.

173. Monfardini C, Milani M, Ostlie N et al. Adoptive protection from experimental myasthenia gravis with T cells from mice treated nasally with acetylcholine receptor epitopes. J Neuroimmunol 2002; 123: 123–134.

174. Ding L, Linsley PS, Huang LY, Germain RN, Shevach EM. IL-10 inhibits macrophage costimulatory activity by selectively inhibiting the up-regulation of B7 expression. J Immunol 1993; 151: 1224–1234.

175. Fiorentino DF, Zlotnik A, Mosmann TR, Howard M, O'Garra A. IL-10 inhibits cytokine production by activated macrophages. J Immunol 1991; 147: 3815–3822.

176. Macatonia SE, Tripp CS, Wolf SF, O'Garra A, Murphy KM. Differential effect of IL-10 on dendritic cell-induced T cell proliferation and IFN-gamma production. Science 1993; 260: 547–549.

177. Enk AH, Angeloni VL, Udey MC, Katz SI. Inhibition of Langerhans cell antigen-presenting function by IL-10. A role for IL-10 in induction of tolerance. J Immunol 1993; 151: 2390–2398.

178. Aste-Amezaga M, Ma X, Sartori A, Trinchieri G. Molecular mechanisms of the induction of IL-12 and its inhibition by IL-10. J Immunol 1998; 160: 5936–5944.

179. de Waal Malefyt R, Yssel H, de Vries JE. Direct effects of IL-10 on subsets of human CD4$^+$ T cell clones and resting T cells. Specific inhibition of IL-2 production and proliferation. J Immunol 1993; 150: 4754–4765.

180. Taga K, Mostowski H, Tosato G. Human interleukin-10 can directly inhibit T-cell growth. Blood 1993; 81: 2964–2971.

181. Groux H, Bigler M, de Vries JE, Roncarolo MG. Interleukin-10 induces a long-term antigen-specific anergic state in human CD4$^+$ T cells. J Exp Med 1996; 184: 19–29.

182. Schwartz RH. Models of T cell anergy: is there a common molecular mechanism? J Exp Med 1996; 184: 1–8.

183. Georgescu L, Vakkalanka RK, Elkon KB, Crow MK. Interleukin-10 promotes activation-induced cell death of SLE lymphocytes mediated by Fas ligand. J Clin Invest 1997; 100: 2622–2633.

184. Bromberg JS. IL-10 immunosuppression in transplantation. Curr Opin Immunol 1995; 7: 639–643.

185. Akdis CA, Blaser K. IL-10-induced anergy in peripheral T cell and reactivation by microenvironmental cytokines: two key steps in specific immunotherapy. FASEB J 1999; 13: 603–609.

186. Hagenbaugh A, Sharma S, Dubinett SM et al. Altered immune responses in interleukin 10 transgenic mice. J Exp Med 1997; 185: 2101–2110.

187. Poussin MA, Goluszko E, Hughes TK, Duchicella SI, Christadoss P. Suppression of experimental autoimmune myasthenia gravis in IL-10 gene-disrupted mice is associated with reduced B cells and serum cytotoxicity on mouse cell line expressing AChR. J Neuroimmunol 2000; 111: 152–160.

188. Llorente L, Zou W, Levy Y et al. Role of interleukin 10 in the B lymphocyte hyperactivity and autoantibody production of human systemic lupus erythematosus. J Exp Med 1995; 181: 839–844.

189. Cross JT, Benton HP. The roles of interleukin-6 and interleukin-10 in B cell hyperactivity in systemic lupus erythematosus. Inflamm Res 1999; 48: 255–261.

190. Huang Y, Yang J, Van Der Meide PH et al. Increased levels of circulating acetylcholine receptor (AChR)-reactive IL-10-secreting cells are characteristic for myasthenia gravis (MG). Clin Exp Immunol 1999; 118: 115–121.

191. Wogensen L, Lee MS, Sarvetnick N. Production of interleukin 10 by islet cells accelerates immune-mediated destruction of beta cells in nonobese diabetic mice. J Exp Med 1994; 179: 1379–1384.

192. Moritani M, Yoshimoto K, Tashiro F et al. Transgenic expression of IL-10 in pancreatic islet A cells accelerates autoimmune insulitis and diabetes in non-obese diabetic mice. Int Immunol 1994; 6: 1927–1936.

193. Pennline KJ, Roque-Gaffney E, Monahan M. Recombinant human IL-10 prevents the onset of diabetes in the nonobese diabetic mouse. Clin Immunol Immunopathol 1994; 71: 169–175.

194. Moritani M, Yoshimoto K, Ii S et al. Prevention of adoptively transferred diabetes in nonobese diabetic mice with IL-10-transduced islet-specific Th1 lymphocytes. A gene therapy model for autoimmune diabetes. J Clin Invest 1996; 98: 1851–1859.

195. Delpy L, Douin-Echinard V, Garidou L, Bruand C, Saoudi A, Guery JC. Estrogen enhances susceptibility to experimental autoimmune myasthenia gravis by promoting type 1-polarized immune responses. J Immunol 2005; 175: 5050–5057.

196. Kong YM, Waldmann H, Cobbold S, Giraldo AA, Fuller BE, Simon LL. Pathogenic mechanisms in murine autoimmune thyroiditis: short- and long-term effects of in vivo depletion of CD4[+] and CD8[+] cells. Clin Exp Immunol 1989; 77: 428–433.

197. Pummerer C, Berger P, Fruhwirth M, Ofner C, Neu N. Cellular infiltrate, major histocompatibility antigen expression and immunopathogenic mechanisms in cardiac myosin-induced myocarditis. Lab Invest 1991; 65: 538–547.

198. Mozes E, Kohn LD, Hakim F, Singer DS. Resistance of MHC class I-deficient mice to experimental systemic lupus erythematosus. Science 1993; 261: 91–93.

199. Rodriguez M, Dunkel AJ, Thiemann RL, Leibowitz J, Zijlstra M, Jaenisch R. Abrogation of resistance to Theiler's virus-induced demyelination in H-2b mice deficient in beta 2-microglobulin. J Immunol 1993; 151: 266–276.

200. Shenoy M, Kaul R, Goluszko E, David C, Christadoss P. Effect of MHC class I and CD8 cell deficiency on experimental autoimmune myasthenia gravis pathogenesis. J Immunol 1994; 153: 5330–5335.

201. Shenoy M, Baron S, Wu B, Goluszko E, Christadoss P. IFN-alpha treatment suppresses the development of experimental autoimmune myasthenia gravis. J Immunol 1995; 154: 6203–6208.

202. Zhang GX, Ma CG, Xiao BG, Bakhiet M, Link H, Olsson T. Depletion of CD8[+] T cells suppresses the development of experimental autoimmune myasthenia gravis in Lewis rats. Eur J Immunol 1995; 25: 1191–1198.

203. Miller AE, Hudson J, Tindall RS. Immune regulation in myasthenia gravis: evidence for an increased suppressor T-cell population. Ann Neurol 1982; 12: 341–347.

204. Skolnik PR, Lisak RP, Zweiman B. Monoclonal antibody analysis of blood T-cell subsets in myasthenia gravis. Ann Neurol 1982; 11: 170–176.

205. Protti MP, Manfredi AA, Straub C, Howard JF, Jr., Conti-Tronconi BM. CD4[+] T cell response to the human acetylcholine receptor alpha subunit in myasthenia gravis. A study with synthetic peptides. J Immunol 1990; 144: 1276–1281.

206. Lisak RP, Laramore C, Levinson AI, Zweiman B, Moskovitz AR. Suppressor T cells in myasthenia gravis and antibodies to acetylcholine receptor. Ann Neurol 1986; 19: 87–89.

207. Lisak RP, Laramore C, Levinson AI, Zweiman B, Moskovitz AR, Witte A. In vitro synthesis of antibodies to acetylcholine receptor by peripheral blood cells: role of suppressor T cells in normal subjects. Neurology 1984; 34: 802–805.

208. Yuen MH, Protti MP, Diethelm-Okita B, Moiola L, Howard JF, Jr., Conti-Fine BM. Immunoregulatory CD8[+] cells recognize antigen-activated CD4[+] cells in myasthenia gravis patients and in healthy controls. J Immunol 1995; 154: 1508–1520.

209. Hohlfeld R, Wekerle H. The thymus in myasthenia gravis. Neurol Clin 1994; 12: 331–342.

210. Levinson AI, Wheatley LM. The thymus and the pathogenesis of myasthenia gravis. Clin Immunol Immunopathol 1996; 78: 1–5.

211. Aharonov A, Tarrab-Hazdai R, Abramsky O, Fuchs S. Immunological relationship between acetylcholine receptor and thymus: a possible significance in myasthenia gravis. Proc Natl Acad Sci USA 1975; 72: 1456–1459.

212. Engel EK, Trotter JL, McFarlin DE, McIntosh CL. Thymic epithelial cell contains acetylcholine receptor. Lancet 1977; 1: 1310–1311.

213. Ueno S, Wada K, Takahashi M, Tarui S. Acetylcholine receptor in rabbit thymus: antigenic similarity between acetylcholine receptors of muscle and thymus. Clin Exp Immunol 1980; 42: 463–469.

214. Schluep M, Willcox N, Vincent A, Dhoot GK, Newsom-Davis J. Acetylcholine receptors in human thymic myoid cells in situ: an immunohistological study. Ann Neurol 1987; 22: 212–222.

215. Kawanami S, Conti-Tronconi B, Racs J, Raftery MA. Isolation and characterization of nicotinic acetylcholine receptor-like protein from fetal calf thymus. J Neurol Sci 1988; 87: 195–209.
216. Kirchner T, Tzartos S, Hoppe F, Schalke B, Wekerle H, Muller-Hermelink HK. Pathogenesis of myasthenia gravis. Acetylcholine receptor-related antigenic determinants in tumor-free thymuses and thymic epithelial tumors. Am J Pathol 1988; 130: 268–280.
217. Kao I, Drachman DB. Thymic muscle cells bear acetylcholine receptors: possible relation to myasthenia gravis. Science 1977; 195: 74–75.
218. Zheng Y, Wheatley LM, Liu T, Levinson AI. Acetylcholine receptor alpha subunit mRNA expression in human thymus: augmented expression in myasthenia gravis and upregulation by interferon-gamma. Clin Immunol 1999; 91: 170–177.
219. Wakkach A, Guyon T, Bruand C, Tzartos S, Cohen-Kaminsky S, Berrih-Aknin S. Expression of acetylcholine receptor genes in human thymic epithelial cells: implications for myasthenia gravis. J Immunol 1996; 157: 3752–3760.
220. Andreetta F, Baggi F, Antozzi C et al. Acetylcholine receptor alpha-subunit isoforms are differentially expressed in thymuses from myasthenic patients. Am J Pathol 1997; 150: 341–348.
221. Navaneetham D, Penn AS, Howard JF, Jr., Conti-Fine BM. Human thymuses express incomplete sets of muscle acetylcholine receptor subunit transcripts that seldom include the delta subunit. Muscle Nerve 2001; 24: 203–210.
222. Marx A, Wilisch A, Gutsche S et al. Low levels of acetylcholine receptor delta subunit message and protein in human thymuses suggest the occurrence of triplet receptors in thymic myoid cells. In: Christadoss P, editor. Myasthenia Gravis: disease mechanisms and immunointervention. New Delhi: Narosa Publishing House; 2000. pp 28–34.
223. Kaminski HJ, Fenstermaker RA, Abdul-Karim FW, Clayman J, Ruff RL. Acetylcholine receptor subunit gene expression in thymic tissue. Muscle Nerve 1993; 16: 1332–1337.
224. Wheatley LM, Urso D, Tumas K, Maltzman J, Loh E, Levinson AI. Molecular evidence for the expression of nicotinic acetylcholine receptor alpha-chain in mouse thymus. J Immunol 1992; 148: 3105–3109.
225. Kornstein MJ, Asher O, Fuchs S. Acetylcholine receptor alpha-subunit and myogenin mRNAs in thymus and thymomas. Am J Pathol 1995; 146: 1320–1324.
226. Hara H, Hayashi K, Ohta K, Itoh N, Ohta M. Nicotinic acetylcholine receptor mRNAs in myasthenic thymuses: association with intrathymic pathogenesis of myasthenia gravis. Biochem Biophys Res Commun 1993; 194: 1269–1275.
227. Wakkach A, Poea S, Chastre E et al. Establishment of a human thymic myoid cell line. Phenotypic and functional characteristics. Am J Pathol 1999; 155: 1229–1240.
228. Wilisch A, Gutsche S, Hoffacker V et al. Association of acetylcholine receptor alpha-subunit gene expression in mixed thymoma with myasthenia gravis. Neurology 1999; 52: 1460–1466.
229. Sine SM, Claudio T. Gamma- and delta-subunits regulate the affinity and the cooperativity of ligand binding to the acetylcholine receptor. J Biol Chem 1991; 266: 19369–19377.
230. Schonbeck S, Chrestel S, Hohlfeld R. Myasthenia gravis: prototype of the antireceptor autoimmune diseases. Int Rev Neurobiol 1990; 32: 175–200.
231. Kirchner T, Hoppe F, Schalke B, Muller-Hermelink HK. Microenvironment of thymic myoid cells in myasthenia gravis. Virchows Arch B cell Pathol 1988; 54: 295–302.
232. Siara J, Rudel R, Marx A. Absence of acetylcholine-induced current in epithelial cells from thymus glands and thymomas of myasthenia gravis patients. Neurology 1991; 41: 128–131.
233. Poea-Guyon S, Christadoss P, Le Panse R et al. Effects of cytokines on acetylcholine receptor expression: implications for myasthenia gravis. J Immunol 2005; 174: 5941–5949.
234. Melms A, Luther C, Stoeckle C et al. Thymus and myasthenia gravis: antigen processing in the human thymus and the consequences for the generation of autoreactive T cells. Acta Neurol Scand Suppl 2006; 183: 12–13.
235. Fujinami RS, Oldstone MB. Amino acid homology between the encephalitogenic site of myelin basic protein and virus: mechanism for autoimmunity. Science 1985; 230: 1043–1045.
236. Gautam AM, Lock CB, Smilek DE, Pearson CI, Steinman L, McDevitt HO. Minimum structural requirements for peptide presentation by major histocompatibility complex class II molecules: implications in induction of autoimmunity. Proc Natl Acad Sci USA 1994; 91: 767–771.
237. Ohno S. Many peptide fragments of alien antigens are homologous with host proteins, thus canalizing T-cell responses. Proc Natl Acad Sci USA 1991; 88: 3065–3068.
238. Farris AD, Keech CL, Gordon TP, McCluskey J. Epitope mimics and determinant spreading: pathways to autoimmunity. Cell Mol Life Sci 2000; 57: 569–578.
239. Conrad B, Weidmann E, Trucco G et al. Evidence for superantigen involvement in insulin-dependent diabetes mellitus aetiology. Nature 1994; 371: 351–355.
240. Paliard X, West SG, Lafferty JA, Clements JR, Kappler JW, Marrack P, Kotzin BL. Evidence for the effects of a superantigen in rheumatoid arthritis. Science 1991; 253: 325–329.
241. Brocke S, Veromaa T, Weissman IL, Gijbels K, Steinman L. Infection and multiple sclerosis: a possible role for superantigens? Trends Microbiol 1994; 2: 250–254.

242. Brocke S, Gaur A, Piercy C, Gautam A, Gijbels K, Fathman CG, Steinman L. Induction of relapsing paralysis in experimental autoimmune encephalomyelitis by bacterial superantigen. Nature 1993; 365: 642–644.
243. Olson JK, Croxford JL, Miller SD. Virus-induced autoimmunity: potential role of viruses in initiation, perpetuation, and progression of T-cell-mediated autoimmune disease. Viral Immunol 2001; 14: 227–250.
244. Matzinger P. Tolerance, danger, and the extended family. Ann Rev Immunol 1994; 12: 991–1045.
245. Blacklaws BA, Nash AA. Immunological memory to herpes simplex virus type 1 glycoproteins B and D in mice. J Gen Virol 1990; 71: 863–871.
246. Eng H, Lefvert AK, Mellstedt H, Osterborg A. Human monoclonal immunoglobulins that bind the human acetylcholine receptor. Eur J Immunol 1987; 17: 1867–1869.

4

Epidemiology and Genetics of Myasthenia Gravis

Jeremiah W. Lanford and Lawrence H. Phillips

1. INTRODUCTION

Epidemiology is the study of disease in a population. The data from epidemiological studies can be used to answer several questions. For example, the age, racial, and geographic distribution of a disease can provide clues to the etiology and risk factors for the disease. This in turn can help guide future research into treatments for disease. The findings from epidemiological studies may also be used to inform public health-care decisions and allocation of health-care resources. Epidemiological studies often cover several years, or even decades, and an overview of trends over time makes it possible to discern changes in the natural history of a disease. The impact of interventions such as newer diagnostic tests, public education, and new treatment modalities can be seen. For these reasons epidemiology remains an important tool in our understanding of the pathophysiology of a disease and its treatment.

MG is an uncommon disease. By current standards its relative rarity makes it an "orphan" disease. Even though individual practitioners may encounter few patients with MG, knowledge of the disease is important for all neurologists and physicians. Since the discovery of auto-antibodies responsible for the disease in the 1970s, MG has been recognized as the prototypical autoimmune disease. We probably know more about the pathophysiology of MG than for any other autoimmune disease. This understanding has helped us to adopt new and aggressive treatments. Unfortunately, there have been very few specific treatment trials for MG. Until recently, the small number of patients with MG has discouraged pharmaceutical companies from investing in large-scale trials of new therapies. Nevertheless most neurologists who care for MG patients feel that current treatments have led to their patients living longer, healthier lives. But is this perception true? We will examine this question through the use of available epidemiological data.

2. EPIDEMIOLOGICAL ISSUES AND MYASTHENIA GRAVIS

2.1 Definition of Terms

Population-based epidemiological studies of chronic diseases rely on measurement of rates of occurrence in a specified population. In order to understand the epidemiology of MG a few terms will be defined. Most of the data presented will be expressed as a rate or fraction of the population, typically in the form of a specific number of individuals affected per 100,000 or 1 million general population. The most commonly used rates are *incidence*, *prevalence*, and *mortality*. Incidence is the number of new cases of a disease that occur in a population during a defined period of time. This rate is most commonly expressed as the number of new cases in a given year, or *average annual incidence*. Prevalence is the number of cases of the disease present in a population at a specified time point, termed *point prevalence*. Mortality is similar to incidence except that it defines the number of deaths of affected individuals over a specified period of time. It is most commonly expressed as the *average annual mortality*.

From: *Current Clinical Neurology*
Myasthenia Gravis and Related Disorders
Edited by: H.J. Kaminski, DOI 10.1007/978-1-59745-156-7_4, © Humana Press, Newyork, NY, 2009

2.2 Epidemiological Studies of Myasthenia Gravis

MG has been the subject of many epidemiological studies for over 50 years. Most studies have been done in North American and European populations. A consistent pattern of a progressive increase in incidence and prevalence over time has been observed *(1)*. As mentioned previously, MG is an uncommon disease. The reported prevalence rates for myasthenia gravis have ranged from 0.5 to 20.4 per 100,000 *(2)*.

In all the populations studied to date, no areas of particularly high or low incidence and prevalence have been identified. No disease "clusters" have been discovered. The natural history of the disease thus appears to be fairly uniform, however, most studies have been done on predominantly Caucasian populations of western European descent. When non-European populations have been studied, there have been some hints that there may be some racial susceptibility differences.

Phillips et al. performed an epidemiological study on MG in central and western Virginia in the mid-1980s. The findings from this study will be discussed as an example of a population-based study of MG. Similar findings have been reported in several other populations. This study took advantage of a referral pattern that resulted in the majority of patients with MG in the state being identified easily from a small number of data sources. The time period for the study encompassed the years between 1980 and 1984. During this time, the prevalence rate increased from 13.4 per 100,000 to 14.2 per 100,000 *(3)*.

The female to male ratio in 1984 was 1.5:1. The median age at onset was 41.7 years for women and 60.3 years for men. The clinical truism that MG is a "disease of young women and old men" seemed to come from the fact that the age-adjusted incidence was particularly high in women in the second and third decades of life. In older age groups, however, the age-adjusted incidence was more or less equal. If the age of 40 years was used as a dividing line between young-onset and older-onset disease, more than 50% of women were in the older-onset group. This resulted in a bimodal age distribution with peaks at 20–29 years and a later steady increase with age greater than 50 years. The bimodal distribution was observed in both genders, but the earlier peak was more pronounced for women. This distribution has been reported in other studies as well *(3, 4)*.

Disease severity was dichotomized into ocular only and generalized forms, and 20–25% of patients had ocular-only disease. Of the patients with generalized myasthenia, 84.8% had serum acetylcholine receptor (AchR) antibodies, but they were found in only 14.3% of patients with ocular-only disease *(3)*.

A particularly interesting finding of this study was that the incidence and prevalence of MG were significantly higher in the African-American population. This study was the first to make such an observation, possibly because most preceding studies had been done in populations that had a relatively small proportion of African-Americans. The observed excess of cases is not unexpected, since it is known that autoimmune diseases in general have higher incidence and prevalence among African-Americans. This finding suggests that there is a common environmental or genetic factor that predisposes African-Americans to a higher incidence of MG as well as other autoimmune diseases *(3)*.

2.3 Current Trends in the Epidemiology of Myasthenia Gravis

In recent epidemiological studies of MG certain trends have become apparent. The most important trend is that the prevalence of the disease has risen progressively since the 1950s. See Fig. 1A *(1)*. There are several possible explanations for this trend. First, the increase may be related to improved diagnosis and heightened awareness of the disease. If this were the sole reason, however, one would expect that incidence rates during the same period would increase in parallel to prevalence. There has been an increase in incidence rates, but the increase in prevalence is substantially greater. See Fig. 1B. Similarly, one might assume that newer and more advanced

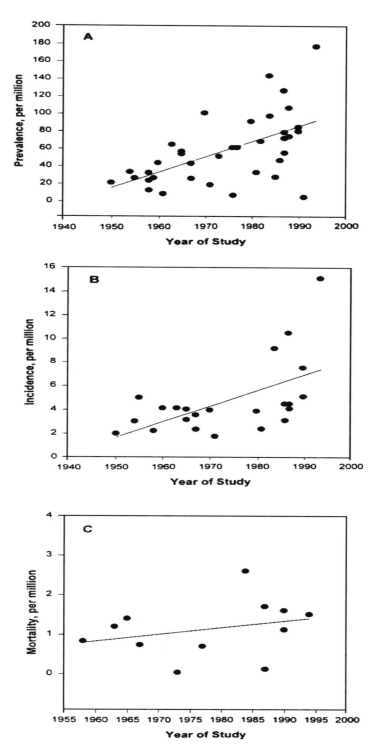

Fig. 1 Regression analysis of reported (**a**) prevalence, (**b**) incidence, and (**c**) mortality for MG between the years of 1950 and 1995. The slopes of the regression lines appear to be similar, but note that the *y*-axes differ

treatments would produce a longer life span and thereby increase prevalence as patients would live longer with the disease. A regression analysis of reported mortality showed that there has been little change in rates over time. See Fig. 1C. Patients with MG seem to be dying at approximately the same rate over the 45 years of available data, but the rates are not specific for cause of death. A general impression of most physicians who care for patients with MG is that the disease is infrequently the cause of death and that patients are living longer with the disease. The increase in prevalence in MG is thus most likely to be due to the combined effects of several factors including improved diagnosis, prolonged survival, and an increase in the population at risk *(1)*.

The latter point is the result of another trend that has been noticeable in the last two decades. The average age of patients with MG appears to be increasing progressively. Increased survival of patients with MG is certainly a factor, since, as more patients with MG grow older, the average age of the entire cohort of patients will certainly increase. Another factor is that the incidence of late-onset MG is increasing. Somnier et al. have reported that the incidence for late-onset MG has risen significantly over the last decade in Denmark, while the incidence for early-onset MG has not changed appreciably *(4)*. Similar findings have been reported by Aragones et al. in Spain and Matsuda. et al. in Japan *(6, 7)*. Within the older-onset MG group the male-to-female ratio is closer to 1:1. Disease severity in the older-onset group appears to be greater than it is in early-onset MG *(8, 9)*. Complications and side effects of treatment are more common in the late-onset MG *(9)*. In addition, the immunological makeup is different in late-onset MG patients, who have different autoantibody profiles and HLA patterns *(8)*.

Three possible etiologies for an increase in late-onset MG have been proposed. First, the increase in the elderly population overall due to increased life expectancy may be responsible, since this increases the population at risk. When incidence rates were standardized based on the growth of the elderly population, the increase in incidence rates was unrelated to the growth in the elderly population *(7)*. Second, increased ability to make an accurate diagnosis of MG has led to an increase in incidence. If this were true, the incidence of both early-onset and late-onset disease would likely have increased at a similar rate. Incidence rates have increased during the past decade where there have been arguably fewer advances in the diagnosis of MG. Finally, an immunological mechanism has been suggested, since there may be a difference in the immunological profile of late-onset MG. This factor is possibly related to an overall increase in autoimmunity in the elderly *(7)*. Regardless of the reason, physicians can expect to see an increasing number of older-onset patients with MG. This has significant implications for treatment, since patients with late-onset disease progress to more severe disease in significantly larger numbers than early-onset MG. Treatment choices will need to be made thoughtfully, since treatment complications are more common, and older patients have more co-morbidities that complicate treatment.

2.4 Seronegative and Muscle-Specific Kinase Antibody-Related Myasthenia Gravis

Since the discovery in 1970 that patients with MG have circulating antibodies to the AChR in their serum, it has become clear that there is a subset of patients who have a similar clinical phenotype but lack AChR antibodies. The term *seronegative MG* (SNMG) has been given to such patients. The proportion of patients with SNMG has been reported to range from 5 to 30% in various studies *(10)*. The distribution is different in ocular versus generalized MG. In patients

Fig. 1 *(Continued)*   for each plot. The slopes of the regression lines for incidence and mortality do not differ significantly ($p = 0.0.354$), but the slope of the prevalence line differs significantly from both incidence (0.0014) and mortality (0.0076). This indicates that prevalence rates have increased much more than either incidence or mortality. Reproduced with permission from: Phillips LH and Torner JC: Epidemiologic evidence for a changing natural history of myasthenia gravis. Neurology, 1996;47:1233–1238.

with ocular MG the distribution of seronegativity is between 50 and 60%, while it is between 20 and 30% in patients with generalized MG *(10)*. In the Virginia study, 84.8% of patients with generalized MG were seropositive *(3, 10)*. The sex ratio of SNMG patients is approximately equal. The age range of SNMG is younger than that of seropositive MG, where patients with SNMG tend to be under 60 years. Clinical symptoms are largely similar, but there is a suggestion that bulbar muscle weakness is greater in SNMG *(10)*.

In 2000, the discovery that some SNMG patients have antibodies to muscle-specific tyrosine kinase (MuSK) has led to a new classification of MG. MuSK-antibody patients were originally included in the seronegative group of MG. Now with the known antibody against MuSK, the clinical features of these patients are being defined. Detailed data about the epidemiology of MuSK is not yet known, but some information can be gleaned from case series.

In studies of SNMG 30–70% have MuSK antibodies *(11, 12)*. The actual prevalence is unknown. There appear to be some racial differences in the distribution of patients with MuSK antibodies. Yeh et al. and Lee et al. have found that the percentage of patients with MuSK antibodies in SNMG ranges from 4% in Chinese patients to 27% in Korean patients *(13, 14)*. The numbers of patients in these studies are relatively small, thus one cannot make any definitive conclusions.

Throughout all studies there is a striking female preponderance for MuSK-antibody-associated MG. In fact, some studies have found only female patients with MuSK antibodies. Patients with MuSK antibodies tend to be younger than the seropositive MG population. The mean age for onset of MuSK-associated MG is between 36 and 44 years *(10)*. Very few patients with onset greater than 60 years have been reported. Three distinct clinical phenotypes have been described: one that is identical to AChR-positive MG; one with severe oculobulbar and facial weakness and atrophy; and one with neck, shoulder, and respiratory muscle weakness *(15)*. Isolated ocular MG has not been described in MuSK-positive MG.

It is clear from the epidemiological data collected so far that MuSK-positive MG is a distinctive subset of the MG population. Separating MuSK-positive MG into a unique clinical entity is important, as the response to treatment and prognosis may differ significantly. Further epidemiological studies should be done to better define the population characteristics of such patients. As there is a suggestion of racial differences, etiological factors might possibly be determined, either genetic or environmental.

2.5 Genetics of MG

Acquired MG is an autoimmune disease of unknown etiology. A variety of etiologies have been proposed, and genetic, environmental, and infectious factors have been proposed. To date, no definite environmental or infectious etiologies have been identified but there is a clear excess of autoimmune diseases in family members of MG patients. There are also numerous, well-documented cases of MG occurring in more than one member of a family. This has led researchers to investigate the possible role that genetics has in the etiology of the disease.

2.6 Proof of Inheritance

Genetic factors have been suggested for autoimmune-generalized MG since the early 1900s. MG is now known to occur in up to 4% of family members of patients with MG *(16)*. The current risk of developing autoimmune-generalized MG is less than 0.01% in the general population. This significant difference suggests that there is a genetic or inherited cause that predisposes people to develop the disease.

Twin studies have provided even more compelling evidence. Namba reviewed a large number of sets of twins with MG and found that 6 out of 21 sets of twins were concordant for the disease. When the zygosity of the twins was determined, 5 of 13 sets of monozygotic twins were concordant for the disease, while none of 7 sets of dizygotic twins were concordant (in one set of twins who were both affected by MG the zygosity was undetermined) *(17)*. The concordance in monozygotic twins gives strong evidence for a genetic etiology rather than an environmental cause.

The mode of inheritance may well be recessive. Evidence for this conclusion comes from the fact that familial MG is most common among siblings (58%) rather than parent and child (15%) *(16)*. Bergoffen and others have reported a family with parental consanguinity in which 5 of 10 siblings had autoimmune MG, suggesting an autosomal-recessive mode of inheritance *(18)*.

3. CANDIDATE GENES

3.1 HLA Complex

Autoimmune MG has been associated with other autoimmune disorders. Kerzin-Storrar et al. studied 44 patients with MG and found that 30% of patients had a positive family history of autoimmune disease *(19)*. This finding suggests there is not only a genetic component for the disease, but a possible overall increase in susceptibility to develop autoimmunity in patients with MG and their family members. This has led researchers to look at the HLA complex as an important factor for the predisposition to develop autoimmunity.

To review briefly, the human leukocyte antigen complex gene products are associated with recognition between self and non-self. The HLA complex is divided into class I, class II, and class III regions. The gene products from class I and class II regions are membrane-bound molecules that present antigenic epitopes to lymphoid cells. Class I gene products are expressed on nearly all nucleated cells, whereas class II gene products are expressed on antigen-presenting cells (macrophages, dendritic cells, and B cells). Class III genes encode proteins associated with the immune process, including complement factors, heat shock proteins, tumor necrosis factor (TNF), and lymphotoxins.

Several studies initially showed an increase in the HLA-B8 and DR3 alleles in patients with MG when compared with the general population *(19, 20)*. This HLA pattern was shown to have certain clinical characteristics that include female preponderance, early age at onset, and thymic hyperplasia. The HLA-DR3 B8 A1 (8.1)-extended ancestral haplotype has now been shown to be reproducibly associated with MG in patients with the above-mentioned characteristics *(21)*. The 8.1 ancestral haplotype's immunological effects and association with other immunological diseases has been demonstrated *(22)*. This suggests that the 8.1 haplotype is associated with a general susceptibility to autoimmunity.

Of particular note, there is a locus of 1.2 Mbase within the 8.1 haplotype, termed *MYSA1*, that is associated with MG and thymic hyperplasia *(21)*. This region has also been shown to be associated with rheumatoid arthritis (RA), which frequently co-exists with MG. Potential gene products in this region include TNF/lymphotoxins. These proteins are known for their role in the production of germinal centers. Other possible gene candidates include IkB-L which has been associated with RA *(23)*.

3.2 AChR as a Self-Antigen and Other HLA Genes

Since the actual antigen involved in autoimmune MG is known, one can ask what there is about the AChR that lends itself to immunogenicity? The α-subunit has been of particular interest, because most autoantibodies are directed against it. Not only is this the region for antibody binding, but the α-subunit is also involved in the direct binding of ACh. This has made the gene, *CHRNA1*, which encodes the α-subunit, a good candidate for influencing genetic susceptibility to MG. Through linkage dysequilibrium analysis a particular allelic variant, located in the first intron of the *CHRNA1* gene, designated *HB*14* has been shown to be closely associated with MG *(24)*. The pathogenic mechanism of *CHRNA-HB*14* in MG is currently unknown.

Further analysis of the association of *CHRNA-HB*14* with MG has allowed determination of whether or not a particular HLA class II gene product is associated with presenting the self-antigen. An association between the HLA *DQA1*0101* gene and *HB*14* loci has been demonstrated. In addition to these two loci, a third locus, the 8.1 ancestral haplotype, has been demonstrated to have an additive effect to the association *(25)*. These findings have given rise to a three-gene model. In this model, the AChR region *HB*14* is presented to immune cells by the HLA class II *DQA1*0101* gene product, and the 8.1 haplotype leads to a non-antigen-specific immune dysregulation *(26)*.

3.3 Other Candidate Genes

Cytotoxic T lymphocyte-associated antigen-4 is expressed on CD4+ and CD8+ T cells. Its role in the immune system is thought to be downregulation of T-cell function. In knockout mice for CTLA-4 there is a profound association with spontaneous autoimmune diseases. This has led investigators to look at CTLA-4 in human diseases. Linkage studies have shown that the CTLA4 3'UTR repeat is associated with type I autoimmune diabetes mellitus *(27)*. This has led to the evaluation of CTLA-4 in other autoimmune diseases as a possible cause for a general increase in autoimmunity. Unfortunately, thus far this gene has not been shown to be highly associated with autoimmune MG.

There have been multiple other genes within the immune system that have been considered as possible candidates for a genetic cause of MG. None to date have proven to have as close a relationship as the above-mentioned three-gene model. Further research into the possible genetic causes of autoimmune MG is needed. With the development of more advanced genotyping methods we will be able to have a better grasp on the genetic cause of autoimmune MG. A better understanding of the genetics of MG may even lead to newer and better therapies for the disease and possibly even its prevention.

REFERENCES

1. Phillips LH, Torner J. Epidemiologic evidence for a changing natural history of myasthenia gravis. Neurology. 1996;47:1233–1238.
2. Phillips LH. The epidemiology of myasthenia gravis. Ann N Y Acad Sci. 2003;998:407–412.
3. Phillips LH, Torner JC, Anderson MS, Cox GM. The epidemiology of myasthenia gravis in central and western Virginia. Neurology, 1992;42:1888–1893.
4. Somnier FE, Keiding N, Paulson OB. Epidemiology of myasthenia gravis in Denmark. A longitudinal and comprehensive population survey. Arch Neurol. 1991;48:733–739.
5. Somnier FE. Increasing incidence of late-onset anti-AChR antibody-seropositive myasthenia gravis. Neurology. 2005;65:928–930.
6. Aragones JM, Bolibar I, Bonfill X et al. Myasthenia gravis: a higher than expected incidence in the elderly. Neurology. 2003;60:1024–1026.
7. Matsuda M, Dohi-Iijima N, Nakamura A et al. Increase in incidence of elderly-onset patients with myasthenia gravis in Nagano Prefecture, Japan. Intern Med. 2005;44:572–577.
8. Aarli JA, Romi F, Skeie GO, Gilhus NE. Myasthenia gravis in individuals over 40. Ann N Y Acad Sci. 2003;998:424–431.
9. Donaldson DH, Ansher M, Horan S et al. The relationship of age to outcome in myasthenia gravis. Neurology. 1990;40:786–790.
10. Vincent A, McConville J, Farrugia ME, Newsom-Davis J. Seronegative myasthenia gravis. Semin Neurol. 2004;24:125–133.
11. Lavrnic D, Losen M, Vujic A et al. The features of myasthenia gravis with autoantibodies to MuSK. J Neurol Neurosurg Psychiatry. 2005;76:1099–1102.
12. Hoch W, McConville J, Helms S et al. Auto-antibodies to the receptor tyrosine kinase MuSK in patients with myasthenia gravis without acetylcholine receptor antibodies. Nat Med. 2001;7:365–368.
13. Yeh JH, Chen WH, Chiu HC, Vincent A. Low frequency of MuSK antibody in generalized seronegative myasthenia gravis among Chinese. Neurology. 2004;62:2131–2132.
14. Lee JY, Sung JJ, Cho JY et al. MuSK antibody-positive, seronegative myasthenia gravis in Korea. J Clin Neurosci. 2006;13:353–5. Epub 2006 Mar 20.

15. Sanders DB, El-Salem K, Massey JM et al. Clinical aspects of MuSK antibody positive seronegative MG. Neurology. 2003;60:1978–1980.

16. Namba T, Brunner NG, Brown SB et al. Familial myasthenia gravis. Report of 27 patients in 12 families and review of 164 patients in 73 families. Arch Neurol. 1971;25:49–60.

17. Namba T, Shapiro MS, Brunner NG, Grob D. Myasthenia gravis occurring in twins. J Neurol Neurosurg Psychiatry. 1971;34:531–534.

18. Bergoffen J, Zmijewski CM, Fischbeck KH. Familial autoimmune myasthenia gravis. Neurology. 1994;44:551–554.

19. Kerzin-Storrar L, Metcalfe RA, Dyer PA et al. Genetic factors in myasthenia gravis: a family study. Neurology. 1988;38:38–42.

20. Evoli A, Batocchi AP, Zelano G et al. Familial autoimmune myasthenia gravis: report of four families. J Neurol Neurosurg Psychiatry. 1995;58:729–731.

21. Giraud M, Beaurain G, Yamamoto AM et al. Linkage of HLA to myasthenia gravis and genetic heterogeneity depending on anti-titin antibodies. Neurology. 2001;57:1555–1560.

22. Price P, Witt C, Allcock R et al. The genetic basis for the association of the 8.1 ancestral haplotype (A1, B8, DR3) with multiple immunopathological diseases. Immunol Rev. 1999;167:257–274.

23. Vandiedonck C, Giraud M, Garchon HJ. Genetics of autoimmune myasthenia gravis: the multifaceted contribution of the HLA complex. J Autoimmun. 2005;25 Suppl:6–11. Epub 2005 Nov 2.

24. Garchon HJ, Djabiri F, Viard JP et al. Involvement of human muscle acetylcholine receptor alpha-subunit gene (*CHRNA*) in susceptibility to myasthenia gravis. Proc Natl Acad Sci U S A. 1994;91:4668–4672.

25. Djabiri F, Caillat-Zucman S, Gajdos P et al. Association of the AChRalpha-subunit gene (*CHRNA*), DQA1*0101, and the DR3 haplotype in myasthenia gravis. Evidence for a three-gene disease model in a subgroup of patients. J Autoimmun. 1997;10:407–413.

26. Garchon HJ. Genetics of autoimmune myasthenia gravis, a model for antibody-mediated autoimmunity in man. J Autoimmun. 2003;21:105–110.

27. Kristiansen OP, Larsen ZM, Pociot F. CTLA-4 in autoimmune diseases–a general susceptibility gene to autoimmunity? Genes Immun. 2000;1:170–84.

5

Clinical Presentation and Epidemiology of Myasthenia Gravis

Jan B.M. Kuks

1. DEFINITION AND CLASSIFICATION

The signs and symptoms of a postsynaptic neuromuscular junction disorder as myasthenia gravis (MG) result from fluctuating strength of the voluntary muscles. The degree of weakness is partly dependent on the exertion of a muscle group but also varies spontaneously over longer periods of time without apparent cause. Remissions occur. All types of voluntary muscles may be involved but usually not to the same extent. In about 15% of the patients with MG, manifestations remain confined to the ocular muscles while among others bulbar symptoms and signs predominate. Usually MG begins with a few isolated signs (Table 1) and extends to other muscles within a few weeks to a few months, but sometimes, even years (see later on in this chapter). Apart from a certain degree of local muscle atrophy, no other neurological abnormalities are expected.

Most patients with fatiguing weakness suffer from the acquired autoimmune disease, myasthenia gravis with antibodies to the acetylcholine receptor (AChR), but other postsynaptic neuromuscular junction disorders do exist. Among these, myasthenia with antibodies to muscle-specific tyrosine kinase (MuSK) at the endplate is now recognized. Acquired MG, developing at any age after birth, should be distinguished from transient neonatal myasthenia, occurring in 10–15% of children of myasthenic mothers and congenital (hereditary) myasthenia, which is usually present at birth but may become apparent in the first years of life, or rarely in adulthood. The author find the analysis of Compston et al. in categorization of patients with acquired MG useful *(1)*: (1) purely ocular MG (weakness restricted to the palpebral levator and the extraocular muscles), (2) early-onset generalized MG (clinical onset prior the age of 40 years), (3) late-onset generalized MG (onset after age of 40 years) and (4) patients with a thymoma, with onset at any age.

2. CLINICAL MANIFESTATIONS

2.1 Ocular

Ocular symptoms are the most frequent manifestations of MG. No consensus is found in the literature about the ocular symptom that most frequently appears first: ptosis or diplopia. Furthermore, any of the extraocular muscles may be involved in isolation or in combination leading to horizontal, vertical or diagonal double vision. In many patients, both ptosis and diplopia are ultimately present. The patient is immediately aware of double vision, as it occurs acutely, but a mild ptosis may escape attention. Some patients appreciate fluctuation of the diplopia from onset and are able to analyze the double images themselves. Others only complain of blurred vision, which becomes normal when looking with one eye. Patients with isolated ocular symptoms during the entire period of illness form a distinct group.

From: *Current Clinical Neurology*
Myasthenia Gravis and Related Disorders
Edited by: H.J. Kaminski, DOI 10.1007/978-1-59745-156-7_5, © Humana Press, Newyork, NY, 2009

Table 1
Clinical Manifestations of Myasthenia Gravis

	Early onset 1–39 years	Late onset 40–85 years	Thymoma	Total
Ocular Diplopia	70 (14)*	84 (36)	17 (1)	171
Ptosis	55 (16)	36 (12)	14	105
Ptosis and Diplopia	64 (10)	47 (18)	15	126
Bulbar Articulation	43	5	6	54
Face	20	2		22
Chewing	1	6	3	10
Swallowing	7	4	1	12
Neck muscles	2	5	4	11
Combined	23	20	16	59
Oculo-bulbar	25	17	30	72
Limbs Arms	15	3	4	22
Hands or fingers	12	1	2	15
Legs	45	6	1	52
Combined	29	5	5	39
Generalized	4	5	16	25
Respiration	4	2	4	10
Total	419 (40)	248 (66)	138 (1)	805

*Initial signs of MG reported by the author's patients. In parentheses is number of the patients who remained purely ocular during the follow-up. This was the case in only one patient with a thymoma, in 9.5% in the early-onset group and 26.6% in the late-onset group. The ocular muscles were involved at onset in 59% of all patients, the bulbar muscles in 30% and the ocular and bulbar muscles combined in 80%. Of the 10 patients with initial respiratory signs, 8 had a prolonged apnea without previous weakness and 2 were children with high fever. The onset with ocular signs was more frequent in the late-onset group than in the early-onset group (74 vs 51%). Limb muscle weakness was more frequent in the early-onset than in the late-onset group (25 vs 8%). Patients with thymomas had the highest incidence of bulbar signs at onset (55 vs 31% in the early-onset and 26% in the late-onset groups).

2.2 Bulbar

In neurological jargon "bulbar muscles" are those innervated by motor neurons originating in the pons and in the medulla oblongata (cranial nerves V, VII, IX, X, XI, XII).

Among bulbar symptoms, speech difficulties manifesting as nasality of the voice or a difficulty in articulation are the most common at onset of MG. More than any other symptom, dysarthria is likely to occur initially under the influence of emotions. At first, it is frequently an isolated and fluctuating symptom that disappears after a "silent period" and may be accompanied by difficulties in swallowing and chewing. If the dysarthria is due to an insufficient function of the palatum, regurgitation of liquids through the nose may occur. A slight insufficiency of the upper pharyngeal muscles results in the sensation that food is sticking in the throat, and this can be documented to be the case by barium swallowing studies. Patients with dysphagia commonly report a preference for cold food. This may relate to improvement in neuromuscular transmission produced by relative cooling of the muscles.

Chewing may be difficult at the end of the meal or the problem may be first appreciated when chewing bubblegum or eating peanuts. If weakness is severe, the jaw sags open and the patient needs to hold the mouth closed. A characteristic posture is seen in Fig. 1A. Most of our patients with chewing problems also had weakness of neck muscles. An important symptom correlating well with the severity of dysphagia is weight loss, and most of our patients with bulbar manifestations lost 5–10 kg of weight in the 3–6 months prior to diagnosis.

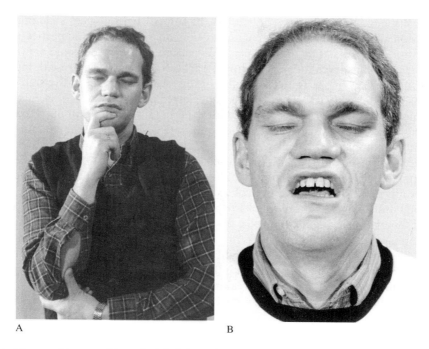

A   B

Fig. 1. A 36-year-old man with typical facial weakness. (**A**) He needs to support his jaw by holding his mouth closed. (**B**) In an attempt to close the eyelids firmly the eyelashes remain visible and the orbicularis oris weakness is evidenced by the straight smile (**B**).

Weakness of neck muscles may result in difficulty in balancing the head, which is particularly troublesome if the patient has to perform work in a bent position. Complaints about stiffness, vague pains in the neck and the back of the head and occasionally paraesthesias are then common, and nearly always set the doctor on the wrong track searching for cervical spine pathology, unless the strength of the neck muscles is formally tested.

Weakness of the facial muscles may occur suddenly – in some patients the initial diagnosis is confused with Bell's palsy – but it is more usual for it to occur insidiously. The patient's complaint is often stiffness, the sensation known from dental anesthesia and sometimes paraesthesias or even hypalgesia. Substantial sensory loss, however, is never due to MG. The facial expression may be changed, especially in emotions. Laughing becomes distorted ("we did not know whether she laughed or cried"). Many patients complain of this change of facial expression and avoid social contacts. People ask them why they always look sad or angry.

Orbicularis oris weakness is first noticed by the inability to whistle or kiss, in sneezing forcefully, in eating soup with a spoon or by the difficulty in pronouncing certain letters (p, f, s). Some patients complain that their tongues are thick and do not fit in their mouths. The time needed for eating a meal increases, and conversation becomes difficult while eating.

Insufficient strength of the orbicularis oculi may cause problems in keeping the eyes closed while washing the hair. Several of our patients complained that they could not close one eye and keep the other open while taking photographs (Fig. 1B). If the eyes are not completely closed during sleeping, they may be irritated on awakening. As all these symptoms may be minor and fluctuating, they are often detected retrospectively and do not cause patients to consult a doctor early in the course of the disease.

The most frequent of the bulbar signs among MG patients with more advanced MG is facial weakness. In its most prominent form it is easily recognizable, but it may be subtle and be of variable severity making it easily overlooked at the time of neurological examination. At rest, facial expression may be unremarkable, but any expression of emotion, and particularly laughing, betrays the loss of normal function. A classical feature is the myasthenic snarl or the "rire verticale" (Fig. 1B). Since weakness of the upper part of the face is also present, laughing makes the eyelids droop, even if ptosis is not noticeable at rest. An early and sensitive sign of orbicularis oris weakness is the inability to whistle or kiss (a neglected part of the neurological examination). Patients cannot blow out their cheeks without air escaping if pressure is applied to the cheeks by the examiner's finger. The closed eyes can be opened with one finger by the examiner, or the eyes cannot be closed completely. Facial weakness may be asymmetrical, but this is rarely as pronounced as that of the ocular signs. The most sensitive functional test of the muscles of articulation is speaking aloud without interruption. An easy test is counting (101, 102, etc.) or reading aloud. A certain degree of quantitation of examination findings is possible by noting when dysarthria or nasal twang is appreciated to the time when the patient becomes unintelligible (Fig. 2). In a minority of patients, the voice first becomes weaker in volume, but not dysarthric. Some hoarseness may occur but aphonia is not a sign of MG. In general, dysarthria and nasality occur while the volume remains normal. Slight palatal weakness may be demonstrated if peak-flow volume measures improve after pinching the nose.

Swallowing difficulties may be due to weakness of the lips, the tongue or pharyngeal muscles and often a combination of these. If observed while eating, some MG patients may be appreciated to support their jaws, underlips and the floor of their mouth in order to counteract gravity and to "chew with their hands". Swallowing in some patients improves if they turn their head, probably thus narrowing their, too wide, throat. Regurgitation of fluids through the nose is a sign of palatal weakness. Coughing after swallowing may be subtle sign of a defective swallowing mechanism. One study demonstrated that elderly patients with bulbar weakness were unable to swallow a 20 cc water bolus *(2)*. Severe impairment of swallowing comes through with drooling of saliva, choking and ventilatory insufficiency.

Weakness of the tongue is apparent by inability of the patients to protrude their tongue and reach the frenulum of the upper lip. If the tongue is pushed against the inner cheek, the degree of weakness can be appreciated directly by the examiner.

Weakness of the masticatory muscles is demonstrated by conventional testing (clenching a tongue depressor). Having patients repeatedly open and close their mouth vigorously until an audible click is heard is a sign of weakness. This maneuver may be easily performed 100 times in 30 s.

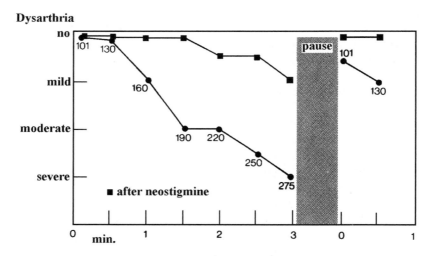

Fig. 2. Counting test. Improvement is appreciated by pausing and with neostigmine.

Weakness of the sternocleidomastoids and the neck muscles may be appreciated on routine testing. A useful functional test is having patients lift their heads in the recumbent position and the individual should be able to look at their toes for 60 s.

Bulbar weakness is not always easily assessed and quantification may be difficult. Dysphagia may be evaluated clinically and electrophysiologically. Weijnen et al. developed several tests and devices that provide quantative measures of bulbar function among MG patients *(3, 4, 5)*.

2.3 Limb, Trunk and Respiratory

Fifteen to twenty percent of the patients complain first of weakness of arms, hands or the legs. In patients under age 30 years, limb weakness, especially of the legs, was the first sign in one third of our patients. This might be explained by younger individuals placing these muscle groups into situations of heavier loading, for example, during sports. Inability to maintain arm position or repetitively elevate the upper extremity, e.g., when hanging laundry, hammering a nail or washing hair, is a common complaint. Isolated weakness of one or more extensors of the fingers, usually the fourth or fifth, may be difficult for the doctor to interpret and sometimes leads to misdiagnosis of a peripheral nerve compression.

Leg weakness frequently leads to sudden falls, and several of our patients had MG diagnosed after a fall from stairs. If the first manifestations are weakness of the limbs or of the trunk muscles, most patients complain of undue fatigability and peculiar feelings of heaviness. If asked about the exact nature of their feelings, they admit that these are different from the normal fatigue after exertion, although some rest is beneficial and restores their normal condition for a variable period of time. Most patients have arm and leg symptoms, but often one predominates, and the differential fatigability may be confirmed by appropriate examination (Fig. 3).

Pain in the back and girdle muscles occurs in some patients and is readily explained as an insufficiency of the postural muscles. It usually disappears at rest or after therapy. Chronic pain is not a feature of MG. Furthermore, many patients with postsynaptic myasthenia report muscle cramps irrespective of the use of anticholinesterases or ciclosporin; this inconvenience may be easily treated with anti-epileptics as phenytoin without worsening the myasthenia.

Weakness of the respiratory and other trunk muscles is rarely the first, isolated sign of the disease, but may be the first manifestation that brings a patient to medical care. We have only seen this in rapidly evolving generalized MG in children with coincident infection or as an effect of

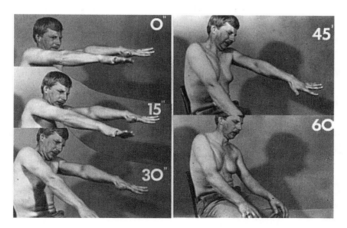

Fig. 3. Extended arm test. The patient is asked to lift his arms with hands in pronation. Normally, this position usually may be sustained for at least 3 min. Myasthenic weakness may lead the patient to drop one or both arms. Weakness of one of the extensor digitorum muscles (especially of the fourth finger) may become manifest with this test.

curare during narcosis ("prolonged apnea"). Some patients report short periods of altered awareness with inspiratory stridor, which are often the forerunners of longer lasting and life-threatening attacks. Rarely, short episodes of unconsciousness occur, which may not be recognized as caused by sudden apneas. Ventilatory symptoms should always lead to a rapid hospitalization. Bilateral vocal fold paralysis was recognized as an initial manifestation of MG in one patient who later developed diplopia, which prompted the evaluation for MG *(6)*.

If a patient is suspected to have MG, it is essential to test muscle strength after exercise. However, we must emphasize that weakness primarily, and not merely "fatigue", is a sign of MG. All tests designed to measure the capacity to do muscle work require the cooperation of the patient. In general this presents no difficulties, but the examiner must discern true weakness from a non-organic failure to generate maximal muscle force. Frank subterfuge may be a cause of such non-organic causes of weakness.

The following procedures in the assessment of muscle strength and exhaustability are useful. The strength of individual muscle groups is tested after rest. Accuracy is improved by using a hand-held dynamometer and the adoption of fixed postures. After moderate exertion, which does not decrease the strength of normal muscles, the strength of the patient is measured again. Moderate exertion may be standardized as follows *(7)*:

1. The arms, as well as the hands and fingers, are stretched horizontally (Fig. 3). This position should be maintained for 3 min without trembling or shaking. The patient may require some encouragement performing the maneuver. Increasing weakness will produce some shaking or a gradual drooping of arms, hands or fingers. If weakness is minimal, it may only be detected by measuring the strength again after 3 min of exertion. This test is very sensitive but not specific. Patients with other neuromuscular diseases may not be able to maintain their arms outstretched for more than a minute, but the strength before and after this effort remains the same.
2. Grip strength on repeated contraction may be measured with a hand-held dynamometer. A simple ergometer can be made out from a blood pressure manometer.
3. Patients should be able to perform knee bends or in older people rise from a standard chair repeatedly without the aid of the arms 20 times.
4. Walking on tiptoes and on the heels at least 30 steps.
5. Straight leg raising to 45° for 1 min in the recumbent position.
6. Vital capacity and peak-flow measurements should give normal values five times in a row. A difficulty with these tests may be the weakness of the lips or of the palate.

In most patients with a generalized MG, but without actual or previous dyspnea, a decrease of the vital capacity and other respiratory parameters is found and even in 40% of pure ocular cases, the vital capacity is decreased *(8)*. Routine lung function tests show that the vital capacity is decreased to a greater extent than the forced expiratory volume. In most patients, peak-flow or vital capacity measurement is a valuable tool in the follow-up and is done easily at any time and circumstance. Most of these tests are quantifiable, and the patient can perform some of them at home, to obtain a reliable picture of diurnal and periodic fluctuation. According to the complaints of the individual patient, other tests may be appropriate. A survey of the tests that are used in our practice is given in Table 2 *(9)*. The data acquired may show a convincing difference between the strength at rest and after exertion. They can also serve as a frame of reference in the evaluation of the effect of anticholinesterases and other treatments.

2.4 Muscle Atrophy

Historically, the observation of focal muscle atrophy has caused much confusion in understanding of MG. On the one hand, otherwise typical patients with MG and localized atrophy were included in early series *(10)*, but on the other hand, muscle atrophy was used as an argument to place such patients in the category of a myasthenic syndrome, with a form of myopathy, neuropathy or ophthalmoplegia. Others referred to patients with MG and atrophy as true MG with myopathy, with neuropathy or with neuromyopathy *(11)*. Osserman *(12)* placed patients with

Table 2
Clinical Investigation of the Patient with Myasthenia Gravis[a]

Inspection of the head at rest
- Note ptosis of the eyelids (usually asymmetrical), ocular deviations or drooping of the head
- Facial weakness may be manifest by disappearance of the nasolabial fold and loss of expression
- Weakness of the neck or cheek muscles may be masked by supporting the jaw
- Ptosis may be compensated by lifting the eyebrows
- There may be tilting or rotation of the head to minimalize diplopia

Ocular functions
- Looking straight toward a bright light provokes ptosis
- Looking aside or upward for 60^b s provokes ptosis, especially on the side of abduction
- Diplopia may occur after sustained lateral/vertical gaze (maximal 60^b s, not further than 45° from the midline)

Bulbar functions
- Test peak expiratory flow or vital lung capacity with and without nose clips to detect palatal weakness
- Repeatedly closing the eyes tightly provokes weakness of the ocular orbicular muscle; eye lashes may remain visible in spite of firmly closing the eyelids.
- Nasal speech may occur after counting 101–199. Note the number where dysarthria or nasality occurs
- Masseter function may be tested by biting on a spatula before and after $100\times$ closing the mouth with click
- Swallowing a glass of water may not be possible without coughing or regurgitation through the nose

Neck musculature (tests in horizontal position)
- Keeping the head raised for 120^b s ("look at your feet")
- Raising the head repeatedly ($20\times$)

Arm (test in sitting position)
- Arms stretched forward in pronate position (90°) for 240^b s
  - Note the beginning of trembling or shaking of the arm/hand
  - Note drooping of individual fingers

Hand
- Inflation of sphygmomanometer until 300 mmHg
- Fist closing/opening with fingers joined together (no digital abduction allowed) ($70\times$)
- Hand grip with a dynamometer (dominant hand $\geq 45^b$ (men) or $\geq 30^b$ (women) kg W, non-dominant hand $\geq 35^b$ and $\geq 25^b$ kg W, respectively)

Leg
- Hip flexion (45° supine in horizontal position) for 100^b s
- Deep knee bends ($20\times$)
- Rising from a standard chair without use of the hands

Ventilatory function
- Vital capacity or peak flow in rested condition
- Vital capacity or peak flow may decrease after repeated testing ($5–10\times$)

[a]Test muscle force directly before and after repetitive activity.
[b]Maximal values according to recommendations of the medical scientific advisory board of the Myasthenia Gravis Foundation of America (Reference 19).

muscular atrophy, comprising 5% of his series, in a separate group and stated that "the histologic changes in the biopsied muscle are indistinguishable from those seen in polymyositis or muscular dystrophy". From a clinical viewpoint, localized muscle atrophy is detectable in about 6–10% of the patients *(11, 12, 13, 14, 15)*, and a still higher percentage is reached if patients with permanent ophthalmoplegia are included. Remarkably few details about the distribution of the atrophy are found in the literature *(11, 14)*, although atrophy of the tongue is particularly appreciated (Fig. 4) *(16)*. In our patients localized muscle atrophy occurred in 9% of early-onset patients, 8% of late-onset patients and 12% of the patients with thymomas *(17)*. In the limbs we found mainly shoulder muscles, forearm muscles (finger extensors) and foot extensor muscles affected. Atrophy

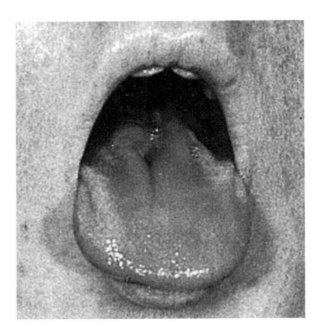

Fig. 4. A 60-year-old man with atrophy of the tongue. Note the presence of one medial and two lateral furrows, which define the "triple furrowed tongue". MG existed from early infancy and was diagnosed at age 52.

in bulbar muscles was more frequently found in our series; part of these patients are now recognized having a-MuSK myasthenia.

2.5 Clinical Classifications and Quantitative Tests

Several scoring systems of myasthenic signs or of the global state of the patient have been proposed *(9)*. The classification of Osserman and Genkins *(18)* is widely used and is in fact a modification of the initial classification of Osserman *(12)* from 1958 and altered again by Oosterhuis *(13)* in 1964. In this classification, juvenile patients and patients with muscle atrophy were excluded as separate entities and class 1 was confined to ocular myasthenia. It may be used retrospectively to classify patients to a certain type, but combines localization, severity and progression and is not well suited to follow individual patients. A Task Force of the Medical Advisory Board of the Myasthenia Gravis Foundation of America proposed an extended modification of this classification scheme in an attempt to provide a standard scheme for use by all investigators *(19)*. Scoring systems exist with categories for ocular, bulbar, limb muscle and respiratory weakness, based on either description or quantified examinations (see Appendix). One system gives a detailed scoring for ocular signs *(20)*.

2.6 Cognitive Involvement

Clinical experience with MG patients does not raise suspicion of central nervous system involvement. Nonetheless, indications for an organic brain syndrome have been found with abnormalities of visual attention and reaction time. Improvement of memory function during plasma exchange in line with the increase of muscle strength was observed in one study *(21)*, but memory defects were not be confirmed in another *(22)*. An auditory vigilance test did not reveal any abnormality *(22)*. Acetylcholine receptor antibodies from MG patients did not bind to the acetylcholine receptors extracted from human brain, which makes it unlikely that central cholinergic receptors are a target of an autoimmune process *(23)*. A review of clinical data regarding cognitive impairment, epilepsy, sleep disturbances, abnormal saccadic eye movements and psychiatric problems is given elsewhere *(24)*. It is our impression that the data on cognitive and memory dysfunction in MG are currently insufficient and inconclusive.

2.7 The Course of the Disease

The initial manifestations of MG have a highly variable pattern and evolution in the individual patient. Patients, when treated and relieved of their symptoms, often appreciate that they have suffered from MG longer than they had initially appreciated. Osserman's impression *(12)* was that MG does not have a classical clinical course. There is a tendency of spread to muscle groups beyond those initially affected, and only in 10% *(25)* to 16% *(26)* of the patients is MG clinically confined to the ocular muscles in the first 3 years (Fig. 5), yet after 3 years generalization occurs in only 3–10% of these purely ocular patients *(12, 25, 26)*. When the disease is restricted to the ocular muscles for a year, there is a likelihood of 84% that it will remain localized *(27)*. If symptoms are localized, but not to the ocular muscles, then spread of weakness to other muscle groups still occurs over 1–3 years. It is very rare that weakness remains confined to the bulbar muscles or to the muscles of the extremities.

In most patients the diagnosis is made in the second year after symptom onset, but a much longer delay is not unusual *(12, 13)*, especially among early-onset women. One reason for this delay that we found in our study of 100 consecutive patients is the more rapid progression of weakness in men than in women in the first 3 months of disease onset *(28)*. An additional factor appears to be a different diagnostic approach to men and women with early-onset disease. No

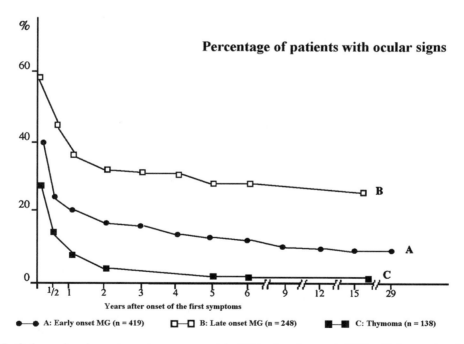

Fig. 5. Only ocular signs at onset were present in 39% of early-onset, 58% of late-onset and 27% of thymoma patients. After 3 years other muscle groups were involved in 84% early-onset, 68% late-onset and 96% thymoma patients. At the end of the follow-up, 9.6% early-onset, 26.6% late-onset and 1% of thymoma patients remained purely ocular *(17)*.

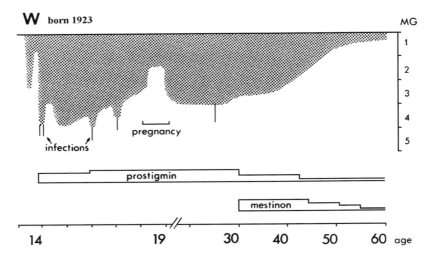

Fig. 6. Pictorial representation of a case history of a patient with MG. Onset of MG occurred with dysarthria during a stressful situation with generalization 6 months later while exacerbations during infections and improvement during pregnancy were observed. During a wartime period, oral prostigmin was replaced by subcutaneous injections (0.5 mg six times a day) because tablets were scarce. The patient improved very gradually after the age of 30, and after the age of 45 only slight weakness was present. Several major operations in the last 10 years were performed without development of weakness. She has been reluctant to discontinue medication (prostigmin 7.5 mg and Mestinon 10 mg three times a day). The severity of symptoms is expressed using a six-point disability score: 0, no symptoms; 5, dependent on artificial ventilation *(35, 37)*.

men underwent psychiatric evaluation compared to 8% of women. Patients with limited or predominant limb muscle weakness had a higher chance of a missed diagnosis. In general, severity of illness increases in the first years of the disease, with a tendency to stabilize, improve or even resolve after many years. A typical case history is schematized in Fig. 6.

Severe exacerbation of MG leading to myasthenic crisis occurs usually in the first years *(26, 29, 30)* of the disease and rarely in the first weeks of onset. One exception is for children with MG with an infection or with rapidly progressive bulbar symptoms that lead to choking and a poor cough. In these patients crises occur more often with gradually increasing generalized weakness during respiratory infections induced by aspiration. Crisis in MG is life threatening and demands aggressive treatment with immunomodulating strategies to minimize the duration.

The mortality of MG has been reduced from 30% to zero in the past decades *(26, 28, 30, 31, 32)*. Spontaneous long-lasting remissions are rare in the first years, but in the long term they are expected in 10–20% of patients *(26, 28, 30)*. Patients with ocular MG have a higher remission rate than those with the generalized disease *(26, 30)*.

2.8 Exacerbating Factors

Certain events are recognized as unfavorably influencing the course of MG and may unmask subclinical MG. Of particular importance are infectious diseases with fever, psychological stress, hyper- or hypothyroidism and certain drugs such as antimalaria drugs (quinine, chloroquine), aminoglycosides, beta-adrenergic receptor-blocking drugs and D-penicillamine *(33)*. Some experts recommend avoidance of vaccinations, but their effect has never been investigated. Patients under immunosuppression are advised by some to stop treatment for 2–3 weeks before immunization *(34)*.

The effect of pregnancy on MG is subject to conflicting reports: either improvement or deterioration or no influence, each one third in all the series *(35)*. Most patients improve in the second part of their pregnancy *(35)*. In a prospective investigation of 64 pregnancies, MG relapsed in 4 of 23 asymptomatic patients who were not on therapy before conception: in patients under treatment, MG improved in 12 of 31 pregnancies, remained unchanged in 13 and deteriorated in six. MG worsened after delivery in 15 of 54 pregnancies *(36)*. Minor exacerbations are experienced 3–10 days before menstruation by at least one third of the women *(37, 13)*.

The effect of ambient temperatures has not been studied extensively. Hot weather in particular is reported to increase weakness *(12, 38)* and may even induce a crisis *(39)*. In our experience, the individual response varies considerably, and several of our patients are considerably better in warm weather.

3. EPIDEMIOLOGY

3.1 Incidence and Prevalence

The annual incidence of MG is 3–5 per million *(40, 41, 42)*; incidental reports mention higher incidences up to 21 *(43, 44)*. The prevalence is about 60 per million *(45, 40, 46, 47, 41, 48, 42)*, in some surveys even higher up to 150 *(49)*. A striking difference of prevalence was appreciated between cities and rural areas *(45)*, but may be due to variation in populations and medical facilities. The prevalence seems to be increasing in the past decades, probably influenced by new diagnostic tests and a decrease of mortality rate. In Virginia the incidence and prevalence were found to be higher in the African-American population than in the corresponding Caucasian population. Ocular MG comprised 25% of this group of patients *(46)*, which was higher than that observed in previous series. In our own patient group we estimate 90% of the generalized patients having a-AChR antibodies and 3% having anti-MuSK *(50)*.

3.2 Age, Gender and Classification

MG may have its onset at any age. In general, women are affected twice as often as men, in the childbearing period even three times as frequently, while the incidence is about equal before puberty and after the age of 40 years *(48, 42, 51, 52)*. The relative incidence is highest in women in the third decade *(40, 18, 42)*, and in some series a late peak is found in older men *(18, 26, 42, 53)*. Several authors mention an increase in incidence of elderly-onset patients *(42, 54, 55)*, and this is explained by population aging and probably environmental factors.

Ocular MG occurs in about 15% of patients and has a higher prevalence in men, especially over the age of 40 *(48, 56)*. Age at onset, gender, type and incidence of thymomas in the author's series

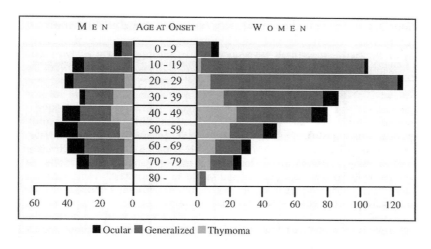

Fig. 7. Clinical and demographic data from 805 patients with MG evaluated between 1960 and 1994 *(17)*.

of 800 patients are given in Fig. 7. A relatively higher onset (22%) in the first decade, or 36% before puberty, is reported from Chinese populations *(57, 58)*, and the disease remains restricted to the ocular muscles in more than 50% of patients. However, the incidence of congenital cases is unknown in these series.

3.3 Acquired Infantile MG

The prevalence of cases with onset in the first decade in European and American series varies from 1 to 3%, which is lower than the overall prevalence in the population. A relatively greater prevalence was found in Japanese *(56)* and Chinese *(58)* populations, with a peak at the ages of 2–3 years for ocular MG. The relatively higher incidence of ocular MG in prepubertal MG was also found in a European series *(59)*.

The prognosis of MG in infancy is usually favorable, although exacerbations may occur with fevers. In sporadic cases, a fulminating onset with life-threatening respiratory insufficiency may occur. The incidence of a-AChR antibodies is lower than that in adults, so that this test cannot be used to differentiate between congenital and autoimmune MG. The outcome of infantile (onset 1–17 years) *(59)* or juvenile MG (onset 1–20 years) *(60)* is comparable with the adult forms, and the same treatment modalities are used although it seems justified to be less aggressive with immunomodulation in prepubertal children *(54)*.

3.4 Myasthenia Gravis in the Elderly

There are several clinical and epidemiological reasons to differentiate between patients with late-onset and early-onset MG (among series the defining age of late onset varies between 40, 45 and 50 years *(47, 61)*). Thymoma is more common in late-onset patients, the response to thymectomy is more uncertain than that in younger patients and antibodies to other muscle components than the acetylcholine receptor are more often found among these patients *(2, 61, 62)*. Although there are no differences between the signs and the symptoms in patients with late-onset MG, progression to severe disease may be more common in MG with onset after the age of 50 years. There is a somewhat increased incidence of weakness of bulbar muscles *(32)*, which cannot be attributed to the increased incidence of a thymoma *(2)*. Immunomodulating therapies are effective but elderly patient may be at increased risk of side effects, in particular for corticosteroids *(2, 17, 32)*.

3.5 Myasthenia Gravis and Thymoma

An incidence for thymoma is about 15% among MG patients *(18, 48)*, with late-onset patients being more commonly affected (Fig. 7). Generally, the course of MG is more severe and crises occur more often than in patients without thymomas *(15, 25, 51)*; however there are contra-dictory studies *(63)*. Patients with thymomas more often have onset of MG with bulbar mani-festations (Table 1). Complications related to the thymoma contribute to a worse outcome, but myasthenic manifestations respond to treatment similarly in thymoma and non-thymoma patients. In our own series *(17)* of 138 patients observed from 1960 to 1994, 16 died from complications of the tumor. In the remaining group 20 patients (16%) died from MG (all before 1984, only two had immunosuppressive therapy) and 55 patients (45%) went into remission with ($n = 37$) or without ($n = 18$) immunosuppression. MG death occurred in 2.8% and full remission in 38% of our 537 non-thymoma patients with generalized disease. Our data also suggest that the natural course of MG with a thymoma is more unfavorable but that the outcome with immuno-modulation is the same as in non-thymoma patients with MG.

3.6 The Incidence of Autoimmune Diseases in MG

The hypothesis of the autoimmune pathogenesis of MG was partly based on the occurrence of other autoimmune disorders among MG patients *(64)*. Many associations of MG and other autoimmune diseases have been reported in large series of MG patients *(18, 25, 40, 47, 42, 65)*. The frequencies in these series vary from 2.3 to 24.2%, with a mean of 12.9% compared with that of a US series of 3.2% *(66)*.

3.7 Genetic Predisposition

Although MG is not considered as a hereditary disease with a definite mode of genetic transmission, familial cases are reported in 1–4% of several series *(12, 51, 67, 68)* with a high of 7.2% *(47)*. In an analysis of 72 familial cases reported up to 1970, 39% belonged to the congenital type (onset before 2 years of age), 22% had occurred between 3 and 18 years and 39% over the age of 18 years. In 76% they occurred in one generation, in 24% in two generations but never in three. In 85% of the families, two members had MG, in 10% three and rarely more *(67)*. In twin studies, 6 out of 15 monozygotic twins both were affected but none out of 9 dizygotic twins *(69)*.

Recently, four separate families each with two affected members were described out of a total of 800 patients *(70)*. In seven of the eight patients a-AChR antibodies were found. No association with a single HLA haplotype was found. In another family, 5 of 10 members in one generation (ages 63–77 years) were described; no genetic abnormalities were found *(71)*. In our own 800 patients *(17)* 14 had a near family member with acquired MG (1.7%). Familial autoimmune MG seems a rarity but can be confused with congenital MG that is more often familial.

3.8 Differential Diagnosis

Making the diagnosis of MG in the patient with a-AChR antibodies is not difficult. If no autoantibodies are found, the history, results of clinical examination, reaction to anticholinesterases or findings on electrodiagnostic testing may strongly point to the diagnosis. Nevertheless, the diagnosis may remain difficult in a small subgroup of patients. In particular, the diagnosis of ocular MG may be difficult. Other myasthenic syndromes, especially the Lambert–Eaton myasthenic syndrome should always be considered in seronegative patients with fluctuating weakness. If the patient has *bulbar symptoms,* one might have reasons to include neuromuscular diseases as amyotrophic lateral sclerosis, oculopharyngeal dystrophy, bulbar spinal atrophy or myotonic dystrophy in the differential diagnosis. Self-evidently other (supra) nuclear brainstem pathology may also lead to bulbar manifestations. Some patients with isolated complaints and signs of dysarthria have no organic disease. Isolated dysphagia may be due to a mechanical impairment or to a disturbance in the parasympathetic innervation of the esophagus (achalasia). Focal dyskinesias may also need to be considered in certain patients.

In the seronegative patient with *predominant limb weakness,* the possibility of Lambert–Eaton myasthenic syndrome surpasses that of MG. Other acquired diseases with more or less fluctuating weakness of limb and trunk muscles may be motor neuron disease, polymyositis, endocrine myopathies and mitochondrial myopathies. It should be realized that there might be a slight benefit appreciated to cholinesterase inhibitor treatment in these diseases. Neck extensor weakness may be a prominent or initial manifestation of MG but also occurs with motor neuron disease and inflammatory myositis. There is also a benign isolated neck extensor myopathy that develops within 3 months without leading to further weakness *(72)*. Finally, some patients just have complaints of muscle pain or muscle fatigue without objective weakness. These symptoms are not suspicious for MG.

REFERENCES

1. Compston DAS, Vincent A, Newsom-Davis J, Batchelor JR. Clinical, pathological, HLA-antigen and immunological evidence for disease heterogeneity in myasthenia gravis. Brain 1980; 103: 579–601
2. Antonini G, Morino S, Gragnani F, Fiorelli M. Myasthenia gravis in the elderly: a hospital based study. Acta Neurol Scand 1996; 93: 260–262

3. Weijnen FG, van der Bilt A, Wokke J, Kuks J, van der Glas HW, Bosman F. What's in a smile? Quantification of the vertical smile of patients with myasthenia gravis. J Neurol Sci 2000;173:124–128.

4. Weijnen FG, Van der Bilt A, Wokke JHJ, Kuks JBM, Van der Glas HW, Bosman F. Maximal bite force and surface EMG in patients with myasthenia gravis. Muscle Nerve 2000;23:1694–1699.

5. Weijnen FG, Kuks JBM, Van der Bilt, Van der Glas HW, Wassenberg MWM, Bosman F: Tongue force in patients with myasthenia gravis. Acta Neurol Scand 2000; 102: 303–308.

6. Osei-Lah A, Reilly V, Capildeo BJ. Bilateral abductor vocal fold paralysis due to myasthenia gravis. J Laryngol Otol 1999;113:678–679

7. Oosterhuis H J G H. Myasthenia gravis. Churchill Livingstone, Edinburgh 1984.

8. Reuther P, Hertel G, Ricker K, Bürkner R. Lungenfunktionsdiagnostik bei myasthenia gravis. Verhandlungen der Deutsche Gesellschaft für Innere Medizin 1978; 84:1579–1582.

9. Kuks JBM, Oosterhuis HJGH. Disorders of the neuromuscular junction: outcome measures and clinical scales. In: Guiloff RJ (ed.): Clinical trials in neurology. Springer, London 2001, pp 485–499.

10. Oppenheim H. Die Myasthenische Paralyse (Bulbärpara1yse ohne anatomischen Befund). Berlin, S Karger 1901.

11. Oosterhuis H J G H, Bethlem J. Neurogenic muscle involvement in myasthenia gravis. J Neurol Neurosurg Psychiatry 1973; 36: 244–254.

12. Osserman KE. Myasthenia Gravis. Grune & Stratton 1958 New York

13. Oosterhuis HJGH 1964 Studies in myasthenia gravis. J Neurol Sci 1964; 1: 512–546

14. Schimrigk K, Samland O. Muskelatrophien bei Myasthenia gravis. Der Nervenarzt 1977; 48: 65–68.

15. Simpson JA. Evaluation of thymectomy in myasthenia gravis. Brain 1958: 81:112–145.

16. Kinnier Wilson S A. Neurology, vol I, London, Edward Arnolds Co 1940, p 1598.

17. Oosterhuis HJGH. Myasthenia Gravis. Groningen Neurological Press. 1997

18. Osserman KE, Genkins G. Studies in myasthenia gravis: review of a twenty year experience in over 1200 patients. Mt Sinai J Med 1971; 34: 497–537.

19. Task Force of the Medical Scientific Advisory Board of the Myasthenia Gravis Foundation of America. Jartezki III A, Barohn RJ, Ernstoff RM, Kaminski HJ, Keesey JC, Penn AS, Sanders DB. Myasthenia Gravis. Recommendations for clinical research standards. Neurology 2000; 55: 16–23

20. Schumm F, Wiethoelter H, Fateh-Moghadam A, Dichgans J. Thymectomy in myasthenia gravis with pure ocular symptoms. J Neurol Neurosurg Psychiat 1985; 48: 332–337.

21. Tucker D M, Roeltgen DP, Wann Ph D Wertheimer RI. Memory dysfunction in myasthenia gravis: evidence for central cholinergic effects. Neurology 1988; 38: 1173–1177 .

22. Lewis S W, Ron M A, Newsom-Davis J. Absence of central functional cholinergic deficits in myasthenia gravis. J Neurol Neurosurg Psychiatry 1989;52:258.

23. Whiting PJ, Cooper J, Lindstrom JM. Antibodies in sera from patients with myasthenia gravis do not bind with nicotinic acetylcholine receptors from human brain. J Neuroimmunol 1987; 16: 205

24. Steiner I, Brenner T, Soffer D, Argov Z. Involvement of sites other than the neuromuscular junction in myasthenia gravis. In Lisak RP (ed.) Handbook of Myasthenia gravis and myasthenic syndromes. Marcel Decker, New York 1994, pp 277–294.

25. Oosterhuis H J G H. Myasthenia gravis. A review. Clin Neurol Neurosurg 1981; 83:105–135.

26. Grob D, Brunner NG, Namba T. The natural course of myasthenia gravis and the effect of various therapeutic measures. Ann NY Acad Sci 1981; 377: 652–669.

27. Grob D. Natural history of myasthenia gravis. In: Engel AG (Ed.). Myasthenia gravis and myasthenic disorders. Oxford University Press, New York, 1999; pp 131–145

28. Beekman R, Kuks J B M, Oosterhuis HJGH. Myasthenia gravis: diagnosis and follow-up of 100 consecutive patients. J Neurol Sci 1997;244:112–118.

29. Cohen MS, Younger D. Aspects of the natural history of myasthenia gravis: crisis and death. Ann NY Acad Sci 1981; 377: 670–677.

30. Oosterhuis H J G H. Observations of the natural history of myasthenia gravis and the effect of thymectomy. Ann NY Acad Sci 1981; 377:678–689.

31. Beghi E, Antozzi C, Batocchi AP et al.: Prognosis of myasthenia gravis: a multicenter follow-up study of 844 patients. J Neurol Sci 1991; 106: 213–220.

32. Donaldson DH, Ansher M, Horan S, Rutherford RB, Ringel SP. The relationship of age to outcome in myasthenia gravis. Neurology 1990; 40: 786–790

33. Wittbrodt ET. Drugs and myasthenia gravis. n update. Arch Int Med 1997; 157: 399–408.

34. Schalke B. Klinische Besonderheiten bei Myasthenia Gravis Patienten. Akt Neurologie 1998; 25: 528–529

35. Plauché W C. Myasthenia gravis in pregnancy: an up date. Am J Ob Gyn 1979; 135:691–697.

36. Batocchi AP, Majolini L, Evoli A, Lino MM, Minisci C, Tonali P. Course and treatment of myasthenia gravis during pregnancy. Neurology 1999; 52: 447–452

37. Leker RR, Karni A, Abramski O. Exacerbation of myasthenia during the menstrual period. J Neurol Sci 1998; 156: 107–111

38. Borenstein S, Desmedt J E. Temperature and weather correlates of myasthenia fatigue. Lancet 1974; ii: 63–66.

39. Guttman L. Heat-induced myasthenic crises. Arch Neurol 1980;37:671–672.

40. Oosterhuis H J G H. The natural course of myasthenia gravis: a longterm follow-up study. J Neurol Neurosurg Psychiatry 1989; 52: 1121–1127.

41. Somnier FE, Keiding N, Paulson OB. Epidemiology of myasthenia gravis in Denmark: a longitudinal and comprehensive population survey. Arch Neurol 1991; 48: 733–739.

42. Storm Mathisen A 1984 Epidemiology of myasthenia gravis in Norway. Acta Neurol Scand 70, 274–284.

43. Aragonès, JM, Bolíbar I, Bonfill X, Bufill E, Mummany A, Alonso F, Illa I. Myasthenia Gravis. A higher than expected incidence in the elderly. Neurology 2003; 60: 1024–1026.

44. Emilio Romagna Study Group on Clinical and Epidemiological Problems in Neurology: Incidence of myasthenia gravis in the Emilia-Romagna region: A prospective multicenter study. Neurology 1998; 51: 255–258.

45. D' Alessandro R, Granieri E, Benassi G, Tola M R, Casmiro M, Mazzanti B, Gamberini G, Caniatti L. Comparative study on the prevalence of myasthenia gravis in the provinces of Bologna and Ferrara, Italy. Acta Neurol Scand 1991; 2: 83–88.

46. Phillips LH, Tomer JC, Anderson MS, Cox GM. The epidemiology of myasthenia gravis in central and western Virginia. Neurology 1992; 42: 1888–1893.

47. Pirskanen R. Genetic aspects in myasthenia gravis, a family study of 264 Finnish patients. Acta Neurol Scand 1977; 56: 365–388

48. Sorensen TT, Holm E B 1989 Myasthenia gravis in the county of Viborg, Denmark. Eur Neurol 29: 177–179.

49. Robertson NP, Deans J, Compston DAS. Myasthenia Gravis: a population based epidemiological study in Cambridgeshire, England. J Neurol Neurosurg Psychiatry 1998; 65: 492–496.

50. Niks EH, Kuks JBM, Verschuuren JJGM. Epidemiology of myasthenia gravis with anti-muscle specific kinase antibodies in The Netherlands. J Neurol Neurosurg Psychiatry. 2007 Apr; 78(4):417–8.

51. Grob D, Arsura E L, Brunner N G, Namba T. The course of myasthenia gravis and therapies affecting outcome. Ann NY Acad Sci 1987; 505:472.

52. Mantegazza R, Beghi E, Pareyson D, AntozziC, Peluchetti D, Sghirlanzoni A, Cosi V, Lombardi M, Piccolo G, Tonali P et al. A multicentre follow-up study of 1152 patients with myasthenia gravis in Italy. J Neurol Sci 1990;237:339–344.

53. Potagas C, Dellatotas G, Tavernarakis A, Molari H, Mourtzouhou P, Koutra H, Matiaks N, Balakas N. Myasthenia Gravis: changes observed in a 30-years retrospective clinical study of a hospital-based population. J Neurol. 2004 Jan;251(1):116–7.

54. Matsuda M, Dohi-Iijima N, Nakamura A, Sekijima Y, Moriat H, Matsuzawa S, Sato S, Yahikozawa H, Tabata K, Yanagawa S, Ideda S. Increase in incidence of elderly-onset patients with myasthenia gravis in Nagano prefecture, Japan. Internal Medicine 2005; 44: 572–577.

55. Somnier FE. Increasing incidence of late-onset anti-AChR antibody-seropositive myasthenia gravis. Neurology 2005; 65: 928–930.

56. Fukuyama Y, Hirayama Y, Osawa M. Epidemiological and clinical features of childhood myasthenia gravis in Japan. Japan Medical Research Foundation University of Tokyo Press, Tokyo 1981 p 19.

57. Hawkins BR, Yu Y L, Wong V, Woo E, IpM S, Dawkins R L. Possible evidence of a variant of myasthenia gravis based on HLA and acetylcholine receptor antibody in Chinese patients. Quart J Med 1989; 70: 235–241.

58. Wong V, Hawkins B R, Yu YL. Myasthenia gravis in Hong Kong Chinese. Acta Neurol Scand 1992; 86: 68–72.

59. Batocchi AP, Evoli A, Palmisani MT, Lo Monaco M, Bartoccioni M, Tonali P. Early onset myasthenia gravis: clinical characteristics and response to therapy. Eur J Pediatr 1990; 150: 66–68

60. Lindner A, Schalke B, Toyka KV. Outcome of juvenile-onset myasthenia gravis: a retrospective study with long term follow-up of 79 patients. J Neurol 1997; 244: 515–520

61. Aarli JA. Late-onset Myasthenia Gravis. Arch Neurol 1999; 56: 25–27.

62. Kuks JBM, Limburg PC, Horst G, Oosterhuis HJGH. Antibodies to skeletal muscle in myasthenia gravis, part 2. Prevalence in non-thymoma patients. J Neurol Sci 1993; 120: 78–81

63. Bril V, Kojic J, Dhanani A. The long-term clinical outcome of myasthenia gravis in patients with thymoma. Neurology 1998; 51: 1198–1200.

64. Simpson JA. Myasthenia gravis, a new hypothesis. Scot Med J 1960; 5: 419–436.

65. Thorlacius S, Aarli JA, Riise T, Matre R, Johnsen HJ. Associated disorders in myasthenia gravis: autoimmune disease and their relation to thymectomy. Acta Neurol Scand 1989;80:290–295.

66. Jacobson DL, Gange SJ, Rose NR, Graham NMH. Epidemiology and estimated population burden of selected autoimmune disease in the United States. Clin Immunol Immunopath 1997;84:223–243

67. Namba T, Brunner N G, Brown S B, Muguruma M, Grob D. Familial myasthenia gravis. Arch Neurol 1971; 25: 49–60.

68. Szobor A. Myasthenia gravis: familial occurrence, a study of 1100 myasthenia gravis patients. Acta Medica Hungarica 1989; 4: 13

69. Behan P O. Immune disease and HLA association with myasthenia gravis. J Neurol Neurosurg Psychiatry 1980; 4: 611–621.
70. Evoli A, Batocchi AP, Zelano G, Uncini A, Palmisani MT, Tonali P. Familial autoimmune myasthenia gravis: report of four families. J Neurol Neurosurg Psychiat 1995; 58: 729–731.
71. Bergoffen J, Zmyewski ChM, Fischbeck KH. Familial autoimmune myasthenia gravis. Neurology 1994; 44: 551–554.
72. Suarez GA, Kelly JJ. The dropped head syndrome. Neurology 1992;42:1625–1627.

6

Ocular Myasthenia

Robert B. Daroff and Michael Benatar

1. INTRODUCTION

Myasthenia gravis (MG) is a generalized disorder that often manifests initially as focal weakness. The most common focal presentation is ocular, with weakness involving the extraocular muscles (EOMs), levators, or orbicularis oculi. Although there may be electrophysiological evidence of myasthenia in lower facial or limb muscles, if the weakness is limited to the ocular muscles, it is designated ocular myasthenia (OM). Eye muscle weakness at the onset of MG is evident in 85–90% of patients, with the percentage of those without clinical evidence of coexisting bulbar or limb weakness at the onset (i.e., pure OM) varying from 18% *(1)* to 59% *(2)* in different series *(3, 4, 5, 6)*.

2. EPIDEMIOLOGY

The epidemiology of MG is discussed in detail in Chapter 15 and here we will discuss it only briefly. Determining the relative frequencies of ocular and generalized MG is compounded by multiple factors. However, using data from several published population-based studies that defined OM on the basis of the modified Osserman classification or the criteria of the Myasthenia Gravis Foundation of America, the pooled estimate of the proportion of all patients with MG who have purely ocular disease is 35% (95% CI, 32–38%) *(7, 8, 9, 10, 11, 12, 13)*. The literature on childhood OM is sparse. Although he does not provide or cite any data, Fenichel states that it rarely begins prior to 6 months of age, and 75% develop after age 10 *(14)*. Pre-pubertal onset is more common in boys and post-pubertal is more common in girls.

3. BASIS FOR OCULAR MUSCLE INVOLVEMENT BY MYASTHENIA GRAVIS

The marked susceptibility of the EOMs in MG may be explained by a variety of factors unique to these muscles *(15)*. One fairly intuitive explanation is that ocular alignment requires a high degree of precision and that a minor reduction in force generation, which would be imperceptible in a limb muscle, leads to misalignment of the visual axis and diplopia. But there are other factors, both anatomic and immunologic, which make EOMs particularly susceptible. One anatomic explanation is that these muscles differ from other skeletal muscles in that they contain both singly and multiply innervated muscle fibers. It is not simply this diversity that predisposes to myasthenia, but rather that the junctional synaptic folds of the singly innervated fibers in EOMs are less complex than those in other skeletal muscles *(16)*. Moreover, multiply innervated muscles are completely devoid of synaptic folds that serve to increase the density of acetylcholine receptors and to focus current into the depths of these folds. Voltage-gated sodium channels are located in the depths of these folds and increase the safety factor for neuromuscular transmission. The lack of synaptic fold complexity in the neuromuscular junctions of EOMs lowers the safety factor, thereby increasing susceptibility to myasthenia. Extraocular muscle also differs immunologically from other skeletal muscle in that "Decay accelerating factor" (Daf1), a membrane-bound complement regulatory gene, is expressed at low levels in extraocular muscle *(17)*. Since complement regulators

From: *Current Clinical Neurology*
Myasthenia Gravis and Related Disorders
Edited by: H.J. Kaminski, DOI 10.1007/978-1-59745-156-7_6, © Humana Press, New York, NY

serve to protect cells from complement deposition and consequent membrane damage, the low level of expression of Daf1 may predispose EOMs to immune-mediated damage. In addition, MG is more likely to be ocular when associated with autoimmune thyroid disease, possibly related to immunological cross-reactivity against shared thyroid and eye muscle antigens *(18)*.

The reasons for preferential involvement of the levator palpebrae muscle are less clear, but may relate to their relatively sparse junctional folds and their constant activation to keep the eyes open, which predisposes the levators to fatigue *(19)*.

4. CLINICAL PRESENTATION

The pattern of muscle involvement with OM ranges from weakness of a single extraocular muscle or lid, to complete external ophthalmoplegia with bilateral ptosis, with any intervening combination. At times, the pattern may mimic a central motility disturbance, such as a gaze palsy or internuclear ophthalmoplegia *(20)*. Although accommodative weakness and fatigue may occur in OM, as may subclinical pupillary abnormalities that can only be detected accurately with quantitative techniques *(21, 22, 23)*, we agree with Glaser and Siatkowski *(24)* that "if pupillary signs are present, another diagnosis must be entertained".

The diagnosis of OM should be strongly suspected by aspects of the *history*, such as (1) episodes of recurrent unilateral, alternating, or bilateral ptosis and (2) progressively increasing ptosis or diplopia during the day, with improvement upon awakening in the morning. In addition, the *examination* may mandate a diagnosis of OM. For example, whereas a third nerve palsy may begin with an isolated ptosis, involvement of EOMs soon follows, and the addition of pain and pupillary dilatation are frequent. Thus, an acquired isolated painless ptosis is almost certainly diagnostic of OM, with the extraordinarily rare exceptions of isolated metastatic carcinoma to the eyelid *(25)*, amyloid infiltration of a levator muscle *(26)*, or unilateral painless lid myositis *(27)*. Similarly, ophthalmoparesis with rapid saccades within the limited range of eye movements only occurs in MG; all other conditions that restrict ocular amplitude also slow saccadic velocity *(20, 28)*. Facial weakness limited to the orbicularis oculi, coexisting with ptosis or extraocular muscle weakness, only occurs in MG. Fatigue of extraocular muscles and lids during sustained upgaze (Fig. 1) also suggests OM, although this, as well as Cogan's "lid twitch" sign,[*] may be secondary to midbrain lesions *(29, 30, 31)*.

Another OM lid sign is "enhanced" ptosis, where passive elevation of a ptotic lid causes the opposite lid to droop. The explanation is Hering's law of equal innervation; for instance, if both eyelids are ptotic due to bilateral levator weakness, and the right is more ptotic than the left, both sides will receive equivalent increased central innervation to elevate the lids. Because of the asymmetric weakness, the right lid will remain more ptotic than the left. But, if the right lid is manually elevated, the increased central innervation ceases and the left lid becomes more ptotic *(22, 32)*.

We must emphasize, however, that no matter how "characteristic" the history and examination is for OM, a dilated pupil or pain negates the diagnosis, unless another disorder coexists with the MG.

5. DIAGNOSTIC TESTING

The methods utilized for the diagnosis of OM include the rest, ice, and Tensilon tests; acetylcholine receptor antibody assays, repetitive nerve stimulation (RNS), and single-fiber electromyography (SFEMG). The methodological quality of the studies examining the diagnostic accuracy of

---

[*]This sign is commonly present in MG patients with unilateral ptosis. The patient is asked to look down, thereby inhibiting the levator muscles. After about 15 seconds, the patient looks up to the examiner's nose or an object at eye level, and if the sign is present, the previously ptotic lid overshoots and is transiently higher than the other lid. The retracted lid then slowly drops to its previous ptotic position.

(a) (b)

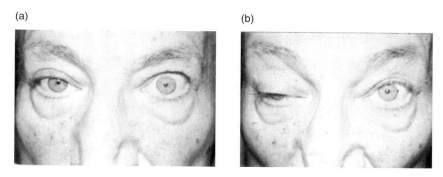

Fig. 1. On the left (**1a**) the patient is attempting to look up, evidenced by the contraction of the frontalis muscle. Note the slight right ptosis and left lid retraction. The right panel (**1b**) demonstrates the lid fatigue that developed during the sustained upward gaze. This was manifested by marked ptosis on the right and lessening of the lid retraction on the left (Photos courtesy of J. Lawton Smith, M.D.).

these tests have varied widely. A common limitation has been the case–control design (in which patients known to have MG are compared to healthy controls or patients with other disorders), rather than a cohort design (in which a series of patients, all of whom were referred for diagnostic evaluation for possible MG, are evaluated against some external reference standard). One of us (MB) critically reviewed the accuracy of the various procedures for the diagnosis of OM *(33)*.

The different techniques suggested for the rest and sleep tests all represent variations of the ptotic eyelid being clinically closed ("rested") for approximately 2 minutes. In the sleep test, the patient is placed in a quiet, darkened room with instructions to close the eyes and try to sleep for about 30 minutes. With both tests, a comparison is made between the degree of ptosis before and after the rest period. A case–control study of the rest test reported a sensitivity of 0.50 and a specificity of 0.97 for the diagnosis of OM *(34)*. A case–control study of the sleep test, albeit of limited methodological quality, reported a sensitivity of 0.99 and a specificity of 0.91 for the diagnosis of OM *(35)*.

Saavedra et al. first described the utility of the ice test for the diagnosis of OM *(36)*, based on the observations by Borsenstein and Desmedt *(37)* that the neuromuscular block in myasthenia is ameliorated when the temperature is lowered. The ice test is typically performed by placing a surgical glove filled with ice over the ptotic eyelid for about 2 minutes. The degree of ptosis is compared before and after the application of ice. Three studies utilized a case–control design to determine the accuracy of the ice test for OM *(38, 39, 40)*. The pooled estimates of sensitivity and specificity for the diagnosis of OM were 0.94 and 0.97, respectively. Consequent to a serious allergic complication experienced by one of RBD's patients, we suggest using latex-free surgical gloves for the ice test.

The Tensilon (edrophonium chloride) test is often regarded as the mainstay of the office diagnosis of OM. The most important technical aspect of the test is the need for an unequivocal endpoint to judge the success or failure of the test. We regard resolution of eyelid ptosis (Fig. 2) or the direct observation of the strengthening of at least a single paretic EOM, as the only reliable endpoints. Thus, patients without ptosis or discernible ophthalmoparesis have no valid endpoints *(41)* (Fig. 3). Others rely heavily on muscle balance endpoints such as prisms, red glass, Maddox rod, and Lancaster red-green torches *(42)*, but these can yield spuriously positive responses if a non-paretic yolk or antagonist muscle becomes weak consequent to the Tensilon *(41)*. When the endpoints are valid, there is no need for a saline placebo control.

The traditional technique for Tensilon testing involved a 2 mg test dose to determine if there were untoward side effects. Following that, 8 mg was intravenously administered. For OM,

(a)   (b)

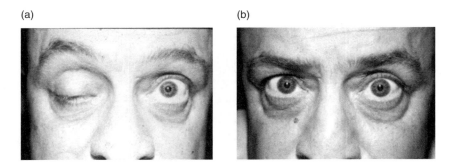

Fig. 2. The left panel (**2a**) depicts an almost complete right ptosis in a patient with ocular myasthenia. The frontalis muscle contraction, elevating the eyebrows, reflects the patient's effort to keep the lids open. Following administration of intravenous edrophonium chloride (Tensilon), the right ptosis is resolved (Fig. **2b**, right) (Photos courtesy of J. Lawton Smith, M.D.).

(a)   (b)

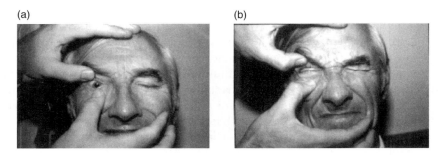

Fig. 3. On the left (**3a**), the patient has weakness of eyelid closure, permitting easy opening by the examiner. On the right (**3b**), the patient shows much stronger eyelid closure following the administration of Tensilon. This patient had myasthenia, demonstrated by EMG and the presence of antibodies. The Tensilon test, by itself, disclosing the increased strength of the orbicularis oculi, would not have been conclusive for the diagnosis of MG, since impaired volition could have explained the orbicularis weakness in **3a**. A Tensilon test with placebo control is unreliable in these circumstances, because IV saline does not cause the same systemic reaction as does Tensilon.

Kupersmith et al. *(43)* recommended injections of no more than 2 mg in 30 s intervals. With this small increment technique, the mean dose for a positive response was 3.3 mg for ptosis and 2.6 mg for extraocular muscle weakness.

Tensilon testing in children should follow the same general technique, except for the dose. Whereas 10 mg is usually regarded as the "full dose" in adults, in children Fenichel *(14)* recommends 0.15 mg/kilogram, with the initial test dose being 10% of the full dose.

A number of studies have utilized oculography with anticholinesterase testing *(44, 45, 46)*, but their expense and specificity do not justify clinical usage. The same holds for oculographic analysis of saccadic waveforms and optokinetic nystagmus, to quantify eye muscle weakness or fatigue.

Some neuro-ophthalmologists and neurologists seem particularly concerned about the complications of Tensilon. A questionnaire mailed to neuro-ophthalmologists yielded almost 200 respondents who had performed a combined total of over 23,000 tests in which only 37 (0.16%) were associated with a serious complication, such as bradyarrhythmia and syncope; yet, 16% of the respondents preferred an alternative diagnostic procedure, such as the sleep or ice test *(47)*.

Although we have never personally observed any serious complications during the performance of a Tensilon test, we do recommend that atropine should always be available (in a syringe) to counter possible muscarinic side effects of Tensilon.

Formal studies of the diagnostic accuracy of the Tensilon test are almost non-existent *(33)*. The only study that we identified is a cohort study of relatively poor methodological quality that reported a sensitivity of 0.92 and a specificity of 0.97 for the diagnosis of OM *(48)*.

Three case–control studies *(49, 50, 51)* and three cohort studies *(48, 52, 53)* reported the utility of acetylcholine receptor antibody testing. The pooled estimate of sensitivity for the diagnosis of OM was lower (0.44) for the higher methodological quality cohort studies *(48, 52, 53)* than for the less well-designed case–control studies (0.66) *(49, 50, 51)*. Specificity in these studies ranged from 0.95 *(52)* to 1.0 *(50)*. The consistently high specificity of elevated titers of acetylcholine receptor antibodies establishes their presence as essentially diagnostic. The utility of testing for muscle-specific receptor tyrosine kinase (MuSK) antibodies is less clear as studies are more limited, but the available data suggest that MuSK antibodies are found in a significant minority of patients with seronegative generalized disease, but are rarely identified in OM *(54, 55)*.

Repetitive nerve stimulation (RNS) for the diagnosis of MG is discussed in Chapter 7. Estimates of the sensitivity of RNS for the diagnosis of OM have ranged from 0.11 (based on stimulation of the radial nerve with recording from anconeus *(56)*), to 0.15 (based on stimulation of the truncus primarius superior with recording from abductor digiti minimi *(52)*), to 0.20 (based on stimulation of the facial nerve, recording from both nasalis and frontalis) *(57)*, to 0.39 (based on stimulation of the facial nerve with recording from orbicularis oculi) *(58)*. Costa et al. found that sensitivity remained around 0.33 when they performed RNS on four different nerve–muscle pairs (facial nerve–nasalis, accessory nerve–trapezius, radial nerve–anconeus, and ulnar nerve–abductor digiti minimi) *(53)*, suggesting that the sensitivity of RNS for the diagnosis of OM is generally poor. Specificity is better, however, ranging from 0.89 *(52)* to 0.97 *(53, 58)*.

As with RNS, the diagnostic accuracy of SFEMG may vary depending upon the individual muscle examined (e.g., orbicularis oculi or frontalis), the recording technique used (e.g., stimulated or volitional), and the type of recording electrode (e.g., single fiber or concentric needle). Studies of the accuracy of SFEMG for the diagnosis of OM all employed a cohort design. Although other methodological issues raise concern, these studies paint a general picture of higher sensitivity for the diagnosis of OM when single-fiber electrodes are used to record jitter from the orbicularis oculi muscle (0.97) *(52, 53)* compared to studies with similar electrodes recording from the frontalis muscle (0.86) *(59, 60)* or concentric needles recording from the frontalis (0.62) *(61)*. The specificity for OM is more variable, ranging from as low as 0.66 *(59)* to as high as 0.98 *(53)*.

6. PROGRESSION TO GENERALIZED MYASTHENIA GRAVIS

A major concern in OM is the potential spread to the generalized form of the disease. Most investigators have reported that approximately 50% of patients progress in this way *(3, 10, 62)*, although some have estimated that the proportion may be as high as 66% *(6)*. Simpson et al. *(5)* reported spread in 85–95% (depending upon which of their two contradictory statements is correct); these figures are at variance with other series and undoubtedly reflect uncertainty as to the definition of OM. In children, rates of spread range from 39 to 49% *(63)*.

The common conviction is that progression is most likely within the first 2 years of disease onset, but the data to support this claim are mixed. One study found that 44 of the 53 patients (89%) who progressed from ocular to generalized MG did so within the first 2 years (3), but other studies found that 50–60% spread in the initial 2 years *(2, 62)*. Indeed, spread may occur up to 24 years after onset *(28)*. Bever et al. (3) found no predictors of spread, including the presence or absence of antibodies or decremental RNS response in uninvolved limb muscles. More recent studies indicate that whereas abnormal limb SFEMG does not predict spread, a normal limb SFEMG provides an 82% likelihood that MG will remain ocular *(64)*.

Twenty percent of children with OM are said to go into a permanent, drug-free remission *(14)*, compared to 10% in Bever et al.'s adult series *(3)*.

7. TREATMENT

The goals of treating OM are to reduce symptoms, and possibly to prevent or delay the progression to generalized MG. The spectrum of treatments includes anticholinesterase drugs, usually pyridostigmine bromide (Mestinon®); steroids; immunosuppressants (azathioprine and cyclosporine A); thymectomy; eye patching (for diplopia); or mechanical measures for bilateral ptosis. In children with OM, the risk of amblyopia, due to ptosis or diplopia, prompts the need for occlusion therapy to ensure the use of both eyes *(65)*. There are proponents and opponents of these various forms of therapy, but the studies evaluating the efficacy of these treatment options are of extremely limited quality. As a result, it is difficult to make evidence-based recommendations.

We uncovered only two randomized controlled drug trials in OM *(66, 67)*, but neither provided useful information for contemporary treatment. The details of these two studies, including the treatments employed and outcome measures used, are summarized in a recent systematic review published in the Cochrane library *(68)*.

Two observational studies, of reasonably good methodological quality, using thymectomy as active therapy *(69, 70)*, reported outcome in terms of improvement in ocular symptoms. Neither study found any benefit from thymectomy. Five observational studies reported the impact of oral steroids on the risk of progression from ocular to generalized MG *(43, 71, 72, 73, 74)*; two of these also examined the effect of azathioprine on the risk *(71, 74)*. The point estimates of the odds ratios in the studies with steroids showed a reduction in the risk of progression in four studies *(43, 71, 72, 74)*, with the confidence interval spanning unity in the only study that did not show a benefit *(73)*. The two studies examining the effects of azathioprine similarly showed a beneficial effect on the risk of progression, with confidence intervals that did not span unity *(71, 74)*. The data from these observational studies suggest that steroids and azathioprine may be of benefit in reducing the risk of progression from ocular to generalized myasthenia.

The paucity of randomized controlled studies makes it impossible to provide evidence-based recommendations on the efficacy of therapy for OM *(75)*. In the absence of evidence, we present our approach to the treatment of these patients. We first try Mestinon for symptomatic relief. Although it often improves ptosis and diplopia, patients may not be satisfied. For instance, patients with complete unilateral ptosis do not have diplopia; Mestinon may relieve the ptosis and unmask the diplopia. Mestinon usually increases the strength of weak extraocular muscles, but may not totally correct the weakness, leaving some diplopia. A small amount of diplopia is more disconcerting than a large separation, since it is easier to ignore a diplopic image that is off in the periphery, than one close to the real image. Thus, eliminating ptosis or substantially reducing diplopia with Mestinon may be more distressing to the patient than their original symptoms.

If we cannot correct diplopia with Mestinon, one of us (RBD) suggests patching an eye with a clip-on spectacle occluder, frosted lenses, or an occluding contact lens. Bilateral ptosis can often be improved with a "crutch" constructed by an optician (Fig. 4).

The two authors of this chapter vary somewhat in their use of steroids for OM. One of us (RBD) uses steroids only as a "last resort" for bilateral total ophthalmoplegia, bilateral ptosis with poor tolerance of crutches, or patient insistence because of overwhelming cosmetic concern *(75)*. The latter refers patient who will "kill myself unless you make me look normal again".

RBD's reluctance to use steroids is based upon the obvious complications, the probable necessity of concomitant prophylactic treatment with a biphosphonate to retard osteoporosis

Fig. 4. Ptosis crutches are thin metallic frames, attached to the rims of eyeglasses. They mechanically elevate the lids, but allow blinking (Photo courtesy of Robert L. Tomsak, M.D., Ph.D.).

(76, 77), and, once started, steroids may be impossible to discontinue without exacerbation of symptoms *(80)*. Although RBD recognizes that steroids may prevent spread to generalized myasthenia, he suspects that they are merely masking the generalized weakness in those patients who do spread.

The other author, MB, prefers a middle ground between those who routinely use steroids in OM and those who do so only rarely. MB will offer steroids to most patients who do not respond to the usual conservative measures described above. Both of us are attentive to the potential complications of steroids, such as diabetes, hypertension, glaucoma, osteoporosis, peptic ulcer disease, and weight gain. We obtain 2 h glucose tolerance tests and bone density scans on all patients prior to initiation of steroids. We provide all patients with gastric prophylaxis in the form of a proton pump inhibitor and prophylaxis against osteoporosis. We advise the patients to follow up regularly with their primary care physicians to monitor blood pressure and glycemic control.

We typically start steroid therapy with prednisone 10 mg every other day, increasing by 10 mg every third dose, until the desired clinical effect is achieved. Because the dose increase is slow, patients often reach the desired effect at 40–60 mg every other day. The dose is maintained for several months and then decreased by 10 mg a month until 30 mg every other day is reached. Any decrease from 30 mg is by increments of 5 mg, and decreases from 20 mg are by increments of 2.5 mg or less. Myasthenics may "crash" with exceedingly small reductions when their alternate day dose is 20 mg or below. A "crash" in an ocular myasthenic is not particularly problematic, but we never know whether the steroids were masking generalized myasthenia, which might manifest dramatically. We do not use azathioprine, cyclosporine, plasma exchange, or intravenous immunoglobulin (IVIg) in OM and do not recommend elective thymectomy.

Rare patients with "fixed" (unchanged for several years) tropias may benefit from corrective strabismus surgery *(79)*. Although lid surgery to improve chronic ptosis is feasible *(80)*, we agree with Sergott *(81)* and do not recommend it. However, blepharoplasty to correct dermatochalasia (redundant, sagging periocular skin) coexisting with MG ptosis may increase the size of the palpebral fissure. Bentley et al. *(79)* used botulinum toxin in a few patients with a fixed tropia to weaken the antagonist of the paretic muscle, thereby straightening the eye. Given the generalized nature of neuromuscular dysfunction in OM and the distant effect of locally injected botulinum *(82, 83)*, the risk seems too great.

REFERENCES

1. Perlo VP, Poskanzer DC, Schwab RS, Viets HR, Osserman KE, Genkins G. Myasthenia gravis: evaluation of treatment in 1,355 patients. Neurology 1966;16:431–439.
2. Ferguson FR, Hutchinson EC, Liversedge LA. Myasthenia gravis; results of medical management. Lancet 1955;269:636–639.
3. Bever CT, Aquino AV, Penn AS, Lovelace RE, Rowland LP. Prognosis of ocular myasthenia. Ann Neurol 1983;14:516–519.

4. Evoli A, Tonali P, Bartoccioni E, Lo Monaco M. Ocular myasthenia: diagnostic and therapeutic problems. Acta Neurol Scand 1988;77:31–35.

5. Simpson JF, Westerberg MR, Magee KR. Myasthenia gravis. An analysis of 295 cases. Acta Neurol Scand 1966;42:Suppl 23 21–27.

6. Grob D, Arsura EL, Brunner NG, Namba T. The course of myasthenia gravis and therapies affecting outcome. Ann N Y Acad Sci 1987;505:472–499.

7. Christensen PB, Jensen TS, Tsiropoulos I, et al. Incidence and prevalence of myasthenia gravis in western Denmark: 1975 to 1989. Neurology 1993;43:1779–1783.

8. Casetta I, Fallica E, Govoni V, Azzini C, Tola M, Granieri E. Incidence of myasthenia gravis in the province of Ferrara: a community-based study. Neuroepidemiology 2004;23:281–284.

9. Lavrnic D, Jarebinski M, Rakocevic-Stojanovic V, et al. Epidemiological and clinical characteristics of myasthenia gravis in Belgrade, Yugoslavia (1983–1992). Acta Neurol Scand 1999;100:168–174.

10. Robertson N, Deans J, Compston D. Myasthenia gravis: a population based epidemiological study in Cambridge-shire, England. J Neurol Neurosurg Psychiatry 1998;65:492–496.

11. Phillips LH, Torner JC, Anderson MS, Cox GM. The epidemiology of myasthenia gravis in central and western Virginia. Neurology 1992;42:1888–1893.

12. Anonymous. Incidence of myasthenia gravis in the Emilia-Romagna region: a prospective multicenter study. Emilia-Romagna Study Group on Clinical and Epidemiological Problems in Neurology. Neurology 1998;51:255–258.

13. Aiello I, Pastorino M, Sotgiu S, et al. Epidemiology of myasthenia gravis in northwestern Sardinia. Neuroepidemiology 1997;16:199–206.

14. Fenichel G. In: Clinical Pediatric Neurology, A Signs and Symptoms Approach. Philadelphia: Elsevier Saunders; 2005;305–307.

15. Yu Wai Man CY, Chinnery PF, Griffiths PG. Extraocular muscles have fundamentally distinct properties that make them selectively vulnerable to certain disorders. Neuromuscul Disord 2005;15:17–23.

16. Spencer RF, Porter JD. Structural organization of the extraocular muscles. Rev Oculomot Res 1988;2:33–79.

17. Porter JD, Khanna S, Kaminski HJ, et al. Extraocular muscle is defined by a fundamentally distinct gene expression profile. Proc Natl Acad Sci U S A 2001;98:12062–12067.

18. Marino M, Barbesino G, Pinchera A, et al. Increased frequency of euthyroid ophthalmopathy in patients with Graves' disease associated with myasthenia gravis. Thyroid 2000;10:799–802.

19. Kaminski HJ, Li Z, Richmonds C, Ruff RL, Kusner L. Susceptibility of ocular tissues to autoimmune diseases. Ann N Y Acad Sci 2003;998:362–374.

20. Leigh R, Zee D. The Neurology of Eye Movements. In. Fourth ed. New York: Oxford University Press; 2006;439–444.

21. Loewenfeld I. The Pupil: Anatomy, Physiology and Clinical Applications. Boston: Butterworth-Heinemann; 1999.

22. Calvert P. Disorder of Neuromuscular Transmission. In: NR M, N N, Biousse V, Kerrison J, eds. Walsh & Hoyt's Clinical Neuro-Ophthalmology. Sixth ed. Philadelphia: Lippincott Williams & Wilkins; 2005;1041–1084.

23. Digre K. Principles and Techniques of Examination of the Pupils, Accommodation, and Lacrimation. In: Miller N, Newman N, Biousse V, Kerrison J, eds. Walsh & Hoyt's Clinical Neuro-Opthalmology. Philadelphia: Lippincott Williams & Wilkins; 2005;715–805.

24. Glaser J, Siatkowski. Infranuclear disorders of eye movement. In: Glaser J, ed. Neuroophthalmology. Third ed. Philadelphia: Lippincott Williams & Wilkins; 1999;405–409.

25. Verhagen WI, Venderink D. Ptosis due to metastasis in the levator palpebrae muscle as first symptom of an adenocarcinoma of the lung. Neuroophthalmology 2001;26:197–199.

26. Liesegang TJ. Amyloid infiltration of the levator palpebrae superioris muscle: case report. Ann Ophthalmol 1983;15:610–613.

27. Ertas FS, Ertas NM, Gulec S, et al. Unrecognized side effect of statin treatment: unilateral blepharoptosis. Ophthal Plast Reconstr Surg 2006;22:222–224.

28. Daroff R. Ocular myasthenia: diagnosis and therapy. In: Glaser J, ed. Neuro-Ophthalmology. St Louis: Mosby; 1980;62–71.

29. Ragge NK, Hoyt WF. Midbrain myasthenia: fatigable ptosis, 'lid twitch' sign, and ophthalmoparesis from a dorsal midbrain glioma. Neurology 1992;42:917–919.

30. Kao YF, Lan MY, Chou MS, Chen WH. Intracranial fatigable ptosis. J Neuroophthalmol 1999;19:257–259.

31. Bhatt A, Haviland J, Black E, Stavern G. Sensitivity and specificity of the Cogan's lid twitch for myasthenia gravis in a prospective series of patients with symptomatic ptosis. Neurology 2005;64:A36.

32. Averbuch-Heller L, Poonyathalang A, von Maydell RD, Remler BF. Hering's law for eyelids: still valid. Neurology 1995;45:1781–1783.

33. Benatar M. A systematic review of diagnostic studies in myasthenia gravis. Neuromuscul Disord 2006;16:459–467.

34. Kubis KC, Danesh-Meyer HV, Savino PJ, Sergott RC. The ice test versus the rest test in myasthenia gravis. Ophthalmology 2000;107:1995–1998.

35. Odel JG, Winterkorn JMS, Behrens MM. The sleep test for myasthenia gravis. A safe alternative to tensilon. J Clin Neuro Ophthalmol 1991;11:288–292.

36. Saavedra J, Femminini R, Kochen S, deZarate JC. A cold test for myasthenia gravis. Neurology 1979;29:1075.

37. Borenstein S, Desmedt JE. Local cooling in myasthenia. Improvement of neuromuscular failure. Arch Neurol 1975;32:152–157.

38. Ellis FD, Hoyt CS, Ellis FJ, Jeffery AR, Sondhi N. Extraocular muscle responses to orbital cooling (ice test) for ocular myasthenia gravis diagnosis. J AAPOS 2000;4:271–281.

39. Ertas M, Arac N, Kumral K, Tuncbay T. Ice test as a simple diagnostic aid for myasthenia gravis. Acta Neurol Scand 1994;89:227–229.

40. Lertchavanakul A, Gamnerdsiri P, Hirunwiwatkul P. Ice test for ocular myasthenia gravis. J Med Assoc Thai 2001;84 Suppl 1:S131–136.

41. Daroff RB. The office Tensilon test for ocular myasthenia gravis. Arch Neurol 1986;43:843–844.

42. Seybold ME. The office Tensilon test for ocular myasthenia gravis. Arch Neurol 1986;43:842–843.

43. Kupersmith MJ, Latkany R, Homel P. Development of generalized disease at 2 years in patients with ocular myasthenia gravis. Arch Neurol 2003;60:243–248.

44. Barton JJ, Fouladvand M. Ocular aspects of myasthenia gravis. Semin Neurol 2000;20:7–20.

45. Tian J, Yang Q, Wei M, Terdiman J, Sun F. Neostigmine-induced alterations in fast phase of optokinetic responses in myasthenic ocular palsies. J Neurol 2002;249:867–874.

46. Toth L, Toth A, Dioszeghy P, Repassy G. Electronystagmographic analysis of optokinetic nystagmus for the evaluation of ocular symptoms in myasthenia gravis. Acta Otolaryngol 1999;119:629–632.

47. Ing EB, Ing SY, Ing T, Ramocki JA. The complication rate of edrophonium testing for suspected myasthenia gravis. Can J Ophthalmol 2000;35:141–144; discussion 145.

48. Nicholson GA, McLeod JG, Griffiths LR. Comparison of diagnostic tests in myasthenia gravis. Clin Exp Neurol 1983;19:45–49.

49. Howard FM, Lennon VA, Finley J, Matsumoto J, Elveback LR. Clinical correlations of antibodies that bind, block or modulate human acetylcholine receptors in myasthenia gravis. Ann N Y Acad Sci 1987;505:526–538.

50. Lefvert A, Bergstrom K, Matell G, Osterman P, Pirskanen R. Determination of acetylcholine receptor antibody in myasthenia gravis: clinical usefulness and pathogenetic implications. J Neurol, Neurosurgery and Psychiatry 1978;41:394–403.

51. Limburg P, The T, Hummel-Tappel E, Oosterhuis H. Anti-acetylcholine receptor antibodies in myasthenia gravis. Part 1: relation to clinical parameters in 250 patients. J Neurol Sci 1983;58:357–370.

52. Padua L, Stalberg E, LoMonaco M, Evoli A, Batocchi A, Tonali P. SFEMG in ocular myasthenia gravis diagnosis. Clin Neurophysiol 2000;111:1203–1207.

53. Costa J, Evangelista T, Conceicao I, de Carvalho M. Repetitive nerve stimulation in myasthenia gravis–relative sensitivity of different muscles. Clin Neurophysiol 2004;115:2776–2782.

54. Bau V, Hanisch F, Hain B, Zierz S. [Ocular involvement in MuSK antibody-positive myasthenia gravis]. Klin Monatsbl Augenheilkd 2006;223:81–83.

55. Bennett DL, Mills KR, Riordan-Eva P, Barnes PR, Rose MR. Anti-MuSK antibodies in a case of ocular myasthenia gravis. J Neurol Neurosurg Psychiatry 2006;77:564–565.

56. Kennett RP, Fawcett PRW. Repetitive nerve stimulation of anconeus in the assessment of neuromuscular transmission disorders. Electroencephalogr Clin Neurophysiol: Electromyogr Motor Control 1993;89:170–176.

57. Zinman LH, O'Connor PW, Dadson KE, Leung RC, Ngo M, Bril V. Sensitivity of repetitive facial-nerve stimulation in patients with myasthenia gravis. Muscle Nerve 2006;33:694–696.

58. Oey PL, Wieneke GH, Hoogenraad TU, van Huffelen AC. Ocular myasthenia gravis: the diagnostic yield of repetitive nerve stimulation and stimulated single fiber EMG of orbicularis oculi muscle and infrared reflection oculography. Muscle Nerve 1993;16:142–149.

59. Rouseev R, Ashby P, Basinski A, Sharpe J. Single fiber EMG in the frontalis muscle in ocular myasthenia: specificity and sensitivity. Muscle Nerve 1992;15:399–403.

60. Ukachoke C, Ashby P, Basinski A, Sharpe J. Usefulness of single fiber EMG for distinguishing neuromuscular weakness from other causes of oculr muscle weakness. Can J Neurol Sci 1994;21:125–128.

61. Benatar M, Hammad M, Doss-Riney H. Concentric-needle single-fiber electromyography for the diagnosis of myasthenia gravis. Muscle Nerve 2006.

62. Weizer JS, Lee AG, Coats DK. Myasthenia gravis with ocular involvement in older patients. Can J Ophthalmol 2001;36:26–33.

63. Mullaney P, Vajsar J, Smith R, Buncic JR. The natural history and ophthalmic involvement in childhood myasthenia gravis at the hospital for sick children. Ophthalmology 2000;107:504–510.

64. Weinberg DH, Rizzo JF, 3rd, Hayes MT, Kneeland MD, Kelly JJ, Jr. Ocular myasthenia gravis: predictive value of single-fiber electromyography. Muscle Nerve 1999;22:1222–1227.

65. Kim JH, Hwang JM, Hwang YS, Kim KJ, Chae J. Childhood ocular myasthenia gravis. Ophthalmology 2003;110:1458–1462.

66. Badrising U, Brandenburg H, van Hilten J, Brietl P, Wintzen A. Intranasal neostigmine as add-on therapy in myasthenia gravis. J Neurol 1996;243:S59.

67. Mount F. Corticotropin in Treatment of Ocular Myasthenia – A Controlled Clinical Trial. Arch Neurol 1964;11:114–124.
68. Benatar M, Kaminski H. Medical and surgical treatments for ocular myasthenia. Cochrane Database Syst Rev 2006.
69. Papatestas AE, Genkins G, Kornfeld P, et al. Effects of thymectomy in myasthenia gravis. Ann Surg 1987;206:79–88.
70. Kawaguchi N, Kuwabara S, Nemoto Y, et al. Treatment and outcome of myasthenia gravis: Retrospective multi-center analysis of 470 Japanese patients, 1999–2000. J Neurol Sci 2004;224:43–47.
71. Mee J, Paine M, Byrne E, King J, Reardon K, O'Day J. Immunotherapy of ocular myasthenia gravis reduces conversion to generalized myasthenia gravis. J Neuroophthalmol 2003;23:251–255.
72. Monsul NT, Patwa HS, Knorr AM, Lesser RL, Goldstein JM. The effect of prednisone on the progression from ocular to generalized myasthenia gravis. J Neurol Sci 2004;217:131–133.
73. Papapetropoulos TH, Ellul J, Tsibri E. Development of generalized myasthenia gravis in patients with ocular myasthenia gravis. Arch Neurol 2003;60:1491–1492.
74. Sommer N, Sigg B, Melms A, et al. Ocular myasthenia gravis: response to long-term immunosuppressant treatment. J Neurol, Neurosurg Psychiatry 1997;62:156–162.
75. Kaminski H, Daroff R. Treatment of ocular myasthenia: steroids only when compelled. Arch Neurol 2000;57:752–753.
76. Lloyd ME, Spector TD, Howard R. Osteoporosis in neurological disorders. J Neurol Neurosurg Psychiatry 2000;68:543–547.
77. Smith GD, Stevens DL, Fuller GN. Myasthenia gravis, corticosteroids and osteoporosis prophylaxis. J Neurol 2001;248:151.
78. Beekman R, Kuks JB, Oosterhuis HJ. Myasthenia gravis: diagnosis and follow-up of 100 consecutive patients. J Neurol 1997;244:112–118.
79. Bentley CR, Dawson E, Lee JP. Active management in patients with ocular manifestations of myasthenia gravis. Eye 2001;15:18–22.
80. Bradley EA, Bartley GB, Chapman KL, Waller RR. Surgical correction of blepharoptosis in patients with myasthenia gravis. Ophthal Plast Reconstr Surg 2001;17:103–110.
81. Sergott RC. Ocular myasthenia. In: Lisak R, ed. Handbook of Myasthenia Gravis and Myasthenic Syndromes. New York: Marcel Dekker; 1994;21–31.
82. Lange DJ, Rubin M, Greene PE, et al. Distant effects of locally injected botulinum toxin: a double-blind study of single fiber EMG changes. Muscle Nerve 1991;14:672–675.
83. Sanders DB, Massey EW, Buckley EG. Botulinum toxin for blepharospasm: single-fiber EMG studies. Neurology 1986;36:545–547.

7

Thymoma-Associated Paraneoplastic Myasthenia Gravis

Philipp Ströbel, Wen-Yu Chuang, and Alexander Marx

1. INTRODUCTION

Since publication of the first edition of this book in 2003, two major new insights into the pathogenesis of thymoma-associated (paraneoplastic) autoimmunities, including myasthenia gravis (MG), have been achieved. First, it has become clear by the analysis of functional single nucleotide polymorphisms that tumor-independent, non-MHC genetic factors contribute to the pathogenesis of paraneoplastic MG at least in a subgroup of patients. Second, thymoma has been identified as a human model of "localized", i.e., tumor-restricted loss of AIRE ("autoimmune regulator") expression. AIRE is a transcriptional regulator that controls immunologic tolerance by the pivotal regulation of "promiscuous" expression of tissue-specific self-antigens in the thymic medulla. The finding of AIRE deficiency in thymoma relates to earlier observations *(1)* that autoantibodies against type I interferons are a conceptionally and diagnostically important hallmark of both thymoma and a germ line-encoded AIRE deficiency syndrome termed **auto-immune polyendocrinopathy syndrome type I (APS I). Of note, anti-IFN autoantibodies and APS I symptoms do not occur in AIRE knockout animals. Therefore, the new findings in thymoma patients have implications beyond paraneoplastic MG and underline the importance of MG(+) compared to MG(−) thymomas as a model system allowing for the analysis of human-specific mechanisms leading to autoimmunity.

MG is a heterogeneous autoimmune disease mediated by autoantibodies that interfere with the function of the neuromuscular junction. Taking the presence or absence of antibodies against the acetylcholine receptor (AChR) as a criterion, MG has been subdivided into seropositive and seronegative types. In seropositive MG, the pathogenic, high-affinity autoantibodies are directed against the nicotinic acetylcholine receptor (AChR) *(2, 3)*, while so-called seronegative MG (SNMG) is now known to be heterogeneous with respect to autoantibody features: depending on ethnic and geographic backgrounds *(4, 5, 6, 7)*, a quite variable percentage of SNMG cases result from autoantibodies (commonly IgG4) to muscle-specific tyrosine kinases (MuSK) at the neuromuscular endplate *(8)*. Furthermore, a third "double-negative" subgroup showing neither anti-AChR nor anti-MuSK autoantibodies exists *(9)*. Whether these patients have autoantibodies against other autoantigen targets or whether their anti-AChR or anti-MuSK autoantibodies cannot be detected by current laboratory tests is the subject of ongoing research.

In 80−90% of cases, MG is associated with pathologic changes of the thymus. In about 10% of anti-AChR-seropositive cases, MG is a paraneoplastic phenomenon caused by a subgroup of epithelial neoplasms of the thymus called thymomas. There are significant associations between

PS and AM are supported by European Union FP6 grants LSHB-CT-2003-503410 (Euro-Thymaide) and 2005105 (EuroMyasthenie), the NIH (International Thymectomy Trial, MGTX) grant MG31506 and grant 10-1740 of the Deutsche Krebshilfe.

From: *Current Clinical Neurology*
Myasthenia Gravis and Related Disorders
Edited by: H.J. Kaminski, DOI 10.1007/978-1-59745-156-7_7, © Humana Press, New York, NY

Table 1
MG Subtypes According to Thymus Pathology. Correlation with Clinical, Epidemiological and Genetic Findings

	EOMG/TFH	Thymoma	LOMG/A
Onset of symptoms age (years)	10–40	15–80	>40
Sex M : F	1 : 3	1 : 1	2 : 1
HLA association	B8; DR3	(DR2, A24)	B7; DR2
Myoid cells	Present	Absent	Present
Intrathymic autoantibody production	Present	Absent	Absent (?)
AIRE expression	Normal	Absent	Normal
TNFA*T1/B*2 homozygosity	Rare	Very frequent	Frequent
TNFA*T2; TNFB1, TNFB*1, C4A*QO, C4B*1, DRB1*03	Frequent	Rare	Rare
CTLA4 +49A/G genotype distribution	Similar to healthy controls	+49A/A associated with pMG	Not available
Autoantibodies against			
AChR (%)	80	>95	90
Striated muscle (%)	10–20	>90	30–60
Titin (%)	<10	>90	30–40
Ryanodine receptor (%)	<5	50–60	20
IL-12, IFN-α, IFN-ω (%)	Infrequent	63–88	Infrequent

EOMG/TFH: early-onset myasthenia gravis/thymic follicular hyperplasia; LOMG/A: late-onset myasthenia gravis/"atrophy"; pMG: paraneoplastic MG.

different thymic alterations and clinicoepidemiological findings *(10, 11, 12, 13)* (Table 1). To highlight peculiarities in thymoma-associated MG, we will shortly review MG with thymic lympho-follicular hyperplasia (TFH) and so-called thymic atrophy as well.

2. HISTOPATHOLOGY OF THE THYMUS IN MG

2.1. *Thymic Lympho-follicular Hyperplasia (TFH)*

TFH is observed in 100% of early-onset anti-AChR-seropositive (EOMG) patients and in most cases of AChR- and MuSK-"double-negative" MG, while it is generally not found in patients with anti-MuSK antibodies *(9, 14, 15)*. TFH is characterized by lymphoid follicles with or without germinal centers that extend around perivascular spaces. Marked TFH is a common finding in MG *(16)*. It may also occur in other autoimmune diseases *(17)* but is rare in healthy persons *(18)*. The hallmark of TFH is the disruption of the basal membrane around perivascular spaces (PVSs) by lymphoid follicles *(19)* resulting in a fusion of the medulla (i.e., a part of the thymus parenchyma) and PVSs (thought to belong to the peripheral (extrathymic) immune system) *(20, 21)*. The inflammatory changes seen in TFH are thought to be directed against a peculiar thymic cell population called myoid cells, as suggested by deposits of complement and autoantibodies on the cell surface of myoid cells in EOMG (Leite, in preparation). Myoid cells are unique non-innervated myoblast-like muscle cells that occur in the thymic medulla outside lymphoid follicles. In the normal thymus and in TFH they are MHC class II negative *(22)* and most probably express both fetal and adult-type AChR *(20, 23, 24, 25)*. Myoid cells in TFH are frequently located in intimate apposition to dendritic cells *(20)* and show increased proliferation and turnover *(26)*. The thymic cortex in TFH shows a normal, age-dependent morphology.

What triggers TFH development in EOMG is not known. However, a contribution of epithelial cell-derived chemokines, particularly CXCL13, has been proposed *(27)*. Moreover, a genetic contribution is likely due to several genetic polymorphisms known to be associated with EOMG,

Table 2
Paraneoplastic Diseases of Presumed Autoimmune Pathogenesis Reported to Be Associated with Thymoma *(43, 79, 101, 102, 122, 123).* **In the majority of these diseases, the immunopathogenesis has not been resolved**

Addison's disease	Neuromyotonia
Agranulocytosis	Panhypopituitarism
Alopecia areata	Pernicious anemia
Aplastic anemia	Polymyositis
Autoimmune colitis (GvHD like)	Pure red cell aplasia
Autoimmune autonomic neuropathy	Rheumatoid arthritis
Autoimmune gastrointestinal dysmotility (intestinal pseudo-obstruction)	Rippling muscle disease
Cushing syndrome	Sarcoidosis
Encephalitis (limbic and/or cortical)	Scleroderma
Hemolytic anemia	Sensory motor neuropathy
Hypogammaglobulinemia (Good syndrome)	Stiff person syndrome
Myasthenia gravis	Systemic lupus erythematosus
Myocarditis	Thyroiditis

such as MHC class I and II *(28, 29)*, IL-1β *(30)*, TNFα *(31, 32, 33)*, IL-10 *(34)* and the AChR alpha- but not beta-subunit *(35, 36, 37)* (Table 1). It is unknown whether hyperexpression of bcl-2 in thymic germinal centers *(38)* or of FAS in mature thymic T cells *(39, 40)* is involved in the etiology of TFH.

2.2. Thymus Histology in Late-Onset MG (LOMG)

For unknown reasons, the incidence of MG diagnosed in persons over 40 years of age has been increasing during the last decades and accounts now for about 60% of all cases diagnosed per year *(41)*. Patients with LOMG are virtually all anti-AChR seropositive. Thymus histology in these patients has traditionally been termed "thymic atrophy". However, morphometric studies have shown that the size of the thymus is essentially normal when compared to age-matched healthy populations, and the number of myoid cells per area of thymic parenchyma is unchanged *(20, 42)*.

2.3. Thymomas and Paraneoplastic MG

Thymomas occur in about 10–15% of MG patients. Although there are rare reports on MG in other tumors *(43)*, a direct pathogenic role is undisputed only for thymomas *(10, 12, 16, 44, 45, 46, 47, 48)*. Although MG is by far the most frequent autoimmune manifestation of thymomas, they can be associated with a broad range of other autoimmune disorders as well (Table 2). The current WHO classification *(49)* distinguishes between thymomas, i.e., thymic epithelial neo-plasms with unique thymus-like organoid features, and thymic carcinomas that resemble malignant tumors elsewhere in the body. The most important thymus-like organoid feature is the capacity to promote intratumorous T-cell development (see below). Importantly, only thymomas but not thymic carcinomas are associated with MG *(43, 50, 51)*. Typical histological findings in MG-associated type A (medullary), AB (mixed), B1 (organoid), B2 (cortical) and B3 thymomas are shown in (Fig. 1 and Color Plate 2, following p. x).

3. PATHOGENETIC CONCEPTS IN SEROPOSITIVE MG

3.1. Pathogenesis of MG in Thymic Lympho-follicular Hyperplasia (TFH)

It is generally accepted that TFH-associated MG results from an intrathymic pathogenesis with acetylcholine receptors (AChR) on thymic myoid cells being primarily involved as (triggering) autoantigens *(52)*. This concept is supported by the following findings:

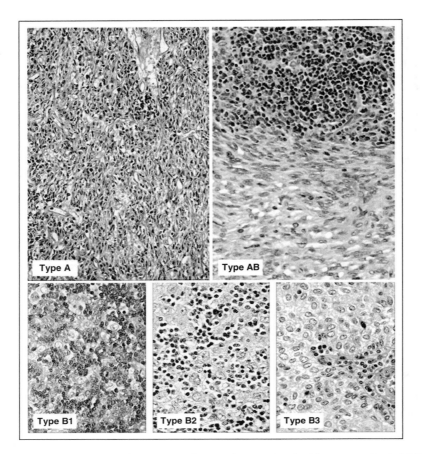

Fig. 1. Morphological spectrum of thymomas that can be associated with paraneoplastic MG. WHO type A (medullary) thymoma with prominent spindle cell features and scarce intermingled lymphocytes, which are mostly mature on flow cytometry. WHO type AB (mixed) thymoma with features of both WHO type A and B (cortical) thymomas. In this case, both areas are clearly separated, whereas in other tumors, the type A and B areas may be indistinguishably interwoven. High numbers of immature T cells, lack of cytologic atypia of the thymic epithelium and presence of numerous Hassall's corpuscles are diagnostic features of type B1 thymomas. Type B2 tumors show somewhat reduced numbers of T cells with marked atypia of the thymic epithelium; occasional Hassall's corpuscles may occur. The picture in type B3 thymomas is clearly dominated by the epithelium with formation of characteristic perivascular spaces and perivascular palisading. Intermingled lymphocytes are mostly of immature phenotype (hematoxylin–eosin). (*see* Color Plate 2, following p. x).

(1) A substantial proportion of autoantibodies in TFH/MG patients recognize the fetal AChR (i.e., AChR with a γ- instead of an ε-subunit) *(53)* that is only expressed on thymic myoid cells and on a few extraocular muscle fibers *(25, 54, 55, 56)*.

(2) Thymus with TFH is the organ where most anti-AChR autoantibodies are produced *(57)*, as suggested by clinical improvement of MG after thymectomy *(58, 59)*.

(3) Extrathymic immunization with AChR can induce experimental autoimmune MG (EAMG) but does not elicit TFH *(60)*.

(4) Transplantation of TFH-affected thymus into SCID mice results in prolonged production of anti-AChR autoantibodies *(61, 62, 63)*.

(5) Abnormal clusters between myoid cells and dendritic cells occur in TFH *(20)*. Since myoid cells remain largely negative for MHC class II in MG they may stimulate pre-activated *(22)* but not naïve AChR-reactive CD4 + T cells *(64)*. Therefore, dendritic cells may take up AChR from destroyed myoid cells

and present AChR peptides to potentially autoreactive T cells that occur as non-tolerized T cells in the normal T-cell repertoire *(65, 66, 67)* and are increased in TFH *(44, 68)*. If activated, such autoreactive T cells may provide help to B cells for autoantibody production inside and outside the thymus *(69, 70)*. Due to increased turnover and apoptosis of myoid cells in TFH *(26)*, the release of AChRs from myoid cells and presentation of AChR peptides by dendritic cells inside the thymus might promote AChR epitope spreading in TFH-associated MG *(22)*. Recent data suggest that not only the adaptive but also the innate immune system might contribute to this breakdown of central tolerance via upregulation of toll-like receptors on thymic epithelial cells *(71)*.

3.2. Pathogenesis of Thymoma-Associated (Paraneoplastic) MG

As summarized in Table 1, there is strong evidence that the pathogenesis of thymoma-associated MG differs from TFH-associated MG *(12, 14, 43, 72)*.

There are important morphological and functional differences between TFH and thymomas:

(1) While the thymic medulla is distorted but principally maintained in TFH, well-defined medullary areas are commonly absent or much smaller in thymomas.
(2) AChR-expressing myoid cells are numerous in the thymic medulla in TFH but absent in thymomas.
(3) MHC class II levels on thymic epithelial cells appear similar in TFH and the normal thymus but are consistently reduced in thymomas *(73, 74, 75, 76)*.
(4) In contrast to prominent intrathymic anti-AChR autoantibody production in TFH, there is no such production in thymomas *(77)*, maybe with very few exceptions *(78)*.
(5) The autoimmune regulator AIRE is always expressed in the thymus and in TFH but absent in the vast majority of thymomas *(79)*. AIRE is expressed most strongly in a small subset of medullary thymic epithelial cells (mTECs) *(80)*, and inactivating *AIRE* mutations are responsible for the autoimmune syndrome APS I or APECED (**a**utoimmune **p**oly**e**ndo**c**rinopathy, **c**andidiasis, **e**ctodermal **d**ystrophy) *(80)*. AIRE is a strong transcriptional activator *(81)* and regulates the "promiscuous" expression of some peripheral tissue antigens (e.g., insulin) in mTECs. Furthermore, AIRE enhances antigen presentation in the thymus *(82)* and influences both expression of co-stimulatory molecules on dendritic cells and their interactions with mature T cells in the periphery *(83)*. In the mouse thymus, the number of AIRE$^+$ mTECs correlates with the presence of thymocytes undergoing negative selection *(84)*, and thymic deletion of autoreactive T cells is apparently *AIRE* gene dose dependent *(85)*.

Unexpectedly, however, in spite of the general absence of AIRE in thymomas, the spectrum of autoantibodies and clinical autoimmune disorders in MG patients with thymomas and TFH is very similar *(79)*, with the important exception of striational (e.g., anti-titin) antibodies and antibodies against IFN-α and IFN-ω, which have not been observed in patients with EOMG/TFH *(1, 79)*. Thus, while autoantibodies against IFN-α and IFN-ω may be a surrogate marker for central (i.e., thymic) AIRE deficiency, these findings argue strongly against a major role for AIRE in the pathogenesis of most thymoma-associated autoimmune disorders including MG.

3.2.1. Genetic Features Contributing to Paraneoplastic MG

In stark contrast to MG in TFH, thymoma-associated MG does not exhibit a clear HLA association (Table 1). Nevertheless, recent findings suggest that both tumor-associated features and the genetic background of a given thymoma patient determine in concert whether there is development of MG. As a paradigm, a germ line single nucleotide polymorphism (SNP) in exon 1 of the T lymphocyte-associated antigen 4 (CTLA4) gene has been shown to exert a prominent predisposing effect to paraneoplastic MG in thymoma patients *(86)*. This finding was surprising since the respective genotype, +49A/A, confers high CTLA4 levels and has been found to be protective against several other autoimmune diseases. CTLA4 is a critical negative regulator of T-cell activation and competitively interferes with the binding of CD28 to B7-1 and B7-2 on antigen-presenting cells *(87)*. The seemingly paradoxical unusual association of paraneoplastic but not other types of MG with a CTLA4^{high} phenotype implies a unique pathogenesis of thymoma-associated MG *(86)*. Specifically, high CTLA4 levels are thought to support the non-tolerogenic selection of CD4 T cells in MG-associated thymomas *(86)* (see below).

3.2.2. Molecular and Functional Features Contributing to Paraneoplastic MG

As mentioned above, the maintenance of thymopoietic activity is now considered a major prerequisite for MG development in the majority of thymomas (maybe not in type A thymomas that often appear to be thymopoietically inactive). Loss of heterozygosity (LOH) of the MHC locus at chromosome 6p21.3 in the neoplastic thymic epithelium appears to be more frequent among MG(+) than MG(−) thymomas. This LOH implies an MHC chimerism between the hemizygous thymoma epithelium (presumed to perform T-cell selection) and MHC heterozygous intratumorous dendritic cells and the peripheral immune system. However, LOH of the MHC locus is observed in only about 50% of MG-associated thymomas, and MHC chimerism might be just one among other mechanisms of non-tolerogenic T-cell selection in a subset of cases *(43, 88)*. Indeed, the consistently reduced levels of MHC class II proteins on thymoma epithelial cells might per se be a feature with autoimmunizing potential *(86)*.

3.2.3. Shared Features Among MG-Associated Thymomas

Although the various histological thymoma subtypes show considerable morphological and functional differences, MG-associated thymomas share the following features:

(1) All MG-associated thymomas to some extent share morphological and functional features with the normal thymus. In particular, they provide signals for the homing of immature hematopoietic precursors and promote their differentiation to apparently mature T cells *(89, 90, 91)*. Intratumorous maturation of mature, naïve CD4+ T cells appears to be of critical importance in >90% of MG(+) thymomas (types AB, B1, B2 and B3) since presence of MG is tightly linked to the generation and subsequent export of high numbers of CD4+CD45RA+-naïve T cells *(92)*, while this subset is reduced in MG(−) thymomas. Mature T cells generated by the thymoma can persist in the periphery for several years after removal of the tumor *(93)*. The number of these T cells increases substantially if the tumor elapses *(93, 94)*.

(2) MG-associated thymomas are enriched for autoreactive T cells with specificity for the AChR α- and ε-subunit *(47, 91, 95, 96, 97)*. There is strong evidence that part of them are generated by intratumorous, non-tolerogenic thymopoiesis *(47, 97)*.

(3) The production of autoreactive T cells in thymomas and their activation outside the thymoma may partly reflect inefficient generation of CD4+CD25+FOXP3+ regulatory T cells that are significantly reduced in thymomas *(98, 99)*.

(4) Concurrent autoimmunity against four apparently unrelated types of autoantigens is highly characteristic of paraneoplastic MG. These autoantigens are (a) the AChR *(48)*, (b) striational muscle antigens, including titin *(100)*, (c) neuronal antigens *(101, 102, 103)* and (d) cytokines IL-12, IFN-α and IFN-ω (see above). Autoimmunity to the ryanodine receptor is also highly characteristic but less frequent *(104, 105)*.

3.3. A Pathogenetic Model of Paraneoplastic Myasthenia Gravis

Taking the above-described features into account, the following model for the development of paraneoplastic MG can be envisaged: in the vast majority (>95%) of thymomas, paraneoplastic MG starts with abnormal, non-tolerogenic T-cell selection inside the thymoma. This step is followed by export of intolerant T cells to the extratumorous immune system, i.e., to lymph nodes, normal thymus, spleen and, maybe, bone marrow. At these sites, T-cell activation, interaction of dendritic cells, T cells and B cells and, finally, the production of autoantibodies occur.

In detail, in the context of low epithelial MHC class II expression (and thus weak T-cell receptor:MHC interactions), a CTLA4^{high} phenotype would favor T-cell survival and tip the balance towards development of MG, while a CTLA4^{low} phenotype would promote negative selection through stronger activation of maturing T cells, resulting in prevention of MG. Furthermore, it has been shown in mice that tolerance induction is influenced by the quantity and quality of MHC/peptide complexes on thymic epithelial cells *(106, 107, 108, 109)*. The MHC level is particularly important when medullary structures (performing negative selection under physiological conditions) are reduced *(110, 111)*. The latter observations might have a bearing for paraneoplastic MG since a reduction of medullary structures is typical for thymomas *(50, 112)*. Epithelial hyperexpression of endogenous proteins together with disturbed T-cell selection might favor the generation of thymoma-derived autoreactive T cells *(44)*. The process of selection and activation of autoreactive, AChR-reactive T cells *(97)* in thymoma patients might not be a random process, as suggested by the observation that these T cells are restricted to the minority HLA isotypes DP14 and DR52a that are infrequent in MG patients without thymoma *(47)*. MG-associated thymomas export mature, naïve CD4+ T cells *(93, 94)* that may not be checked by adequate numbers of regulatory T cells of appropriate specificities *(98)*. Export of naïve T cells might gradually replace the normally tolerant, thymus-derived T-cell repertoire by a "chimeric" autoimmunity-prone T-cell repertoire derived from both the thymoma and the non-neoplastic residual thymus. In accordance with this model, MG can rarely develop after surgical removal of the thymoma *(113)*. In order to be pathogenically relevant, the potentially autoantigen-reactive T cells have to become activated to provide help for autoantibody-producing B cells outside the thymoma *(46)* (Fig. 2). At this stage, it can be hypothesized that anti-IL-12 or anti-IFN-α autoantibodies *(114)* facilitate the CD4 T-cell-dependent production of anti-AChR autoantibodies.

3.4. Etiological Triggers of Paraneoplastic MG

The mechanisms that activate the autoimmunity-prone T-cell repertoire, i.e., the etiologies triggering the MG-provoking autoimmune cascade, remain yet to be defined. We have observed

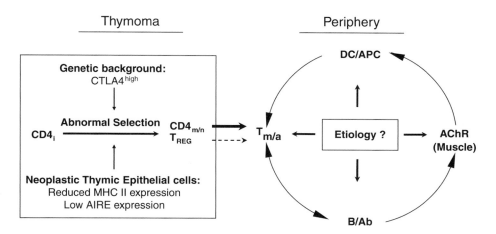

Fig. 2. Simplistic unifying model for the pathogenesis of paraneoplastic MG. MG-associated thymomas promote T-cell maturation from immature CD4+ precursors (CD4$_i$) to fully mature, naïve CD4 T cells (CD4$_{m/n}$). The genetic background, particularly a CTLA4^{high} phenotype in the context of low expression of MHC class II molecules on the tumor epithelium, may favor abnormal T-cell selection with survival of potentially autoreactive T cells. After export to the "periphery" (residual thymus, lymph nodes, bone marrow and eventually inflamed muscle) and after an yet unknown trigger, naïve T cells are activated (T$_{m/a}$) by antigen-presenting dendritic cells (DC/APC) that provide help to autoantibody (Ab)-producing B cells. Whether the AChR in skeletal muscle or thymic myoid cells is the primary triggering autoantigen or becomes secondarily involved is unknown.

paraneoplastic MG following pregnancy, unspecific infectious or traumatic "stimuli", but in most cases no major event can be identified *(43)*. The "periphery" outside the thymoma where emigrant T cells become activated can clearly be the residual thymus *(115)* that is frequently enriched in autoreactive T cells and lymphoid follicles, resulting in a TFH-like morphology in many MG-associated thymoma cases (unpublished). However, other lymphoid organs have to play a role in this process, given that even complete surgical removal of thymoma plus residual thymus is often not followed by a decline of autoantibody titers *(116)*.

3.5. Major Unresolved Questions in Paraneoplastic MG

The above model still awaits reconciliation with the following open points:

(1) It has remained unclear why MG is so predominant among the autoimmune diseases associated with thymomas. Abnormal proteins that contain AChR-, titin- or ryanodine receptor (RyR)-like epitopes *(97, 105, 117, 118)* have been described in thymomas, but it is not known whether these proteins are crucial for the development of anti-AChR, anti-titin or anti-RyR autoimmunity. Alternatively, we speculate that the lack of myoid cells in thymomas might play a role in this respect.

(2) In spite of absence of AIRE in thymomas and despite the high frequency of various autoantibodies in the serum of thymoma patients, clinical autoimmune *diseases* other than MG are relatively rare (about 10% of cases) *(79, 119)* (Table 3). A possible explanation could be that clinical symptoms in MG may be elicited by autoantibodies and complement at even nanomolar concentrations *(41)*, while other autoimmune diseases like type I diabetes may not be autoantibody mediated but follow a more complex attack of cytotoxic T cells *(120)*.

Table 3
Autoimmune Diseases with Defined Autoantigens, Pathogenic Relevance of Autoantibodies or Autoreactive T cells in Thymoma Patients and Relationship to Histological Thymoma Subtype
(43, 48, 124, 125)

Autoimmune disease	Autoantigen	Pathogenic relevance of autoantibodies or autoreactive T cells	Preferred thymoma subtype (WHO)
Myasthenia gravis	AChR[1]	Yes	WHO type B > AB > A
	StrA[2]	Probably no	
	IFN-α, IL-12	Probably yes	
Neuromyotonia	VGKC[3]	Yes	Not established
Peripheral neuropathy	VGKC[3]	Yes	Not established
Stiff person syndrome	GAD[4]	Unknown	Not established
Rippling muscle disease	Neuronal AChR	Probably yes	WHO type AB, B
Intestinal pseudo-obstruction	Neuronal AChR	Probably yes	Not established
Limbic encephalitis	Neuronal nuclear antigens, glial antigens	Probably yes	Not established
Red blood cell aplasia	Unknown	Yes (T cells) (Auto-Abs?)	WHO type A > B
Neutropenia, pancytopenia	Unknown	(Auto-Abs?)	Not established
Polymyositis, dermatomyositis	Unknown	Unknown	WHO type B, C

[1]AChR, Acetylcholine receptor.
[2]StrA, striational antigens (myosin, titin and ryanodine receptor).
[3]VGKC, voltage-gated potassium channel.
[4]GAD, glutamic acid decarboxylase.

(3) Another enigma is the occurrence of MG in patients with type A thymomas. Since this thymoma subtype usually exhibits minimal or virtually no intratumorous thymopoiesis in terms of immature CD4 + CD8 + and naïve CD4 + T cells *(90, 94)*, the above pathogenetic model (Fig. 2) may not apply to this subtype. However, type A thymomas are frequently associated with pure red cell aplasia that is believed to result from an attack of cytotoxic T cells on red cell precursors in the bone marrow. Therefore, it might be speculated that CD8 + and not CD4 + T cells might be relevant in MG associated with type A thymoma.

(4) Finally, it has not been elucidated which autoantigens maintain the prolonged autoantibody response after thymoma resection during which the adjacent residual non-neoplastic thymus with its myoid cells is usually removed as well. An obvious candidate is the AChR itself: AChR could be released from skeletal muscle endplates following destruction by autoantibodies or cytotoxic T cells. Hypothetically, released AChR and striational antigens may be presented to autoreactive T cells by the intramuscular inflammatory infiltrate *(121)* or by antigen-presenting cells in regional lymph nodes *(12, 112)*.

REFERENCES

1. Meager A, Visvalingam K, Peterson P, et al. Anti-interferon autoantibodies in autoimmune polyendocrinopathy syndrome type 1. PLoS Med. 2006;3:e289.
2. Lindstrom JM, Seybold ME, Lennon VA, Whittingham S, Duane DD. Antibody to acetylcholine receptor in myasthenia gravis: prevalence, clinical correlates, and diagnostic value. 1975 [classical article]. Neurology. 1998;51:933–939.
3. Tzartos SJ, Barkas T, Cung MT, et al. Anatomy of the antigenic structure of a large membrane autoantigen, the muscle-type nicotinic acetylcholine receptor. Immunol Rev. 1998;163:89–120.
4. Lee JY, Sung JJ, Cho JY, et al. MuSK antibody-positive, seronegative myasthenia gravis in Korea. J Clin Neurosci. 2006;13:353–355.
5. Vincent A, Leite MI. Neuromuscular junction autoimmune disease: muscle specific kinase antibodies and treatments for myasthenia gravis. Curr Opin Neurol. 2005;18:519–525.
6. Yeh JH, Chen WH, Chiu HC, Vincent A. Low frequency of MuSK antibody in generalized seronegative myasthenia gravis among Chinese. Neurology. 2004;62:2131–2132.
7. Bartoccioni E, Marino M, Evoli A, Ruegg MA, Scuderi F, Provenzano C. Identification of disease-specific auto-antibodies in seronegative myasthenia gravis. Ann N Y Acad Sci. 2003;998:356–358.
8. Hoch W, McConville J, Helms S, Newsom-Davis J, Melms A, Vincent A. Auto-antibodies to the receptor tyrosine kinase MuSK in patients with myasthenia gravis without acetylcholine receptor antibodies. Nat Med. 2001;7:365–368.
9. Leite MI, Strobel P, Jones M, et al. Fewer thymic changes in MuSK antibody-positive than in MuSK antibody-negative MG. Ann Neurol. 2005;57:444–448.
10. Willcox N. Myasthenia gravis. Curr Opin Immunol. 1993;5:910–917.
11. Drachman DB. Myasthenia gravis. N Engl J Med. 1994;330:1797–1810.
12. Marx A, Wilisch A, Schultz A, Gattenlohner S, Nenninger R, Muller-Hermelink HK. Pathogenesis of myasthenia gravis. Virchows Arch. 1997;430:355–364.
13. Chuang WY, Strobel P, Gold R, et al. A CTLA4high genotype is associated with myasthenia gravis in thymoma patients. Ann Neurol. 2005;58:644–648.
14. Willcox N, Schluep M, Ritter MA, Newsom-Davis J. The thymus in seronegative myasthenia gravis patients. J Neurol. 1991;238:256–261.
15. Williams CL, Hay JE, Huiatt TW, Lennon VA. Paraneoplastic IgG striational autoantibodies produced by clonal thymic B cells and in serum of patients with myasthenia gravis and thymoma react with titin. Lab Invest. 1992;66:331–336.
16. Vincent A. Aetiological factors in development of myasthenia gravis. Adv Neuroimmunol. 1994;4:355–371.
17. Muller-Hermelink HK, Marx A, T K. Thymus. In: Damjanov I LJ, ed. Anderson's Pathology (ed 10). St. Louis: Mosby; 1996:1218–1243.
18. Middleton G, Schoch EM. The prevalence of human thymic lymphoid follicles is lower in suicides. Virchows Arch. 2000;436:127–130.
19. Roxanis I, Micklem K, Willcox N. True epithelial hyperplasia in the thymus of early-onset myasthenia gravis patients: implications for immunopathogenesis. J Neuroimmunol. 2001;112:163–173.
20. Kirchner T, Schalke B, Melms A, von Kugelgen T, Muller-Hermelink HK. Immunohistological patterns of non-neoplastic changes in the thymus in Myasthenia gravis. Virchows Arch B Cell Pathol Incl Mol Pathol. 1986;52:237–257.
21. Flores KG, Li J, Sempowski GD, Haynes BF, Hale LP. Analysis of the human thymic perivascular space during aging. J Clin Invest. 1999;104:1031–1039.

22. Curnow J, Corlett L, Willcox N, Vincent A. Presentation by myoblasts of an epitope from endogenous acetylcholine receptor indicates a potential role in the spreading of the immune response. J Neuroimmunol. 2001;115:127–134.

23. Navaneetham D, Penn AS, Howard JF, Jr., Conti-Fine BM. Human thymuses express incomplete sets of muscle acetylcholine receptor subunit transcripts that seldom include the delta subunit. Muscle Nerve. 2001;24:203–210.

24. Geuder KI, Marx A, Witzemann V, et al. Pathogenetic significance of fetal-type acetylcholine receptors on thymic myoid cells in myasthenia gravis. Dev Immunol. 1992;2:69–75.

25. Schluep M, Willcox N, Vincent A, Dhoot GK, Newsom-Davis J. Acetylcholine receptors in human thymic myoid cells in situ: an immunohistological study. Ann Neurol. 1987;22:212–222.

26. Bornemann A, Kirchner T. An immuno-electron-microscopic study of human thymic B cells. Cell Tissue Res. 1996;284:481–487.

27. Meraouna A, Cizeron-Clairac G, Panse RL, et al. The chemokine CXCL13 is a key molecule in autoimmune myasthenia gravis. Blood. 2006;108:432–440.

28. Compston DA, Vincent A, Newsom-Davis J, Batchelor JR. Clinical, pathological, HLA antigen and immunological evidence for disease heterogeneity in myasthenia gravis. Brain. 1980;103:579–601.

29. Degli-Esposti MA, Andreas A, Christiansen FT, Schalke B, Albert E, Dawkins RL. An approach to the localization of the susceptibility genes for generalized myasthenia gravis by mapping recombinant ancestral haplotypes. Immunogenetics. 1992;35:355–364.

30. Huang D, Shi FD, Giscombe R, Zhou Y, Ljunggren HG, Lefvert AK. Disruption of the IL-1beta gene diminishes acetylcholine receptor-induced immune responses in a murine model of myasthenia gravis. Eur J Immunol. 2001;31:225–232.

31. Skeie GO, Pandey JP, Aarli JA, Gilhus NE. TNFA and TNFB polymorphisms in myasthenia gravis. Arch Neurol. 1999;56:457–461.

32. Huang DR, Pirskanen R, Matell G, Lefvert AK. Tumour necrosis factor-alpha polymorphism and secretion in myasthenia gravis. J Neuroimmunol. 1999;94:165–171.

33. Franciotta D, Cuccia M, Dondi E, Piccolo G, Cosi V. Polymorphic markers in MHC class II/III region: a study on Italian patients with myasthenia gravis. J Neurol Sci. 2001;190:11–16.

34. Huang DR, Zhou YH, Xia SQ, Liu L, Pirskanen R, Lefvert AK. Markers in the promoter region of interleukin-10 (IL-10) gene in myasthenia gravis: implications of diverse effects of IL-10 in the pathogenesis of the disease. J Neuroimmunol. 1999;94:82–87.

35. Garchon HJ, Djabiri F, Viard JP, Gajdos P, Bach JF. Involvement of human muscle acetylcholine receptor alpha-subunit gene (CHRNA) in susceptibility to myasthenia gravis. Proc Natl Acad Sci U S A. 1994;91:4668–4672.

36. Djabiri F, Gajdos P, Eymard B, Gomez L, Bach JF, Garchon HJ. No evidence for an association of AChR beta-subunit gene (CHRNB1) with myasthenia gravis. J Neuroimmunol. 1997;78:86–89.

37. Giraud M, Beaurain G, Yamamoto AM, et al. Linkage of HLA to myasthenia gravis and genetic heterogeneity depending on anti-titin antibodies. Neurology. 2001;57:1555–1560.

38. Onodera J, Nakamura S, Nagano I, et al. Upregulation of Bcl-2 protein in the myasthenic thymus. Ann Neurol. 1996;39:521–528.

39. Masunaga A, Arai T, Yoshitake T, Itoyama S, Sugawara I. Reduced expression of apoptosis-related antigens in thymuses from patients with myasthenia gravis. Immunol Lett. 1994;39:169–172.

40. Moulian N, Bidault J, Truffault F, Yamamoto AM, Levasseur P, Berrih-Aknin S. Thymocyte Fas expression is dysregulated in myasthenia gravis patients with anti-acetylcholine receptor antibody. Blood. 1997;89:3287–3295.

41. Vincent A, Palace J, Hilton-Jones D. Myasthenia gravis. Lancet. 2001;357:2122–2128.

42. Sempowski GD, Hale LP, Sundy JS, et al. Leukemia inhibitory factor, oncostatin M, IL-6, and stem cell factor mRNA expression in human thymus increases with age and is associated with thymic atrophy. J Immunol. 2000;164:2180–2187.

43. Muller-Hermelink HK, Marx A. Thymoma. Curr Opin Oncol. 2000;12:426–433.

44. Sommer N, Willcox N, Harcourt GC, Newsom-Davis J. Myasthenic thymus and thymoma are selectively enriched in acetylcholine receptor-reactive T cells. Ann Neurol. 1990;28:312–319.

45. Kirchner T, Schalke B, Buchwald J, Ritter M, Marx A, Muller-Hermelink HK. Well-differentiated thymic carcinoma. An organotypical low-grade carcinoma with relationship to cortical thymoma. Am J Surg Pathol. 1992;16:1153–1169.

46. Marx A, Schultz A, Wilisch A, Helmreich M, Nenninger R, Muller-Hermelink HK. Paraneoplastic autoimmunity in thymus tumors. Dev Immunol. 1998;6:129–140.

47. Nagvekar N, Moody AM, Moss P, et al. A pathogenetic role for the thymoma in myasthenia gravis. Autosensitization of IL-4- producing T cell clones recognizing extracellular acetylcholine receptor epitopes presented by minority class II isotypes. J Clin Invest. 1998;101:2268–2277.

48. Vincent A. Antibodies to ion channels in paraneoplastic disorders. Brain Pathol. 1999;9:285–291.

49. Müller-Hermelink HK, Engel P, Kuo TT, et al. Tumours of the thymus: Introduction. In: Travis MD, Brambilla E, Müller-Hermelink HK, Harris CC, eds. World Health Organization Classification of Tumours Pathology and Genetics of Tumours of the Lung, Thymus and Heart. Vol. 7. Lyon: IARC Press; 2004;145–151.

50. Marx A, Strobel P, Zettl A, et al. Thymomas. In: Travis MD, Brambilla E, Müller-Hermelink HK, Harris CC, eds. World Health Organization Classification of Tumours Pathology and Genetics of Tumours of the Lung, Thymus and Heart. Vol. 7. Lyon: IARC Press; 2004;152–153.

51. Rosai J. Histological typing of tumours of the thymus (2nd ed.). Berlin and Heidelberg: Springer-Verlag; 1999.

52. Wekerle H, Ketelsen UP. Intrathymic pathogenesis and dual genetic control of myasthenia gravis. Lancet. 1977;1:678–680.

53. Weinberg CB, Hall ZW. Antibodies from patients with myasthenia gravis recognize determinants unique to extra-junctional acetylcholine receptors. Proc Natl Acad Sci U S A. 1979;76:504–508.

54. Geuder KI, Marx A, Witzemann V, Schalke B, Kirchner T, Muller-Hermelink HK. Genomic organization and lack of transcription of the nicotinic acetylcholine receptor subunit genes in myasthenia gravis-associated thymoma. Lab Invest. 1992;66:452–458.

55. Kaminski HJ, Kusner LL, Block CH. Expression of acetylcholine receptor isoforms at extraocular muscle endplates [published erratum appears in Invest Ophthalmol Vis Sci 1996 Jul;37(8):6A]. Invest Ophthalmol Vis Sci. 1996;37:345–351.

56. MacLennan C, Beeson D, Buijs AM, Vincent A, Newsom-Davis J. Acetylcholine receptor expression in human extraocular muscles and their susceptibility to myasthenia gravis [see comments]. Ann Neurol. 1997;41:423–431.

57. Scadding GK, Vincent A, Newsom-Davis J, Henry K. Acetylcholine receptor antibody synthesis by thymic lympho-cytes: correlation with thymic histology. Neurology. 1981;31:935–943.

58. Nieto IP, Robledo JP, Pajuelo MC, et al. Prognostic factors for myasthenia gravis treated by thymectomy: review of 61 cases [see comments]. Ann Thorac Surg. 1999;67:1568–1571.

59. Tsuchida M, Yamato Y, Souma T, et al. Efficacy and safety of extended thymectomy for elderly patients with myasthenia gravis. Ann Thorac Surg. 1999;67:1563–1567.

60. Meinl E, Klinkert WE, Wekerle H. The thymus in myasthenia gravis. Changes typical for the human disease are absent in experimental autoimmune myasthenia gravis of the Lewis rat. Am J Pathol. 1991;139:995–1008.

61. Schonbeck S, Padberg F, Marx A, Hohlfeld R, Wekerle H. Transplantation of myasthenia gravis thymus to SCID mice. Ann N Y Acad Sci. 1993;681:66–73.

62. Spuler S, Marx A, Kirchner T, Hohlfeld R, Wekerle H. Myogenesis in thymic transplants in the severe combined immunodeficient mouse model of myasthenia gravis. Differentiation of thymic myoid cells into striated muscle cells. Am J Pathol. 1994;145:766–770.

63. Spuler S, Sarropoulos A, Marx A, Hohlfeld R, Wekerle H. Thymoma-associated myasthenia gravis. Transplantation of thymoma and extrathymomal thymic tissue into SCID mice. Am J Pathol. 1996;148:1359–1365.

64. Baggi F, Nicolle M, Vincent A, Matsuo H, Willcox N, Newsom-Davis J. Presentation of endogenous acetylcholine receptor epitope by an MHC class II-transfected human muscle cell line to a specific CD4+ T cell clone from a myasthenia gravis patient. J Neuroimmunol. 1993;46:57–65.

65. Melms A, Malcherek G, Gern U, et al. T cells from normal and myasthenic individuals recognize the human acetylcholine receptor: heterogeneity of antigenic sites on the alpha- subunit. Ann Neurol. 1992;31:311–318.

66. Jermy A, Beeson D, Vincent A. Pathogenic autoimmunity to affinity-purified mouse acetylcholine receptor induced without adjuvant in BALB/c mice. Eur J Immunol. 1993;23:973–976.

67. Salmon AM, Bruand C, Cardona A, Changeux JP, Berrih-Aknin S. An acetylcholine receptor alpha subunit promoter confers intrathymic expression in transgenic mice. Implications for tolerance of a transgenic self-antigen and for autoreactivity in myasthenia gravis. J Clin Invest. 1998;101:2340–2350.

68. Melms A, Schalke BC, Kirchner T, Muller-Hermelink HK, Albert E, Wekerle H. Thymus in myasthenia gravis. Isolation of T-lymphocyte lines specific for the nicotinic acetylcholine receptor from thymuses of myasthenic patients. J Clin Invest. 1988;81:902–908.

69. Hohlfeld R, Wekerle H. The immunopathogenesis of myasthenia gravis. In: Angel AG, ed. Myasthenia gravis and myasthenic disorders. Oxford: Oxford University Press; 1999:87–110.

70. Sims GP, Shiono H, Willcox N, Stott DI. Somatic hypermutation and selection of b cells in thymic germinal centers responding to acetylcholine receptor in myasthenia gravis. J Immunol. 2001;167:1935–1944.

71. Bernasconi P, Barberis M, Baggi F, et al. Increased toll-like receptor 4 expression in thymus of myasthenic patients with thymitis and thymic involution. Am J Pathol. 2005;167:129–139.

72. Vincent A, Willcox N, Hill M, Curnow J, MacLennan C, Beeson D. Determinant spreading and immune responses to acetylcholine receptors in myasthenia gravis. Immunol Rev. 1998;164:157–168.

73. Willcox N, Schluep M, Ritter MA, Schuurman HJ, Newsom-Davis J, Christensson B. Myasthenic and nonmyasthenic thymoma. An expansion of a minor cortical epithelial cell subset? Am J Pathol. 1987;127:447–460.

74. Inoue M, Fujii Y, Okumura M, et al. T-cell development in human thymoma. Pathol Res Pract. 1999;195:541–547.

75. Kadota Y, Okumura M, Miyoshi S, et al. Altered T cell development in human thymoma is related to impairment of MHC class II transactivator expression induced by interferon-gamma (IFN- gamma). Clin Exp Immunol. 2000;121:59–68.

76. Strobel P, Helmreich M, Kalbacher H, Muller-Hermelink HK, Marx A. Evidence for distinct mechanisms in the shaping of the CD4 T cell repertoire in histologically distinct myasthenia gravis-associated thymomas. Dev Immunol. 2001;8:279–290.

77. Newsom-Davis J, Willcox N, Schluep M, et al. Immunological heterogeneity and cellular mechanisms in myasthenia gravis. Ann N Y Acad Sci. 1987;505:12–26.

78. Fujii Y, Monden Y, Nakahara K, Hashimoto J, Kawashima Y. Antibody to acetylcholine receptor in myasthenia gravis: production by lymphocytes from thymus or thymoma. Neurology. 1984;34:1182–1186.

79. Ströbel P, Murumagi A, Klein R, et al. Deficiency of the autoimmune regulator AIRE in thymomas is insufficient to elicit autoimmune polyendocrinopathy syndrome type I (APS-1). J Pathol. 2007;211:563–571.

80. Heino M, Peterson P, Kudoh J, et al. Autoimmune regulator is expressed in the cells regulating immune tolerance in thymus medulla. Biochem Biophys Res Commun. 1999;257:821–825.

81. Bjorses P, Halonen M, Palvimo JJ, et al. Mutations in the AIRE gene: effects on subcellular location and transactivation function of the autoimmune polyendocrinopathy-candidiasis-ectodermal dystrophy protein. Am J Hum Genet. 2000;66:378–392.

82. Anderson MS, Venanzi ES, Chen Z, Berzins SP, Benoist C, Mathis D. The cellular mechanism of Aire control of T cell tolerance. Immunity. 2005;23:227–239.

83. Ramsey C, Hassler S, Marits P, et al. Increased antigen presenting cell-mediated T cell activation in mice and patients without the autoimmune regulator. Eur J Immunol. 2006;36:305–317.

84. Zuklys S, Balciunaite G, Agarwal A, Fasler-Kan E, Palmer E, Hollander GA. Normal thymic architecture and negative selection are associated with Aire expression, the gene defective in the autoimmune- polyendocrinopathy-candidiasis-ectodermal dystrophy (APECED). J Immunol. 2000;165:1976–1983.

85. Liston A, Lesage S, Wilson J, Peltonen L, Goodnow CC. Aire regulates negative selection of organ-specific T cells. Nat Immunol. 2003;4:350–354.

86. Chuang W, Strobel P, Gold R, et al. A CTLA4high Genotype Is Associated with Myasthenia Gravis in Thymoma Patients. Ann Neurol. 2005;58:644–648.

87. Carreno BM, Bennett F, Chau TA, et al. CTLA-4 (CD152) can inhibit T cell activation by two different mechanisms depending on its level of cell surface expression. J Immunol. 2000;165:1352–1356.

88. Zettl A, Ströbel P, Wagner K, et al. Recurrent genetic aberrations in thymoma and thymic carcinoma. Am J Pathol. 2000;157:257–266.

89. Takeuchi Y, Fujii Y, Okumura M, Inada K, Nakahara K, Matsuda H. Accumulation of immature CD3-CD4 + CD8- single-positive cells that lack CD69 in epithelial cell tumors of the human thymus. Cell Immunol. 1995;161:181–187.

90. Nenninger R, Schultz A, Vandekerckhove B, Hünig T, Müller-Hermelink HK, Marx A. Abnormal T lymphocyte development in myasthenia gravis-associated thymomas. New York, London: Plenum Press; 1997.

91. Nenninger R, Schultz A, Hoffacker V, et al. Abnormal thymocyte development and generation of autoreactive T cells in mixed and cortical thymomas. Lab Invest. 1998;78:743–753.

92. Strobel P, Helmreich M, Menioudakis G, et al. Paraneoplastic myasthenia gravis correlates with generation of mature naive CD4(+) T cells in thymomas. Blood. 2002;100:159–166.

93. Buckley C, Douek D, Newsom-Davis J, Vincent A, Willcox N. Mature, long-lived CD4 + and CD8 + T cells are generated by the thymoma in myasthenia gravis. Ann Neurol. 2001;50:64–72.

94. Hoffacker V, Schultz A, Tiesinga JJ, et al. Thymomas alter the T-cell subset composition in the blood: a potential mechanism for thymoma-associated autoimmune disease [In Process Citation]. Blood. 2000;96:3872–3879.

95. Sommer N, Harcourt GC, Willcox N, Beeson D, Newsom-Davis J. Acetylcholine receptor-reactive T lymphocytes from healthy subjects and myasthenia gravis patients. Neurology. 1991;41:1270–1276.

96. Conti-Fine BM, Navaneetham D, Karachunski PI, et al. T cell recognition of the acetylcholine receptor in myasthenia gravis. Ann N Y Acad Sci. 1998;841:283–308.

97. Schultz A, Hoffacker V, Wilisch A, et al. Neurofilament is an autoantigenic determinant in myasthenia gravis. Ann Neurol. 1999;46:167–175.

98. Strobel P, Rosenwald A, Beyersdorf N, et al. Selective loss of regulatory T cells in thymomas. Ann Neurol. 2004;56:901–904.

99. Fattorossi A, Battaglia A, Buzzonetti A, Ciaraffa F, Scambia G, Evoli A. Circulating and thymic CD4 CD25 T regulatory cells in myasthenia gravis: effect of immunosuppressive treatment. Immunology. 2005;116:134–141.

100. Aarli JA, Skeie GO, Mygland A, Gilhus NE. Muscle striation antibodies in myasthenia gravis. Diagnostic and functional significance. Ann N Y Acad Sci. 1998;841:505–515.

101. Etienne M, Weimer LH. Immune-mediated autonomic neuropathies. Curr Neurol Neurosci Rep. 2006;6:57–64.

102. Vernino S, Lennon VA. Autoantibody profiles and neurological correlations of thymoma. Clin Cancer Res. 2004;10:7270–7275.

103. Marx A, Kirchner T, Greiner A, Muller-Hermelink HK, Schalke B, Osborn M. Neurofilament epitopes in thymoma and antiaxonal autoantibodies in myasthenia gravis. Lancet. 1992;339:707–708.

104. Mygland A, Aarli JA, Matre R, Gilhus NE. Ryanodine receptor antibodies related to severity of thymoma associated myasthenia gravis. J Neurol Neurosurg Psychiatry. 1994;57:843–846.

105. Mygland A, Kuwajima G, Mikoshiba K, Tysnes OB, Aarli JA, Gilhus NE. Thymomas express epitopes shared by the ryanodine receptor. J Neuroimmunol. 1995;62:79–83.

106. Fukui Y, Ishimoto T, Utsuyama M, et al. Positive and negative CD4 + thymocyte selection by a single MHC class II/peptide ligand affected by its expression level in the thymus. Immunity. 1997;6:401–410.

107. Ashton-Rickardt PG, Tonegawa S. A differential-avidity model for T-cell selection. Immunol Today. 1994;15:362–366.

108. Hogquist KA, Jameson SC, Heath WR, Howard JL, Bevan MJ, Carbone FR. T cell receptor antagonist peptides induce positive selection. Cell. 1994;76:17–27.

109. Barton GM, Rudensky AY. Requirement for diverse, low-abundance peptides in positive selection of T cells. Science. 1999;283:67–70.

110. Laufer TM, Fan L, Glimcher LH. Self-reactive T cells selected on thymic cortical epithelium are polyclonal and are pathogenic in vivo. J Immunol. 1999;162:5078–5084.

111. van Meerwijk JP, MacDonald HR. In vivo T-lymphocyte tolerance in the absence of thymic clonal deletion mediated by hematopoietic cells. Blood. 1999;93:3856–3862.

112. Muller-Hermelink HK, Wilisch A, Schultz A, Marx A. Characterization of the human thymic microenvironment: lymphoepithelial interaction in normal thymus and thymoma. Arch Histol Cytol. 1997;60:9–28.

113. Vincent A, Willcox N. The role of T-cells in the initiation of autoantibody responses in thymoma patients. Pathol Res Pract. 1999;195:535–540.

114. Meager A, Vincent A, Newsom-Davis J, Willcox N. Spontaneous neutralising antibodies to interferon–alpha and interleukin-12 in thymoma-associated autoimmune disease [letter]. Lancet. 1997;350:1596–1597.

115. Conti-Tronconi BM, McLane KE, Raftery MA, Grando SA, Protti MP. The nicotinic acetylcholine receptor: structure and autoimmune pathology. Crit Rev Biochem Mol Biol. 1994;29:69–123.

116. Somnier FE. Exacerbation of myasthenia gravis after removal of thymomas [see comments]. Acta Neurol Scand. 1994;90:56–66.

117. Marx A, O'Connor R, Geuder KI, et al. Characterization of a protein with an acetylcholine receptor epitope from myasthenia gravis-associated thymomas [see comments]. Lab Invest. 1990;62:279–286.

118. Marx A, Wilisch A, Schultz A, et al. Expression of neurofilaments and of a titin epitope in thymic epithelial tumors. Implications for the pathogenesis of myasthenia gravis. Am J Pathol. 1996;148:1839–1850.

119. Strobel P, Bauer A, Puppe B, et al. Tumor recurrence and survival in patients treated for thymomas and thymic squamous cell carcinomas: a retrospective analysis. J Clin Oncol. 2004;22:1501–1509.

120. Lang KS, Recher M, Junt T, et al. Toll-like receptor engagement converts T-cell autoreactivity into overt autoimmune disease. Nat Med. 2005;11:138–145.

121. Maselli RA, Richman DP, Wollmann RL. Inflammation at the neuromuscular junction in myasthenia gravis. Neurology. 1991;41:1497–1504.

122. Evoli A, Minisci C, Di Schino C, et al. Thymoma in patients with MG: characteristics and long-term outcome. Neurology. 2002;59:1844–1850.

123. Rickman OB, Parisi JE, Yu Z, Lennon VA, Vernino S. Fulminant autoimmune cortical encephalitis associated with thymoma treated with plasma exchange. Mayo Clin Proc. 2000;75:1321–1326.

124. Vernino S, Auger RG, Emslie-Smith AM, Harper CM, Lennon VA. Myasthenia, thymoma, presynaptic antibodies, and a continuum of neuromuscular hyperexcitability. Neurology. 1999;53:1233–1239.

125. Pande R, Leis AA. Myasthenia gravis, thymoma, intestinal pseudo-obstruction, and neuronal nicotinic acetylcholine receptor antibody. Muscle Nerve. 1999;22:1600–1602.

8

Electrodiagnosis of Neuromuscular Junction Disorders

Bashar Katirji

1. INTRODUCTION

The electrodiagnostic (EDX) examination in patients with suspected neuromuscular junction (NMJ) disorders requires a good comprehension of the physiology and pathophysiology of neuromuscular transmission. The EDX studies that are useful in the evaluation of such patients include (1) motor nerve conduction studies (NCSs); (2) conventional needle electromyography (EMG); (3) repetitive nerve stimulation (RNS); and (4) single-fiber EMG. This chapter reviews the basic knowledge of neuromuscular transmission as it relates to the EDX studies and discusses in detail the EDX studies and findings in various NMJ disorders.

2. BASIC CONCEPTS OF NEUROMUSCULAR TRANSMISSION

Performing EDX studies in NMJ disorders depends on the understanding of few important concepts inherent in neuromuscular transmission. These physiologic facts dictate the type of RNS and single-fiber EMG needed for the accurate diagnosis of NMJ disorders. The physiology of neuromuscular transmission is discussed in detail in other chapters and is addressed briefly here *(1, 2, 3, 4)*.

2.1. Quantum

A quantum is the amount of acetylcholine (ACh) packaged in a single vesicle, which is approximately 5000–10,000 ACh molecules. Each quantum (vesicle) released results in a 1 mV change of postsynaptic membrane potential. This occurs spontaneously during rest and forms the basis of miniature end-plate potential (MEPP).

The number of quanta released after a nerve action potential depends on the number of quanta in the *immediately available (primary) store* and the probability of release, i.e., $m = p \times n$, where m is the number of quanta released during each stimulation, p the probability of release (effectively proportional to the concentration of calcium and typically about 0.2, or 20%), and n the number of quanta in the immediately available store. Under normal conditions, a single nerve action potential triggers the release of 50–300 vesicles (quanta) with an average of about 60 quanta (60 vesicles).

In addition to the immediately available store of ACh-containing synaptic vesicles located beneath the presynaptic nerve terminal membrane, *a secondary (or mobilization) store* starts to replenish the immediately available store after 1–2 s of repetitive nerve action potentials. A large *tertiary (or reserve) store* is also available in the axon and cell body.

2.2. End-Plate Potential

The end-plate potential (EPP) is the potential generated at the postsynaptic membrane following a nerve action potential and neuromuscular transmission. Since each released vesicle (quanta) results in a 1 mV change in the postsynaptic membrane potential, the ACh release after a nerve action potential results in about a 60 mV change in the amplitude of the membrane potential.

From: *Current Clinical Neurology*
Myasthenia Gravis and Related Disorders
Edited by: H.J. Kaminski, DOI 10.1007/978-1-59745-156-7_8, © Humana Press, New York, NY

2.3. Safety Factor

Under normal conditions, the number of quanta (vesicles) released at the NMJ by the presynaptic terminal (about 60 vesicles) far exceeds the postsynaptic membrane potential change required to reach the *threshold* needed to generate a postsynaptic muscle action potential (7–20 mV). The safety factor produces an EPP that always reaches threshold, results in an all-or-none muscle fiber action potential (MFAP), and prevents neuromuscular transmission failure despite repetitive action potentials. In addition to quantal release, several other factors contribute to the safety factor and EPP including ACh receptor conduction properties, ACh receptor density, ACh-esterase activity, synaptic architecture, and sodium channel density at the NMJ.

2.4. Calcium Influx into the Terminal Axon

Following depolarization of the presynaptic terminal, voltage-gated calcium channels (VGCCs) open leading to calcium influx. Through a calcium-dependent intracellular cascade, vesicles are docked at active release sites (called active zones) and release ACh molecules. Calcium then diffuses slowly away from the vesicle release site in 100–200 ms. The rate at which motor nerves are repetitively stimulated in the EDX laboratory dictates whether calcium accumulation plays a role in enhancing the release of ACh or not.

2.5. Compound Muscle Action Potential

The compound muscle action potential (CMAP) is obtained, during motor NCS, with supramaximal stimulation while recording via surface electrode placed over the belly of a muscle. The CMAP represents the summation of all MFAPs generated in a muscle following stimulation of all motor axons.

3. ELECTRODIAGNOSTIC TESTS IN NEUROMUSCULAR JUNCTION DISORDERS

3.1. Routine Motor Nerve Conduction Studies

Sensory NCSs are always normal in NMJ disorders. Motor NCSs are, however, helpful in the evaluation of all disorders affecting the motor unit. In NMJ disorders, CMAP amplitude is the most useful parameter analyzed since motor distal latencies, conduction velocities, F waves' minimal latencies, and H-reflexes are normal. The CMAP amplitudes are usually normal in postsynaptic disorders (such as MG) due to the effect of the safety factor: after a single supramaximal stimulus, and despite blockade of ACh receptors, EPPs achieve thresholds and generate MFAPs in all muscle fibers resulting in normal CMAP. Occasionally, such as in a myasthenic crisis, the CMAP amplitudes may be borderline or slightly diminished due to severe postsynaptic neuromuscular blockade. In contrast to postsynaptic disorders, the CMAP amplitudes on routine NCSs are often low in presynaptic disorders (such as Lambert–Eaton syndrome, LES) since many EPPs do not reach threshold, and many muscle fibers do not fire.

3.2. Conventional Needle Electromyography

The needle EMG is usually normal in NMJ disorders. However, nonspecific changes, more commonly encountered in myopathies or neurogenic disorders, may occasionally be associated with NMJ disorders, particularly when chronic and severe.

3.2.1. Moment-to-Moment Variation Instability of Motor Unit Action Potentials

In healthy subjects, individual motor unit action potential (MUAP) amplitude, duration, and phases are stable with little, if any, morphology variation. However, individual MUAP amplitude and morphology in NMJ disorders may vary significantly during activation due to intermittent end-plate blockade, slowing, or both. During needle EMG recording, moment-to-moment variation instability should be distinguished from MUAP overlap. This can be achieved by always recording from a single MUAP at a time.

3.2.2. Short-Duration, Low-Amplitude, and Polyphasic MUAPs

These MUAPs are seen primarily in proximal muscles and are similar in morphology to those seen in myopathies. In NMJ disorders, "myopathic" MUAPs are caused by physiological blocking and slowing of neuromuscular transmission at end plates during voluntary activation. This leads to exclusion of MFAPs from the MUAP (hence the short duration and low amplitude) and asynchrony of neuromuscular transmission of muscle fibers (hence the polyphasia).

3.2.3. Fibrillation Potentials

Fibrillation potentials are rarely encountered in NMJ disorders *(5, 6)*. They are usually inconspicuous and present mostly in proximal muscles. The mechanism of fibrillation potentials in NMJ disorders is not clear but may be related to chronic neuromuscular transmission blockade or loss of end plates, resulting in "effective" denervation of individual muscle fibers. Since fibrillation potentials are rare in NMJ disorders, their presence should always raise the suspicion of an alternate diagnosis or associated illness.

3.3. Repetitive Nerve Stimulation

3.3.1. Principles

Motor nerves may be repetitively stimulated and the resultant CMAPs evaluated for evidence of amplitude decrement or increment. The rate at which motor nerves are stimulated in the EDX laboratory dictates whether calcium accumulation plays a role in enhancing the release of ACh. After a single stimulus, calcium diffuses out of the presynaptic terminal in about 100–200 ms. Hence, at a slow rate of RNS (i.e., a stimulus every 200 ms or more, or a stimulation rate of <5 Hz), calcium's role in ACh release is not enhanced and subsequent nerve action potentials reach the nerve terminal long after calcium has dispersed. In contrast, with rapid RNS (i.e., a stimulus every 100 ms or less, or stimulation rate >10 Hz), calcium influx is greatly enhanced and the probability of release of ACh quanta increases.

3.3.2. Techniques

In the EMG laboratory, RNSs often follow the performance of motor NCSs. Electromyographers and EDX technologists should master the various motor NCS and RNS techniques to avoid technical factors that may result in false positive and false negative studies. There are certain prerequisites that are essential for performing reliable and reproducible RNSs.

- *Acetylcholinesterase inhibitors* Patients on acetylcholinesterase inhibitors (such as pyridostigmine) should be asked to withhold their medication, preferably for 12–24 h before RNS, if medically not contraindicated. These agents improve neuromuscular transmission and may mask a CMAP decrement resulting in a false negative RNS.
- *Limb temperature control* Limb temperature should be maintained at around 33°C at the recording site. A cool limb enhances neuromuscular transmission and may mask a CMAP decrement resulting in a false negative RNS. The exact reason for this effect is not well understood, but it may be due to decreased functional effectiveness of acetylcholinesterase resulting in increased amount of ACh available at the NMJ.
- *Limb immobilization* The limb tested should be immobilized as best as possible to prevent movement, particularly at the stimulation or recording sites. These sites should be well secured with tape, and the limb could be secured to a board using tape or Velcro. Movement at either site may result in an

apparent CMAP amplitude decay or increment, potentially leading to a false diagnosis of a NMJ disorder.

- *Stimulation intensity* Supramaximal stimulation (i.e., 10–20% above the intensity level needed for a maximal response) is needed to assure that all nerve fibers are activated and supramaximal CMAP obtained. Unnecessary high-intensity or long-duration stimuli should be avoided to prevent movement artifacts and excessive pain.
- *Stimulation frequency and number of stimuli* The stimulus rate and number of stimuli applied during RNS depend on the clinical problem and the working diagnosis.
  a. Slow RNS is usually done at a rate of 2–3 Hz; this rate is low enough to prevent calcium accumulation but high enough to deplete the quanta in the immediately available stores before the secondary (mobilization) stores start to replenish it. Three to five stimuli are adequate since the maximal decrease in ACh release occur after the first four stimuli. There is nothing to be gained in exceeding 9–10 stimuli.
  b. Rapid RNS is done with a frequency of 20–50 Hz to ensure accumulation of calcium in the presynaptic terminal. Since this is extremely painful, a train of 3–5 s is sufficient. A brief (10 s) period of maximal isometric voluntary exercise has the same effect as rapid RNS at 20–50 Hz. This is much less painful than rapid RNS and, hence, may be done on multiple motor nerves. Brief exercise should substitute for rapid RNS in most cooperative subjects. However, rapid RNS is necessary in patients who cannot produce a strong isometric exercise (e.g., young children, comatose patients, or patients with severe weakness).
- *Nerve selection* The choice of nerve to be stimulated and the muscle to be recorded from depends on the patient's manifestations. Useful nerves for RNS are the median and ulnar nerves, recording abductor pollicis brevis and abductor digiti minimi, respectively. Since the upper limb is easily immobilized, these RNSs are well tolerated and accompanied by minimal movement artifact. However, since distal muscles are often spared in postsynaptic NMJ disorders (such as MG), recording from a proximal muscle is often necessary. Slow RNS of the spinal accessory nerve, recording the trapezius muscle, is the most common study of a proximal nerve. It is relatively well tolerated and subject to less movement artifact when compared to RNS of other proximal nerves such as the musculocutaneous or axillary nerves, recording the biceps or deltoid muscles, respectively. Finally, facial slow RNS is indicated in patients with ocular or facial weakness when MG is suspected and other RNSs are normal or equivocal. However, the facial CMAP is low in amplitude and often plagued by large stimulation artifacts. This renders measurement of decrement difficult and subject to error.

3.3.3. Findings

Slow RNS does not abolish any MFAPs in healthy subjects. Although the subsequent EPPs fall in amplitude (due to the relative decrease in ACh release), the EPPs remain above threshold (due to the safety factor) and ensure generation of MFAPs with each stimulation (Fig. 1A). In addition, the secondary store begins to replace the depleted quanta after the first few seconds with a subsequent stabilization or rise in the EPP amplitude. Thus, all muscle fibers generate MFAPs with slow RNS and the CMAP does not change. In *postsynaptic disorders* (such as MG), the safety factor is reduced since there are fewer ACh receptors available for binding to ACh. Hence, the baseline EPPs are reduced but usually still above threshold. Slow RNS, however, results in a variable number of EPPs falling below threshold at many end plates and a failure to generate MFAPs at these neuromuscular junctions. The decline in the number of MFAPs is reflected by a decrement of the CMAP amplitude *(7, 8)*. In *presynaptic disorders* (such as LES), EPPs are low at baseline due to impaired ACh release, with many end plates often not reaching threshold, and many muscle fibers do not fire. This results in low baseline CMAP. Slow RNS causes a CMAP decrement, due to further decline of ACh release with subsequent stimuli, resulting in further loss of many MFAPs *(9, 10)*.

Rapid RNS in healthy subjects increases EPP amplitude. This has no ultimate effect on the number of MFAPs or the size of CMAP since all EPPs are and remain above threshold (Fig. 1B). In patients with *presynaptic disorders* (such as LES), the baseline CMAP is low in amplitude since many muscle fibers do not reach threshold due to inadequate release of quanta after a single

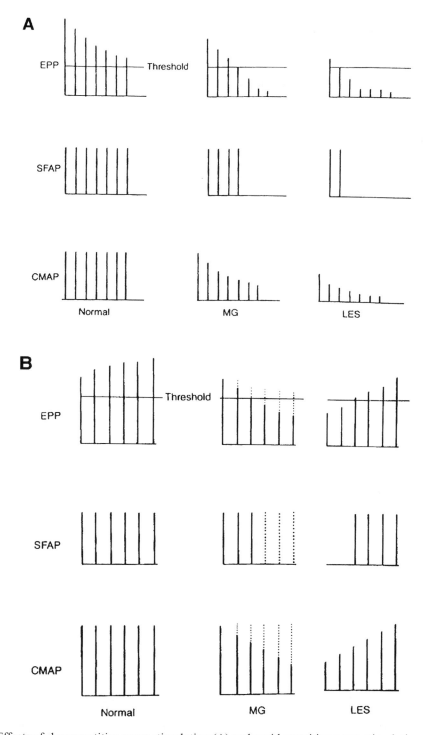

Fig. 1. Effects of slow repetitive nerve stimulation (**A**) and rapid repetitive nerve stimulation (**B**) on EPP, SFAP, MFAP and CMAP in normal subjects and in patients with MG and LES (marked ELS) (from Oh S. Clinical electromyography, neuromuscular transmission studies, Baltimore, 1988, Williams and Wilkins, with permission).

stimulus. However, calcium influx is greatly enhanced with rapid RNS resulting in larger releases of quanta and larger EPPs. This leads to increasingly more muscle fibers reaching threshold required for the generation of MFAPs. Thus, more MFAPs are generated and hence the CMAP increment *(9, 10, 11)*. In patients with *postsynaptic disorders*, the depleted stores are compensated by the calcium influx induced by rapid RNS, resulting usually in no change of CMAP. In severe postsynaptic defect (such as during myasthenic crisis), the increased quantal release cannot compensate for the marked NMJ blockade resulting in a drop in EPP amplitude. Hence, fewer MFAPs are generated resulting in CMAP decrement.

3.3.4. Measurements

After establishing a supramaximal CMAP, slow RNS is usually performed by applying three to five stimuli to a mixed or motor nerve at a rate of 2–3 Hz. Calculation of the decrement with slow RNS is accomplished by comparing the baseline (first) CMAP amplitude to the lowest CMAP amplitude. In NMJ disorders, the CMAP decrement is usually maximal at the third or fourth CMAP which then plateaus or begins to improve by the fifth or sixth response due to the mobilization store re-supplying the immediately available store (Fig. 2). The CMAP decrement is expressed as a percentage and calculated as follows:

$$\% \ decrement = \frac{Amplitude(1st \ response) - Amplitude(3rd \ or \ 4th \ response)}{Amplitude(1st \ response)} \times 100$$

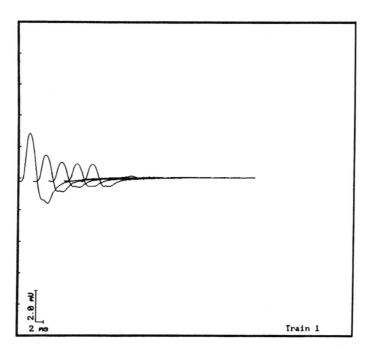

Fig. 2. Slow repetitive nerve stimulation (3 Hz) of the median nerve (recording thenar muscles) in a patient with myasthenic crisis. The largest decrement is between the first and the second CMAPs (45%) while the maximal decrement is between the first and third CMAPs (65%). Note that the CMAP amplitude plateaus after the third response and that the baseline (first) CMAP amplitude is slightly diminished during myasthenic crisis (see text).

Slow RNS at rest should be repeated after an interval of 1–2 min to confirm a normal or abnormal response. A reproducible decrement of more than 10% is considered abnormal and eliminates false positives. Decrement less than 10% is equivocal and not diagnostic. If there is a reproducible decrement at rest (≥10%), slow RNS should be repeated after the patient exercises for 10 s to demonstrate repair of the decrement ("post-exercise facilitation"). If there is no or equivocal decrement (<10%) with slow RNS at rest, the patient should perform maximal voluntary exercise for 1 min (exercise for 30 s, rest for 5 s, and exercise for another 30 s). Immediately after exercise, slow RNS is repeated at 1, 2, 3, 4, and 5 min later. Since the amount of ACh released with each stimulus is at its minimum 2–5 min after exercise, slow RNS after exercise provides a high chance of detecting a defect in NMJ by demonstrating a worsening CMAP decrement (*post-exercise exhaustion*).

Rapid RNS is most useful in patients with suspected presynaptic NMJ disorders such as LES or botulism. The optimal frequency is 20–50 Hz for 2–10 s (Fig. 3). A typical rapid RNS applies 200 stimuli at a rate of 50 Hz (i.e., 50 Hz for 4 s). Calculation of CMAP increment after rapid RNS is as follows:

$$\% \ increment = \frac{Amplitude(Highest \ response) - Amplitude(1st \ response)}{Amplitude(1st \ response)} \times 100$$

A CMAP increment of more than 50–100% is considered abnormal. A modest increment of 25–40% may occur in normal individuals. This is likely caused by increased synchrony of MFAPs following tetanic stimulation (*physiologic post-tetanic facilitation or pseudofacilitation*) *(3, 4)*. Maximal voluntary isometric exercise for brief periods (10 s) has the same effect as rapid RNS, is much less painful, and is a good substitute in cooperative subjects *(11)*. A single supramaximal stimulus is applied to generate a baseline CMAP; then the patient performs a 10 s maximal isometric voluntary contraction which is followed by another stimulus that produces a

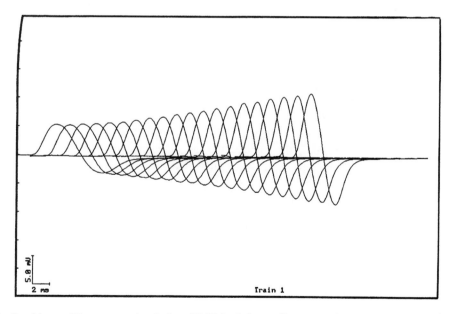

Fig. 3. Rapid repetitive nerve stimulation (30 Hz) of the median nerve (recording thenar muscles) in an infant with botulism. There is a 110% increment of CMAP amplitude between the baseline (first) and last (twentieth) responses.

post-exercise CMAP. Calculation of CMAP increment after brief (10 s) voluntary contraction is similar to the calculation of the increment following rapid RNS, which is given as follows:

$$\% \ increment = \frac{Amplitude \ of \ post \ exercise \ response - Amplitude \ of \ preexercise \ response}{Amplitude \ of \ preexercise \ response} \times 100$$

3.4. Single-Fiber EMG

3.4.1. Principles

Single-fiber EMG (SFEMG) is the selective recording of a small number (usually two or three) of MFAPs innervated by a single motor unit. The study aims to analyze the effects of the variation in the time it takes the EPPs to reach threshold on MFAPs generation. There is an increased variation in the time taken to attain an EPP capable of reaching threshold. SFEMG recording requires special expertise and understanding of the micro-environment of motor unit physiology. Although the examination may be applied to many neuromuscular disorders, SFEMG jitter study is most useful in clinical practice in the diagnosis of NMJ disorders, particularly myasthenia gravis *(12, 13, 14)*.

3.4.2. Techniques

SFEMG jitter study has specific requirements, which are essential for the completion and accurate interpretation of data *(4, 12)*.

1. SFEMG is performed by inserting a single-fiber concentric needle into a muscle. This electrode has a small recording surface (25 μm), which restricts the number of recordable MFAPs and results in an effective recording area of 300 μm³, as compared to a concentric needle electrode that records from approximately 1 cm³(Fig. 4). The use of a reusable single-fiber electrode for recording SFEMG raises concerns regarding the risk of transmission of infectious agents such as prion disease. Concentric needle electrodes, however, have the advantages of being less costly and disposable, decreasing the possibility and concern of infection. Recent investigations have shown that assessment of neuromuscular transmission with a concentric needle electrode has the same sensitivity and specificity as when assessed with a

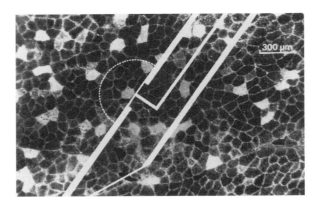

Fig. 4. Cross-section of a muscle stained for glycogen after stimulating an isolated motor axon to show the muscle fiber distribution of one motor unit (the fibers of the motor unit become depleted of glycogen and therefore appear pale). A single-fiber EMG electrode is superimposed to show the uptake area. The strategy of recording jitter is to position the electrode (as shown) in order to record from two muscle fibers belonging to the same motor unit (from Stålberg E, Trontelj JV. Single Fiber Electromyography. Studies in Healthy and Diseased Muscle. 2nd ed. New York: Raven Press, 1994, with permission).

single-fiber electrode *(15, 16)*. However, the potential recorded with a concentric electrode may not always represent a single fiber, but may be generated by two or three muscle fibers.

2. A 500 Hz low-frequency filter attenuates signals from distant fibers (more than 500 μm from the electrode). Filter settings should be set at 500 Hz for the high-pass filter and 10–20 kHz for the low-pass filter.

3. Selected single MFAPs should have a rise time of 300 μs and preferable peak-to-peak amplitude of 200 μV or more.

4. An amplitude threshold allows recording from a single MFAP by triggering on it on a screen with a delay line capability.

5. Computerized equipment assists in calculating individual and mean interpotential intervals (IPIs) and jitters (see below).

Volitional (recruitment) SFEMG is a common method for recording MFAPs, where the patient activates and maintains the firing rate of the motor unit. This technique is not possible if the patient cannot cooperate (e.g., a child or a patient with dementia, encephalopathy, coma, or severe weakness) and is difficult if the patient is unable to maintain a constant and stable voluntary muscle contraction (e.g., in patients with parkinsonism, tremor, dystonia, or spasticity). The small recording area of a SFEMG needle only registers signals from muscle fibers in its immediate vicinity, usually five to seven fibers at any time. Given the normal mosaic distribution of muscle fibers within motor units, these fibers may be innervated by up to five to seven different motor units. By using an amplitude threshold to trigger the oscilloscope trace on the closest MFAP (the action potential with the sharpest rise time and the greatest amplitude), MFAPs from other motor units are excluded from the oscilloscope screen. With minimal voluntary activation, the needle is positioned until two muscle potentials (a pair) from a single motor unit are recognized (Fig. 4). When a muscle fiber pair is identified, one fiber triggers the oscilloscope (triggering potential) and the second precedes or follows the first (slave potential). After recording multiple consecutive firings of these two muscle fibers (usually about 50–100 consecutive discharges), the electromyographer determines, usually with the aid of a computerized system, the consecutive interpotential intervals (IPIs) and calculates the difference between consecutive IPIs (Fig. 5). Comparison of IPIs illustrates the slight variation in transmission at the NMJ, named the neuromuscular jitter. Jitter is most accurately determined by calculating a *mean consecutive difference* (MCD) (see Measurements). Although jitter analysis may be obtained from any skeletal muscle, the most common muscles examined by voluntary SFEMG are the extensor digitorum communis, frontalis, and orbicularis oculi. These muscles are ideal because of their frequent involvement in NMJ disorders and the ability of most patients to control and sustain their voluntary activity to a minimum as required for the test.

Axonal stimulation SFEMG is an alternative method of motor unit activation. It is performed by inserting another monopolar needle electrode near the intramuscular nerve twigs and stimulating at a low current and constant rate. Then, the SFEMG is moved slightly until one or more MFAPs are recorded (Fig. 6). This technique requires that the electromyographer manipulates two electrodes, a stimulating and recording electrode. It has the advantage of not requiring patient participation and, thus, may be performed on children, uncooperative, or comatose patients. Another advantage of this technique is that the rate of stimulation may be adjusted from slow rate (2–3 Hz) to a rapid rate (20–50 Hz). This is helpful in differentiating presynaptic from postsynaptic disorders since the jitter improves significantly with rapid RNS in LES while it does not change or worsens in MG (see below) *(17, 18)*.

In volitional SFEMG, the jitter is calculated as the variation in IPIs between two MFAPs; one potential is time-locked by the trigger and all the variation of both end plates is expressed by the jitter of the slave potential. In contrast, the IPI in stimulated SFEMG is measured as the latency between the stimulus artifact and a single MFAP and, thus, only one end plate is involved in the analysis. Hence, stimulation SFEMG jitter normal values are smaller than their volitional counterparts.

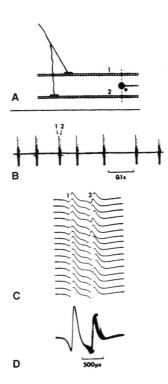

Fig. 5. Principle of voluntary single-fiber jitter recording. (**A**) The single-fiber EMG electrode is positioned by the electromyographer until it is possible to record from muscle fiber pair (1 and 2) innervated by the same motor axon (see also Fig. 4). (**B**) Muscle fiber action potentials firing at a low degree of voluntary activation. (**C**) As in B, but with a faster sweep speed, a sweep triggered by the first potential (the triggering potential) and showing successive discharges of the pair in a raster mode. (**D**) As in C, but shown in a superimposed mode illustrating the variability in the interpotential intervals (IPIs) which reflect the neuromuscular jitter (from Stålberg E, Trontelj JV. Single Fiber Electromyography. Studies in Healthy and Diseased Muscle. 2nd ed. New York: Raven Press, 1994).

3.4.3. Measurements

The jitter is best expressed as the mean consecutive difference (MCD) of all interpotential intervals (IPI) recorded of the muscle pair *(12, 19)*. It is calculated as follows:

$$MCD = \frac{(IPI1 - IPI2) + (IPI2 - IPI3) + ... + (IPIN - 1 - IPIN)}{N - 1}$$

where IPI1 is the interpotential interval of the first discharge, IPI2 of the second discharge, etc., and N is the number of discharges recorded. The mean jitter of the muscle sampled is reported after analyzing 10–20 muscle fiber pairs.

Normal values for jitter differ between muscles and tend to increase with age (Table 1) *(20, 21)*. As jitter values obtained by stimulation SFEMG are calculated on the basis of one end plate, the normal values are lower than those obtained by volitional activation. To calculate the normal stimulation jitter value, it is recommended that the reference data for volitional SFEMG is multiplied by 0.80. Blocking is measured as the percentage of discharges of a motor unit in which a single-fiber potential does not fire. For example, in 100 discharges of the pair, if a single potential is missing 30 times, the blocking is 30%. In general, blocking occurs when the jitter values are significantly abnormal.

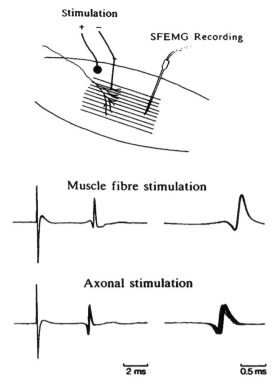

Fig. 6. Principle of stimulation single-fiber jitter recording. A monopolar needle stimulating cathode is inserted into the muscle near the motor point while the anode is a surface electrode. Lower tracing discloses normal jitter with axonal stimulation while the upper tracing discloses a very low jitter (<4 µs) resulting from direct muscle stimulation (from Stålberg E, Trontelj JV. Single Fiber Electromyography. Studies in Healthy and Diseased Muscle. 2nd ed. New York: Raven Press, 1994, with permission).

The results of SFEMG jitter study are expressed by (1) the mean jitter of all muscle pairs studied, (2) the percentage of pairs with impulse blocking, and (3) the percentage of pairs with normal jitter *(4, 12, 13)*. The jitter is considered abnormal when one or more of the following criteria are met (for individuals above 60 years, the second criterion should be more than 20%):

1. Mean jitter value exceeds the normal limit.
2. More than 10% of pairs have jitter greater than the upper limit of normal.
3. Impulse blocking is frequently seen in the majority of fiber pairs in a muscle.

3.4.4. Findings

Neuromuscular jitter is defined as the random variability of the time interval between two MFAPs innervated by the same motor unit. In normal subjects, there is a slight variability in the amount of acetylcholine released at the synaptic junction from one moment to another. Though a nerve action potential results in a muscle action potential at all times, the rise of end-plate potential is variable, resulting in a small variation of the muscle pair's IPI. *Impulse blocking* is defined as the failure of neuromuscular transmission of one of the potentials and represents the most extreme abnormality of the jitter.

Table 1
Reference Values for Jitter Measurements During Voluntary Muscle Activation (μs): 95% Confidence Limits for Mean Jitter/95% Confidence Limits for Jitter Values of Upper Limit of Individual Fiber Pairs (Adapted from Reference 20)

Muscle/age	10 years	20 years	30 years	40 years	50 years	60 years	70 years	80 years	90 years
Frontalis	33.6/ 49.7	33.9/ 50.1	34.4/ 51.3	35.5/ 53.5	37.3/ 57.5	40.0/ 63.9	43.8/ 74.1		
Orbicularis oculi	39.8/ 54.6	39.8/ 54.7	40.0/ 54.7	40.4/ 54.8	40.9/ 55.0	41.8/ 55.3	43.0/ 55.8		
Orbicularis oris	34.7/ 52.5	34.7/ 52.7	34.9/ 53.2	35.3/ 54.1	36.0/ 55.7	37.0/ 58.2	38.3/ 61.8	40.2/ 67.0	42.5/ 74.2
Tongue	32.8/ 48.6	33.0/ 49.0	33.6/ 50.2	34.8/ 52.5	36.8/ 56.3	39.8/ 62.0	44.0/ 70.0		
Sternocleidomastoid	29.1/ 45.4	29.3/ 45.8	29.8/ 46.8	30.8/ 48.8	32.5/ 52.4	34.9/ 58.2	38.4/ 62.3		
Deltoid	32.9/ 44.4	32.9/ 44.5	32.9/ 44.5	32.9/ 44.6	33.0/ 44.8	33.0/ 45.1	33.1/ 45.6	33.2/ 46.1	33.3/ 46.9
Biceps	29.5/ 45.2	29.6/ 45.2	29.6/ 45.4	29.8/ 45.7	30.1/ 46.2	30.5/ 46.9	31.0/ 48.0		
Extensor digitorum communis	34.9/ 50.0	34.9/ 50.1	35.1/ 50.5	35.4/ 51.3	35.9/ 52.5	36.6/ 54.4	37.7/ 57.2	39.1/ 61.1	40.9/ 66.5
Abductor digiti minimi	44.4/ 63.5	44.7/ 64.0	45.2/ 65.5	46.4/ 68.6	48.2/ 73.9	51.0/ 82.7	54.8/ 96.6		
Quadriceps	35.9/ 47.9	36.0/ 48.0	36.5/ 48.2	37.5/ 48.5	39.0/ 49.1	41.3/ 50.0	44.6/ 51.2		
Tibialis anterior	49.4/ 80.0	49.3/ 79.8	49.2/ 79.3	48.9/ 78.3	48.5/ 76.8	47.9/ 74.5	47.0/ 71.4	45.8/ 67.5	44.3/ 62.9

Jitter analysis is highly sensitive but not specific. It is frequently abnormal in MG and other NMJ disorders, but may also be abnormal in a variety of neuromuscular disorders including neuropathies, myopathies, and anterior horn cell disorders. Thus, a diagnosis of a NMJ disorder, such as myasthenia gravis, by jitter analysis should always be interpreted only in the contest of the patient's clinical manifestations, nerve conduction studies, and needle EMG findings.

4. NEUROMUSCULAR DEFECT FINDINGS IN NEUROMUSCULAR JUNCTION DISORDERS

4.1. Myasthenia Gravis

Myasthenia gravis (MG) is the best understood and thoroughly studied of all human organ-specific autoimmune diseases. MG is caused by an antibody-mediated attack on the postsynaptic nicotinic ACh receptors in the neuromuscular junction. It is characterized by a reduction of skeletal muscle postsynaptic ACh receptors resulting in a decrease in the EPP amplitude necessary for action potential generation. Up to 85% of patients with MG have an elevated ACh receptor antibody (seropositive MG). Antibodies to muscle-specific tyrosine kinase (MuSK) are present in sera of 25–50% of MG patients who have negative ACh receptor antibody. The term "seronegative MG" should hence be reserved to patients who have negative antibodies to ACh receptor and MuSK.

The EDX study is an essential component of the available armament in the diagnosis of MG. The EDX studies are most useful in seronegative patients with suspected MG and in patients with negative or equivocal tensilon test or neurological findings. Individual EDX studies of patients with suspected MG should be tailored to the patient's symptomatology.

4.1.1. Baseline Motor Nerve Conduction Studies

Routine motor nerve conduction studies (NCSs), including CMAP amplitudes, are normal in MG. A single supramaximal stimulus to a motor nerve results in ACh release and postsynaptic EPPs which reach threshold despite ACh receptor blockade because of the presence of the safety factor. Hence, MFAPs are generated in all muscle fibers resulting in a normal CMAP. Occasionally such as in the midst of a myasthenic crisis, the CMAPs may be borderline or slightly diminished due to severe postsynaptic neuromuscular blockade (See Fig. 2). Also, in patients taking large quantities of cholinesterase inhibitors, such as pyridostigmine, there may be a tendency to record multiple CMAPs after a single stimulus applied to the nerve.

4.1.2. Slow Repetitive Nerve Stimulation

Slow RNS results in a decrease in quantal release due to the depletion of the immediate ACh stores. This causes that many EPPs do not reach threshold needed to generate action potentials. The loss of many MFAPs is reflected as a decremental CMAP on slow RNS. The greatest decrease in CMAP amplitude and area in MG occurs between the first and the second responses, but the decrement usually continues to the third or fourth response *(2, 3, 4, 7, 8)*. Often, after the fifth or sixth stimulus, the secondary stores are mobilized and no further loss of MFAP occurs (See Fig. 2). This results in stabilization, or sometimes slight improvement of the CMAP after the fifth or sixth stimulus, giving the characteristic "U-shaped" decrement. A reproducible CMAP decrement of >10% is abnormal and eliminates false positives; CMAP decrement of 5–10% is considered equivocal. The diagnostic yield of slow RNS in MG is increased by the following:

1. *Recording from clinically weakened muscles, such as proximal and facial muscles.* This often means recording from a proximal muscle in generalized MG (such as the trapezius) or from a facial muscle in ocular MG (such as the orbicularis oris or nasalis) (Fig. 7). This strategy increases the diagnostic sensitivity of slow RNS in the diagnosis of MG *(22)*. Compared to recording distal muscles (such as the abductor digiti minimi), slow RNS of the facial and spinal accessory nerves, recording from orbicularis oculi and trapezius muscles, respectively, increases the diagnostic sensitivity of the study by about 5–20% *(23)*. Though often technically challenging, a decremental response on slow RNS of the facial nerve is highly specific for the diagnosis of MG *(24)* and is common in MuSK-positive MG patients *(25)*.
2. *Obtaining slow RNS following exercise* looking for post-exercise exhaustion. After performing slow RNS at rest, the patient is asked to exercise the tested muscle for 1 min. Then, slow RNSs are repeated every 30–60 s for 4–5 min. Post-exercise exhaustion usually occurs 4–6 min after exercise and is particularly useful in patients with suspected MG and equivocal CMAP decrement at rest. However, post-exercise studies only increase the yield of RNS by about 5–7% *(26)*.

4.1.3. Rapid Repetitive Nerve Stimulation and Post-exercise Facilitation

Rapid RNS is not useful or necessary in the diagnosis of MG. It should only be considered when a presynaptic NMJ disorder (such as LES or botulism) is clinically suspected and needs to be excluded, or if the baseline CMAPs are borderline or low in amplitudes. Often, CMAP evaluation after a brief (10 s) exercise is sufficient in these situations and eliminates the severe discomfort induced by rapid RNS. With rapid RNS (or following brief exercise), the depleted ACh stores are usually compensated by increased ACh release due to the accumulation of presynaptic calcium resulting in no change of CMAP amplitude. In severe myasthenics, rapid RNS may cause a decrement when the increased ACh release cannot compensate for the marked postsynaptic neuromuscular blockade. Hence, the findings in MG (no CMAP change or decrement) are in contrast to the impressive CMAP increment seen in patients with a presynaptic NMJ disorder such as LES or botulism (see below).

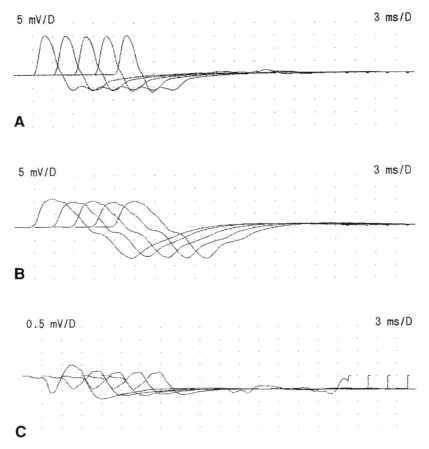

Fig. 7. Slow repetitive stimulation in a 35-year-old man with binocular diplopia and a history of ptosis, revealing (**A**) no decrement simulating median nerve (recording thenar muscles), (**B**) mild and equivocal decrement (10%) stimulating spinal accessory nerve (recording upper trapezius muscle), and (**C**) significant decrement (40%) simulating the facial nerve (recording orbicularis oculi muscle).

4.1.4. Single-Fiber EMG

Evaluation of neuromuscular transmission in patients with suspected MG is the most common indication for performing SFEMG. Commonly tested muscles in patients with suspected MG are the extensor digitorum communis, orbicularis oculi, and frontalis.

In patients with MG, abnormal jitter values are common (Fig. 8). This is frequently accompanied by neuromuscular impulse blocking, which reflects the failure of one of the muscle fibers to transmit an action potential due to the inability of EPP to reach threshold. SFEMG is extremely sensitive in detecting MG; a normal SFEMG jitter study in a weak muscle virtually excludes the diagnosis of MG. The published diagnostic sensitivity of SFEMG ranges from 90 to 99% (Fig. 9) *(12, 13, 14, 22)*. This makes SFEMG the most sensitive diagnostic study in MG *(22)*. However, abnormal jitter is nonspecific since it is often seen in a variety of neuromuscular disorders. Hence, SFEMG jitter findings should be correlated with the history, examination, and conventional needle EMG.

SFEMG sensitivity and specificity is equal in MG patients with ACh receptor antibody positive or seronegative patients. However, SFEMG of limb muscles, such as the extensor

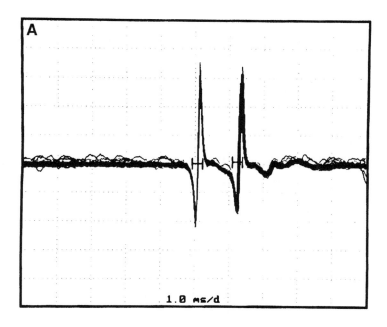

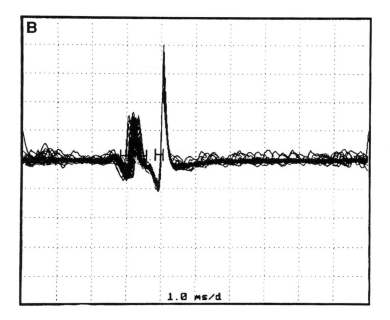

Fig. 8. Neuromuscular jitter recording frontalis muscle with voluntary activation in a 22-year-old patient with ptosis (shown in a superimposed mode – see Fig. 5 also). (**A**) Normal jitter (mean consecutive discharge (MCD) = 18.5 μs), (**B**) abnormal jitter (MCD = 65 μs).

digitorum communis, are often normal in MuSK-positive MG patients who frequently have prominent facial and bulbar muscle weakness *(27)*. In comparison, SFEMG of facial muscles, such as the orbicularis oculi, has a similar sensitivity in MuSK-positive patients compared to ACh receptor-positive patients *(27)*.

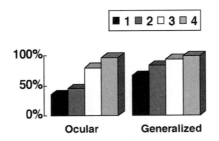

1 = *Slow RNS recording distal (hand) muscle*

2 = *Slow RNS recording proximal (shoulder) muscle*

3 = *Single Fiber EMG of forearm muscle (extensor digitorum communis)*

4 = *Single Fiber EMG of facial muscle (frontalis)*

Fig. 9. The diagnostic sensitivity of slow repetitive nerve stimulation and single-fiber EMG in the diagnosis of myasthenia gravis (Data compiled from Oh SJ, Kim DE, Kuruoglu R, Bradley RJ, Dwyer D. Diagnostic sensitivity of the laboratory tests in myasthenia gravis. Muscle Nerve. 1992;15:720–724 and from Howard, JF, Sanders DB, and Massey JM. The electrodiagnosis of myasthenia gravis and the Lambert–Eaton syndrome. Neurol Clin 1994;12:305–330).

4.1.5. Conventional Needle EMG

The needle EMG examination in evaluation of MG serves mostly in excluding other neuromuscular causes of weakness such as motor neuron disease, neuropathies, or myopathies. Though the needle EMG is usually normal in patients with MG, certain findings are occasionally present and not inconsistent with the diagnosis. These findings usually occur in patients with moderate or severe MG and are most evident in proximal muscles. Findings on needle EMG that are compatible with MG include (1) moment-to-moment variation of MUAP morphology (MUAP instability), (2) short duration, low amplitude and polyphasic MUAPs, (3) fibrillation potentials, and fasciculation potentials. *MUAP instability* occurs when individual muscle fibers are either blocked or come to action potential at varying intervals, leading to an MUAP that changes in configuration (amplitude and/or number of phases) from impulse to impulse. *Short duration, low amplitude, polyphasic MUAPs* are seen when blocking is severe, resulting in dropout of muscle fibers, thereby reducing the number of muscle fibers per motor unit. *Fibrillation potentials*, rarely seen in MG, are due to effective denervation of muscle fibers due to chronic neuromuscular blockade or loss of end plates *(5)*. *Fasciculation potentials* may be encountered in patients treated with large doses of cholinesterase inhibitors, such as pyridostigmine. All these findings are, however, not specific and more often seen in myopathies or denervating disorders.

4.2. Lambert—Eaton Syndrome

Lambert–Eaton syndrome (LES) is a rare autoimmune disorder of the neuromuscular junction associated with small cell lung cancer (SCLC) in approximately 40–50% of patients *(9, 10, 28, 29)*. A significant predictor for developing SCLC in patients with LES is smoking at the time of diagnosis. SCLC is commonly detected soon after the onset of LES symptoms and is rarely diagnosed beyond 5 years. The neuromuscular junction disorder is caused by antibodies against the presynaptic P/Q type voltage-gated calcium channels (VGCC) detected in three-fourths of patients with LES *(30)*. The block of VGCCs results in a decrease in calcium influx during depolarization of the presynaptic membrane and interferes with the calcium-dependent release of ACh from its stores in vesicles into the synaptic cleft. The EDX abnormalities encountered in LES constitute the mainstay of diagnosis as described originally by Lambert and Eaton *(9, 10, 28, 29)*.

4.2.1. Baseline Motor Nerve Conduction Studies

The CMAPs, obtained during routine motor NCSs, are low in amplitudes in patients with LES, since many end plates do not reach threshold after a single stimulus due to inadequate release of synaptic vesicles. Thus, a low number of MFAPs is generated leading to low-amplitude CMAPs usually present in all muscles. Low-amplitude CMAPs may be the first clue that a patient with weakness may have LES, and the electromyographer may be the first to diagnose LES in the EMG laboratory by evaluating post-exercise CMAP in patients with universally low CMAP amplitude referred for a variety of reasons *(4)*.

4.2.2. Rapid Repetitive Nerve Stimulation and Post-exercise Facilitation

Rapid RNS (usually 20–50 Hz) or brief (10 s) exercise enhances calcium influx into the presynaptic terminal, which results in larger releases of quanta and larger EPPs. While many EPPs do not reach threshold after the first stimulus, many muscle fibers achieve threshold required for the generation of MFAPs with the subsequent stimuli. Hence, rapid RNS or brief (10 s) exercise results in a CMAP amplitude increment (Fig. 10). This post-tetanic facilitation exceeds 60% in all muscles and 100% in at least one muscle in 80% of LES patients *(30, 31)*. It is also common to have robust increments exceeding 200% in many LES patients *(9, 10, 11, 28, 29, 32, 33)*, particularly those with positive VGCC serology *(34)*.

4.2.3. Slow Repetitive Nerve Stimulation

Slow RNS (usually 2–3 Hz) results in >10% CMAP decrement in all patients with LES *(30)*. With slow RNS, ACh release is reduced further because of the depletion of the immediately available stores, and at this slow rate of stimulation, calcium does not accumulate in the presynaptic terminal. The end result is further loss of many MFAPs and a decrement of CMAP amplitude. However, slow RNS is not useful in LES diagnosis since it cannot distinguish it from the decrement seen in postsynaptic disorders including MG.

4.2.4. Single-Fiber EMG

SFEMG jitter is abnormal in LES, similar to the findings in MG *(4, 12, 13, 17, 35)*. However, axonal stimulation SFEMG technique may help distinguish these two disorders. With this method, the jitter improves significantly with rapid rate stimulation (20–50 Hz) in LES, due to enhancement of calcium influx which results in larger releases of quanta. In contrast, rapid rate of axonal stimulation does not change or worsen the jitter in MG.

4.2.5. Conventional Needle EMG

The findings on needle EMG in LES are very similar to MG, although the changes are often more conspicuous and only present in severe cases.

4.3. Botulism

Botulism is a rare but serious and potentially fatal illness. It is caused by the toxicity of botulinum toxin which is produced by the anaerobic bacterium, *Clostridium botulinum*. The toxin has a significant affinity to both muscarinic and nicotinic cholinergic nerve terminals resulting in autonomic failure and skeletal muscle paralysis. Botulinum toxin results in failure of ACh release from the presynaptic terminal and ultimately leads to destruction of the presynaptic terminal. Botulinum toxin first attaches irreversibly to the axonal terminal, and interferes with the calcium-dependent intracellular cascade that is responsible for ACh release, by cleaving proteins essential for docking and fusion of the presynaptic vesicles at the presynaptic active zones. Depending on the mode of entry of the toxin into the bloodstream, botulism is classified into four clinically distinct forms: Food-borne (classical) botulism, infant botulism, wound botulism, and iatrogenic (inadvertent) botulism.

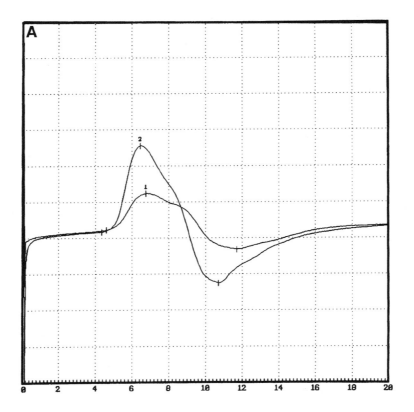

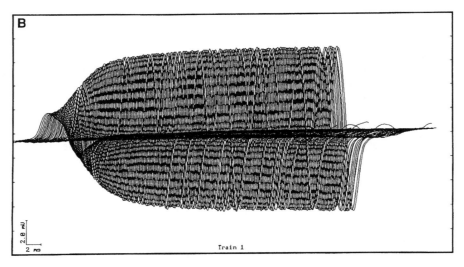

Fig. 10. Post-tetanic compound muscle action potential (CMAP) evaluation in a patient with LES stimulating the ulnar nerve (recording hypothenar muscles). (**A**) Baseline low amplitude CMAP at 2.0 mV (waveform 1). Following 10 s of voluntary maximal isometric exercise, there is a significant CMAP increment (waveform 2) to 4.8 mV (increment = 140%) (sensitivity 2 mV/division). (**B**) Rapid (50 Hz) stimulation revealing marked increment of CMAP amplitude (250%).

The EDX studies in botulism are a mainstay in diagnosis. The study is a rapid and readily available method of diagnosis for patients with suspected botulism, while definitive bioassays for botulinum toxin or stool cultures for *C. botulinum* are usually delayed and may be negative. The EDX findings in botulism are compatible with a presynaptic defect of neuromuscular transmission and are somewhat similar to the findings in LES *(6, 36, 37, 38, 39)*. However, the results of EDX studies are more variable and dependent on the amount of toxic exposure and the timing of the study. During the early course of the disease, it is common for the EDX results to change significantly from day to day.

4.3.1. Baseline Motor Nerve Conduction Studies

Low CMAP amplitudes, obtained during routine motor NCSs, are the most consistent finding present in 85% of patients, particularly when recording from weak muscles (usually proximal) *(6, 38)*. Low CMAP is the result of the inadequate release of quanta (vesicles), which in turn render many EPPs unable to reach threshold after a single stimulus. In contrast to CMAP amplitudes, motor distal latencies and conduction velocities are normal.

4.3.2. Rapid Repetitive Nerve Stimulation and Post-exercise Facilitation

Rapid RNS, or CMAP following brief (10 s) of isometric exercise, in patients with botulism results in CMAP increment. With tetanic stimulation, Ca^2 influx is greatly enhanced resulting in larger releases of quanta. This leads to increasing number of muscle fibers reaching the threshold required for the generation of muscle action potentials. The increment in botulism is modest, ranging between 50 and 100% (Fig. 3), when compared to the increment in LES, which often surpasses 200% (Table 2) *(6, 38, 39)*. The increment may occasionally be absent, especially in severe cases such as those caused by type A toxin, presumably due to the degree of presynaptic blockade *(40)*.

4.3.3. Slow Repetitive Nerve Stimulation

Slow RNS in patients with botulism may cause a decrement of CMAP amplitude. However, this is infrequent and mild not exceeding 8–10% of baseline. It is caused by the progressive depletion of the immediately available ACh stores without accumulation of Ca^2 in the presynaptic terminals.

4.3.4. Single-Fiber EMG

Increased jitter with blocking may be observed by single-fiber EMG *(36, 37)*. Axonal stimulation jitter usually improves during rapid stimulation.

Table 2
Baseline CMAP and Repetitive Stimulation Findings in Common Neuromuscular Junction Disorders

NMJ disorder	NMJ defect	CMAP amplitude	Slow RNS	Rapid RNS or post-exercise facilitation
Myasthenia gravis	Postsynaptic	Normal	Decrement	Normal or decrement
Lambert–Eaton syndrome	Presynaptic	Low in all muscles	Decrement	Marked increment: >100% in at least one muscle and >60% in all muscles
Botulism	Presynaptic	Low in proximal and weak muscles	Decrement	Modest increment: in weak muscles (50–100%)

NMJ, neuromuscular junction; CMAP, compound muscle action potential; RNS, repetitive nerve stimulation.

4.3.5. Conventional Needle EMG

The needle EMG in botulism is variable and depends on the amount of toxic exposure. The needle EMG may be normal or sometimes reveals increase in the number of short-duration, low-amplitude, and polyphasic MUAPs, occasional fibrillation potentials, and MUAP instability. In severe cases, rapid recruitment of only few MUAPs may be apparent.

4.4. Congenital Myasthenic Syndromes

Congenital myasthenic syndromes (CMS) are caused by genetic defects of the presynaptic or postsynaptic apparatus. Most cases of CMS, regardless of primary etiology, demonstrate a degeneration of the postsynaptic region and simplification of junctional folds often with a concomitant reduction of ACh receptors *(41)*.

Given the similar anatomic pathology of CMSs, slow RNS often produces a decremental response, which may be absent during rest and only elicited after several minutes of exercise. In cholinesterase deficiency and slow-channel syndrome, the best characterized CMSs, a single electrical stimulation leads to repetitive CMAPs, similar to the findings in organophosphate poisoning and in patients taking large doses of cholinesterase inhibitors.

4.5. Electrodiagnostic Strategy in a Patient with Suspected Neuromuscular Junction Disorder

The EDX study of a patient with suspected NMJ disorder should start with a detailed history and comprehensive neurological examination. Sensory and motor NCSs in two limbs (preferably an upper and a lower extremity) should be the initial studies. The clinical situation and baseline CMAP amplitudes dictate the next steps (Fig. 11) *(2, 3)*.

- If the CMAP amplitudes are low, a presynaptic defect should be suspected and ruled out.
- If the diagnosis of LES is clinically suspected, baseline and post-exercise CMAP of at least two distal motor nerves is a sufficient screening test. In LES, CMAP increment is usually robust and may surpass 200%. A rapid RNS of one distal nerve for 3–5 s, which is extremely painful, is sufficient for confirmation if there are definite CMAP increments after brief exercise.

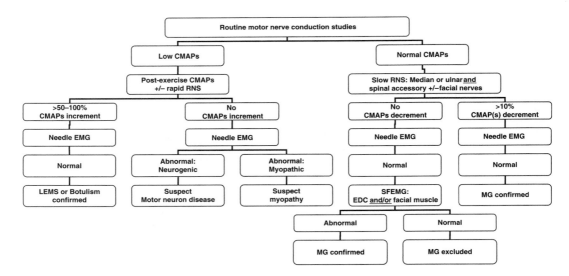

Fig. 11. Electrodiagnostic strategy in patients with suspected neuromuscular junction disorder.

- If the diagnosis of botulism is considered, the choice of muscle should include clinically weak muscles. In botulism, the CMAP increment is usually 50–100%. Also, the study may be repeated in 1–2 days, particularly if the initial evaluation was done during the early phase of the illness.
- If the diagnosis of MG is clinically suspected, slow RNS at rest and for 4–5 min following a 1-min exercise should be performed on at least two motor nerves. The selection of nerves and muscles is dependent on the clinical manifestations with the goal to record from clinically weak muscles. One should start by performing slow RNS on a distal hand muscle (such as the abductor digiti minimi or abductor pollicis brevis) and move on to a proximal muscle such as the upper trapezius. Facial RNS should be reserved for patients with oculobulbar manifestations and normal slow RNS recording from distal and proximal muscles (see Figure 7). SFEMG of one or two muscles (such as the frontalis, orbicularis oculi, or extensor digitorum communis) should be considered if the diagnosis of MG is still considered despite normal RNS studies and absence of serum ACh receptor antibodies.
- If the CMAPs obtained on motor NCSs in a patient with suspected MG are low or borderline, post-exercise CMAP screening should always be done to exclude LES. A misdiagnosis of MG is often made if post-exercise CMAP evaluation is not done and a slow RNS shows a CMAP decrement.
- Post-exercise CMAP screening is recommended for patients with weakness associated with a malignancy (particularly a small cell lung cancer) or if the clinical situation does not clearly differentiate between LES and MG.

5. NEUROMUSCULAR DEFECT FINDINGS IN OTHER NEUROMUSCULAR DISORDERS

5.1. Amyotrophic Lateral Sclerosis

Amyotrophic lateral sclerosis (ALS) is a relentlessly progressive, fatal disease caused by degeneration of both upper and lower motor neurons. Slow rates of RNS result in a decrementing response in at least 50% of ALS patients, suggesting functional alterations of the neuromuscular junction that accompany this disease *(42, 43)*. The decrement is usually inversely proportional to CMAP amplitude. Although many authors favor a presynaptic origin to the neuromuscular junction defect in ALS patients, none of these patients exhibit any significant post-exercise facilitation of the baseline low CMAP amplitude. Single-fiber EMG jitter, which is abnormal in the majority of patients with ALS, does not differentiate presynaptic from postsynaptic neuromuscular junction defects. The neuromuscular transmission defect is best explained by reinnervation of partial or completely denervated motor end plates with reduction of ACh release parameters.

5.2. Miller-Fisher Syndrome

The Miller-Fisher syndrome, considered a variant of Guillain–Barre syndrome, usually consists of a clinical triad of ataxia, ophthalmoplegia, and areflexia. In more than 90% of patients, serum antibodies to GQ1b ganglioside are detected. Single-fiber EMG studies in many patients with Miller-Fisher syndrome show abnormal neuromuscular transmission in cranial innervated as well as limb muscles *(44, 45, 46)*. These changes are more common in GQ1b-positive patients and are consistent with a terminal motor axon dysfunction. Sera from GQ1b-positive patients bind to the presynaptic terminal causing an increase in miniature end-plate potential frequency. In vitro studies found both presynaptic and postsynaptic reversible blocking effects to GQ1b-positive sera *(47)*. Other studies found that anti-GQ1b antibodies induced irreversible quantal release of ACh from nerve terminals, eventually causing neuromuscular blockade *(48)*.

5.3. Guillian—Barré Syndrome

Early on, it was recognized that increased jitter occurs in patients with early Guillian–Barré syndrome (GBS). This was felt to be due to unstable peripheral nerve conduction during the demyelination phase. Reversible blocking at the NMJ was detected in vitro by sera from GBS patients who are seropositive or seronegative to anti-GM1 antibodies *(49)*. Dysfunction at the NMJ was documented by distinctly abnormal jitter values. However, intermittent genuine impulse blocking without correspondingly increased jitter seen in GBS patients is a likely sign of axolemmal dysfunction *(50)*

REFERENCES

1. Boonyapisit K, Kaminski HJ, Ruff RL. The molecular basis of neuromuscular transmission disorders. Am J Med 1999;06:97–113.
2. KatirjiB. Electromyography in Clinical Practice. A Case Study Approach. Second edition, St Louis, Mosby/Elsevier, 2007.
3. Katirji B, Kaminski HJ. An electrodiagnostic approach to the patient with neuromuscular junction disorder. Neuro Clin. 2003;20:557–586.
4. Oh SJ. Electromyography: Neuromuscular Transmission Studies. Baltimore, Williams and Wilkins, 1988, pp. 87–110, 243–253.
5. Barberi S, Weiss GM, Daube JR. Fibrillation potentials in myasthenia gravis. Muscle Nerve 1982;5:S50.
6. Cherrington M. Botulism. In: Katirji B, Kaminski HJ, Preston DC, Ruff RL, Shapiro EB, eds. Neuromuscular disorders in Clinical Practice. Boston, Butterworth-Heinemann, 2002: 942–952.
7. Desmedt JE, Borenstein S. Diagnosis of myasthenia gravis by nerve stimulation. Ann NY Acad Sci 1976; 74:174–188.
8. Ozdemir C, Young RR. The results to be expected from electrical testing in the diagnosis of myasthenia gravis. Ann NY Acad Sci 1976;74:203–222.
9. Eaton LM, Lambert EH. Electromyography and electric stimulation of nerves in diseases of motor unit: observations on myasthenic syndrome associated with malignant tumors. JAMA 1957; 163:1117–1124.
10. Lambert EH, Eaton LM, Rooke ED. Defect of neuromuscular conduction associated with malignant neoplasms. Am J Physiol 1956; 187: 612–613.
11. Tim RW, Sanders DB. Repetitive nerve stimulation studies in the Lambert-Eaton myasthenic syndrome. Muscle Nerve 1994; 17: 995–1001.
12. Stålberg E, Trontelj JV. Single Fiber Electromyography. Studies in Healthy and Diseased Muscle. 2nd ed. New York, Raven Press, 1994.
13. Sanders DB, Stålberg EV. Single fiber Electromyography. Muscle Nerve 1996;19:1069–1083.
14. Sanders DB, Howard JF. Single fiber EMG in myasthenia gravis. Muscle Nerve 1986;9:809–819.
15. Sarrigiannis PG, Kennett RP, Read S, Farrugia ME. Single-fiber EMG with concentric needle electrode: validation in myasthenia gravis. Muscle Nerve 2006;33:61–65.
16. Benatar M, Hammad M, Doss-Riney H. Concentric-needle single-fiber Electromyography for the diagnosis of myasthenia gravis. Muscle Nerve 2006;33: 61–65.
17. Trontelj JV, Stålberg E. Single motor endplates in myasthenia gravis and Lambert Eaton myasthenic syndrome at different firing rates. Muscle Nerve 1990;14:226–232.
18. Sanders DB. The effect of firing rate on neuromuscular jitter in Lambert-Eaton myasthenic syndrome. Muscle Nerve 1992;15:256–258.
19. Ekstedt J, Nilsson G, Stålberg E. Calculation of the electromyographic jitter. J Neurol Neurosurg Psychiatry 1974;37:526–539.
20. Gilchrist JM. Ad hoc committee of the AAEM special interest group on SFEMG. Single fiber EMG reference values: a collaborative effort. Muscle Nerve 1992;15:151–161.
21. Gilchrist JM. Single fiber EMG. In: Katirji B, Kaminski HJ, Preston DC, Ruff RL, Shapiro EB, eds. Neuromuscular disorders in Clinical Practice. Boston, Butterworth-Heinemann, 2002:141–150.
22. Oh SH, Kim DE, Kuruoglu R, Bradley RJ, Dwyer D. Diagnostic sensitivity of the laboratory tests in Myasthenia gravis. Muscle Nerve 1992;5:720–724.
23. Costa J, Evangelista T, Conceicao I, de Carvalho M. Repetitive nerve stimulation in myasthenia gravis–relative sensitivity of different muscles. Clin Neurophysiol 2004;115(12):2776–2782.
24. Zinman LH, O'connor PW, Dadson KE, et al. Sensitivity of repetitive facial-nerve stimulation in patients with myasthenia gravis. Muscle Nerve 2006;33: 694–696.
25. Oh SJ, Hatanaka Y, Hemmi S, et al. Repetitive nerve stimulation of facial muscles in MuSK antibody-positive myasthenia gravis. Muscle Nerve 2006;33: 500–504.
26. Rubin DI, Hentschel K. Is exercise necessary with repetitive nerve stimulation in evaluating patients with suspected myasthenia gravis? Muscle Nerve 2007;35: 103–106.
27. Farrugia ME, Kennett RP, Newsom-Davis J, et al. Single-fiber electromyography in limb and facial muscles in muscle-specific kinase antibody and acetylcholine receptor antibody myasthenia gravis. Muscle Nerve 2006;33: 568–570.

28. Lambert EH, Rooke ED, Eaton LM, et al. Myasthenic syndrome occasionally associated with bronchial neoplasm: neurophysiologic studies. In Viets HR ed. Myasthenia Gravis: The second International Symposium Proceedings 1959, Springfield, IL, Charles C. Thomas, 1961: 362–410.

29. O'Neil JH, Murray NMF, Newsom-Davis J. The Lambert -Eaton myasthenic syndrome: a review of 50 cases. Brain 1988;111:577–596.

30. Juel VC, Massey JM, Sanders DB. Lambert-Eaton Myasthenic syndrome. Findings in 97 patients. Muscle Nerve 2006;34:543.

31. Oh SJ, Kurokawa K, Claussen GC, Ryan JR HF. Electrophysiological diagnostic criteria of Lambert-Eaton myasthenic syndrome. Muscle Nerve 32: 515–520, 2005.

32. Maddison P, Newsom-Davis J, Mills KR. Distribution of electrophysiological abnormality in Lambert-Eaton myasthenic syndrome. J Neurol Neurosurg Psychiatry 1998;5: 213–217.

33. Maddison P, Newsom-Davis J. The Lambert-Eaton myasthenic syndrome. In: Katirji B, Kaminski HJ, Preston DC, Ruff RL, Shapiro EB, eds. Neuromuscular disorders in Clinical Practice. Boston, Butterworth-Heinemann, 2002: 931–941.

34. Oh SJ, Hatanaka Y, Claussen GC, Sher E. Electrophysiological differences in seropositive and seronegative Lambert-Eaton myasthenic syndrome. Muscle Nerve 2007; 35:178–183.

35. Oh SJ, Hurwitz EL, Lee KW, Chang CW, Cho HK. The single-fiber EMG in the Lambert-Eaton myasthenic syndrome. Muscle Nerve 1989;12:159–161.

36. Schiller HH, Stålberg E. Human botulism studied with single-fiber electromyography. Arch Neurol 1978;35:346–349.

37. Padua L, Aprile I, Monaco ML, et al. Neurophysiological assessment in the diagnosis of botulism: usefulness of single-fiber EMG. Muscle Nerve 1999;22:1388–1392.

38. Cherrington M. Electrophysiologic methods as an aid in diagnosis of botulism: a review. Muscle Nerve 1982;5:528–529.

39. Cornblath DR, Sladky JT, Sumner AJ. Clinical electrophysiology of infantile botulism. Muscle Nerve 1983;6:448–652.

40. Souayah N, Karim H, Kamin SS, et al. Severe botulism after focal injection of botulinum toxis. Neurology 2006;67:1855–1856.

41. Engel AG. Congenital myasthenic syndromes. In: Katirji B, Kaminski HJ, Preston DC, Ruff RL, Shapiro EB, eds. Neuromuscular disorders in Clinical Practice. Boston, Butterworth-Heinemann, 2002:953–963.

42. Maselli RA, Wollman RB, Leung C et al. Neuromuscular transmission in amyotrophic lateral sclerosis. Muscle Nerve 1993;16:1193–1203.

43. Killian JK, Wilfong AA, Burnett L, Appel SH, Boland D. Decremental motor responses to repetitive nerve stimulation in ALS. Muscle Nerve 17:747–754 1994.

44. Lo YL, Chan LL, Pan A, Ratnagopal P. Acute ophthalmoparesis in the anti-GQ1b antibody syndrome: electrophysiological evidence of neuromuscular transmission defect in the orbicularis oculi. J Neurol Neurosurg Psychiatry 2004;75:436–440.

45. Sartucci F, Cafforio G, Borghetti D, Domenici L, Orlandi G, Murri L. Electrophysiological evidence by single fiber electromyography of neuromuscular transmission impairment in a case of Miller Fisher syndrome. Neurol Sci 2005;26:125–128.

46. Lange D J, Deangelis T, Sivak MA. Single-fiver electromyography shows terminal axon dysfunction in Miller Fisher syndrome. Muscle Nerve 2006;34:232–234.

47. Buchwald B, Bufler J, Carpo M, et al. Combined pre- and postsynaptic action of IgG antibodies in Miller Fisher syndrome. Neurology 2001;56:67–74.

48. Plomp JJ, Molenaar PC, O'Hanlon GM, et al. Miller Fisher anti-GQ1b antibodies: $\forall$-latrotoxin–like effects on motor endplates. Ann Neurol 1999;45:189 –199.

49. Buchwald B, Toyka KV, Zielasek J, et al. Neuromuscular blockade by IgG antibodies from patients with Guillain–Barré syndrome: a macropatch-clamp study. Ann Neurol 1998;44:913–922.

50. Spaans F, Vredeveld JW, Morre HH, Jacobs BC, De Baets MH. Dysfunction at the motor end-plate and axon membrane in Guillain–Barré syndrome: a single-fiber EMG study. Muscle Nerve 2003;27:426–434.

9

Autoantibody Testing in the Diagnosis and Management of Autoimmune Disorders of Neuromuscular Transmission and Related Diseases

Mark A. Agius, David P. Richman, and Angela Vincent

1. INTRODUCTION

The neuromuscular junction (NMJ), the synapse between the motor nerve and the muscle fiber, includes the motor nerve terminal, the synaptic cleft, and the "endplate" region of the postsynaptic muscle fiber. A number of molecules, which include ion channels and other proteins at the NMJ, may be targeted by the immune system resulting in disordered neuromuscular transmission. Antibodies to acetylcholine receptors (AChR), resulting in the clinical picture of acquired myasthenia gravis (MG), were the first to be demonstrated to be pathogenic. More recently, additional targets in both the nerve terminal and the muscle cell have been identified, including calcium and potassium ion channels in the motor nerve terminal, and components of the AChR clustering mechanisms at the NMJ, as well as proteins located intracellularly within the muscle fiber. In most cases, the antibodies that bind to these proteins appear to be pathogenic and their identification allows for specific treatments and, in certain situations, may suggest modifications to current treatment.

Thus one of the most satisfying aspects of current understanding of the role of antibodies in MG and other diseases of the NMJ is the ability to measure these antibodies effectively in the laboratory. This is largely due to the availability of various neurotoxins that bind specifically and with high affinity to the ion channels that are the targets for these antibodies. Alternative approaches, including new cellular techniques, are needed for autoantibody detection against other proteins, for example the muscle-specific kinase (MuSK) and various intracellular proteins. Here we start by discussing some general principles and then go on to describe the clinical syndromes and the methods that can be used to investigate the pathogenic role of autoantibodies.

2. SPECTRUM OF ANTIBODIES TO TARGETS AT THE NMJ AND ASSOCIATED MOLECULES

Most of the proteins targeted in autoimmune myasthenic disorders and related conditions are located in the cell membrane of the motor nerve or muscle endplate with an extracellular domain that is accessible to circulating substances. A functional blood–nerve barrier, composed of endothelial cells and gap junctions, is not present at the NMJ exposing it to the circulation. In some instances, however, the autoimmune targets appear to be intracellular. Whereas a pathogenic role for autoimmunity against NMJ proteins with an extracellular domain is well characterized, it is still unclear as to whether an immune attack against intracellular components is significant in the pathophysiology of these conditions.

The molecules identified to be targets in the NMJ are illustrated (Fig. 1 and Color Plate 3, following p. 313). These include calcium and potassium channels in the motor nerve terminal,

From: *Current Clinical Neurology*
Myasthenia Gravis and Related Disorders
Edited by: H.J. Kaminski, DOI 10.1007/978-1-59745-156-7_9, © Humana Press, New York, NY

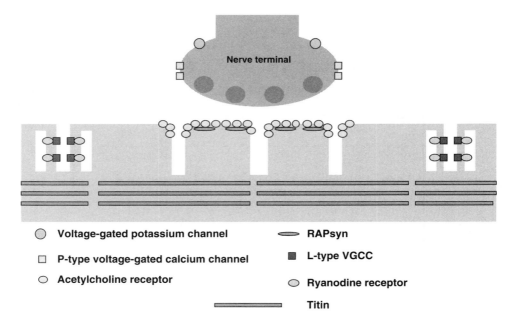

Fig. 1. Diagram of the neuromuscular junction with proteins that are targeted for autoimmune attack. (*see* Color Plate 3, following p. 313)

AChRs and MuSK in the postsynaptic muscle endplate, and rapsyn, titin, and the ryanodine receptor intracellularly in the muscle cell.

3. IDENTIFYING A PATHOGENIC ROLE FOR ANTIBODIES IN ANTIBODY-MEDIATED DISORDERS

The classic studies that demonstrate a role for autoantibodies in the disorders discussed in this volume are described elsewhere. Here we shall just draw attention to the important principles underlying these observations. In MG, four observations were crucial in leading to the identification of the autoimmune basis of the condition. First, active immunization of animals with purified AChR led to the development of experimental autoimmune myasthenia gravis (EAMG) *(1)*. Second, there was a reduction in AChRs at the motor endplates of patients *(2)*. Third, patients with MG had antibodies to AChR *(3)*, and fourth, it was possible to transfer the signs of the disease by injecting IgG of MG patients into mice *(4)*. These observations made the case that MG was an autoimmune disease so compelling that it was justifiable to attempt plasma exchange as an experimental treatment *(5)*.

In Lambert–Eaton syndrome (LES) and neuromyotonia (NMT), by contrast, active immunization has still not been carried out satisfactorily. However, the combination of clinical response to plasma exchange and passive transfer of immunoglobulins to mice *(6, 7, 8)*, or occasionally other species, has shown that these conditions are due to autoantibodies, and the identification of the target antigens (or some of them) has followed.

4. DISEASE PHENOTYPES IN AUTOIMMUNE MYASTHENIC DISORDERS

In immune-mediated diseases, the clinical syndrome generated is related to the target antigen and its role in the function of the NMJ and the muscle fiber. At the same time, overlapping phenotypes occur as a consequence of immune attacks directed at distinct molecular targets. For

Table 1
Autoantigens in Myasthenia Gravis Subtypes

	AChR	Titin	MuSK	Thymic pathology	Age of onset (years)
Early-onset MG	+ + +[1]	−	−	Germinal centers	< 50
Thymoma MG	+	+	−	Thymoma	Any usually 30–60
Late-onset MG[2]	+	−	−		>40
Late-onset MG with titin antibodies	+	+	−	Atrophy	> 60
MuSK MG	−	−	+ +	−	Mostly <50
Seronegative MG	+[3]	−	−	Some germinal centers	Any

[1] + + +: high serum antibody concentration: +: low serum antibody concentration.
[2] Late-onset MG is classified into titin positive and negative patients.
[3] Only detected by binding to clusters of high-density AChRs.

instance, both AChR and MuSK antibodies can be associated with different forms of MG and are likely to disturb neuromuscular transmission by different mechanisms. Identification of these mechanisms may allow for improved diagnosis and treatment.

Patients with acquired MG with AChR antibodies fall into three distinct groups: early-onset MG without thymoma, patients with thymoma, and patients with late-onset MG without thymoma (Table 1) *(9)*. The late-onset group, furthermore, may include at least two subpopulations of patients.

5. HETEROGENEITY AND PATHOPHYSIOLOGICAL EFFECTS OF AChR ANTIBODIES

Serum AChR antibodies provide a specific marker for MG. When present in humans, they invariably mediate AChR loss, simplification of the postsynaptic membrane, dysfunction of NMJ transmission, and clinical disease. In patients with MG, plasma exchange-induced reduction in serum anti-AChR levels is generally associated with clinical improvement *(10)* that is often marked *(11)*. The duration of benefit as in other antibody-mediated diseases in general may last from 1 to 3 months. In individual patients, there is often a rough correlation between severity of disease and the concentration of serum AChR antibodies *(12)*.

In order to understand the effects of antibodies on NMJ function, it is necessary to refer to the animal model, EAMG, that results from active immunization against purified AChR. The majority of anti-AChR antibodies in EAMG and many antibodies in MG bind to a conformational determinant, known as the main immunogenic region (MIR), on the extracellular aspect of the AChR alpha subunit. This region includes the alpha subunit 67–76 amino acid sequence *(13)*. Complement activation, leading to the deposition of membrane attack complex, by these antibodies is pivotal to the development of AChR loss in EAMG. C4-deficient guinea pigs do not develop EAMG and prevention of complement activation prevents EAMG *(14)*. The antibodies cross-link the AChRs in the membrane and their Fc regions mediate C1q binding and macrophage activation. However, since efficient complement activation requires binding of C1q by cross-linked adjacent immunoglobulin molecules, this may be reduced or prevented by decreased AChR density or distortion of the postsynaptic membrane. In fact, the macrophage infiltration induced by injection of AChR monoclonal antibodies correlates with AChR density *(15)*.

Human MG is characterized by postsynaptic membrane simplification and reduced AChR content. This pathology is induced and maintained by anti-AChR antibodies and is also mediated by complement activation *(16)*. Although many antibodies are against the MIR, some patients have predominantly antibodies to other subunits. In addition, macrophages invade the postsynaptic membrane in human MG, albeit in relatively few numbers *(17)*. Anti-AChR antibodies may have additional effects on the target molecules by blocking ion channel function *(18)*. An additional mechanism is the increased internalization, or antigenic modulation, of the AChR molecules by specific antibody *(19)*. Passive transfer of these modulating antibodies does not cause overt EAMG when complement activation is prevented *(20)*.

The lack of a precise correlation between serum AChR antibody concentrations and clinical weakness in patients with MG, therefore, is likely to be dependent on several confounding factors. These include variables related to the structure and function of the antibodies themselves, including the nature of the variable region forming the binding site and which, consequently, determines the AChR site to which it binds. Effector functions of the constant regions also likely influence clinical expression. These are a function of the class and subclass of the antibodies and their interactions with components of the innate immune system, particularly complement and macrophages. Most AChR antibodies in human MG belong to the IgG1 and IgG3 subclasses and are strong activators of complement. Nevertheless, the polyclonal B-cell response means that changes in subpopulations of pathogenic antibodies may result in changes in the clinical manifestations, independent of the total level of the antibody. The relationship between the serum antibody concentration and pathology also involves equilibration between serum and tissue concentrations. The density and organization of the target epitopes are additional factors, as are the presence of co-occurring antibodies directed at additional targets.

5.1. Binding Antibodies Measured by Immunoprecipitation

This is the assay that tends to be most useful in clinical practice. It uses extracted human skeletal muscle AChR labeled with radio-iodinated alpha-bungarotoxin (α-BgTx) (Table 2). It detects AChR antibodies in 85% of patients with generalized MG *(21, 22)* and 50–75% of patients with purely ocular MG *(22)*. The patients with early-onset MG with generalized weakness tend to have the highest concentrations of serum AChR antibody levels when compared with other patients with MG (Table 1). The AChR antibody-positive, or seropositive, patients with ocular MG tend to have serum AChR concentrations that are lower than seropositive patients with generalized disease. This suggests that the clinical manifestations may, at least in part, be a function of the absolute AChR antibody concentration. A particular susceptibility of ocular muscle NMJs for clinical manifestations of NMJ dysfunction is also suggested by the occurrence of ocular muscle involvement early in the course of disease in many patients with botulism *(23)* or with other neurotoxins. It is also possible that the antigenic targets in the NMJs of extraocular muscles and levator palpebrae provide distinct epitopes for the autoimmune process *(24)*. The patients with generalized MG, who are consistently negative for AChR antibodies, appear to manifest a different disease mechanism (see below).

Table 2
Neurotoxins Used to Quantitative Assays for Ion Channel Autoantibodies

Antigen	Source	Neurotoxin	Source	Use
AChR	Muscle or muscle cell lines (TE671 cells)	α-Bungarotoxin	*Bungarus multicinctus*	Diagnosis of MG
VGCC	Human or rabbit cerebellum or neuronal or small cell lung cancer lines	ω-conotoxin MVIIC	*Conus magus*	Diagnosis of LES and investigation of some CNS disorders
VGKC	Human or rabbit cortex	Dendrotoxins	Dendroaspis species	Diagnosis of NMT and investigation of some CNS disorders

5.2. Blocking Antibodies

Blocking antibodies that inhibit the binding of radiolabeled α-BgTx to the AChR probably compete for binding to the ACh binding site *(25)*, although in theory they may be non-competitive and result in allosteric inhibition. While AChR blocking antibodies may be detected in many patients with MG, they appear to represent a minority of the AChR antibodies and usually occur in association with AChR binding antibodies. They are suggested to be important pathogenically in instances where there is an acute severe exacerbation of MG, but blocking antibodies do not necessarily interfere with AChR function. Antibodies that do inhibit function can be detected by measuring ion flux through the AChR or ACh-induced currents *(26, 27)*.

5.3. Assays to Detect Modulating Antibodies

The application of serum from patients with MG to muscle cell lines can interfere not only in function but also with AChR expression *(19)*. This is because the antibodies cross-link the AChR in the membrane and increase their rate of degradation. The AChR are measured by binding of radiolabeled-BgTx after exposure to the serum for 16 h. It is a relatively non-specific test and the results should be compared with suitable control sera. The detection of these antibodies is most useful when the AChR binding test yields a negative result *(25)*. It is possible that in some instances, however, the decrease in AChR concentration occurs as a consequence of antibodies directed to postsynaptic targets adjacent to the AChR, rather than to the AChR itself. In the majority of cases, modulating antibodies directed at the AChR also tend to be positive in immunoprecipitation assays.

5.4. Other Autoantibodies

Apart from antibodies to molecular targets located at the NMJ, patients with MG may possess other autoimmune antibodies. These include antibodies directed against thyroglobulin, thyroid microsomes, parietal cells and intrinsic factor, and nuclear antigens. Other autoantibodies are less common *(25)*. The association between these autoantibodies, or other autoimmune diseases, and generalized or ocular MG appears to lie in thymic hyperplasia. Thymic hyperplasia with intramedullary germinal centers is invariably present in patients with early-onset MG. It has also been described in patients with thyroid autoimmune disease and in other autoimmune conditions including Addison's disease and pemphigus. Thyrotropin appears to be expressed in the thymus, and the association between thymic hyperplasia and thyroid disease may also be related to loss of tolerance in the thymus *(28)*.

6. AUTOIMMUNE MG IN ASSOCIATION WITH THYMOMA

One hundred percent of patients with thymoma and MG have detectable serum AChR antibodies *(29)*. In addition, patients with thymoma without clinically detectable MG may sometimes have AChR antibodies suggesting the presence of subclinical disease. The antibodies and their epitopes may differ in their fine specificities from those with early-onset autoimmune MG without thymoma *(30)*. The serum concentrations of AChR antibodies in thymoma cases tend to be lower than in those with generalized disease without thymoma.

Complement-activating, striated muscle antibodies were initially described in 1960 by immunofluorescence in patients with MG *(31)*. The majority of the striational antibodies are directed against the large structural protein, titin *(32)*. Ryanodine receptor (RyR) and possibly other proteins constitute additional autoimmune targets *(33)*. Ninety-five percent of patients with thymoma and MG also have titin antibodies, whereas 70% of patients with thymoma and MG possess anti-RyR antibodies *(34)*. Patients with detectable RyR antibodies invariably also possess titin antibodies. The clinical pattern of weakness and fatigability of patients with MG and thymoma is usually indistinguishable from those with MG without thymoma. However, some differences between patients in these two groups may represent the effects of distinct antibodies. Thus, some patients with thymoma, titin, and RyR antibodies may have a myocarditis or a focal myositis *(29)*. The involvement of skeletal muscle and heart muscle is suspected to be a consequence of the striated muscle antibodies providing additional disease-producing potential. In addition, the MG patients with titin antibodies appear to have a more severe course despite lower AChR antibody concentrations compared to those without detectable titin antibodies *(34)*. A more severe course tends to occur in patients with titin antibodies in the presence or in the absence of thymoma.

Patients with thymoma require thymectomy to prevent complications stemming from local extension of the tumor (including the superior vena cava syndrome). Thymectomy in these patients, however, may not be followed by clinical improvement. Furthermore, the course of MG may be more severe after thymectomy and on occasions MG may be precipitated in the postoperative period. It is suggested that the more severe clinical course of MG often character-izing patients with thymoma may be related to pathogenic effects of titin and RyR antibodies *(34, 35, 36)*. Patients with thymoma occasionally have been reported to possess antibodies to other autoimmune targets. These include potassium channels *(37)*, calcium channels, as well as rapsyn *(38, 39)* and other proteins including glutamic acid decarboxylase *(40)*. Titin and RyR anti-bodies, and antibodies to the cytokines interferon-alpha and interleukin 12 *(41)*, are also present in a high proportion of the subgroup of patients with late-onset MG who do not have a thymoma *((34, 41)*, see below).

6.1. *Striated Muscle Antibody Assays*

Titin and RyR antibodies may be detected by enzyme-linked immunosorbent assay (ELISA) or immunoblot. A single epitope of titin, MGT30, reacts with 90% of titin antibodies. This epitope has been employed for the titin assays, and a commercial version is available. Striated muscle antibody assays detected by immunofluorescence are also commercially available and may be used in lieu of specific titin and RyR antibody assays. The detection of striated antibodies may be useful diagnostically in patients when AChR antibodies are negative. However, this situation is uncommon. A more important diagnostic role for striated muscle antibodies is to suggest the presence of a thymoma when present in patients under the age of 60 years *(41)*. The sensitivity and specificity of RyR antibodies for thymoma in this age group is 70% *(34)*.

7. LATE-ONSET MG NOT ASSOCIATED WITH A DETECTABLE THYMOMA

The mean AChR antibody concentration in these patients tends to be significantly lower than the early-onset MG group without thymoma *(34)*. The use of antibody profiles utilizing AChR, titin, and RyR antibodies has characterized two subpopulations of patients in this group. Patients who do not have detectable anti-striated muscle antibodies may resemble the early-onset MG group and usually present before 60 years of age *(41)*, whereas the patients with titin antibodies present later. Similarly to patients with thymoma, patients with late-onset MG with titin or RyR antibodies tend to have a poorer response to thymectomy than those with late-onset MG without detectable titin or RyR antibodies *(36)*.

8. ROLE OF ANTIBODIES IN DIAGNOSIS AND TREATMENT IN PATIENTS WITH MG

Patients with AChR serum antibodies almost invariably have MG. These antibodies provide a valuable diagnostic tool and indicate a mechanism of disease pathogenesis that is more precise than can be obtained by electrophysiological studies. At this time, the usefulness of following serum concentrations of AChR antibodies in order to monitor treatment in individual patients is questioned. Identification of subsets of pathogenically important antibodies, including antibodies to additional targets, and measuring their concentration could prove to be important.

The presence of antibodies to targets other than the AChR may also influence treatment decisions. Since titin antibodies appear to correlate with disease severity, their presence in patients with thymoma may suggest a role for more aggressive management early in the course of the disease prior to thymectomy. Similarly, the presence of titin antibodies in patients with onset of disease above the age of 50 years may suggest a more aggressive immunosuppression regimen early in the course of the disease and the avoidance of thymectomy.

9. SERONEGATIVE (SN) MG

The observations that indicated that SNMG was a distinct antibody-mediated condition are of some interest and are briefly summarized.

9.1. Plasma Exchange and Passive Transfer

In seronegative MG, as in LES and NMT, the clinical and experimental evidence for antibody-mediated disease preceded the identification of a target antigen. Several authors showed that the patients responded well to plasma exchange and that passive transfer to mice produced some electrophysiological evidence of MG *(42, 43)*.

9.2. Functional and Binding Studies

The identification of the target antigen(s), however, has been difficult. The possibilities are several. There could be low-affinity antibodies to AChRs that are not detected by the RIA, but which nevertheless bind effectively in vivo. That is now thought to be the case, as shown by binding of SNMG IgG to cell lines expressing human AChR (see below). Antibodies from seropositive MG patients bound, but those from AChR antibody negative patients did not *(44)*. Moreover, when SNMG plasma was injected into mice, the AChR extracted from the mouse muscles were not complexed with IgG *(42)*.

Another explanation is the presence of antibodies binding to another muscle surface antigen. This is supported by the evidence that both AChR antibody-positive and negative sera bound to TE671 cells, which are derived from a rhabodmyosarcoma and express muscle antigens in addition to AChR *(44)*. This suggests that, if not binding to the AChR itself, the serum antibodies must be binding to other muscle proteins.

This was confirmed by showing that application of plasma or serum from SNMG patients to TE6712 cells leads to a rapid reduction in AChR function as demonstrated by ion flux studies or by patch clamp *(45, 46)*. In order to affect AChR function, without binding directly to it, the antibodies would need to alter the function of some other membrane protein, perhaps one that stimulated an intracellular pathway. There are various intracellular pathways in muscle that could affect AChR function. In particular, the membrane receptors that stimulate protein kinase A or protein kinase C, such as CGRP and ATP receptors, are thought to be present at the neuromuscular junction and could be targets for SNMG antibodies. However, these pathways do not appear to be involved in SNMG, because inhibitors of either protein kinase A or C do not prevent the effects of the antibodies on AChR function *(46)*.

9.3. Antibodies to a Candidate Antigen

These functional studies suggested that the antibodies might be binding to another type of receptor. An attractive candidate was the muscle-specific kinase, MuSK. The kinase is known to be present at the NMJ, and to be a receptor tyrosine kinase that causes autophosphorylation followed by phosphorylation of rapsyn and AChR. Therefore, it is clearly closely associated with AChR at the mature as well as the developing neuromuscular junction. In fact, SNMG IgG was

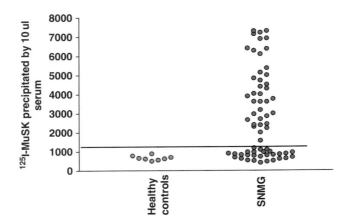

Fig. 2. Immunoprecipitation of ^{125}I-MuSK by sera from patients with seronegative MG. Recombinant extracellular domain of rat MuSK was provided by Dr. Werner Hoch (University of Bristol). The SNMG sera tested were provided by Professor John Newsom-Davis, University of Oxford, and Dr. Amelia Evoli, Catholic University of Rome. (*see* Color Plate 4, following p. x)

found to bind strongly to COS cells expressing recombinant MuSK *(47)*. To provide a screening assay for MuSK antibodies, Hoch et al. *(47)* used the soluble extracellular domain of recombinant MuSK. Antibodies from SNMG patients with well-established disease bound specifically to MuSK and positive results were found in about 70%.

ELISA is not very easy to use if the concentration of human serum required is relatively high. Non-specific binding to the plates or to other antigens contaminating the purified recombinant protein can make analysis of the results difficult. Immunoprecipitation of directly labeled rat MuSK is considerably more sensitive, and the results correlate strongly with the ELISA data (Fig. 2 and Color Plate 4, following p. x). This technique is now available commercially and is used widely, but the proportion of patients who are positive differs worldwide, apparently correlating with geographical latitude *(48a)*.

MG in the absence of detectable AChR antibodies, in turn, does not appear to represent a homogenous disease entity. It is likely that patients with SNMG, who do not possess detectable antibodies to MuSK, may possess antibodies directed at other proteins at the NMJ. These targets include other proteins involved in the clustering of AChRs at the endplate including intracellular proteins such as rapsyn *(39)*. Furthermore, some MG patients, who are seronegative for AChR antibodies at initial presentation, develop detectable serum anti-AChR antibodies over time. It is possible that at presentation many patients have serum concentrations of AChR antibodies that are below the assay sensitivity. A recent development, which should allow better detection of AChR antibodies at low titre or of low-affinity antibodies, is the use of cell lines expressing high-density clusters of AChRs *(48a, 48b)*.

10. LAMBERT–EATON SYNDROME AND CEREBELLAR ATAXIA

10.1. Plasma Exchange and Passive Transfer

Early studies showed that LES was associated with certain autoantibodies and other autoimmune diseases *(49)* and that plasma exchange was an effective treatment *(50)*. Particularly important was the demonstration that injection of patient IgG into mice resulted in a reduction in the quantal release of ACh and an altered number and distribution of active zone particles. The results mirrored the findings in patients *(51, 52)*.

10.2. Functional and Binding Studies

In order to begin to define the nature of the antigen, functional studies on cell lines were used. These began by looking for a candidate antigen, voltage-gated calcium channel (VGCC), on small cell lung cancer cells, and by showing that the LES IgGs, when applied for several days in culture, were able to downregulate the VGCC numbers *(53)*. A number of small cell lung cancer cells have been used subsequently, and in all cases VGCCs have been demonstrated and the effects of LEMS antibodies confirmed *(54, 55)*. The main target of the LEMS antibodies appears to be the P/Q-type VGCC, which contains the alpha 1a subunit. These VGCCs are expressed at the motor nerve terminal *(56)* and also on the Purkinje cells in the cerebellum *(57)*.

10.3. Radioimmunoprecipitation Assays

The radioimmunoprecipitation assay for VGCC antibodies was made possible by the discovery in the late 1980s of the cone snail toxins (Table 2) *(58)*. The various species of cone snails live on the seabed and produce a rich variety of neurotoxins that are specific for either pre- or postsynaptic ion channels. The first snail toxin to be used to identify VGCCs in LEMS sera was from *Conus geographus*. This neurotoxin binds to the N-type VGCC, which is expressed widely in the nervous and neuroendocrine systems. Using VGCCs extracted from human cerebellum or cortex labeled by ^{125}I-omega-conotoxin GVIA, about 40% of LEMS patients are positive *(59)*. However, using the P/Q-type selective omega-conotoxin MVIIC under very similar conditions, up to 90% of LEMS patients are positive *(60, 61)*. These results emphasize the exquisite specificity of these toxins. By contrast, the source of tissue is probably not so important. Similar results are found with rabbit or human cerebellum, indicating that the VGCCs are highly conserved in evolution and antigenically similar between species *(62)*.

The radioimmunoassay has been used extensively as a diagnostic test, and some studies report serial estimations during treatment. For instance, VGCC antibodies are found to fall by 30% on average in patients treated with intravenous immunoglobulin *(63)*. The fall does not start until 2 weeks after the treatment, suggesting that it is not a direct result of "blocking" antibodies. The VGCC levels return to pretreatment levels by about 8 weeks, roughly correlating with clinical manifestations.

Although principally in use as a diagnostic test for LEMS, there are other applications for measuring VGCC antibodies. It has been established that a proportion of patients with small cell lung cancer have LEMS and that VGCC antibodies are present in such patients even without clear clinical evidence of LEMS *(64)*. Moreover, some patients with cerebellar ataxia with and without clinical evidence of LEMS have VGCC antibodies *(65)*. Thus serum from patients who are at risk of SCLC and who present with cerebellar or a myasthenic syndrome should be tested for VGCC antibodies.

11. ACQUIRED NEUROMYOTONIA

11.1. Plasma Exchange and Passive Transfer

Like LEMS, acquired NMT was not thought to be autoimmune until clinical studies showed a response of some patients to plasma exchange or intravenous immunoglobulin therapies. In one of the few reports of passive transfer, an enhancement of neuromuscular transmission with resistance to the action of curare was induced in the mouse nerve–muscle preparations after passive transfer of NMT IgG *(66)*.

11.2. Functional and Binding Assays

A functional effect of NMT IgG on neuronal cell lines was successfully demonstrated by using dorsal root ganglion cells, which developed repetitive action potentials, very similar to those found in the presence of 4-diamino-pyridine, a potassium channel blocker *(7)*. In the PC-12 and NB-1 neuroblastoma cell lines, incubating the cells overnight in the NMT IgG led to a reduction in the voltage-gated potassium channel (VGKC) currents without any observable change in the current/voltage density. These alterations, which were not seen with briefer incubation times, suggest that the antibodies are not blocking channel function directly but rather reducing the number of VGKC *(67, 68, 69)*.

Using cell lines expressing VGKC, it is possible to detect binding of some antibodies from NMT patients, but the sensitivity is not sufficient for routine analysis (unpublished data). Ideally one would use cell lines expressing different VGKC isotypes so as to test the involvement of different targets in the disease process. Immunohistochemistry on *Xenopus* oocytes expressing some VGKC subtypes demonstrated antibody binding *(70)*, but this approach is time consuming and has not been evaluated further (Section 10.3).

There are numerous subtypes of VGKC, particularly of the shaker type. Present nomenclature identifies these as Kv1.1–1.9. To date, it has only been possible to identify autoantibody binding to Kv1.1, 1.2, and 1.6 (and perhaps 1.3) using the dendroaspis neurotoxins that bind specifically to these subtypes (Table 2) *(70)*.

The sensitivity of the assay depends on using high specific activity dendrotoxin to label VGKC extracted from human or rabbit cortex, but with this product it is not possible to saturate binding sites. About 40% of NMT patients are positive by this approach *(7, 71)*. The antibodies do not appear to inhibit binding of dendrotoxin to the VGKC. Few serial studies of VGKC antibodies have thus far been performed. Perhaps the most striking report is that of a woman with MG and thymoma who following a thymoma recurrence developed an episode of "limbic encephalitis" *(72)*. This responded strikingly to plasma exchange, strongly implying an antibody-mediated basis for the central disease. In a retrospective analysis of VGKC antibodies in over 80 sera collected from this patient over 13 years, a single peak of VGKC antibodies was found, corresponding closely to the time of appearance of her limbic manifestations (Fig. 3 and Color Plate 5, following p. x) *(72)*. VGKC antibodies and a clinical response to plasma exchange were found in a patient with classical Morvan's syndrome *(73)*, which includes neuromyotonia with sleep disorder,

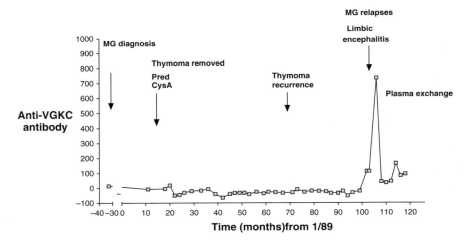

Fig. 3. VGKC antibody levels in a patient with a 13-year history of thymoma-associated MG. She developed a single episode of limbic signs (memory loss, confusion, disorientation, agitation) associated with a peak of antibodies to VGKC. The antibody levels fell following plasma exchange and her clinical state improved. From Buckley et al. *(72)*. (*see* Color Plate 4, following p. x)

limbic manifestations, and autonomic dysfunction. Subsequently many patients with VGKC antibodies have been recognized. Most have either peripheral symptoms or limbic symptoms, but a few have a broad combination of peripheral hyperexcitability, autonomic dysfunction, and cognitive and sleep disturbance. The relationship between antibody specificity and clinical phenotype is not well established (see Vincent and Hart Chapter 14).

12. STIFF PERSON SYNDROME

Autoimmune disorders involving components of central nervous system synapses have been identified with some similarity to the peripheral disorders discussed above. Stiff person syndrome (SPS), a condition characterized by muscle rigidity and spasms, was suspected to have an autoimmune basis because of its association with autoantibodies and other autoimmune conditions. Furthermore, patients with SPS respond to immune treatments including plasma exchange *(74)*. The molecular autoimmune targets have been identified to be components of CNS synapses at which γ-amino butyric acid (GABA) is a neurotransmitter. GABA is the major inhibitory neurotransmitter in the CNS. Ninety percent of patients with SPS possess antibodies directed at glutamic acid decarboxylase (GAD), the rate-limiting enzyme involved in the synthesis of GABA. High serum GAD antibodies are specific for SPS, and CSF GAD-specific antibodies also occur in SPS *(75)*. Diabetes mellitus is often present and GAD antibodies are also present in patients with insulin-dependent diabetes mellitus in the absence of SPS. Patients with diabetes mellitus without SPS may have lower serum concentrations than those with SPS and tend to have antibodies to the 65 kD isoform whereas patients with SPS may have antibodies to both the 65 kD and the 67 kD isoforms *(76)*. Some patients with SPS in association with cancer are reported to not have detectable GAD antibodies but have serum antibodies to synapse-related proteins including amphiphysin *(77)* and, in one case, gephyrin, a protein that anchors GABA receptors *(78)*. Thus, autoimmune SPS may represent a disease phenotype or syndrome that may result as a consequence of distinct immunopathogenesis and etiology.

13. CONCLUSIONS

Acquired MG may be considered the prototypic autoimmune disorder. Observations pertaining to MG are also pertinent to other autoimmune disorders of the NMJ and related disorders. Thus in MG, phenotypic variations are at least in part a function of different mechanisms of disease. The antibody target, the concentration of antibodies, and the fraction of antibodies that are complement activating are all factors likely to influence the clinical presentation and the course. Serological testing is of value in the diagnosis as well as in the treatment of individual patients. AChR is the main autoantigen in most patients with acquired MG. However it is not the only potential target for treatment. We are developing a clearer picture of a broader spectrum of antigenic targets and the means by which the antibodies to these targets may influence the disease course and treatment. While in most instances the diseases do not overlap with each other, e.g., MG and LES or MG and SNMG, in other instances the antibody syndromes occur together particularly in MG in association with thymoma. Patients with AChR antibody-positive MG with early-onset appear to have disease mechanisms that are distinct from those with those with thymoma and MG. Seronegative MG represents a separate condition. Further identification of additional targets of the autoimmune disease is likely to lead to better understanding of the etiology of these disorders.

REFERENCES

1. Patrick J, Lindstrom J. Autoimmune response to acetylcholine receptor. Science 1973;180:871–72.
2. Fambrough DM, Drachman DB, Satyamurti S. Neuromuscular junction in myasthenia gravis: decreased acetylcholine receptors. Science 1973;182:293–295.

3. Lindstrom JM, Seybold ME, Lennon VA, Whittingham S, Duane DD. Antibody to acetylcholine receptor in myasthenia gravis: Prevalence, clinical correlates, and diagnostic value. Neurology 1976;26:1054–9.

4. Toyka, KV, Drachman, DB, Griffin DE, Pestronk A, Winkelstein JA, Fischbeck KH, Kao I. Myasthenia gravis: study of humoral immune mechanisms by passive transfer to mice. N Eng J Med 1977; 296:125–131.

5. Pinching AJ, Peters DK, Newsom-Davis JN. Remission of myasthenia gravis following plasma exchange. Lancet 1976;2:1373–1376.

6. Lang B, Newsom-Davis J, Prior C, Wray D. Antibodies to motor nerve terminals: an electrophysiological study of a human myasthenic syndrome transferred to a mouse. J Physiol (Lond) 1983;344: 335–345.

7. Shillito P, Molenaar PC, Vincent A, et al. Acquired neuromyotonia: Evidence for autoantibodies directed against K^+ channels of peripheral nerves. Ann Neurol 1995;38:714–722.

8. Mossman S, Vincent A, Newsom-Davis J. Passive transfer of myasthenia gravis by immunoglobulins: lack of correlation between antibody bound, acetylcholine receptor loss and transmission defect. J Neurol Sci 1988;84:15–28.

9. Compston DA, Vincent A, Newsom-Davis J, Batchelor JR. Clinical, pathological, HLA antigen and immunological evidence for disease heterogeneity in myasthenia gravis. Brain 1980;103:579–601.

10. Newsom-Davis J, Pinching AJ, Vincent A, Wilson SG. Function of circulating antibody to acetycholine receptor in myasthenia gravis: investigated by plasma exchange. Neurology 1978;28:266–272.

11. Dau, PC. Plasmpheresis therapy in mysthenia gravis. Muscle Nerve 1980;3:468–482.

12. Oosterhuis HJGH, Limburg PC, Hummel-Tappel E. Anti-acetylcholine receptor antibodies in myasthenia gravis. Part 2. Clinical and serological follow-up of individual patients. J Neurol Sci 1983;58:371–385.

13. Tzartos SJ, Loutrari HV, Tang F, et al. Main immunogenic region of torpedo electroplax and human muscle acetylcholine receptor: localization and microheterogeneity revealed by the use of synthetic peptides. J Neurochem 1990;54:51–61.

14. Lennon VA, Seybold ME, Lindstrom JM, Cochrane C, Ulevitch R. Role of complement in the pathogenesis of experimental autoimmune myasthenia gravis. J Exp Med 1978;147:973–983.

15. Corey AL, Richman DP, Agius MA, Wollmann RL. Refractoriness to a second episode of experimental myasthenia gravis: correlation with AChR concentration and morphologic appearance of the postsynaptic membrane. J Immunol 1987;138:3269–3275.

16. Gomez CM, Richman DP. Chronic experimental autoimmune myasthenia gravis induced by monoclonal antibody to acetycholine receptor: biochemical and electrophysiological criteria. J Immunol 1987;139:73–76.

17. Maselli RA, Richman DP, Wollmann RL. Inflammation at the neuromuscular junction in myasthenia gravis. Neurology 1991;41:1497–1504.

18. Gomez CM, Richman DP. Anti-acetylcholine receptor antibodies directed against the alpha bungarotoxin binding site induce a unique form of experimental myasthenia. Proc Natl Acad Sci (USA) 1983;80:4089–4093.

19. Drachman DB, Adams RN, Josifek LF, Self SG. Functional activities of autoantibodies to acetylcholine receptors and the clinical severity of myasthenia gravis. N Engl J Med 1982;307:769–75.

20. Dupont BL, Twaddle GM, Richman DP. Suppression of passive transfer acute experimental myasthenia by F(Ab')2 fragments. J Neuroimmunol 1987;16:47 Abstract.

21. Howard FM, Lennon VA, Finley J, Matsumoto J, Elveback LR. Clinical correlations of antibodies that bind, block or modulate human acetylcholine receptors in myasthenia gravis. Ann NY Acad Sci 1987;505:526–538.

22. Vincent A, Newsom-Davis J. Anti-acetycholine receptor antibodies. J Neurol Neurosurg Psychiatry. 1980;43:590–600.

23. Maselli RA, Ellis W, Mandler RN, Sheikh F, Senton G, Knox S, Salari-Namin H, Agius M, Wollmann RL, Richman DP. Cluster of wound botulism in California: clinical, electrophysiologic, and pathologic study. Muscle Nerve 1997;20:1284–1295.

24. Kaminski, HJ. Acetylcholine receptor epitopes in ocular myasthenia. Ann NY Acad Sci. 1998;841:309–19.

25. Lennon VA. Serological diagnosis of myasthenia gravis and the Lambert Eaton Myasthenic syndrome. In Lisak RP, ed. Handbook of Myasthenia Gravis and Myasthenic Syndromes. Marcel Dekker, Inc. New York. 1994: 149–164.

26. Lang B, Richardson G, Rees J, Vincent A, Newsom-Davis J. Plasma from myasthenia gravis patients reduces acetylcholine receptor agonist-induced Na + flux into TE671 cell line. J Neuroimmunol 1988;19:141–8.

27. Bufler J, Pitz R, Czep M, Wick M, Franke C. Purified IgG from seropositive and seronegative patients with myasthenia gravis reversibly blocks currents through nicotinic acetylcholine receptor channels. Ann Neurol 1998;43:458–64.

28. Murakami M, Hosoi Y, Araki O, Morimura T, Imamura M, Ogiwara T, Mizuma H, Mori M. Expression of thyrotropin receptors in rat thymus. Life Sci 2001; 68:2781–7.

29. Aarli J. Myasthenia gravis and thymoma. In Lisak RP, ed Handbook of Myasthenia Gravis and Myasthenic Syndromes. Marcel Dekker, Inc. New York. 1994:207–224.

30. Lefvert AK, Bergstrom K, Mattel G, Osterman PO, Pirksanen R. Determination of acetylcholine receptor antibody in myasthenia gravis: clinical usefulness and pathogenetic implications. J Neurol Neurosurg Psychiatry 1978;41:394–403.

31. Strauss AJ, Seegal BC, Hsu JC, Burkholder PM, Nastuk WL, Osserman KE. Immunofluorescence demonstration of a muscle binding, complement fixing serum globulin fraction in myasthenia gravis. Proc Soc Exp Biol (NY) 1960;195:184–191.

32. Aarli JA, Stefansson K, Marton LSG, Wollmann RL. Patients with myasthenia gravis and thymoma have in their sera IgG autoantibodies against titin. Clin Exp Immunol 1990;82:284–288.

33. Skeie GO, Lunde PK, Sejersted OM, Mygland A, Aarli JA, Gilhus NE. Autoimmunity against the ryanodine receptor in myasthenia gravis. Acta Physiol Scand 2001;171:379–384.

34. Romi F, Skeie GO, Aarli JA, Gilhus NE. Muscle autoantibodies in subgroups of myasthenia gravis patients. J Neurol 2000;247:369–375.

35. Romi F, Skeie GO, Aarli JA, Gilhus NE. The severity of myasthenia gravis correlates with the serum concentration of titin and ryanodine receptor antibodies. Arch Neurol 2000;57:1596–600.

36. Romi F. Muscle autoantibodies in myasthenia gravis (MG): clinical, immunological and therapeutic implications. 2001. Thesis, University of Bergen.

37. Mygland A, Vincent A, Newsom-Davis J, Kaminski H, Zorzato F, Agius M, Gilhus NE, Aarli JA. Autoantibodies in thymoma-associated myasthenia gravis with myositis or neuromyotonia. Arch Neurol 2000;57:527–531

38. Lee E-K, Maselli RA, Ellis WG, Agius MA. Morvan's fibrillary chorea: a paraneoplastic manifestation of thymoma. J Neurol Neurosurg Psych 1998;65:857–862.

39. Agius MA, Zhu S, Kirvan CA, Schafer A, Lin MY, Fairclough RH, Oger JJ-F, Aziz T, Aarli JA. Rapsyn antibodies in myasthenia gravis. Ann NY Acad Sci 841:516–524.

40. Hagiwara H, Enomoto-Nakatani S, Sakai K, Ugawa Y, Kusunoki S, Kanazawa IJ. Stiff-person syndrome associated with invasive thymoma: a case report. Neurol Sci 2001;193:59–62

41. Buckley C, Newsom-Davis J, Willcox N, Vincent A. Do titin and cytokine antibodies in MG patients predict thymoma or thymoma recurrence? Neurology 2001;57:1579–82.

42. Mossman S, Vincent A, Newsom-Davis J. Myasthenia gravis without acetylcholine-receptor antibody: a distinct disease entity. Lancet 1986;1:116–9.

43. Evoli A, Batocchi AP, Lo Monaco M, Servidei S, Padua L, Majolini L, Tonali P. Clinical heterogeneity of seronegative myasthenia gravis. Neuromuscul Disord 1996;6:155–61.

44. Blaes F, Beeson D, Plested P, Lang B, Vincent A. IgG from "seronegative" myasthenia gravis patients binds to a muscle cell line, TE671, but not to human acetylcholine receptor. Ann Neurol 2000;47:504–510.

45. Yamamoto T, Vincent A. Ciulla TA, Lang B, Johnston I. Newsom-Davis J. Seronegative myasthenia gravis: a plasma factor inhibiting agonist-induced acetylcholine receptor function copurifies with IgM. Ann Neurol 1991;30:550–557.

46. Plested CP, Newsom-Davis J, Vincent A. Seronegative myasthenia plasmas and non-IgG fractions transiently inhibit nAChR function. Ann NY Acad Sci 1998;841:501–4.

47. Hoch W, McConville J, Helms S, Newsom-Davis J, Melms A, Vincent A. Autoantibodies to the receptor tyrosine kinase MuSK in patients with myasthenia gravis without acetylcholine receptor antibodies. Nat Med 2001;7:365–368.

48a. Vincent A, Leite MI, Farrugia ME, Jacob S, Viegas S, Shiraishi H, Benveniste O, Morgan BP, Hilton-Jones D, Newsom-Davis J, Beeson D, Willcox N. Myasthenia gravis seronegative for acetylcholine receptor antibodies. Ann NY Acad Sci. 2008;1132:84–92.

48b. Leite MI, Jacob S, Viegas S, Cossins J, Clover L, Morgan BP, Beeson D, Willcox N, Vincent A. IgG1 antibodies to acetylcholine receptors in "seronegative" MG. Brain 2008;131:1940–1952.

49. Lennon VA, Lambert EH, Whittingham S, Fairbanks V. Autoimmunity in the Lambert-Eaton myasthenic syndrome. Muscle Nerve 1982;5:S21–5.

50. Newsom-Davis J, Murray NM. Plasma exchange and immunosuppressive drug treatment in the Lambert-Eaton myasthenic syndrome. Neurology 1984;34: 480–485.

51. Fukunaga H, Engel AG, Osame M, Lambert EH. Paucity and disorganization of presynaptic membrane active zones in the Lambert-Eaton myasthenic syndrome. Muscle Nerve 1982;5:686–697.

52. Fukuoka T, Engel AG, Lang B, Newsom-Davis J, Prior C, Wray DW. Lambert-Eaton myasthenic syndrome: I. Early morphological effects of IgG on the presynaptic membrane active zones. Ann Neurol 1987;22:139–199.

53. Roberts A, Perera S, Lang B, Vincent A, Newsom-Davis J. Paraneoplastic myasthenic syndrome IgG inhibits 45Ca2+ flux in a human small cell carcinoma line. Nature 1985;317:737–9.

54. Johnston I, Lang B, Leys K, Newsom-Davis J. Heterogeneity of calcium channel autoantibodies detected using a small cell lung cancer line derived from a Lambert-Eaton syndrome patient. Neurology 1994;44:334–338.

55. Viglione MP, O'Shaughnessy TJ, Kim Y. Inhibition of calcium currents and exocytosis by Lambert-Eaton myasthenic syndrome antibodies in human lung cancer cells. J Physiol Lond 1995;488:303–17.

56. Protti DA, Reisen R, MacKinley TA, Uchitel OD. Calcium channel blockers and transmitter release at the normal human neuromuscular junction. Neurology 1996;46:1391–1396.

57. Pinto A, Gillard S, Moss F, Whyte K, Brust P, Williams M, Stauderman K, Harpold M, Lang B, Newsom-Davis J, Bleakman D, Lodge D, Boot J. Human autoantibodies specific for α_{1A} calcium channel subunit reduce both P-type and Q-type calcium currents in cerebellar neurons. Proc Natl Acad Sci 1998;95:8328–8333.

58. Espiritu DJ, Watkins M, Dia-Monje V, Cartier GE, Cruz LJ, Olivera BM. Venomous cone snails: molecular phylogeny and the generation of toxin diversity. Toxicon 2001;39:1899–916.

59. Sher E, Comola M, Nemni R, Canal N, Clementi F. Calcium channel autoantibody and non-small-cell lung cancer in patients with Lambert-Eaton syndrome. Lancet 1990;335:413.

60. Motomura M, Johnston I, Lang B, Vincent A, Newsom-Davis J. An improved diagnostic assay for Lambert-Eaton myasthenic syndrome. J Neurol Neurosurg Psychiatry 1995;58:85–87.

61. Lennon VA, Kryzer TJ, Greismann GE, O'Suilleabhain PE, Windebank AJ, Woppmann A, Miljanich GP, Lambert E. Calcium channel antibodies in the Lambert-Eaton myasthenic syndrome and other paraneoplastic syndromes. N Engl J Med 1995;332:1467–74.

62. Motomura M, Lang B, Johnston I, Palace J, Vincent A, Newsom-Davis J. Incidence of serum anti-P/O-type and anti-N-type calcium channel autoantibodies in the Lambert-Eaton myasthenic syndrome. J Neurol Sci 1997;147:35–42.

63. Bain PG, Motomura M, Newsom-Davis J, Misbah SA, Chapel HM, Lee ML, Vincent A, Lang B. Effects of intravenous immunoglobulin on muscle weakness and calcium channel antibodies in the Lambert-Eaton myasthenic syndrome. Neurology 1996;47:678–683.

64. Mason WP, Graus F, Lang B, Honnorat J, Delatrre J-Y, Vallderiola F, Antoine C, Rosenblum MK, Rosenfeld MR, Newsom-Davis J, Posner JB, Dalmau J. Paraneoplastic cerebellar degeneration and small-cell carcinoma. Brain 1997;120:1279–1300.

65. Trivedi R, Mundanthanan G, Amyes L, Lang B, Vincent A. Which antibodies are worth testing in subacute cerebellar ataxia? Lancet 2000;356:565–566.

66. Sinha S, Newsom-Davis J, Mills K, et al. Autoimmune aetiology for acquired neuromyotonia (Isaacs' syndrome). Lancet 1991;338, 75–77.

67. Sonoda, Y., Arimura, K., Kurono, A. et al. Serum of Isaacs' syndrome suppresses potassium channels in PC-12 cell lines. Muscle Nerve 1996;19:1439–1446.

68. Arimura K, Watanabe O, Kitajima I. et al. Antibodies to potassium channels of PC12 in serum of Isaacs' syndrome: Western blot and immunohistochemical studies. Muscle Nerve 1997;20:299–305.

69. Nagado T, Arimura K, Sonoda Y, et al. Potassium current suppression in patients with peripheral nerve hyperexcitability. Brain 1999;122:2057–2066.

70. Harvey AL, Rowan EG, Vatanpour H, Fatehi M, Castaneda O, Karlsson E. Potassium channel toxins and transmitter release. Ann NY Acad Sci 1994;710:1–10.

71. Hart IK, Waters C, Vincent A, et al. Autoantibodies detected to expressed K^+ channels are implicated in neuromyotonia. Ann Neurol 1997;41:238–46.

72. Buckley C, Oger J, Clover L, Tuzun E, Carpenter K, Jackson M, Vincent Potassium channel antibodies in two patients with reversible limbic encephalitis. Ann Neurol 2001;50:73–8.

73. Liguori R, Vincent A, Clover L, Avoni P, Plazzi G, Cortelli P, Baruzzi A, Carey T, Gambetti P, Lugaresi E, Montagna P. Morvan's syndrome: peripheral and central nervous system and cardiac involvement with antibodies to voltage-gated potassium channels. Brain 2001;124:2417–2426.

74. Levy LM, Dalakas MC, Floeter MK. The stiff-person syndrome: an autoimmune disorder affecting neurotransmission of gamma-aminobutyric acid. Ann Intern Med 1999;131:522–30.

75. Dalakas, MC, Li M, Fujii M, Jacobowitz DM. Quantification, specificity, and intrathecal synthesis of GAD_{65} antibodies. Neurology 2001;57:780–784.

76. Sohnlein P, Muller M, Syren K, Hartmann U, Bohm BO, Meinck HM, Knip M, Akerblom HK, Richter W. Epitope spreading and a varying but not disease-specific GAD65 antibody response in Type I diabetes. The Childhood Diabetes in Finland Study Group. Diabetologia 2000;43:210–7.

77. Schmierer K, Valdueza JM, Bender A, DeCamilli P, David C, Solimena M, Zschenderlein R. Atypical stiff-person syndrome with spinal MRI findings, amphiphysin autoantibodies, and immunosuppression. Neurology 1998;51:250–2.

78. Butler MH, Hayashi A, Ohkoshi N, Villmann C, Becker CM, Feng G, De Camilli P, Solimena M. Autoimmunity to gephyrin in Stiff-Man syndrome. Neuron 2000 May;26(2):307–12.

10

Treatment of Myasthenia Gravis

Henry J. Kaminski

1. INTRODUCTION

In 1895, Jolly used the Greek terms for muscle and weakness forming "myasthenia" and the Latin "gravis" for "severe" to describe a condition manifesting with fatiguing strength that often leads to death *(1, 2)*. Treatment for myasthenia gravis (MG) began in the 1930s. Edgeworth, a physician with MG, treated herself with ephedrine describing its benefit in a placebo-controlled trial *(3, 4)*. However, its primary effect is as a central nervous system stimulant with only minor influences on neuromuscular transmission *(5)*. Walker, as a house officer at St. Alfege's in the United Kingdom, appreciated the similarity of MG to curare poisoning, which was treated with physostigmine. She administered physostigmine to a patient with MG, leading to prompt improvement in ptosis *(6, 7)*. Lazar Remen had published a report in 1932 *(8)*, 2 years prior to Walker's work, demonstrating the beneficial effects of neostigmine in a patient with MG, but his discovery has largely gone unnoticed *(9)*.

In the late 1800s, it had been appreciated that patients with myasthenia gravis often had pathology of the thymic gland *(10)*. Blalock, a cardiothoracic surgeon at Johns Hopkins, encountered a 19-year-old woman who had suffered for years with worsening relapsing and remitting weakness consistent with MG *(11, 12)*. A chest radiograph demonstrated a mediastinal mass. Blalock removed a thymic cyst, and postoperatively, the patient improved significantly. Later, he reported the observations from six patients, who all improved postoperatively *(13, 14)*. With these reports thymectomy became an established treatment, although its efficacy remains debated *(15, 16)* and is the subject of a NIH-funded trial (see Chapter 12) *(17)*.

Initial results of adrenocorticotropic hormone treatment of MG were not impressive and widespread use of prednisone did not occur until the 1970s *(15)*. Several immunosuppressive agents were used to treat MG in the 1960s with azathioprine becoming an established treatment during the 1970s. In 1976, plasma exchange was identified as an effective acute treatment for severe MG and also served to support the presence of circulating factors as a cause of MG *(18)*. Current common treatments of intravenous immunoglobulin (IVIg), cyclosporine, and mycophenolate mofetil came into use in the final two decades of the last century.

In the first edition of this text, I commented that "the rationale for all therapies for MG is compromised by the lack of well-designed investigations with appropriate statistical power to draw firm conclusions of efficacy" *(19)*. I specified that challenges for rigorous trial development included the disorder's variable presentation and clinical course as well as its relatively low incidence and prevalence *(20, 21)*. I mentioned that there was a lack of agreed-upon clinical scales to measure outcome and poor cooperation in clinical trials among major centers that care for patients with MG. As you will read in Chapter 18, the MG research community has taken significant steps to utilize a common set of clinical measures *(20)*, and over 60 centers from across the world are now participating in the MGTX study *(17)*. Euromyasthenia (www.euromyasthenia.eu) has been formed which knits together European investigators. The Myasthenia Gravis Foundation of Argentina and research groups in Japan are also forging coordinated collaborations. The following summary reflects my personal bias regarding treatment recommendations,

From: *Current Clinical Neurology*
Myasthenia Gravis and Related Disorders
Edited by: H.J. Kaminski, DOI 10.1007/978-1-59745-156-7_10, © Humana Press, New York, NY

but at the time of this writing MG is making great strides in improving the rigor of evidence-based therapeutic recommendations *(22)*.

2. PATIENT EDUCATION AND EVALUATION

Education is the most important factor in achieving optimal health for patients with MG. Care of the patient with MG should begin with a thorough discussion of the natural history, treatment options, and disease pathogenesis. The patient, family, and the other interested parties should all participate in this discussion and benefit from the appreciation of the chronic nature of the illness with a variable prognosis which often requires treatments with poor side effect profiles. Psychological adjustment is difficult to an illness in which definite predictions of outcome cannot be made and fear exists regarding the development of incapacitating weakness. Poor psychological adjustment may limit functional improvement despite significant resolution of myasthenic weakness *(23)*. The neurologist should attempt to strike a balance between the difficulties in MG treatment and the hope of complete remission. In the United States, treatment decisions need to be made with cost (Table 1) and insurance considerations, but of course, patients across the globe work within the economic constraints of their own nations.

Patients should be directed to the MG Foundation of America and other national MG organizations for additional information. The MG Foundation of America maintains an Internet site (www.myasthenia.org), which provides excellent resources on various aspects of the disease. Support group services provided by local chapters of the Foundation are usually of particular benefit.

A list of medications, which may exacerbate myasthenic weakness, should be given to each patient. Those medications, which are particularly prone to compromise patient strength, are listed in Table 2. The MG Foundation of America maintains a complete, current list on their web site. However, counseling must indicate that these medications may be used, if indicated for other medical reasons, but that caution needs to be exercised. The patient should inform other healthcare professionals of the MG diagnosis. Because of these patient's special needs and the unfamiliarity of many therapists, physicians, and nurses with MG, the neurologist may wish to provide the patient's referring physicians with general information regarding MG.

Computed tomography (CT) or magnetic resonance imaging of the chest for detection of thymoma is indicated at the time of diagnosis. Because CT is less expensive and there is no indication that magnetic resonance imaging is more sensitive or specific, I prefer CT *(24)*. Seybold argues strongly that plain chest x-rays are best for pediatric age group because of the extreme rarity of thymoma in this population *(25)*. However, I would argue that the test with greater sensitivity is preferred when screening a population with a low likelihood of disease. It must be expected that false positives and negatives *(26)* will occur; however, since the majority of patients undergo therapeutic thymectomy, a false positive result is not of clinical significance.

Table 1
Cost Comparisons of Oral Myasthenia Gravis Treatments

Treatment	Cost from Epocrates data bank*
Prednisone 20 mg	100 tablets for $26.63
Azathioprine (150 mg/day)	$84/month
Cyclosporine (200 mg/day)	$318/month of Neoral®
Mycophenolate mofetil (2 g/day)	$780/month of Cellcept®
Tacrolimus (3 mg/day)	$360/month of Prograf®
Pyridostigmine (60 mg, 120 mg/day)	$72/month

*Searched March 2008.

Table 2
Drugs That Frequently Exacerbate Myasthenia Gravis

Antibiotics and antimicrobials	Central nervous system drugs
Neomycin, Kanamycin, Streptomycin, Gentamicin, Tobramycin, Amikacin, Polymyxin B, Colistins, Tetracyclines, Lincomycin, Clindamycin, Erythromycin, Ampicillin, Imipenem, Clarithromycin, Emetine, Ciprofloxacin	Diphenylhydantoin, Trimethadione*, Lithium, Chlorpromazine, Trihexyphenidyl, gabapentin
Cardiovascular drugs	**Antirheumatic agents**
β-blockers, Quinidine, Procainamide, Verapamil, Trimethaphan	D-penicillamine*, Chloroquine*, Prednisone
Other drugs	
Procaine/Lidocaine, d,l-Carnitine, Lactate, Methoxyflurane, Magnesium, Contrast agents, Citrate anticoagulation, Trasylol, Ritonavir, Levonorgestrel, Desferrioxamine, Interferon-α*, Pyrantel pamoate	

*Induction of autoimmune disorder.

However, not identifying a thymoma in a patient who chooses not to undergo thymectomy is deleterious. Although striational antibodies are associated with thymoma, they are not of adequate sensitivity or specificity to eliminate the need for imaging *(27)*. Patients with positive striational antibodies who choose not to undergo thymectomy or are beyond the age typically treated with thymectomy may require serial imaging for the detection of a thymoma.

I routinely screen for thyroid dysfunction because of high frequency of coincidence with MG and the benefit to myasthenic manifestations that patients may derive in treatment of thyroid abnormalities. Systemic lupus erythematosis, rheumatoid arthritis, and vitamin B12 deficiency are associated with MG, but screening should be performed if clinically indicated. IgA deficiency is estimated to have a prevalence as high as 1 per 1,000, and treatment of such patients with IVIg may lead to life-threatening allergic reactions *(28)*. This has led to the recommendation that all patients be screened for its presence prior to IVIg treatment. Consideration may be given for screening of IgA deficiency in anticipation of IVIg therapy to avoid delay in initiation of treatment in emergency situations. Tuberculin skin tests may be performed in anticipation of future corticosteroid treatment because of the possibility that such treatment could activate dormant tuberculosis.

3. TREATMENTS

Treatment choices should be individualized based on the severity of the disease, patient's lifestyle and career, associated medical conditions, as well as the patient's assessment of risk and benefit of various therapies. For example, a young woman with an active career may require immunosuppressive treatment as long as she appreciates cosmetic and reproductive concerns. An older individual with a sedentary lifestyle and mild general weakness may do well with cholinesterase inhibitors. Generalized MG is no longer the grave disease it once was, with life spans of patients approaching normal *(29)*. This has occurred from the combination of modern critical care and immunosuppressive therapy. However, quality of life remains compromised *(30, 31, 32)*, and there remains a need to define disability and refine quality of life assessments among patients with MG, even those assessed as doing well by their treating physicians.

The complications of treatment need thorough explanation, and the patient and physician must work together to optimize the care plan. As mentioned, cost (Table 1) influences treatment selection. If a patient has insurance with a drug benefit, the physician is not limited, and in my experience, all insurance companies approve use of the medications listed in Table 1. Nearly all pharmaceutical corporations also have established programs to provide medications at reduced cost for patients who document financial need. The web site, www.needymeds.com, maintains contact information for such programs.

3.1. Cholinesterase Inhibitors

Acetylcholinesterase (AChE) inhibitors, which retard the hydrolysis of acetylcholine at the neuromuscular junction *(33)*, are usually the initial treatments for MG. Pyridostigmine bromide is the most commonly used agent. Initial therapy begins at doses of 30–60 mg of pyridostigmine, and dosing intervals of pyridostigmine are set at 3–6 hours depending on symptoms, individual doses of greater than 180 mg rarely being effective. Because of erratic absorption, variability in patient activity levels, severity of disease, and degrees of muscarinic side effects, strict dosing guidelines cannot be recommended. Once a patient becomes aware of the effects of pyridostig- mine on their symptoms and the adverse reactions to the medication, they should be able to modify the dose and timing of administration within limits set by the physician. Timed-release forms are available but tend to have erratic absorption; however, rare patients demonstrate a preference for these, particularly when given at bedtime for nocturnal symptoms. I tend not to use timed-release preparations. An elixir form of pyridostigmine is available, which may be helpful for some patients with difficulty swallowing and is easier to administer for patients with small nasogastric tubes. The manufacturer recommends crushing a tablet with a small amount of fruit juice or applesauce to provide a liquid or easy-to-swallow dose (Mestinon 60 mg tablet is a dose equivalent to one teaspoon of Mestinon syrup). Prostigmine bromide (neostigmine) or mytelase chloride (ambenonium) is used but tends to have greater side effects and shorter period of action (Table 3).

In experimental MG produced in animals and with cholinesterase inhibition, a splice variant of the AChE is produced, which is soluble and more effectively eliminates acetylcholine from the synapse *(34, 35)*. Patients and physicians often appreciate a diminution of the beneficial effect of cholinesterase inhibition over time especially with effective immunosuppressive therapy, and it is likely the splice variant is responsible for this clinical observation. Therefore, AChE inhibitor treatment may be tapered when clinical benefit is lost. Patients can always restart the medication if they identify development of symptoms during the taper.

AChE inhibitor treatment is generally safe, but significant side effects may occur. If a patient has prominent bulbar muscle weakness, administration of AChE inhibitors will often not reverse weakness to an extent that a patient will be safe to swallow. Swallowing difficulties may be further worsened by excess and thick saliva, especially when concomitant treatment with anti-muscarinic medications is used. Such apparent paradoxical responses should be appreciated and a reduction of AChE inhibitors may improve symptoms. AChE inhibitors do not reverse respiratory muscle weakness reliably. Respiratory secretions may be increased, which complicates treatment of patients with pulmonary diseases and may actually worsen breathing. Reductions or discontinuation of

Table 3
Equivalent Doses of Cholinesterase Inhibitor Medications

Medication	Oral	Intramuscular	Intravenous
Pyridostigmine bromide	60 mg	2.0 mg	0.7 mg
Neostigmine bromide	15 mg	None	None
Neostigmine methylsulfate	None	1.5	0.5 mg
Ambenonium	7.5 mg	None	None

AChE inhibitors may improve respiratory symptoms *(36)*. Rare patients develop significant bradycardia necessitating discontinuation of the medication *(37)*.

Gastrointestinal complaints of nausea, vomiting, diarrhea, and abdominal cramps are most common but may be controlled with administration of atropine and glycopyrrolate. Muscle twitching, fasciculations, and cramps may be particularly bothersome to patients. Stretching exercises help cramps, while reassurance alleviates concerns regarding abnormal muscle movements. Rarely, patients may develop confusion from AChE inhibitor therapy.

AChE inhibitor-induced weakness ("cholinergic crisis") is frequently discussed but is rarely encountered and some experts do not think it exists (Lewis Rowland, personal communication). The use of intravenous edrophonium to determine if muscle weakness is secondary to MG or AChE treatment is unreliable *(15, 38, 39)*. If cholinergic weakness is seriously considered, then AChE therapy should be temporarily discontinued and the patient watched for improvement. In patients with myasthenic crisis, discontinuation of AChE inhibitors is recommended to limit secretions while on artificial ventilation (Chapter 10) *(40)*.

3.2. Corticosteroids

Corticosteroids are the least expensive (Table 1), most reliable, and rapid acting of maintenance therapies for MG, and this statement is based on general clinical experience rather than controlled clinical trials *(41)*. Their use is complicated by a myriad of adverse effects (Table 4). Corticosteroids are recommended for patients who are compromised by generalized weakness that limits their ability to function and which cannot be adequately improved by AChE inhibitors and moderation of activity *(42, 43)*. Patients with respiratory or bulbar weakness generally require immunosuppression and usually will need corticosteroid therapy. Patients need to be made aware of expected complications and agree to institution of the medication. Some patients may not accept steroid therapy, and treatment with the slower acting agents (see below) often coupled with IVIg or plasma exchange may be considered as an alternative.

The optimal manner to initiate corticosteroids is dependent on the severity of weakness and subject to prescriber's personal preference *(44)*. With initiation of 60–80 mg per day, up to 50% of patients develop a worsening of strength in the first month, usually within the first few days, of treatment *(15, 45, 46, 47, 48, 49)*. Sustained improvement usually begins within 2 weeks and in the majority, within 1 month, but rare patients have a delay of 2 months. Maximal improvement usually occurs at 6 months. For patients with severe weakness, I admit the patient and institute a high-dose regimen or a rapidly increasing dosage coupled with an acute therapy, usually plasma exchange. In my opinion, plasma exchange limits the severity of the corticosteroid-associated increase in weakness, but this method has not been evaluated formally. Gradual initiation of

Table 4
Immunomodulating Drugs for Myasthenia Gravis and Common Adverse Effects

Treatment	Major adverse effects
Prednisone	Osteoporosis, weight gain with central obesity, glaucoma, cataracts, hypertension, peripheral edema, psychiatric changes (depression, mania, personality alterations), sleep disturbance, easy bruising, glucose intolerance
Azathioprine	Idiosyncratic flu-like reaction, leukopenia, hepatotoxicity, alopecia, teratogen, possible risk of neoplasia
Cyclosporine	Renal insufficiency, hypertension, gingival hyperplasia, numerous drug interactions
Mycophenolate mofetil	Anemia, leukopenia, gastrointestinal discomfort, diarrhea
Tacrolimus	Tremor, headache, diarrhea, hypertension, nausea, renal insufficiency, hyperkalemia, hypomagnesemia

corticosteroids beginning with 20 mg and increasing 5 mg every 3 days until a dose of 60–80 mg is reached decreases the frequency of exacerbation but delays the onset of improvement *(48, 50, 51)*. Such a regimen may be considered in patients with moderate weakness and allows outpatient therapy. Regardless of treatment initiation, ultimately converting to a q.o.d. dosing schedule is desired to limit steroid complications *(46)*. Some advocate an every-other-day treatment with 100 mg of prednisone at onset of therapy *(52)*, and in a multinational, collaborative study of thymectomy that is ongoing at the time of this writing, the participating investigators agreed that an alternate day treatment regimen was the preferred approach to corticosteroid treatment. Regardless of regimen, the corticosteroid should be given as a single morning dose in an attempt to mimic the morning peak of cortisol production. High-dose solumedrol has been used, but there is no reason to believe that such treatment is of greater benefit than the treatment with standard regimens *(53)*. Before common use of "steroid-sparing" adjunct therapy, slightly less than 20% of patients were able to stop corticosteroid treatment *(46)*.

Monitoring serum potassium and glucose as well as blood pressure and treatments to limit osteoporosis is usually necessary. Calcium, vitamin D supplement, and bisphosphonates are indicated for most patients *(54)*. Ophthalmological evaluation for cataracts and glaucoma should be performed on a yearly basis. Treatments for gastric ulcer prophylaxis need not be instituted unless patients have a history of ulcer disease or develop symptoms of gastric irritation. The emotional effects of corticosteroids are common and for patients often the most immediate and deleterious *(55, 56)*. Despite the clinical observation, formal studies of the behavioral consequences of chronic corticosteroid are essentially nonexistent.

Mechanism of Action. Corticosteroids have numerous effects on the immune system leading to general immunosuppression *(57, 58, 59)*. The therapeutic effects for MG appear to be related to (1) reductions of lymphocyte differentiation and proliferation, (2) redistribution of lymphocytes from the circulation into tissues that remove them from sites of immunoreactivity, (3) alterations of lymphokine function (primarily tumor necrosis factor, interleukin-1, and interleukin-2), and (4) inhibition of macrophage function, in particular antigen processing and presentation. Acetylcholine receptor (AChR) antibody levels decrease in the first few months of therapy. Also, corticosteroids may exert their benefit in MG by increasing the muscle's AChR synthesis *(33)*.

3.3. Azathioprine

Azathioprine is used alone or in combination with corticosteroids *(40, 60, 61, 62, 63)*. Although azathioprine is often utilized for its "corticosteroid-sparing" effect, there are limited data that address its relative benefit over corticosteroids. Palace and colleagues did demonstrate in a small, placebo-controlled trial a steroid-sparing effect, but only after more than 1 year of treatment *(64)*. Both the physician and the patient need to be cognizant that improvement is extremely gradual. It is critical for the patient to receive a dose based on total body weight (1–3 mg/kg/day in divided doses) since underdosing may lead to the false impression of a lack of efficacy. Improvement appears to correlate with elevations of mean red blood cell volume *(61)* and reduction of white blood cell count. If either of these is not observed, then the dose may be increased to its maximum. If white blood cell counts fall below 3,000–3,500 white blood cells/mm^3, the dose needs to be reduced and adjusted to maintain counts above 3,500. Levels below 1,000 white blood cells/mm^3 require discontinuation of the drug *(40, 65)*.

In the United States, 10% of patients in the first 2 weeks of initiation of azathioprine develop fever and flu-like symptoms necessitating discontinuation of the medication. This reaction is not prominently observed in Europe and is perhaps related to differences in formulation of the tablets. Mild hepatotoxicity may manifest with elevations of serum transaminase levels, but usually without clinically evident liver disease. Azathioprine needs to be discontinued if transaminase levels double. Monitoring of liver function tests and complete blood counts are necessary throughout the treatment course. Pancreatitis rarely occurs with azathioprine treatment. Allopurinol reduces metabolism of azathioprine necessitating reduction of dose by one-third to avoid complications. There is an increased risk of neoplasm, primarily lymphoma, in patients taking azathioprine for organ rejection and some autoimmune disorders *(66)* but not confirmed among patients with MG, but this risk needs an explanation to the patient *(60)*. The agent also has teratogenic potential, and it is generally advised for women and men planning a pregnancy to discontinue its use *(52)*. However, one study of MG patients receiving azathioprine during pregnancy did not report birth defects *(67)*.

Azathioprine is a purine analogue, which inhibits synthesis of nucleic acids. The drug is converted to 6-mercaptopurine and metabolized to 6-thioguanine nucleotides, which are the cytotoxic agent *(68,69)*. Thiopurine *S*-methyltransferase and xanthine oxidase oppose formation of 6-thioguanine nucleotides. Patients with thiopurine *S*-methyltransferase deficiency are at risk for severe reactions and should be identified by red cell thiopurine *S*-methyltransferase enzyme activity prior to initiation of azathioprine therapy.

Mechanism of Action. The primary effect on the immune system is interference with T- and B-cell proliferation and has been presumed to be related to its direct effect on nucleic acid synthesis. Azathioprine metabolites are now being identified as having immunosuppressive effect *(70)*. Coupled with plasma exchange, azathioprine may offer relative selectivity for interference with the anti-AChR autoimmune response *(71)*. This is because depletion in immunoglobulin after plasma exchange would induce a compensatory increase in antibody-producing B cells. Activated anti-AChR B cells would respond vigorously and therefore may be preferentially affected by azathioprine.

3.4. Cyclosporine

Cyclosporine has been used to treat MG patients, with severe MG responding poorly to corticosteroids and thymectomy *(22, 72, 73, 74)*. Doses of 5 mg/kg/day in two divided doses have been recommended. Although cyclosporine works more quickly than azathioprine, the patient and physician still cannot expect a response for several months of starting treatment. In a retrospective review, clinical improvement in 55 of 57 patients who took cyclosporine for an average of 3.5 years usually occurred in less than 7 months. Corticosteroids were discontinued or decreased in nearly all the patients taking them *(73)*. Five percent could not afford or tolerate the drug *(73)*. I have also found that lower doses may be initiated (2–3 mg/kg/day) with similar benefit. However, such an approach requires rigorous evaluation. Cyclosporine may be used alone and this approach does have the advantage of avoiding corticosteroid side effects *(43)*, although it should be appreciated that the cyclosporine does promote osteoporosis *(75)*. Some advocate monitoring of drug levels to achieve a trough level of 75–150 ng/ml *(65)*, but it is not clear that a specific level needs to be maintained. Monitoring of drug levels may only be of benefit to assure compliance and in patients on medications that affect cyclosporine metabolism. No study exists that evaluates the relative benefits of long-term cyclosporine versus prednisone for MG treatment.

The primary side effects of cyclosporine are renal insufficiency and hypertension. Creatinine measures need to be monitored throughout treatment with cyclosporine. In the study of Ciafaloni et al., 16 patients had serum creatinine elevations of 30–70% above pretreatment values (mean 48%) *(73)*. Five patients required discontinuation of cyclosporine because the creatinine value did not fall despite dose reduction. These patients were older than 62 years and the renal toxicity occurred 3–11 years after initiation of therapy. Cyclosporine has many drug interactions, and it is critical to check updated drug databases for interactions that may lower or elevate cyclosporine levels or lead to specific toxicities.

Mechanism of Action. Cyclosporine is derived from a fungus and is a cyclic undecapeptide with actions directed exclusively on T cells. Cyclosporine blocks T-helper cell synthesis of cytokines, in particular interleukin-2 and interleukin-2 surface receptors. The agent also interferes with transcription of genes critical in T-cell function. Cyclosporine suppresses T-helper cell-dependent function, maintains tolerance in transplant patients, and prevents experimental autoimmune MG *(76)*.

3.5. Mycophenolate Mofetil

In this century, mycophenolate mofetil (MM, Cellcept®) has become an increasingly popular agent for treatment of MG in the United States. Small studies showed that MM leads to a reduction of corticosteroid dose, improved strength, and lowering of AChR antibody levels *(77, 78, 79)*. A study of patients with purely ocular myasthenia has shown it to be effective and well tolerated *(80)*. Two randomized, placebo-controlled studies have recently been completed which did not demonstrate efficacy of MM in steroid-sparing end points *(81, 82)*. They were both compromised by their short duration of 6 months. The slow onset of biological action of MM would not have been expected to show efficacy in such a short time frame. I hope that these negative studies do not compromise the use of what I consider an effective medication.

MM is given at a standard dose of 1 g twice daily and no data exist regarding whether higher doses are more effective. In studies of patients with MG, MM is well tolerated and adverse effects are limited to gastrointestinal upset and anemia. Leukopenia may occur. Although no firm information in humans regarding the risk of birth defects is available, animal studies indicate a teratogenic potential and it is recommended that women of childbearing age be informed of this and birth control offered. Experience from studies of transplant patients indicates that there may be an increased risk of cancer with MM treatment *(83)*, and case reports have appeared of MM treatment associated with lymphoma *(84)*.

Mechanism of Action. MM is another agent commonly used to treat transplant rejection with effects primarily on T- and B-cell proliferation through inhibition of guanosine nucleotide synthesis *(76, 85, 86)*. MM is hydrolyzed to mycophenolic acid, which inhibits inosine monophosphate dehydrogenase, the rate-limiting enzyme in the synthesis of guanosine nucleotides. Mycophenolic acid is a more potent inhibitor of the dehydrogenase isoform expressed in activated lymphocytes leading to its preferential immunosuppressant properties. Additional mechanisms of immunosuppression include (1) apoptosis of activated T lymphocytes, (2) alterations in cell adhesion molecule expression decreasing recruitment of lymphocytes to sites of inflammation, and (3) reduction of inducible nitric oxide synthase activity.

3.6. Tacrolimus

Throughout the world, but less so in the United States, tacrolimus (FK 506) *(87, 88, 89)* is being utilized for MG. The largest study utilized a tacrolimus dose (0.1 mg/kg/day) in over 200 patients who were steroid dependent and receiving cyclosporine. Strength assessments and AChR antibody levels were improved at the end of the observation time and the study found no major adverse effects, although about 5% of patients did withdraw secondary to medication effects. Substitution of tacrolimus for cyclosporine appeared to reduce the overall frequency of medication-related complications *(87, 88)*. Common side effects are gastrointestinal discomfort and paresthesias. Tacrolimus is nephrotoxic and hepatotoxic and may worsen hypertension and diabetes.

Mechanism of Action. Tacrolimus is a macrolide with several activities that may be therapeutic in MG patients. It modulates the activity of T cells, increases their apoptosis, and may enhance T regulatory cells *(90)*. Tacrolimus is also reported to enhance muscle contraction by modulation of intracellular calcium-release channels *(91)* and appears to increase intracellular concentration of glucocorticoid by blocking their export *(92)*.

3.7. Plasma Exchange

Plasma exchange (plasmapheresis) produces rapid improvement in weakness regardless of whether patients are seronegative or seropositive for anti-AChR antibody *(93, 94, 95)*. It is used in myasthenic crisis as well as to optimize muscle function prior to surgery, including thymectomy. The exchanges may be performed in an outpatient setting and, depending on the skill of the exchange technicians, often through peripheral venous catheters. Typically exchanges are done to remove one to two plasma volumes three times per week for up to six exchanges. In my experience, more than 6 exchanges during a single course are not beneficial, but others describe benefit after 14 exchanges *(65)*. Patients may show improvement within 48 hours of the first or second exchange and continue to improve during an exchange course. Treatment reduces immunoglobulin levels rapidly; however, rebound may occur in weeks, leading to clinical worsening if concomitant immunosuppressive treatment is not used. In rare patients, plasma exchange is used as a chronic therapy, but this has not been formally studied *(93)*. The exchange is usually performed with resins that remove proteins at certain molecular weights and studies of immunoabsorbent resins do not demonstrate any superiority to standard resins *(96, 97)*, but preclinical work suggests that selective exchange may ultimately prove to be more effective *(98)*.

The general usefulness of plasma exchange is limited by its restriction to major medical centers and the frequent need for large bore intravenous catheters. During the infusion, patients may complain of paresthesias from citrate-induced hypocalcemia and hypotension may occur at initiation of the exchange *(99)*. Some patients have nausea and vomiting related to fluid shifts and electrolyte alterations during the exchange. Infectious and thrombotic complications related to venous access occur *(100, 101)*. The exchange process reduces coagulation factors, and concomitant use of heparin in the case of venous catheters may reduce platelet levels leading to bleeding tendencies. Thrombotic complications also occur *(99)*.

Mechanism of Action. Removal of circulating pathogenic antibody and in all likelihood other factors, such as complement proteins, produce the clinical benefits of plasma exchange *(33)*. The removal of antibodies that block AChR function may be primary in mediating the improvement that is sometimes observed within hours of plasma exchange. Therefore, it is likely that other mechanisms are important in mediating the therapeutic effect of plasma exchange.

3.8. Intravenous Immunoglobulin

IVIg is used in similar circumstances as plasma exchange, but its use is based on limited evidence *(102)*. A randomized study demonstrated that IVIg is superior to placebo in treatment of MG exacerbation *(103)*, but the important question is whether it is superior to other active treatments for MG. The standard treatment regimen is a 4- to 6-hour infusion of 400 mg/kg/day dose for 5 days, but some advocate a 1 g/kg/day dose for 2 days *(28)*. The latter can be administered without a higher rate of complications; however, in older patients with cardiac disease the lower dose regimen may be prudent. IVIg therapy improves strength rapidly, often within 5 days of initiation, and lowers AChR antibodies, but the response is short lived and not uniformly observed. A retrospective study comparing IVIg and plasma exchange for treatment of myasthenic crisis showed patients had a better respiratory and functional outcome with plasma exchange, but IVIg had fewer complications *(100)*. A randomized trial compared three exchanges to either 3 or 5 days of IVIg (400 mg/kg/day) and showed a similar efficacy with again IVIg having less adverse effects *(102)*. IVIg may also be used as a maintenance therapy to limit corticosteroid dosage, while other agents, such as cyclosporine or azathioprine, are taking effect.

Adverse effects with IVIg occur commonly but most are minor *(28, 104)*. Headache occurs frequently during infusion. Mild headaches may be treated with acetaminophen or nonsteroidal anti-inflammatory drugs (since MG patients are frequently receiving corticosteroids, nonsteroidals should be used with caution to limit gastrointestinal irritation). More severe migraine headaches occur, particularly in patients with a history of migraine. In appropriate individuals, triptan agents may be used for their treatment. Aseptic meningitis may occur and recur with subsequent treatments. Some claim that prophylaxis with nonsteroidal anti-inflammatory drugs may be of benefit, but others disagree *(28, 104)*. Chills, myalgia, or chest discomfort may occur early during the infusion and usually resolves on stopping the infusion for 30 min and resuming at a slower rate. Flu-like symptoms may occur in the days following treatment and some patients are described as having a transient worsening of strength. Urticaria, lichenoid cutaneous lesions, pruritus of the palms, and petechiae may occur within days or weeks of treatment and resolve over weeks to months.

Rarely, significant complications occur with IVIg therapy. Anaphylactic reactions occur in patients with immunoglobulin A deficiency, which may be present in 1 in 1,000 people *(28)*. This deficiency should be screened for prior to infusion and consideration may be given to provide an IVIg preparation depleted of immunoglobulin A to deficient individuals *(105)*. Acute renal tubular necrosis, which is usually reversible, occurs in patients who have renal insufficiency. Age over 65 years, diabetes, and dehydration increase the risk of renal injury. Renal failure is associated with high concentration of sucrose in one proprietary IVIg product but occurs with other preparations *(104)*. Diluting the IVIg preparation and slowing the rate of infusion decreases the chance of renal damage. Close monitoring of creatinine and blood urea nitrogen is essential if the agent needs to be given in patients with renal insufficiency. Deep vein thrombosis, cerebral infarction, and myocardial infarction rarely occur *(106, 107)*. For the immobilized MG patient, physicians should be vigilant for the signs of deep vein thrombosis and appropriate prophylactic treatments administered.

The choice of IVIg or plasma exchange for a particular patient with myasthenic exacerbation lies in assessment of potential risks of each therapy. Since plasma exchange is offered only at specialized centers, IVIg has the advantage of being more accessible, although shortages of IVIg do occur, and vascular access is usually not difficult for administration of IVIg. In patients with renal insufficiency or risks of thrombotic complications, plasma exchange may be preferred. Cost of a single course of IVIg is on the order of $10,000, while outpatient plasma exchange is slightly less *(42)*.

Mechanism of Action. IVIg impacts on the autoimmune process by several mechanisms including inhibition of cytokines, competition with autoantibodies, inhibition of complement deposition, interference with Fc receptor binding on macrophages or the immunoglobulins on B cells, and interference with antigen recognition by sensitized T cells. Anti-idiotype mechanism appears to be the most critical in clinical efficacy *(108, 109)*.

3.9. Other Immunosuppressive Treatments

Intravenous and oral cyclophosphamide has been used to treat MG resistant to corticosteroids and one study found that half of the patients were asymptomatic after 1 year *(22, 110, 111, 112)*. Nearly all patients developed alopecia. Additional complications include diarrhea, nausea, vomiting, and hemorrhagic cystitis, which may be severe. The drug has carcinogenic and teratogenic potential as well as a likelihood of producing infertility *(66)*. Rarely, interstitial pneumonitis and hepatic injury occur.

High-dose cyclophosphamide alone is not myeloablative, which allows a patient's endogenous stem cells to repopulate the hematopoietic/immune systems. Such an approach has been applied to patients with treatment-resistant autoimmune diseases, including MG *(113)*. In the limited number of patients studied, several have been placed in remission.

Individual reports and small series describe the response of MG patients to rituxmab *(114, 115, 116, 117)*, a monoclonal antibody against the B-cell surface membrane marker CD20 *(118)*. A controlled trial has been organized to rigorously determine its efficacy.

4. SPECIFIC CLINICAL SITUATIONS

4.1. Myasthenia Gravis and Pregnancy

MG preferentially affects women of childbearing age leading to particular therapeutic issues. The effect of pregnancy on the clinical course of MG is highly variable among women, and each pregnancy may affect an individual woman differently *(67)*. A patient appears to have an equal probability of worsening, staying the same, or improving during the pregnancy or postpartum period *(67, 119)*. The normal weight gain and other effects of pregnancy would be expected to exacerbate the fatigue of the MG patient. Labor is an exhausting event and should be expected to weaken the symptomatic MG patient; however, the majority of women are able to deliver vaginally without complications *(67, 120, 121)*. If pre-eclampsia or eclampsia is present, then the obstetrician should be made aware that intravenous magnesium sulfate is contraindicated because of its deleterious effects on neuromuscular transmission *(122)*. MG should not be considered a contraindication to pregnancy for most women.

MG treatments may affect the fetus *(67, 120)*. Azathioprine, MM, and cyclosporine are potential teratogens, and every attempt should be made to discontinue the medications prior to conception. Mestinon and prednisone are safe. There are no large studies of the effect of plasma exchange or IVIg in pregnant women but they appear to be safe. Thymectomy should be delayed until after delivery, since there is no reason to believe the delay would be deleterious compared to the potential complications of performing a major surgical procedure during pregnancy.

Approximately a third of the infants of MG patients have neonatal myasthenia *(67, 120)*. This is a transient disorder manifested by general weakness, difficulty feeding, and respiratory insufficiency in some. Such infants require supportive care and cholinesterase treatment. Weakness resolves over weeks without any permanent weakness or risk of development of MG later in life. Development of neonatal MG is independent of the clinical status of the mother and the mother may be asymptomatic. AChR antibody levels of the mother also do not predict occurrence of MG. An increased ratio of antibodies against the fetal AChR compared to the antibodies directed only against the adult AChR isoform correlates with the development of neonatal myasthenia *(123)*. Case reports exist of infants with arthrogryposis multiplex that was due to passage of antibodies against the fetal AChR, which severely compromised fetal muscle development *(124)*. If a woman with MG has a history children with arthrogryposis multiplex or recurrent fetal loss, consideration should be given to more intensive therapy of MG during pregnancy *(125, 126, 127)*.

4.2. Treatment of the Myasthenia Gravis Patient with Thymoma

Ten percent of patients with MG have a thymoma, and Chapter 7 details thymoma-associated MG. The treatment of patients with thymoma does not differ from those patients without thymoma but is complicated by the necessity to treat the neoplasm. Most, but not all, studies suggest that thymoma-associated MG has a more severe clinical picture *(128, 129)*. Once a thymoma is detected it should be removed along with removal of the remaining thymus, although the patient should be aware that the surgery is not to cure MG. If the tumor has broken through its capsule to involve adjacent tissue, local irradiation is necessary. However, some thymomas are aggressive and produce widespread metastases. Patients require lifelong monitoring for recurrence of tumor *(130)*. In the elderly patient with a contraindication to surgery, consideration may be given to monitor tumor growth by computed tomography since many tumors are very slow growing.

4.3. Juvenile Myasthenia Gravis

The pathophysiology of children younger than 18 years presenting with MG does not differ from the adult disorder *(131, 132)*. Differential diagnosis and diagnostic methods are identical to adults, although congenital myasthenic syndromes are more likely. Treatment considerations are more difficult. Prepubertal patients appear to have higher rates of spontaneous remission, which prompts some to consider a delay in thymectomy. Corticosteroids retard growth and increase the severity of osteoporosis later in life. Treatment with immunosuppressive drugs has the potential of total dose-dependent risk of neoplasia. As in adults, treatment must be individualized and the parents need to appreciate the complications of therapy *(133)*.

4.4. The Treatment-Resistant Patient

Patients referred to me for a poor treatment response come in three varieties: (1) those who never had MG, (2) those with complications related to treatment, and (3) those with severe MG.

When faced with a patient who continues to complain of weakness, it is imperative to confirm that the diagnosis of MG is based on unequivocal clinical manifestations and supported by serological or electrophysiological studies. Care is further complicated if the patients have been treated for MG with immunosuppressive treatments and developed complications, which produce weakness or fatigue. If on review of previous evaluation the physician concludes MG is not present, then the patient must be informed of the determination using the utmost sensitivity. Medications for MG will need to be tapered. Even if an alternate medical diagnosis is identified as a cause of symptoms, patients will have been traumatized and require expert psychological counseling to assist them in recovering from misdiagnosis.

Corticosteroids produce a legion of side effects that may lead patients to complain of "fatigue", "tiredness", and "weakness", which often do not distinguish from their presenting complaints of MG. Similar complaints could occur with immunosuppressants that produce anemia. The neurologist, therefore, must take a detailed history to clearly define that symptoms, corroborated by examination, are related to MG. Sleep disturbances, including sleep apnea, may complicate MG and be exacerbated by weight gain related to steroid therapy, which lead to excessive somnolence, misinterpreted as fatigue *(134, 135)*. I have seen several patients with MG who have developed steroid myopathy after several months of daily, high-dose therapy, who were also treated aggressively with plasma exchange and other immunosuppressive treatments. These patients have proximal muscle weakness and obvious systemic complications of steroids without MG manifestations *(136)*. Taper of steroids, in particular to an every-other day regimen, in concert with physical therapy, leads to significant functional improvement after several months.

Fortunately, MG patients with persistent weakness despite prolonged therapy (usually this means thymectomy, corticosteroids and at least one additional modality) are rare. Appropriate dosing of medications must be confirmed and enough time from treatment initiation needs to pass to be sure an agent would be expected to have taken effect. If these are true, then alternative treatments need to be considered. If a patient has moderate to severe weakness, then I use IVIg or plasma exchange in the hopes of bringing about relatively rapid improvement, which may then be provided monthly until oral medications take effect. If corticosteroids are at low doses, then an increase may also be helpful. Coincident with these treatments, a change to another immunosuppressant is made.

4.5. Experimental Treatments for Myasthenia Gravis

In spite of the success of the current therapy for MG, there remains a need for specific treatments to eliminate the autoimmune reaction leaving other immune functions intact and without systemic side effects. Various methods have been attempted: (1) administration of autoantigen or the parts of its sequence to induce tolerance, (2) depletion of AChR-specific B cells, (3) interruption of the complex between major histocompatibility class II molecules, epitope peptide, T-cell receptor, and $CD4^+$ molecule, (4) depletion of AChR-specific T cells, (5) complement inhibition, and (6) enhancement of cholinergic activity *(33)*.

REFERENCES

1. Keesey J. Myasthenia Gravis. An Illustrated History. Roseville, CA: Publisher's Design Group, 2002.
2. Pascuzzi R. The history of myasthenia gravis. Neurol Clin 1994;12:231–242.
3. Edgeworth H. The effect of ephedrine in the treatment of myasthenia gravis, second report. J Am Med Assoc 1933;100:1401.
4. Edgeworth H. A report of progress on the use of ephedrine in a case of myasthenia gravi. J Am Med Assoc 1930;94:1136.
5. Sieb J, Engel A. Ephedrine: effects on neuromuscular transmission. Brain Res 1993;623:167–171.
6. Walker MB. Treatment of myasthenia gravis with physostigmine. Lancet 1934;1:1200–1201.
7. Keesey JC. Contemporary opinions about Mary Walker: a shy pioneer of therapeutic neurology. Neurology 1998;51:1433–1439.
8. Remen L. Zur Pathogenese und Therapie der Myasthenia gravis pseudoparalytica. Dtsch Ztschr f Nervenheilkunde 1932;128:66–78.
9. Viets HR. Introductory remarks. In: Ossermann KE, ed. Myasthenia Gravis. New York: New York Academy of Sciences, 1966: 5–7.
10. Bell ET. Tumors of the thymus in myasthenia gravis. J Nerv Ment Dis 1917;45:130–143.
11. Blalock A, Mason MF, Morgan HJ, Riven SS. Myasthenia gravis and tumors of the thymic region. Report of a case in which the tumor was removed. Ann Surg 1939;110:544–561.
12. Kirschner PA. The history of surgery of the thymus gland. Chest Surg Clin N Am 2000;10:153–165.
13. Blalock A. Thymectomy in treatment of myasthenia gravis. J Thoracic Surg 1944;13:316.
14. Blalock A, Harvey AM, Ford FF, Lilienthal JL. The treatment of myasthenia gravis by removal of the thymus gland: preliminary report. JAMA 1941;117:1529.
15. Rowland LP. Controversies about the treatment of myasthenia gravis. J Neurol Neurosurg Psychiatry 1980;43:644–659.
16. Gronseth GS, Barohn RJ. Practice parameter: thymectomy for autoimmune myasthenia gravis (an evidence-based review): report of the Quality Standards Subcommittee of the American Academy of Neurology [see comments]. Neurology 2000;55:7–15.
17. Wolfe GI, Kaminski HJ, Jaretzki A, 3rd, Aban A, Cutter G, Newsom-Davis J. Status of the Thymectomy Trial for Non-thymomatous Myasthenia Gravis Patients by Receiving Prednisone. In: Kaminski HJ, Barohn RJ, eds. Myasthenia Gravis and Related Disorders: 11th International Conference. New York: New York Academy of Sciences, 2008;1132:344–347.
18. Pinching AJ, Peters DK, Newsom-Davis J. Remission of myasthenia gravis following plasma exchange. Lancet 1976;2:1373–1376.
19. Kaminski HJ. Treatment of myasthenia gravis. In: Kaminski HJ, ed. Myasthenia Gravis and Related Disorders. Totowa, NJ: Humana Press, Inc, 2003: 197–221.
20. Jaretzki A, 3rd, Barohn RJ, Ernstoff RM, et al. Myasthenia gravis: recommendations for clinical research standards. Task Force of the Medical Scientific Advisory Board of the Myasthenia Gravis Foundation of America [see comments]. Neurology 2000;55:16–23.
21. Kissel JT, Franklin GM. Treatment of myasthenia gravis: a call to arms. Neurology 2000;55:3–4.
22. Hart IK, Sathasivam S, Sharshar T. Immunosuppressive agents for myasthenia gravis. Cochrane Database Syst Rev 2007 Oct 17;(4):CD005224.
23. Kulaksizoglu IB. Mood and anxiety disorders in patients with myasthenia gravis: aetiology, diagnosis and treatment. CNS Drugs 2007;21:473–481.
24. Emskotter T, Trampe H, Lachenmayer L. Magnetic resonance imaging in myasthenia gravis. An alternative to mediastinal computerized tomography? Dtsch Med Wochenschr 1988;113:1508–1510.
25. Seybold ME. Diagnosis of myasthenia gravis. In: Engel AG, ed. Myasthenia Gravis and Myasthenic Disorders. New York: Oxford University Press, 1999: 146–166.
26. de Kraker M, Kluin J, Renken N, Maat AP, Bogers AJ. CT and myasthenia gravis: correlation between mediastinal imaging and histopathological findings. Interact Cardiovasc Thorac Surg 2005;4:267–271.
27. Kaminski HJ, Santillan C, Wolfe GI. Autoantibody testing in neuromuscular disorders. Neuromuscular junction and muscle disorders. J Clin Neuromusc Dis 2000;2:96–105.
28. Dalakas MC. Immunotherapies in the treatment of neuromuscular disorders. In: Katirji B, Kaminski H, Preston D, Ruff R, Shapiro B, eds. Neuromuscular Disorders in Clinical Practice. Boston: Butterworth Heinemann, 2002: 364–383.
29. Phillips LH, Torner JC. Epidemiologic evidence for a changing natural history of myasthenia gravis. Neurology 1996;47:1233–1238.
30. Paul RH, Cohen RA, Goldstein JM, Gilchrist JM. Fatigue and its impact on patients with myasthenia gravis. Muscle Nerve 2000;23:1402–1406.
31. Padua L, Evoli A, Aprile I, et al. Quality of life in patients with myasthenia gravis. Muscle Nerve 2002;25:466–467.

32. Padua L, Evoli A, Aprile I, et al. Health-related quality of life in patients with myasthenia gravis and the relationship between patient-oriented assessment and conventional measurements. Neurol Sci 2001;22:363–369.

33. Conti-Fine BM, Milani M, Kaminski HJ. Myasthenia gravis: past, present, and future. J Clin Invest 2006; 116:2843–2854.

34. Soreq H, Seidman S. Acetylcholinesterase-new roles for an old actor. Nat Rev Neurosci 2001;2:294–302.

35. Brenner T, Hamra-Amitay Y, Evron T, Boneva N, Seidman S, Soreq H. The role of readthrough acetylcholinesterase in the pathophysiology of myasthenia gravis. FASEB J. 2003;17:214–222.

36. Liggett SB, Daughaday CC, Senior RM. Ipratropium in patients with COPD receiving cholinesterase inhibitors. Chest 1988;94:210–212.

37. Arsura EL, Brunner NG, Namba T, Grob D. Adverse cardiovascular effects of anticholinesterase medications. Am J Med Sci 1987;293:18–23.

38. Daroff RB. Ocular myasthenia: diagnosis and therapy. In: Glaser J, ed. Neruo-opthalmology. St. Louis, MO: C.V. Mosby, 1980: 62–71.

39. Daroff RB. The office tensilon test for ocular myasthenia gravis. Arch Neurol 1986;43:843–844.

40. Juel VC, Massey JM. Autoimmune myasthenia gravis: recommendations for treatment and immunologic modulation. Curr Treat Opt Neurol 2005;7:3–14.

41. Schneider-Gold C, Gajdos P, Toyka KV, Hohlfeld RR. Corticosteroids for myasthenia gravis. Cochrane Database Syst Rev 2005;(2):CD002828.

42. Bedlack RS, Sanders DB. Steroids have an important role. Muscle Nerve 2002;25:117–121.

43. Rivner MH. Steroids are overutilized. Muscle Nerve 2002;25:115–117.

44. Hilton-Jones D. When the patient fails to respond to treatment: myasthenia gravis. Pract Neurol 2007;7:405–411.

45. Bae JS, Go SM, Kim BJ. Clinical predictors of steroid-induced exacerbation in myasthenia gravis. J Clin Neurosci 2006;13:1006–1010.

46. Pascuzzi RM, Coslett HB, Johns TR. Long-term corticosteroid treatment of myasthenia gravis: report of 116 patients. Ann Neurol 1984;15:291–298.

47. Beghi E, Antozzi C, Batocchi AP, et al. Prognosis of myasthenia gravis: a multicenter follow-up study. J Neurol Sci 1991;106:213–220.

48. Seybold M, Drachman D. Gradually increasing doses of prednisone in myasthenia gravis: reducing the hazards of treatment. N Engl J Med 1974;290:81–84.

49. Grob D, Arsura EL, Brunner NG, Namba T. The course of myasthenia gravis and therapies affecting outcome. Ann NY Acad Sci 1987;505:472–499.

50. Durelli L, Maggi G, Casadio C, Ferri R, Rendine S, Bergamini L. Actuarial analysis of the occurrence of remissions following thymectomy for myasthenia gravis in 400 patients. J Neurol Neurosurg Psychiatry 1991;54:406–411.

51. Drachman DB. Myasthenia gravis. New Engl J Med 1994;330:1797–1810.

52. Seybold ME. Treatment of myasthenia gravis. In: Engel AG, ed. Myasthenia Gravis and Myasthenic Disorders. Oxford: Oxford University Press, 1999: 167–204.

53. Arsura E, Brunner N, Namba T, Grob D. High-dose intravenous methylprednisolone in myasthenia gravis. Arch Neurol 1985;42:1149–1153.

54. Lewis SJ, Smith PE. Osteoporosis prevention in myasthenia gravis: a reminder. Acta Neurol Scand 2001;103:320–322.

55. Brown ES, Vera E, Frol AB, Woolston DJ, Johnson B. Effects of chronic prednisone therapy on mood and memory. J Affect Disord 2007;99:279–283.

56. Perantie DC, Brown ES. Corticosteroids, immune suppression, and psychosis. Curr Psychiatry Rep 2002;4:171–176.

57. Kirwan J, Power L. Glucocorticoids: action and new therapeutic insights in rheumatoid arthritis. Curr Opin Rheumatol 2007;19:233–237.

58. Chinenov Y, Rogatsky I. Glucocorticoids and the innate immune system: crosstalk with the toll-like receptor signaling network. Mol Cell Endocrinol 2007;275:30–42.

59. Angeli A, Masera RG, Sartori ML, et al. Modulation by cytokines of glucocorticoid action. Ann NY Acad Sci 1999;876:210–220.

60. Hohlfeld R, Michels M, Heininger K, Besinger U, Toyka KV. Azathioprine toxicity during long-term immunosuppression of generalized myasthenia gravis. Neurology 1988;38:258–261.

61. Witte AS, Cornblath DR, Parry GJ, Lisak RP, Schatz NJ. Azathioprine in the treatment of myasthenia gravis. Ann Neurol 1984;15:602–605.

62. Kuks J, Djojoatmodjo S, Oosterhuis H. Azathioprine in myasthenia gravis: observations in 41 patients and a review of literature. Neuromusc Disord 1991;1:423–431.

63. Group MGCS. A randomised clinical trial comparing prednisone and azathioprine in myasthenia gravis. Results of the second interim analysis. Myasthenia Gravis Clinical Study Group. J Neurol Neurosurg Psychiatry 1993;56:1157–1163.

64. Palace J, Newsom-Davis J, Lecky B. A randomized double-blind trial of prednisolone alone or with azathioprine in myasthenia gravis. Myasthenia Gravis Study Group. Neurology 1998;50:1778–1783.

65. Sanders D, Howard F, Jr. Disorders of neuromuscular transmission. In: Bradley W, Daroff R, Fenichel G, Marsden C, eds. Neurology in Clinical Practice. Boston: Butterworth Heinemann, 2000: 2167–2185.

66. Machkhas H, Harati Y, Rolak LA. Clinical pharmacology of immunosuppressants: guidelines for neuroimmunotherapy. In: Rolak LA, Harati Y, eds. Neuroimmunology for the Clinician. Boston: Butterworth Heinemann, 1997: 77–104.

67. Batocchi AP, Majolini L, Evoli A, Lino MM, Minisci C, Tonali P. Course and treatment of myasthenia gravis during pregnancy. Neurology 1999;52:447–452.

68. Stolk JN, Boerbooms AM, de Abreu RA, et al. Reduced thiopurine methyltransferase activity and development of side effects of azathioprine treatment in patients with rheumatoid arthritis. Arthritis Rheum 1998;41:1858–1866.

69. Armstrong VW, Oellerich M. New developments in the immunosuppressive drug monitoring of cyclosporine, tacrolimus, and azathioprine. Clin Biochem 2001;34:9–16.

70. Cara CJ, Pena AS, Sans M, et al. Reviewing the mechanism of action of thiopurine drugs: towards a new paradigm in clinical practice. Med Sci Monit 2004;10:RA247–RA254.

71. Conti-Fine B, Bellone M, Howard JJ, Protti M. Myasthenia gravis. The immunobiology of an autoimmune disease. In: Neuroscience Intelligence Unit. Austin, TX: R.G. Landes, 1997: 230pp.

72. Tindall RSA, Phillips JT, Rollins JA, Wells L, Hall K. A clinical therapeutic trial of cyclosporine in myasthenia gravis. Ann NY Acad Sci 1993;681:539–551.

73. Ciafaloni E, Nikhar NK, Massey JM, Sanders DB. Retrospective analysis of the use of cyclosporine in myasthenia gravis. Neurology 2000;55:448–450.

74. Lavrnic D, Vujic A, Rakocevic-Stojanovic V, et al. Cyclosporine in the treatment of myasthenia gravis. Acta Neurol Scand 2005;111:247–252.

75. Tamler R, Epstein S. Nonsteroid immune modulators and bone disease. Ann NY Acad Sci 2006;1068:284–296.

76. Marder W, McCune WJ. Advances in immunosuppressive therapy. Semin Respir Crit Care Med 2007;28:398–417.

77. Ciafaloni E, Massey JM, Tucker-Lipscomb B, Sanders DB. Mycophenolate mofetil for myasthenia gravis: an open-label pilot study. Neurology 2001;56:97–99.

78. Chaudhry V, Cornblath DR, Griffin JW, O'Brien R, Drachman DB. Mycophenolate mofetil: a safe and promising immunosuppressant in neuromuscular diseases. Neurology 2001;56:94–96.

79. Meriggioli MN, Rowin J, Richman JG, Leurgans S. Mycophenolate mofetil for myasthenia gravis: a double-blind, placebo-controlled pilot study. Ann NY Acad Sci 2003;998:494–499.

80. Chan JW. Mycophenolate mofetil for ocular myasthenia. J Neurol 2008.

81. Sanders DB, Hart K, Mantegazza R, et al. An international, phase III, randomized trial of mycophenolate mofetil in myasthenia gravis. Neurology in press.

82. Group TMS. A trial of mycophenolate mofetil with prednisone as initial immunotherapy in myasthenia gravis1. Neurology in press.

83. O'Neill JO, Edwards LB, Taylor DO. Mycophenolate mofetil and risk of developing malignancy after orthotopic heart transplantation: analysis of the transplant registry of the International Society for Heart and Lung Transplantation. J Heart Lung Transplant 2006;25:1186–1191.

84. Vernino S, Salomao DR, Habermann TM, O'Neill BP. Primary CNS lymphoma complicating treatment of myasthenia gravis with mycophenolate mofetil. Neurology 2005;65:639–641.

85. Hood KA, Zarembski DG. Mycophenolate mofetil: a unique immunosuppressive agent. Am J Health Syst Pharm 1997;54:285–294.

86. Allison AC, Eugui EM. Mycophenolate mofetil and its mechanisms of action. Immunopharmacology 2000;47:85–118.

87. Ponseti JM, Azem J, Fort JM, Codina A, Montoro JB, Armengol M. Benefits of FK506 (tacrolimus) for residual, cyclosporin- and prednisone-resistant myasthenia gravis: one-year follow-up of an open-label study. Clin Neurol Neurosurg 2005;107:187–190.

88. Ponseti JM, Azem J, Fort JM, et al. Long-term results of tacrolimus in cyclosporine- and prednisone-dependent myasthenia gravis. Neurology 2005;64:1641–1643.

89. Ponseti JM, Gamez J, Azem J, Manuel LC, Vilallonga R, Armengol M. Tacrolimus for myasthenia gravis: a clinical study of 212 patients. Ann NY Acad Sci 2007.

90. Furukawa Y, Yoshikawa H, Iwasa K, Yamada M. Clinical efficacy and cytokine network-modulating effects of tacrolimus in myasthenia gravis. J Neuroimmunol 2008.

91. Lamb GD, Stephenson DG. Effects of FK506 and rapamycin on excitation-contraction coupling in skeletal muscle fibres of the rat. J Physiol 1996;494 (Part 2):569–576.

92. Croxall JD, Paul-Clark M, Van Hal PT. Differential modulation of glucocorticoid action by FK506 in A549 cells. Biochem J 2003;376:285–290.

93. Gajdos P, Chevret S, Toyka K. Plasma exchange for myasthenia gravis. Cochrane Database Syst Rev 2002:CD002275.

94. Chiu HC, Chen WH, Yeh JH. The six year experience of plasmapheresis in patients with myasthenia gravis. Ther Apher 2000;4:291–295.

95. Carandina-Maffeis R, Nucci A, Marques JF, Jr. et al. Plasmapheresis in the treatment of myasthenia gravis: retrospective study of 26 patients. Arq Neuropsiquiatr 2004;62:391–395.
96. Haupt WF, Rosenow F, van der Ven C, Birkmann C. Immunoadsorption in Guillain-Barre syndrome and myasthenia gravis. Ther Apher 2000;4:195–197.
97. Grob D, Simpson D, Mitsumoto H, et al. Treatment of myasthenia gravis by immunoadsorption of plasma. Neurology 1995;45:338–344.
98. Kostelidou K, Trakas N, Tzartos SJ. Extracellular domains of the beta, gamma and epsilon subunits of the human acetylcholine receptor as immunoadsorbents for myasthenic autoantibodies: a combination of immunoadsorbents results in increased efficiency. J Neuroimmunol 2007;190:44–52.
99. Shemin D, Briggs D, Greenan M. Complications of therapeutic plasma exchange: a prospective study of 1,727 procedures. J Clin Apher 2007;22:270–276.
100. Qureshi AI, Choudhry MA, Akbar MS, et al. Plasma exchange versus intravenous immunoglobulin treatment in myasthenic crisis. Neurology 1999;52:629–632.
101. Gajdos P, Chevret S, Clair B, Tranchant C, Chastang C. Clinical trial of plasma exchange and high-dose intravenous immunoglobulin in myasthenia gravis. Myasthenia Gravis Clinical Study Group. Ann Neurol 1997;41:789–796.
102. Gajdos P, Chevret S, Toyka K. Intravenous immunoglobulin for myasthenia gravis. Cochrane Database Syst Rev 2008:CD002277.
103. Zinman L, Ng E, Bril V. IV immunoglobulin in patients with myasthenia gravis: a randomized controlled trial. Neurology 2007;68:837–841.
104. Kazatchkine MD, Kaveri SV. Immunomodulation of autoimmune and inflammatory diseases with intravenous immune globulin. N Engl J Med 2001;345:747–755.
105. Cunningham-Rundles C, Zhou Z, Mankarious S, Courter S. Long-term use of IgA-depleted intravenous immunoglobulin in immunodeficient subjects with anti-IgA antibodies. J Clin Immunol 1993;13:272–278.
106. Paran D, Herishanu Y, Elkayam O, Shopin L, Ben-Ami R. Venous and arterial thrombosis following administration of intravenous immunoglobulins. Blood Coagul Fibrinolysis 2005;16:313–318.
107. Vucic S, Chong PS, Dawson KT, Cudkowicz M, Cros D. Thromboembolic complications of intravenous immunoglobulin treatment. Eur Neurol 2004;52:141–144.
108. Zhu KY, Feferman T, Maiti PK, Souroujon MC, Fuchs S. Intravenous immunoglobulin suppresses experimental myasthenia gravis: immunological mechanisms. J Neuroimmunol 2006;176:187–197.
109. Fuchs S, Feferman T, Meidler R, et al. A disease-specific fraction isolated from IVIG is essential for the immunosuppressive effect of IVIG in experimental autoimmune myasthenia gravis. J Neuroimmunol 2008;194:89–96.
110. Niakan E, Harati Y, Rolak LA. Immunosuppressive drug therapy in myasthenia gravis. Arch Neurol 1986;43:155–156.
111. Perez M, Buot WL, Mercado-Danguilan C, Bagabaldo ZG, Renales LD. Stable remissions in myasthenia gravis. Neurology 1981;31:32–37.
112. Lin PT, Martin BA, Weinacker AB, So YT. High-dose cyclophosphamide in refractory myasthenia gravis with MuSK antibodies. Muscle Nerve 2006;33:433–435.
113. Drachman DB, Brodsky RA. High-dose therapy for autoimmune neurologic diseases. Curr Opin Oncol 2005;17:83–88.
114. Baek WS, Bashey A, Sheean GL. Complete remission induced by rituximab in refractory, seronegative, muscle-specific, kinase-positive myasthenia gravis. J Neurol Neurosurg Psychiatry 2007;78:771.
115. Hain B, Jordan K, Deschauer M, Zierz S. Successful treatment of MuSK antibody-positive myasthenia gravis with rituximab. Muscle Nerve 2006;33:575–580.
116. Gajra A, Vajpayee N, Grethlein SJ. Response of myasthenia gravis to rituximab in a patient with non-Hodgkin lymphoma. Am J Hematol 2004;77:196–197.
117. Wylam ME, Anderson PM, Kuntz NL, Rodriguez V. Successful treatment of refractory myasthenia gravis using rituximab: a pediatric case report. J Pediatr 2003;143:674–677.
118. Pescovitz MD. Rituximab, an anti-cd20 monoclonal antibody: history and mechanism of action. Am J Transplant 2006;6:859–866.
119. Plauché WC. Myasthenia gravis in mothers and their newborns. Clin Obstet Gynecol 1991;34:82–99.
120. Hoff JM, Daltveit AK, Gilhus NE. Myasthenia gravis in pregnancy and birth: identifying risk factors, optimising care. Eur J Neurol 2007;14:38–43.
121. Ciafaloni E, Massey JM. The management of myasthenia gravis in pregnancy. Semin Neurol 2004;24:95–100.
122. Kalidindi M, Ganpot S, Tahmesebi F, Govind A, Okolo S, Yoong W. Myasthenia gravis and pregnancy. J Obstet Gynaecol 2007;27:30–32.
123. Gardnerova M, Eymard B, Morel E, et al. The fetal/adult acetylcholine receptor antibody ratio in mothers with myasthenia gravis as a marker for transfer of the disease to the newborn. Neurology 1997;48:50–54.
124. Hoff JM, Daltveit AK, Gilhus NE. Artrogryposis multiplex congenita – a rare fetal condition caused by maternal myasthenia gravis. Acta Neurol Scand Suppl 2006;183:26–27.

125. Riemersma S, Vincent A, Beeson D, et al. Association of arthrogryposis multiplex congenita with maternal antibodies inhibiting fetal acetylcholine receptor function. J Clin Invest 1996;98:2358–2363.

126. Polizzi A, Huson SM, Vincent A. Teratogen update: maternal myasthenia gravis as a cause of congenital arthrogryposis. Teratology 2000;62:332–341.

127. Brueton LA, Huson SM, Cox PM, et al. Asymptomatic maternal myasthenia as a cause of the Pena-Shokeir phenotype. Am J Med Genet 2000;92:1–6.

128. Romi F, Gilhus NE, Aarli JA. Myasthenia gravis: disease severity and prognosis. Acta Neurol Scand Suppl 2006;183:24–25.

129. Bril V, Kojic J, Dhanani A. The long-term clinical outcome of myasthenia gravis in patients with thymoma. Neurology 1998;51:1198–1200.

130. Evoli A, Minisci C, Di Schino C, et al. Thymoma in patients with MG: characteristics and long-term outcome. Neurology 2002;59:1844–1850.

131. Parr JR, Jayawant S. Childhood myasthenia: clinical subtypes and practical management. Dev Med Child Neurol 2007;49:629–635.

132. Andrews PI. Autoimmune myasthenia gravis in childhood. Semin Neurol 2004;24:101–110.

133. Andrews I. A treatment algorithm for autoimmune myasthenia gravis in childhood. NY Acad Sci 1998;841:789–802.

134. Prudlo J, Koenig J, Ermert S, Juhasz J. Sleep disordered breathing in medically stable patients with myasthenia gravis. Eur J Neurol 2007;14:321–326.

135. Nicolle MW, Rask S, Koopman WJ, George CF, Adams J, Wiebe S. Sleep apnea in patients with myasthenia gravis. Neurology 2006;67:140–142.

136. Al-Shekhlee A, Kaminski HJ, Ruff RL. Endocrine myopathies and muscle disorders related to electrolyte disturbance. In: Katirji B, Kaminski HJ, Preston D, Shapiro B, Ruff RL, eds. Neuromuscular Disorders in Clinical Practice. Boston: Butterworth Heinemann, 2002: 1187–1204.

11

Neurocritical Care of Myasthenic Crisis

J. Americo Fernandes Filho and Jose I. Suarez

1. INTRODUCTION

Myasthenic crisis (MC) is defined as an exacerbation of myasthenia gravis (MG) weakness provoking an acute episode of respiratory failure that leads to institution of mechanical ventilation (MV). Weakness may involve respiratory, altering respiratory mechanics, or bulbar muscles, which compromise airway protection and patency. MC is the most dangerous complication of MG and is a life-threatening condition that requires immediate recognition and patient care in an intensive care unit setting for its optimal management *(1, 2)*.

MC usually occurs within the first 2 years after MG onset (74% of patients) *(3)*, and 15–20% of patients with MG will experience at least one episode of crisis. Its mortality has declined from 40% in the early 1960s to 5% in the 1970s, which most likely reflects improvement of ventilatory and general medical management of these patients in intensive care units. However, despite the decrease in mortality, the duration of MC has not changed significantly and continues to average 2 weeks *(3)*. This chapter reviews the most important aspects of MC including its pathophysiology, etiology, and management.

2. PATHOPHYSIOLOGY

2.1. Precipitants

MC is most commonly precipitated by infections (30–40% of cases) *(4, 5)*. The most frequent are respiratory tract infections caused by viral or bacterial agents. Aspiration pneumonia occurs in about 10% of MC patients *(4, 5)* and may be more prevalent in patients with oropharyngeal weakness as manifested by difficulty in swallowing and chewing, altered facial expression, and dysarthria *(6)*. Certain medications exacerbate MG and lead to MC (Table 1) *(7, 8)*. It is crucial to obtain a reliable history of current medication use including specific inquiry of the patient, family, and friends of nonprescription drugs, including the so-called alternative medicines. Telithromycin, an antibiotic of the ketolide class, has recently been reported to cause myasthenia exacerbation or crisis, or unmasking of myasthenia gravis shortly after initial intake (minutes to a few days) which may be confused with an anaphylactic reaction. Extreme caution should be used when prescribing this medication to myasthenic patients *(9, 10)*. When treatment of MG was limited to anticholinesterase medications, overdose could lead to MC, but now is rare and its importance previously may have been overstated *(4, 11)*. Other factors which may lead to MC include botulinum toxin injection *(12)*, recent surgical procedures (including thymectomy), and trauma. In 30–40% of the patients no trigger can be identified *(5)*. The presence of thymoma seems to be a risk factor for MC. MG patients with thymoma may have a more severe disease course (see Chapter 7), and thymoma patients are identified twice as often among patients in crisis compared to patients without thymoma (30 vs. 15%) *(4)*.

From: *Current Clinical Neurology*
Myasthenia Gravis and Related Disorders
Edited by: H.J. Kaminski, DOI 10.1007/978-1-59745-156-7_11, © Humana Press, New York, NY

Table 1
Drugs which Exacerbate Myasthenia Gravis

Definite association	Corticosteroids
Probable association	*Antibiotics*: aminoglycosides, ciprofloxacin, clindamycin, telithromycin
	Antiarrhythmics: procainamide, propranolol, timolol
	Neuropsychiatric: phenytoin, trimethadione, lithium
Possible association	*Antibiotics*: ampicillin, imipenen/cilastatin, erythromycin
	Antiarrhythmics: propafenone, verapamil, quinidine,
	Miscellaneous: trihexyphenidyl, chloroquine, neuromuscular-blocking drugs, carbamazepine, oral contraceptives, transdermal nicotine

2.2. Respiratory Abnormalities

As weakness of respiratory and oropharyngeal muscles progresses, there is a decrease in lung expansion along with the efficiency of cough reflex to clear the airway. Patients' forced vital capacity (FVC) progressively decreases, making them prone to developing atelectasis and, eventually, to respiratory failure *(4)*. It is proposed that patients follow a stereotyped sequence of events leading to MC directly related to FVC *(13)*:

1. Normal respiratory function is present with FVC of 65 ml/kg.
2. Poor cough is evident with accumulation of secretions with FVC of 30 ml/kg.
3. With FVC of 20 ml/kg, the sigh mechanism is impaired with development of atelectasis and hypoxia.
4. Sigh is lost with FVC of 15 ml/kg, and atelectasis and shunting appear.
5. Hypoventilation starts with FVC of 10 ml/kg.
6. Hypercapnia develops with FVC between 5 and 10 ml/kg.

2.3. Oropharyngeal Dysfunction

The oropharyngeal muscles maintain the patency of the upper airway by regulation of its cross-sectional area and their dysfunction increases resistance to airflow *(14)*. Weakness of laryngeal muscles causes the vocal cords to remain adducted during inspiration instead of the normal abduction *(14, 15)*. This creates the so-called "sail phenomenon". Paralyzed vocal cords, because of their position and curvature, catch the airflow during inspiration being pulled inward to midline like the sail of a boat *(14)*. Weakness of the tongue obstructs the oropharyngeal cavity. In addition to the mechanical obstruction of the airway, oropharyngeal dysfunction leads to an inability to protect from aspiration *(6, 14)*.

2.4. Clinical Presentation and Evaluation

MC may occur in patients previously diagnosed but also may be the presenting event of MG. Patients with the definite diagnosis of MG that complain of worsening weakness or shortness of breath need to be assessed for dysphagia, stridor, and adequacy of ventilation. During the examination, the respiratory pattern requires assessment. Rapid, shallow breathing indicates respiratory muscle fatigue, which should not be confused with psychogenic hyperventilation *(4)*. Diaphragmatic function is assessed by observation of abdominal movements during the respiratory cycle. With severe weakness, a paradoxical respiratory pattern develops with inward movement of the abdomen during inspiration. The strength of neck muscles correlates with diaphragm strength and weakness of these muscles should alert the clinician to possible respiratory compromise. A simple way to evaluate ventilatory reserve is by asking the patient to count from 1 to 25 in a single breath *(4)*. Patients with significant limitation should undergo measures of respiratory parameters.

Assessment of oropharyngeal muscles begins with inquiries of difficulty swallowing, episodes of choking, or coughing while eating. A wet, gurgling voice or stridor may indicate the need for intubation for airway protection *(4)*. Swallowing may be tested at bedside by observing the patient swallow three ounces of water and observing for coughing or choking *(16)*. However, caution should be observed in performing this assessment, and it need not be done in patients with clear signs of oropharyngeal weakness.

Patients may present with worsening of respiratory status because of vocal cord paralysis. In these patients, flexible laryngoscopy or flow volume loops should be performed *(14, 15)*. The finding of vocal cords in the adducted position should alert the physician for the possible need of a cricothomy in the case that conventional endotracheal intubation is not feasible *(15)*.

In one series of 63 MC, initial manifestations were generalized weakness in 76% of patients, focal bulbar weakness in 19%, and weakness of respiratory muscles in 5%. The majority of the patients (68%) required MV between 1 and 3 days after initial myasthenic exacerbation *(1)*. Some patients may present with respiratory failure without any evidence of generalized weakness *(3, 14)*.

The respiratory status of patients at risk for MC should be monitored closely. Bedside measurements of forced vital capacity (FVC, normal ≥ 60 ml/kg), negative inspiratory force (NIF, normal > 70 cm H_2O), and positive expiratory force (PEF, normal > 100 cm H_2O) should be done serially. Arterial blood gases are also important, and pulse oximetry is not a substitute for this measurement. Blood gas assessments are abnormal (showing hypoxia and hypercarbia) with advanced respiratory dysfunction, but patients may have normal oxygenation by pulse oximetry while arterial blood gases show developing hypercarbia, a sign of impending respiratory arrest.

Values of FVC ≤ 1.0 liter (less than 15 ml/kg body weight), NIF < 20 cm H_2O and PEF < 40 cm H_2O are indications for institution of MV *(4, 5, 17)*. However, each patient should be considered individually, with assessment of comfort level, heart rate, respiratory rate, arterial blood gas values, and the ability to protect and maintain their airway. Respiratory dysfunction may progress rapidly and should be promptly recognized. Some independent predictors identified for the need to institute MV include abnormal chest roentgenogram on hospital admission (pneumonia or atelectasis, Fig. 1.) and complications during hospitalization such as atelectasis,

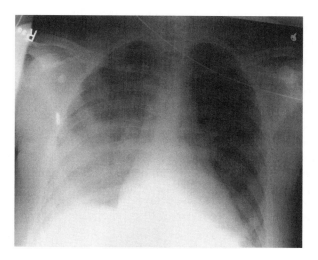

Fig. 1. Anteroposterior chest X-ray of a 33-year-old male with known MG for 1 year, who presented with worsening of generalized weakness, and dysphagia, followed by tachypnea and fever. The X-ray on admission revealed right lower lung pneumonia and left lower lobe atelectasis. The patient required mechanical ventilation within 12 hours of admission.

Table 2
Neuromuscular Disorders in Critical Care Patients

Condition	Key presenting features
Acute intermittent porphyria	Asymmetric limb weakness progressing to quadriplegia after several attacks
Botulism	Nausea and vomiting preceding muscle weakness. Blurred vision, dysphagia, dysarthria, descending muscle paralysis, dilated pupils, dry mouth, constipation and urinary retention
Critical illness myopathy	Patient with COPD or asthma requiring mechanical ventilation and use of neuromuscular blockers and corticosteroids; patients with sepsis, "critical" systemic illness, transplant patients
Critical illness polyneuropathy	Patient with sepsis and difficulty weaning from the ventilator, diminished or absent reflexes
Electrolyte imbalance	Generalized muscle weakness, cardiac arrhythmias with or without rhabdomyolisis
Guillain–Barré syndrome	Preceded by upper respiratory or gastrointestinal infection; ascending paralysis, arreflexia
Lambert–Eaton myasthenic syndrome	Symmetric proximal muscle weakness, hypoactive or absent deep tendon reflexes, dry mouth, blurred vision, orthostatic hypotension
Lead poisoning	Pure motor weakness, initially of extensor muscles, fasciculations, abdominal pain, constipation, anemia, renal failure
Motor neuron disease	Weakness, wasting, fasciculations
Organophosphate poisoning	Exposure to insecticides, petroleum additives, and modifiers of plastics followed by acute cholinergic crisis (muscle weakness, myosis, abdominal cramping)
Polymyositis	Proximal, symmetrical muscle weakness, elevated creatine kinase
Prolonged neuromuscular blocking	Patient with impaired renal or hepatic failure who had been on continuous neuromuscular-blocking agents

cardiac arrhythmia, and anemia requiring transfusion *(18)*. Early MV is preferable to prevent worsening atelectasis. Patients with low FVC but not to the level to warrant prompt intubation, and those with independent predictors of MV should be admitted to an intensive care unit for serial measurements of the spirometry parameters. Cardiac arrhythmias are common among patients with MC (14–17% of patients) and are an additional reason for intensive care monitoring of patients with MC and evidence of impending MC *(1, 19)*.

2.5. Differential Diagnosis

In any patient who is difficult to wean from a ventilator, MG should be considered; other possible disorders are Lambert–Eaton myasthenic syndrome, botulism, Guillain–Barré syndrome, polymyositis, motor neuron disease, critical illness myopathy/polyneuropathy, and organophosphate poisoning among others (Table 2) *(20, 21, 22, 23, 24)*. Electrodiagnostic studies (Chapter 8) are instrumental in differentiating among these conditions and may guide additional specific evaluation as well as provide prognostic information *(20)*.

3. MANAGEMENT

3.1. General

To assure recovery of patients from MC, their overall medical condition must be optimized. Identified triggers should be removed or corrected. All patients should undergo extensive evaluation for infection including cultures of sputum, urine, and blood as well as other sites as clinically indicated. Empiric use of antibiotics needs to be carefully tailored to the clinical situation because of the potential for some antibiotics to further impair neuromuscular transmission further as well as the development of bacterial resistance. Avoidance of the unnecessary use of antibiotics is further emphasized by the observation that *Clostridium difficile* colitis complicates prolonged crisis *(3, 4)*. Nutritional assessment needs to be made early in the course of MC. A determination is made for the necessity of short-term naso-gastric tube feeding, peripheral hyperalimentation, or a gastrostomy tube with the expectation that weakness may not resolve quickly. Electrolyte imbalance will compromise neuromuscular function, and close monitoring with appropriate correction is required. Intermittent positive pressure breathing may be useful in preventing development of atelectasis and if the patient is already intubated or with respiratory tract infection, aggressive pulmonary toilet is necessary. Deep venous thrombosis prophylaxis measures should be undertaken with the use of compression stockings, sequential compression devices, and subcutaneous heparin. Coagulation status may be deranged from heparin use when plasma exchange is performed, and intravenous immunoglobulin therapy may predispose to hypercoagulation. Gastrointestinal prophylaxis is given either with sulcrafate, histamine receptor blockers, or proton pump inhibitors to prevent stress ulcers and gastrointestinal bleeding. Psychological support for the patient and family is also important with emphasis on the ultimate ability to return most patients to an excellent functional level.

3.2. Ventilatory Management

Noninvasive bilevel-positive pressure ventilation (BiPAP) is used for patients with chronic ventilatory insufficiency caused by neuromuscular disorders like muscular dystrophies, amyotrophic lateral sclerosis, spinal muscular atrophy. It has also been shown to be useful in patients with acute respiratory insufficiency caused by disorders like MG. In one series, endotracheal intubation during episode of myasthenia crisis was successfully avoided in 70% of trials. In this series, the presence of hypercapnia ($PaCO_2 > 50$ mm Hg) predicted BiPAP failure and need for intubation. However, patients with severe ventilatory impairment and bulbar dysfunction should be excluded from a trial of noninvasive ventilation *(25, 26)*.

Once the decision for MV is made, rapid sequence intubation should be performed *(17)*. This entails bag-masking the patient to obtain arterial oxygen saturation of greater than 97%, administration of free-running intravenous normal saline, continuous blood pressure monitoring, and bolus infusion of sedative medications (usually etomidate at 0.2–0.3 mg/kg). If short acting muscle relaxants are needed (preferably they are to be avoided), non-depolarizing agents like vecuronium should be used. Oral intubation should be undertaken whenever possible.

Once the patient is intubated, a mode of ventilation needs to be chosen, but there is no perfect selection. The assist-control (AC) and synchronized intermittent mandatory ventilation (SIMV) are commonly used, each one with its advantages and disadvantages. Muscle fatigability with weakness and poor lung expansibility are the main determinants in the development of MC. Thus, the initial objectives of mechanical ventilation should be to promote rest and expand the lungs. It was thought that, as long as the patient does not have primary lung disease compromising lung compliance, large tidal volumes (15 ml/kg) combined with lower rates to maintain a normal minute ventilation, with positive end expiratory pressure at levels of 5–15, could be used, provided that peak airway pressures are maintained within acceptable limits (<40 cm H_2O) *(3)*. However, recent literature suggests that smaller tidal volumes (7–8 ml/kg) with faster respiratory rates (12–16 breaths/min) should be used to avoid lung injury adding intermittent sighs (1.5 × tidal volume, 3–4 times every hour) to avoid atelectasis *(27)*.

3.3. Ventilator Weaning

Several parameters may be helpful in when to initiate weaning from MV: FVC >15 ml/kg, NIF ≤ 30, PEF ≥ 40 cm H_2O, and minute ventilation < 15 L/min *(4, 17, 28)*. However, these parameters have limited predictive power, and other general conditions should be satisfied before weaning is attempted *(17)*: (1) The patient needs to be adequately oxygenating with a $PO_2 >$ 60 mmHg with FiO_2 40% and PEEP < 5 cm H_2O. (2) The patient also needs to have intact respiratory drive (in patients with MG, respiratory drive is found not to be impaired *(29)*), to be able to protect their airway and have an adequate cough reflex. (3) The hemodynamic status should be stable, electrolyte levels normal, and nutritional status adequate. (4) The patient should be free of infection or other significant medical complications. (5) The need for respiratory suctioning should be less than every 2–3 hours.

For MG patients, one major and initial indicator of timing for MV weaning is the improvement of general muscle strength by objective physical examination. The rapid shallow breathing index is thought to be the best predictor of successful weaning *(17)* and is calculated by division of the tidal volume by the respiratory rate of the patient while temporarily off the ventilator. Patients with index values greater than 100 have a 95% likelihood of failing a weaning trial *(28)*.

Once patients meet criteria for extubation, the method chosen for a weaning trial varies according to an individual physician's clinical experience. One method is a daily trial of continuous positive airway pressure (CPAP) and pressure support levels of 5–15 cm H_2O. If the patient remains comfortable after 1–4 hours, the level of pressure support can be decreased by 1–3 cm H_2O each day. Patients should be returned to the original ventilatory mode overnight or when signs of fatigue appear *(4)*.

The SIMV mode is another method of weaning. It is accomplished by decreasing the ventilatory rate by 2–3 breaths once or more times a day based on patient comfort level *(17)*. An increase in respiratory rate, fall in tidal volumes, agitation, and tachycardia may all be indicators of fatigue *(4)*. Once a patient has demonstrated good endurance (the number of hours patient tolerates the weaning mode with minimal support, usually more than 2 hours) and general conditions are adequate, the patient may be extubated.

Extubation should be performed early in the day. Stridor may occur immediately to 1 hour after extubation, and aerosolized racemic epinephrine may reverse the condition, but reintubation might be necessary. Laryngospasm is less common but is a life-threatening condition. Lidocaine administered intravenously (2 mg/kg) may significantly reduce laryngospasm, if used several minutes prior to extubation *(28)*.

3.4. Treatment of Neuromuscular Dysfunction

Therapies may be divided into those that improve strength rapidly with a short duration of action and those that improve strength slowly with a more permanent response. In the first category, there are acetylcholinesterase inhibitors (AchI), plasmapheresis, and intravenous immunoglobulin (IVIg) (Chapter 10). The second category comprises immunosuppressive agents such as corticosteroids, azathioprine, cyclophosphamide, cyclosporine, mycophenolate, and tacrolimus *(30)*. The latter are not used in rescuing patients from MC but in long-term management. Corticosteroid treatment during MC is debated. Some advocate their use during crisis in patients only with refractory weakness, others suggest that they be started while the patient is treated with IVIg or plasmaphereis *(4, 5, 31)*.

Cholinesterase inhibitors provide symptomatic improvement by means of decreasing the degradation of acetylcholine at the neuromuscular junction. The continuous intravenous infusion of pyridostigmine is reported as an effective treatment in MC compared to plasmapheresis as measured by mortality, duration of ventilation, and outcome. Its use has been defended especially when MC is triggered by an infection *(1)*. However, it is common practice to discontinue these agents in mechanically ventilated patients due to the propensity for promoting excess respiratory secretions and mucous plugging *(4)*. Another concern is the reported occurrence of cardiac arrhythmias in MC patients, including those receiving intravenous pyridostigmine *(1, 19)*. Cholinesterase inhibitors could increase cholinergic activity at cardiac muscarinic synapses leading to arrhythmias *(19)*.

Color Plates

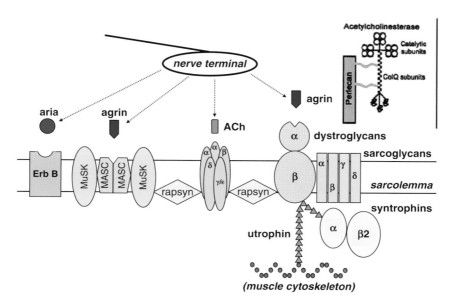

Color Plate 1. Schematic representation of postsynaptic region of the neuromuscular junction. Ach, acetylcholine; ARIA, acetylcholine receptor-inducing activity; MASC, myotube-associated specificity component; MuSK, muscle-specific kinase. (*see* discussion on p. *5*)

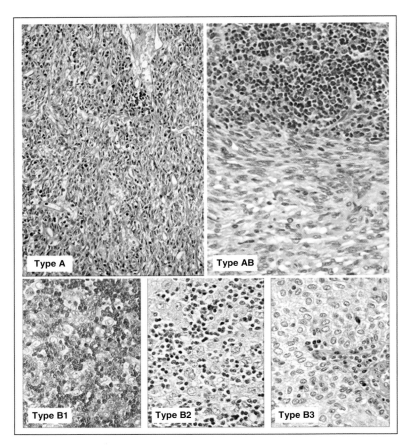

Color Plate 2. Morphological spectrum of thymomas that can be associated with paraneoplastic MG. WHO type A (medullary) thymoma with prominent spindle cell features and scarce intermingled lymphocytes, which are mostly mature on flow cytometry. WHO type AB (mixed) thymoma with features of both WHO type A and B (cortical) thymomas. In this case, both areas are clearly separated, whereas in other tumors, the type A and B areas may be indistinguishably interwoven. High numbers of immature T cells, lack of cytologic atypia of the thymic epithelium and presence of numerous Hassall's corpuscles are diagnostic features of type B1 thymomas. Type B2 tumors show somewhat reduced numbers of T cells with marked atypia of the thymic epithelium; occasional Hassall's corpuscles may occur. The picture in type B3 thymomas is clearly dominated by the epithelium with formation of characteristic perivascular spaces and perivascular palisading. Intermingled lymphocytes are mostly of immature phenotype (hematoxylin–eosin). (*see* discussion on p. *107*)

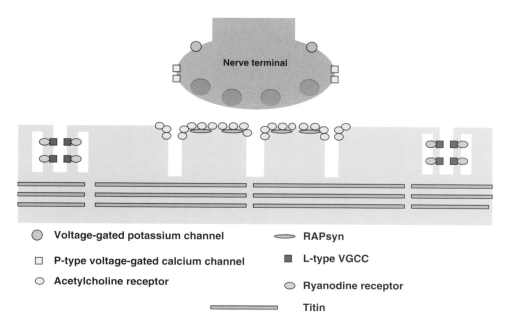

Voltage-gated potassium channel   **RAPsyn**

P-type voltage-gated calcium channel   **L-type VGCC**

Acetylcholine receptor   **Ryanodine receptor**

Titin

Color Plate 3. Diagram of the neuromuscular junction with proteins that are targeted for autoimmune attack. (*see* discussion on p. *143*)

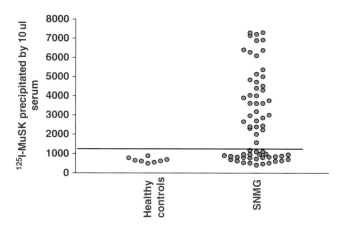

Color Plate 4. Immunoprecipitation of ^{125}I-MuSK by sera from patients with seronegative MG. Recombinant extracellular domain of rat MuSK was provided by Dr. Werner Hoch (University of Bristol). The SNMG sera tested were provided by Professor John Newsom-Davis, University of Oxford, and Dr. Amelia Evoli, Catholic University of Rome. (*see* discussion on p. *150*)

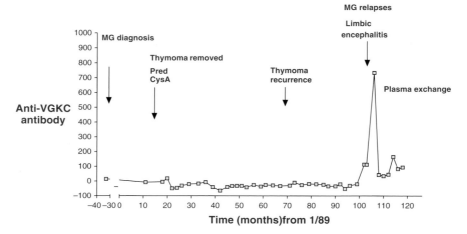

Color Plate 5. VGKC antibody levels in a patient with a 13-year history of thymoma-associated MG. She developed a single episode of limbic signs (memory loss, confusion, disorientation, agitation) associated with a peak of antibodies to VGKC. The antibody levels fell following plasma exchange and her clinical state improved. From Buckley et al. *(72)*. (*see* discussion on p. *152*)

Plasmapheresis is the preferred treatment in managing MC, with a reported efficacy of 75% *(4)*. Its mechanism of action is related to the removal of circulating factors, in addition to acetylcholine receptor (AChR) antibodies. Even though AChR antibody titers do not correlate well with MG severity, the decrease of AChR antibodies titers correlates with clinical improvement. There is no standard protocol for plasmapheresis. The usual regimen is exchange of 1–1.5 plasma volumes every day or every other day for 5–6 treatments. Clinical improvement may be seen as early as 24 hours, but in most patients the first effect appears after 2–3 sessions of plasmapheresis. Some patients may have an initial worsening, thought to be due to reduction in plasma concentration of cholinesterase. The duration of effect is usually less than 10 weeks if other immunosuppressive treatment is not instituted *(30, 32)*.

The most frequent complications of plasmapheresis are hypotension, electrolyte imbalance (reduction of calcium, potassium, and magnesium), depletion of clotting factors, and thrombocytopenia. Obtaining vascular access may at times be difficult and may be complicated by pneumothorax, thrombosis, and infection *(4, 30, 32)*. When a large volume of plasma is removed, intravascular volume should be replaced with albumin in saline solution *(32)*. Any electrolyte imbalance should be corrected to prevent exacerbating weakness. The coagulopathy is usually mild but should be kept in mind when utilizing subcutaneous heparin for deep venous thrombosis prophylaxis. Reducing anti-hypertensive drug dosages and administering intravenous fluids before the procedure may avoid hypotension.

Another option for specific treatment of MC is IVIg. A standard course consists of 400 mg/kg/day of IVIg for 5 days *(30)*. Response to IVIg is usually observed 5 days after treatment initiation. Some report IVIg to have a similar efficacy and complication rate as plasmapheresis *(33)*. However, others find patients treated with plasmapheresis to be at a superior respiratory status at 2 weeks (ability to extubate) and a better functional outcome at 1 month compared to IVIG *(34)*. Furthermore, some patients may not respond to IVIg but have improvement after subsequent plasmapheresis *(35)*. IVIg may be a better choice in patients with hemodynamic instability, vascular access problems, or with a poor response to plasmapheresis. Adverse effects occur in fewer than 10% of patients and most commonly include headache, chills and fever, fluid overload, and rarely renal failure *(36, 37)*. Anaphylaxis may occur to IgA components in patients with IgA deficiency *(37)*. IgA levels and baseline renal function should be obtained before treatment is initiated.

The use of preoperative plasma exchange is advocated in patients with generalized MG and preoperative symptoms of oropharyngeal or respiratory weakness or in patients with symptoms controlled by use of cholinesterase inhibitors, in order to prevent delayed extubation and reintubation. IVIg has also been used preoperatively, but the variability in reaching maximal response (3–19 days) should be kept in mind *(2)*.

3.5. Outcome of Myasthenic Crisis

One study identified three main independent predictors of prolonged MV: age > 50, pre-intubation serum bicarbonate ≥ 30, and peak vital capacity < 25 ml/Kg at days 1–6 post-intubation. The proportion of patients intubated for more than 2 weeks was 88% in patients with three risk factors, 46% with two, 21% with one, and zero for patients with no risk factors. The same study also revealed that atelectasis, anemia requiring transfusion, *Clostridium difficile* infection, and congestive heart failure were complications associated with prolonged intubation *(3)*. Tracheostomy is usually performed after 2 weeks of intubation, but early tracheostomy is recommended for patients expected to require prolonged MV. Tracheostomy is more comfortable for the patient, reduces the risk of tracheolaryngeal stenosis, permits more effective suctioning of tracheal secretions, and facilitates weaning from MV due to reduction in dead space and in resistance to air flow from endotracheal tube *(4, 28)*. In general, 25% of patients will be extubated by 1 week, 50% by 2 weeks, and 75% by 1 month. Intubation and MV for more than 2 weeks is associated with a threefold increase in hospital stay (median 63 days) and a twofold increase in likelihood of functional dependency at discharge *(3)*. A third of patients who survive the first crisis will have a second episode *(4)*.

Although crisis is potentially a lethal event and is always devastating for the patient and loved ones, optimizing critical care and long-term treatment of MG has prevented the event from being "grave", as it once was.

REFERENCES

1. Berrouschot J, Baumann I, Kalischewski P, Sterker M, Schneider D. Therapy of myasthenic crisis. Crit Care Med 1997;25:1228–35.
2. Vern JC. Myasthenia gravis: Management of myasthenic crisis and preoperative care. Semin Neurol 2004;24:75–81.
3. Thomas CE, Mayer SA, Gungor Y, Swarup R, Webster EA, Chang I, et al. Myasthenic crisis: Clinical features, mortality, complications, and risk factors for prolonged intubation. Neurology 1997;48:1253–60.
4. Mayer SA. Intensive care of the myasthenic patient. Neurology 1997; 48(Suppl 5):S70–S75.
5. Bedlack RS, Sanders DB. How to handle myasthenic crisis. Postgrad Med 2000;107(4):211–4, 220–2.
6. Weijnen FG, Van Der Bilt A, Wokke JHJ, Wassenberg MWM, Oudenaarde I. Oral functions of patients with myasthenia gravis. Ann NY Acad Sci 1998;841:773–6.
7. Wittbrodt, ET. Drugs and myasthenia gravis. Arch Intern Med 1997;157:399–408.
8. Pascuzzi RM. Medications and myasthenia gravis: a reference for health care professionals. www.myathenia.org/drugs/reference.htm.
9. Perrot X, Bernard N, Vial C, Antoine JC, Laurent H, Vial T, et al. Myasthenia gravis exacerbation or unmasking associated with telithromycin treatment. Neurology 2006;67:1–3.
10. Jennet AM, Bali D, Jasti P, Shah B, Browning LA. Telithromycin and myasthenia crisis. Clin Infect Dis 2006;43:1621–22.
11. Grob D. Natural history of myasthenia gravis. In: Engel AG, ed. Myathenia Gravis and Myasthenic Disorders. New York: Oxford University Press, 1999: 131–45.
12. Borodic G. Myasthenic crisis after botulinum toxin. Lancet 1998;352:1832.
13. Ropper AH. The Guillain-Barré syndrome. N Engl J Med 1992;326:1130–36.
14. Putman MT, Wise RA. Myasthenia gravis and upper airway obstruction. Chest 1996;109:400–04.
15. Cridge PB, Allegra J, Gerhard H. Myasthenic crisis presenting as isolated vocal cord paralysis. Am J Emerg Med 2000;18:232–3.
16. DePippo KL, Holas MA, Reding MJ. Validation of the 3 oz water swallow test for aspiration after stroke. Arch Neurol 1992;49:1259–61.
17. McDonagh DL, Borel CO. Ventilatory management in the neurosciences critical care unit. In: Suarez JI, ed. Critical Care Neurology and Neurosurgery. Totowa: Humana Press, 2004:151–166.
18. Suarez JI, Boonyapisit K, Zaidat OO, Kaminski HJ, Ruff RL. Predictors of mechanical ventilation in patients with myasthenia gravis exacerbation [abstract]. J Neurol Sci 2001;187:S464.
19. Mayer SA, Thomas CE. Therapy of myasthenic crisis. Crit Care Med 1998;26:1136.
20. Maher J, Rutledge F, Remtulla H, Parkes A, Bernardi L, Bolton CF. Neuromuscular disorders associated with failure to wean from the ventilator. Intensive Care Med 1995; 21:737–43.
21. Vita G, Girlanda P, Puglisi RM, Marabello L, Messina C. Cardiovascular-reflex testing and single fiber electromyography in botulism. Arch Neurol 1987 44:202–206.
22. Suarez JI, Cohen ML, Larkins J, Kernich CA, Hricick DE, Daroff RB. Acute intermittent porphyria. Clinicopathologic correlation. Report of a case and review of the literature. Neurology 1997 48:1678–83.
23. Torbey MT, Suarez JI, Geocadin R. Less common causes of quadriparesis and respiratory failure. In: Suarez JI, ed. Critical Care Neurology and Neurosurgery. Totowa: Humana Press, 2004:493–513.
24. Bosch EP, Smith BE. Disorders of peripheral nerves. In: Bradley WG, Daroff RB, Fenichel GM, Marsden CD, eds. Neurology in Clinical Practice, 3rd ed., vol. 2. Boston: Butterworth-Heineman, 2000:2045–2130.
25. Rabinstein A, Wijdicks FM. BiPAP in acute respiratory failure due to myasthenic crisis may prevent intubation. Neurology 2002;59:1647–49.
26. Agarwal R, Reddy C, Gupta D. Noninvasive ventilation in acute neuromuscular respiratory failure due to myasthenic crisis: Case report and review of literature. Emerg Med J 2006:e6.
27. The acute respiratory distress syndrome network. Ventilation with lower tidal volumes as compared with traditional tidal volumes for acute lung injury and the acute respiratory distress syndrome. N Engl J Med 2000;342:1301–08.
28. Wijdicks FM. Management of airway and mechanical ventilation. In: Wijdicks FM, ed. The Clinical Practice of Critical Care Neurology. Philadelphia: Lippincott-Raven, 1997: 25–45.

29. Borel CO, Teitelbaum JS, Hanley DF. Ventilatory drive and carbon dioxide response in ventilatory failure due to myasthenia gravis and Guillain-Barré syndrome. Crit Care Med 1993;21:1717–26.
30. Kokontis L, Gutmann L. Current treatment of neuromuscular diseases. Arch Neurol 2000;57:939–43.
31. Seybold ME. Treatment of myasthenia gravis. In: Engel AG, ed. Myasthenia Gravis and Myasthenic Disorders. New York: Oxford University Press, 1999: 167–91.
32. Batocchi AP, Evoli A, Di Schino C, Tonali P. Therapeutic apheresis in myasthenia gravis. Ther Apher 2000;4:275–79.
33. Gajdos P, Chevret S, Clair B, Tranchant C, Chastang C. Clinical trial of plasma exchange and high-dose intravenous immunoglobulin in myasthenia gravis. Ann Neurol 1997;41:789–96.
34. Qureshi AI, Choudhry MA, Akbar MS, Mohammad Y, Chua HC, Yahia AM, et al. Plasma exchange versus intravenous immunoglobulin treatment in myasthenic crisis. Neurology 1999;52:629–32.
35. Stricker RB, Kwiatkowska BJ, Habis JA, Kiprov DD. Myasthenic crisis. Response to plasmapheresis following failure of intravenous γ-globulin. Arch Neurol 1993;50:837–40.
36. Hamrock DJ. Adverse events associated with intravenous immunoglobulin therapy. Int Immunopharmacol 2006;6:535–542.
37. Dalakas, MC. Intravenous immunoglobulin in the treatment of autoimmune neuromuscular diseases: Present status and practical therapeutic guidelines. Muscle Nerve 1999;22:1479–97.

12

Thymectomy for Non-thymomatous MG

Alfred Jaretzki III and Joshua R. Sonett

1. INTRODUCTION

There continues to be a debate regarding the effectiveness of thymectomy in the treatment of non-thymomatous myasthenia gravis (MG) and, when undertaken, which surgical technique is the procedure of choice. Additionally, critical issues remain unanswered including patient selection, preparation for surgery, timing of surgery, and optimal peri-operative management.

The debate persists primarily because of the lack of controlled prospective studies. In addition, analysis has been complicated by the absence, until very recently, of accepted objective definitions of severity of the illness and response to therapy as well as variable patient selection, timing of surgery, type of surgery, and methods of analysis of results. Without resolution of these issues by properly controlled prospective studies, there can be no unequivocal determination of the effectiveness of thymectomy or valid comparison of the various thymectomy techniques.

In this review, based on previous analyses (1), attempts will be made to clarify some of the controversial issues concerning thymectomy for non-thymomatous MG and make limited recommendations based on the best available evidence.

2. TOTAL THYMECTOMY IS INDICATED

Even before the role of the thymus in MG was understood, "total" thymectomy was considered the goal of surgery. In 1941 Blalock wrote "complete removal of all thymus tissue offers the best chance of altering the course of the disease" (2) and this has been reiterated by most leaders in the field (3, 4, 5, 6, 7, 8, 9, 10, 11, 12, 13, 14, 15, 16). It has since been clearly demonstrated that the thymus plays a central role in the autoimmune pathogenesis of MG. Pathologic and immunologic studies support this thesis (17, 18, 19).

The concept that the entire thymus should be removed, when thymectomy is undertaken in the treatment of non-thymomatous MG, is supported by animal models and by the clinical experience. In an animal model of myasthenia gravis, complete neonatal thymectomy in rabbits prevents experimental autoimmune myasthenia gravis, whereas incomplete removal does not (20). In humans, both incomplete transcervical and incomplete transsternal resections have been followed by persistent symptoms that were later relieved by a more extensive re-operation with the finding of residual thymus (21, 22, 23, 24, 25, 26, 27). And removal of as little as 2 g of residual thymus has been therapeutic (25). In addition, several studies comparing aggressive thymic resections with limited resections support the premise that entire thymus should be removed (1, 28, 29).

Accordingly, although total thymectomy appears indicated in the treatment of non-thymomatous MG, the optimal approach that balances extent of resection, morbidity, patient acceptance, and results remains controversial.

From: *Current Clinical Neurology*
Myasthenia Gravis and Related Disorders
Edited by: H.J. Kaminski, DOI 10.1007/978-1-59745-156-7_12, © Humana Press, New York, NY

3. SURGICAL ANATOMY OF THE THYMUS

Since complete removal of the thymus appears indicated when a thymectomy is performed in the treatment of non-thymomatous MG, its anatomy should be understood by all those involved in the treatment of these patients and in the analysis of the results of the surgery. Importantly, the thymus is not "two well-defined lobes that appear almost as distinct as do the two lobes of the thyroid" as described by Blalock *(2)*.

Detailed surgical-anatomical studies (Fig. 1) have demonstrated that the thymus frequently consists of multiple lobes in the neck and mediastinum, often separately encapsulated, and *these lobes may not be contiguous*. In addition, *unencapsulated lobules* of thymus and *microscopic foci* of thymus may be *widely* and *invisibly* distributed in the pretracheal and anterior mediastinal fat from the level of the thyroid, and occasionally above, to the diaphragm and bilaterally from beyond each phrenic nerve *(30, 31)*. Occasionally, microscopic foci of thymus have been found in the subcarinal fat *(32)*.

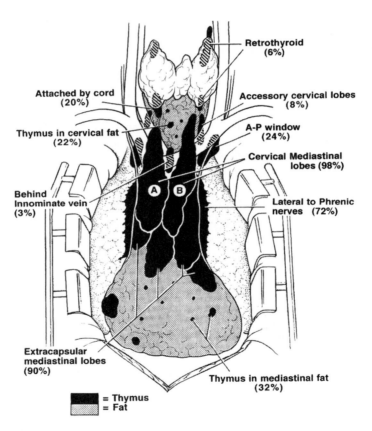

Fig. 1. Composite Anatomy of the Thymus. This illustration represents what is now accepted as the surgical anatomy of the thymus *(31)*. The frequencies (percentages of occurrence) of the variations are noted. Thymus was found outside the confines of the two classical cervical-mediastinal lobes (**A** and **B**) in the neck in 32% of the specimens, in the mediastinum in 98% (Black = thymus. gray = fat, which may contain islands of thymus and microscopic thymus) (Reprinted with permission from *Neurology* **48**-Suppl 5: S52–S63, 1997. Lippincott-Raven Publishers, Philadelphia, PA).

4. RESECTIONAL POTENTIAL OF THE SURGICAL TECHNIQUES

In evaluating the results of surgery and in determining if total thymectomy should remain the goal, an understanding of how much gross and microscopic thymus each resectional technique removes is necessary.

The following presents *estimates* of what each type of resection can accomplish. These estimates are based on published reports, drawings, photographs of resected specimens, videos of the procedures when available, and our personal experience. The videos taken at operation and photographs of the resected specimens are frequently the most revealing. A review of what appears to be the potential of each thymectomy technique strongly suggests that all resections are not equal in extent (Fig. 2) *(1)*.

Regardless of the selected technique, however, when complete removal of all thymic tissue is the goal of the thymectomy technique employed, the actual extent of each individual resection is determined in part by the operating surgeon's conviction that as much thymus that can be removed safely should be removed, the surgeon's commitment to take the time to do so meticulously, and the surgeon's experience with the technique employed. This is certainly true in the use of the trans*cervical* and the video-assisted thoracoscopic approaches, perhaps less so in the open trans*sternal* approaches.

4.1. Combined Transcervical and Transsternal Thymectomy

Transcervical–transsternal maximal thymectomy (31) is also known as *extended cervicomediastinal thymectomy (33)*. Under direct vision these procedures employ wide exposure in the neck and a complete median sternotomy. An en bloc resection is used removing in a single specimen all gross thymus, suspected cervical-mediastinal thymus, and anterior cervical-mediastinal fat. The resections include removal of both sheets of mediastinal pleura, with the specimen and sharp dissection on the pericardium to ensure no microscopic thymus is left on these structures. The resections are exenteration in extent and are "performed as if it were an en bloc dissection for a malignant tumor" *(34)*. Extreme care is taken to protect the recurrent laryngeal, left vagus, and phrenic nerves. A photograph of a typical specimen illustrates the extent of these resections (Fig. 3). A video illustrating the details of this procedure is available *(35)*. Lennquist has described a similar procedure, but with a less extensive resection in the neck and mediastinum *(36)*.

Comment. The *maximal* thymectomy has been described as the benchmark against which the other resectional procedures should be measured *(11)*. Its design is based on the surgical anatomy of the thymus. It predictably removes all surgically available thymus in the neck and mediastinum, including the unencapsulated lobes and the lobules of thymus and microscopic thymus in the pre-cervical and anterior mediastinal fat that are not visible to the naked eye. The en bloc technique is to ensure that islands of thymus are not left behind and to guard against the potential of seeding of thymus in the wound. Piecemeal removal of the thymus may "herald a poorer prognosis than clean removal" *(37)*. The sharp dissection on the pericardium (with division of the fingers of pericardium that enter the thymus) and the removal of the mediastinal pleural sheets with the specimen (with the confirmed adherent microscopic thymus) are an integral part of the procedure. Blunt dissection has been demonstrated to leave microscopic foci of thymus on the pericardium and pleura.

4.2. Transsternal Thymectomies

The *extended transsternal thymectomy (7, 28, 38, 39, 40, 41, 42, 43, 44, 45, 46)* is also known as *aggressive transsternal thymectomy (47)* or *transsternal radical thymectomy (48)*. The extent of these resections varies *in the mediastinum*, may or may not include all the mediastinal tissue removed by the "maximal" technique, and frequently do not. The cervical extensions are removed

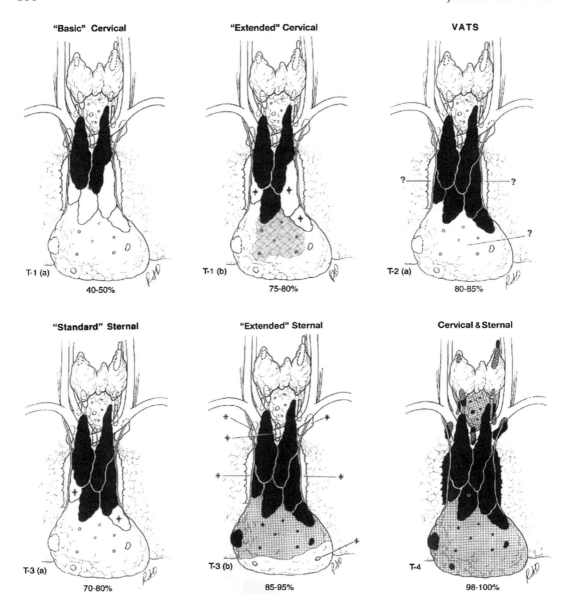

Fig. 2. Estimated Resectional Extent of Six Thymectomy Techniques. See text for details. The VATET resection is not illustrated (Black = thymus; gray = fat, removed by the designated operation – it may contain islands of thymus and microscopic thymus; * = thymic lobes or thymo-fatty tissue that may or may not be removed; white = tissue not removed by the designated procedure; and ? = thymo-fatty tissue that may or may not be removed by the designated procedure) (Modified from *Neurology* **48**-Suppl 5: S52–S63, 1997. Lippincott-Raven Publishers, Philadelphia, PA).

from below, with or without some additional cervical tissue, but without a formal neck dissection. Mulder describes in detail an aggressive mediastinal dissection approximating the dissection of the "maximal" technique *(46, 49)*. A video illustrating the details of a very extensive mediastinal dissection is also available *(50)*.

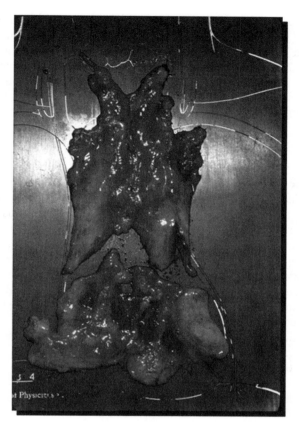

Fig. 3. Photograph of a Typical Transcervical–Transsternal "Maximal" Thymectomy Specimen. The specimen consists of the en bloc resected cervical and mediastinal thymus, suspected thymus, and pretracheal and anterior mediastinal fat from thyroid isthmus to diaphragm with adherent mediastinal pleural sheets bilaterally. The specimen was divided after removal from the patient to indicate the division between the grossly visible thymus and the distal anterior mediastinal fat (Reprinted with permission from *General Thoracic Surgery*, fifth edition, Lippincott Williams & Wilkins).

Comment. These transsternal resections are identical in the mediastinum to the mediastinal dissection of the "maximal" procedure *when* they are exenteration in extent, use sharp dissection on the pericardium, *include* removal of both mediastinal pleural sheets, and remove the other tissue in the mediastinum defined in the "maximal" technique. They do, however, remove less tissue in the neck where a small amount of thymus is present in approximately 30% of the specimens. Mulder has expressed the view, however, that the risk to the recurrent laryngeal nerves in performing the extensive neck dissection of the combined transcervical–transsternal maximal thymectomy is not "justified by the small potential gain" *(49)*.

The *standard transsternal thymectomy* used by the pioneers Blalock, Keynes, and Clagett *(2, 3, 51, 52)* was originally limited to removal of the well-defined cervical-mediastinal lobes that were thought to be the entire gland *(53)*. Currently, although a complete *(54)* or partial *(55, 56, 57)* sternotomy may be performed, the resection is more extensive than originally described and includes removal of all visible mediastinal thymic lobes. Mediastinal fat, varying in extent, may or may not be removed. The cervical extensions of the thymus are removed from below, with or without some adjacent cervical fat. Variations of this technique include a video-assisted technique using a complete median sternotomy via a limited lower transverse sternal skin incision *(58)*.

Comment. This resection, as illustrated in surgical textbooks *(59, 60, 61, 62)*, falls short of total thymectomy. Residual thymus has been found in the neck and in the mediastinum at re-operation following a "standard" trans*sternal* thymectomy *(24, 25, 26)*. Accordingly, most MG centers since the early 1980s *(25, 33, 38, 39, 40, 41, 42, 43, 44, 47, 48, 63)*, with some exceptions *(61, 64)*, have considered the *standard* trans*sternal* thymectomy to be incomplete and no longer use it.

4.3. Transcervical Thymectomies

The *extended transcervical thymectomy (65, 66, 67)* employs a special manubrial retractor for improved exposure of the mediastinum. The mediastinal dissection is extra-capsular and includes resection of the visible mediastinal thymus and mediastinal fat. Sharp dissection may or may not be performed on the pericardium. The mediastinal pleural sheets may be included in the resection but usually are not. The neck exploration and dissections vary in extent and may or may not be limited to exploration and removal of the cervical-mediastinal extensions of the thymus. Variations of this procedure include the associated use of mediastinoscopy *(68)*, partial median sternotomy *(69, 70)*, trans*cervical* video-assisted thoracoscopy *(71, 72)*, and a combined trans-cervical-*subzyphoid* approach with reported ability to achieve a thymic resections as extensive as the combined transcervical–transsternal "maximal" thymectomy procedure *(73)*.

Comment. The *extended transcervical thymectomy* resection, as warned by Cooper *(65)*, when performed by others may be less extensive than the procedure described by him. This concern appears to be supported by a textbook description of the technique *(74)*. Although the incision is in the neck, the resection may or may not include the occasional accessory thymic lobes in the neck, retro-thyroid thymus, or thymus in the pretracheal fat.

It is not clear that the addition of mediastinoscopy, a partial sternal split, or a transcervical video-assisted thoracoscopy improve the extent of the resection, although the latter deserves close scrutiny in view of the apparent results (Table 2). However, the addition of a bilateral subzyphoid approach described in one report, based on the author's description of the procedure, is intriguing *(73)*. A video review of the procedure and comparative follow-up results will be of interest.

The *basic transcervical thymectomy* employs an *intra*capsular *extraction* of the mediastinal thymus via a small cervical incision and is limited to the removal of the *intracapsular* portion of the central cervical-mediastinal lobes. *No* other tissue is removed in the neck or mediastinum *(5, 75, 76)*.

Comment. The *basic transcervical thymectomy*, although originally considered to be a "total thymectomy" *(5)*, is unequivocally a limited resection, incomplete in both the neck and the mediastinum. The limited extent is evident in descriptions of the technique *(75, 76)*, the surgical anatomy of the thymus in MG *(31)*, and the findings of residual thymus at many re-operations after an initial trans*cervical* resection *(21, 22, 23, 24, 25, 26, 27)*. In addition, Henze *(23)* re-operated on 27% of his "basic" transcervical series of 95 patients, because of persistent or progressively severe symptoms, finding 10–60 g of residual thymus in all. And a computer-matched study, comparing the "basic" trans*cervical* to the "standard" trans*sternal* operation, demonstrated that twice as much thymus was removed by the trans*sternal* than in the trans*cervical* procedure, even though the trans*sternal* operation was the more limited "standard" type *(29)*. The *basic transcervical thymectomy* not only routinely fails to remove accessory thymus present in the neck and mediastinum but encapsulated thymic lobes in the mediastinum as well. Accordingly, we believe that it should be abandoned.

4.4. Videoscopic-Assisted Thymectomies

Variations in videoscopic thymectomy are continually being developed. Their resectional potential and the results are still in the investigative phase. Two basic techniques have been employed with several additional approaches recently being introduced.

The video-assisted thoracic surgery (VATS) thymectomy employs unilateral videoscopic exposure of the mediastinum (right or left) with removal of the grossly identifiable thymus and variable amounts of anterior mediastinal fat. Sharp dissection may or may not be used on the pericardium and the mediastinal pleural sheets are not routinely removed with the specimen. Complete removal of the mediastinal and diaphragmatic fat on the operative side is routinely performed. The cervical extensions of the thymus are usually removed from below *(16, 77, 78, 79, 80, 81)*. Illustrations and photographs of the technique and specimens are available *(16, 80, 81, 82)*.

The *video-assisted thoracoscopic extended thymectomy* (VATET) uses bilateral thoracoscopic exposure of the mediastinum for improved visualization of both sides of the mediastinum. Extensive removal of the mediastinal thymus and peri-thymic fat is described, the thymus and fat being removed separately. Sharp dissection may or may not be used on the pericardium. The mediastinal pleural sheets are usually not removed. A cervical incision is performed with exposure of the recurrent laryngeal nerves and reported removal of the cervical thymic lobes and pretracheal fat under direct vision *(83, 84)*. A more formal and extensive transcervical resection has been employed *(85)*.

The *robotic-aided video-assisted thoracoscopic thymectomy*, unilateral and bilateral, is in the developmental stage *(86, 87, 88, 89)*. The instrumentation offers enhanced optics with three-dimensiona visualization and ×12 magnification; the surgical arms allow precise tissue dissection. Therefore the technique has the potential for extensive, yet safe, resection of the thymus and anterior mediastinal fat as well as exploration of the neck, if considered indicated. The unilateral approach appears to have the disadvantages of the unilateral VATS thymectomy. Accordingly, the bilateral procedure has the appeal of the VATET thymectomy *(87)*. Despite improved optics, the clinical advantage of the robotic technique versus a thoracoscopic approach is not clearly evident and certainly adds significant cost and time to the procedure.

The *infrasternal thymectomies* – These include a *video-assisted thoracoscopic subxyphoid thymectomy (90)*, an *infrasternal mediastinoscopy (91)*, and a combined *transcervical-subxyphoid thymectomy* described as a thymic resection as "extensive as the combined transcervical–transsternal 'maximal' thymectomy" *(73)*.

Comment. The VATS resections, based on the published reports, vary in extent in the mediastinum. They appear to be more extensive than the standard transsternal but less than the extended transsternal resections. No attempt is made to remove the accessory thymus in the neck. Since it is unilateral, left or right, the contralateral side of the mediastinum does not appear to be as well visualized as the operative side. The VATET operation is conceptually more complete than the VATS since it offers excellent visualization of both sides of the mediastinum and includes a neck dissection as well. It may well offer the same clinical response to a thymic resection as the extended transsternal thymectomy.

As warned by Mack, however, "thymectomy is an advanced *VATS* procedure that should only be undertaken by surgeons. . .well versed in simpler VATS operations and who have the enthusiasm and patience to pursue minimally invasive techniques" *(81)*. Clearly this caveat applies to the VATET thymectomy as well as any form of video-assisted thoracoscopic thymectomy approach.

We have had no experience with, nor have seen operative videos of, the subxyphoid approaches. Accordingly, we must reserve judgment.

5. PROBLEMS IN THE ANALYSIS OF THYMECTOMY FOR MG

5.1. Data Analysis

Inappropriate statistical analysis, including the comparison of unrelated statistical techniques, has led in many instances to incorrect conclusions concerning the relative merits of the thymectomy techniques. The following is a brief review of material previously reviewed and analyzed in detail *(1, 92)*.

Life table analysis, using the Kaplan–Meier method, is considered the preferred statistical technique for the analysis of remissions following thymectomy *(93)*. It provides comparative analysis using all follow-up information accumulated to the date of assessment, including information on patients subsequently lost to follow-up and on those who have not yet reached the date of assessment *(69, 94)*. This analysis should be supplemented by multivariable analysis to identify and correct for significant variables. "Hazard Rates" (remissions per 1,000 patient-months) corrects for length of follow and "censors" patients lost to follow-up. However, these rates depend on the "risk" (remissions per unit of time) being constant *(95)*, which may not be the case in MG.

Uncorrected crude rates, the number of remissions divided by the number of patients operated upon *or* sometimes divided by only the number of patients followed (potentially a very different denominator), have been the primary form of analysis in the comparative evaluation of remissions and improvement following thymectomy. This is unfortunate since this form of analysis does not include in the evaluation all the follow-up information accumulated to the date of assessment. In addition, patients evaluated many years after surgery may appear to do as well or better than patients with a shorter follow-up. And even the differing denominators in the two subsets (*patients operated upon versus patients followed*) are not comparable but have frequently been compared without comment. Accordingly, uncorrected crude data should have no place in the comparative analysis of results of thymectomy *and for this reason has been omitted from this presentation.*

Although correcting crude data for *mean* length of follow-up enables a *rough* comparison of the *uncorrected* crude data remission rates and confirms the fallacy of comparing uncorrected crude data, it should not be used as a substitute for life table analysis and *therefore is also not included in this presentation.*

5.2. Other Pitfalls of the Analysis of Results

Unfortunately, in addition to the problems with statistical analysis, most of the reports are flawed because of the "absence of standardized methods for assessing patient status, both before and after surgery" *(48)*. The following is a brief review of these issues *(1)*.

The studies are primarily retrospective. Accordingly, they do not address the variability and unpredictability of MG, the differing response to treatment among patients with different severities and sub-types of MG, nor the bias inherent in the selection of patients for thymectomy. Clinical classifications have been used to stratify patients and evaluate results, yet there have been at least 15, often conflicting, classifications, and these clinical classification systems are not quantitative, do not accurately describe changes in disease severity, and even when standardized are not a reliable measure of response. In addition, patients are frequently not examined, the evaluation being by phone.

Although "remission" is considered the measurement of choice in defining results following thymectomy *(96, 97, 98)*, until recently there has been no uniform definition of the remission status. "Improvement" and changes in "mean grade", widely used as determinants of success, are unreliable measurements without quantitative and objective criteria of evaluation that have been rarely applied. In addition, when immunosuppression is included in the preoperative or postoperative thymectomy regimen, the patients are not compared to a control group and do not follow a predefined schedule of medications and dose reduction that is required in assessing the additive benefit from thymectomy and immunosuppression. *Thus, under these circumstances, it is not possible to infer retrospectively the effects of thymectomy itself (99).*

In comparing results of different techniques, other confounding factors are also frequently ignored and may conceal the disadvantages of the procedure being touted *(1)*. These include (1) failure to assess or define the length of illness preoperatively; (2) failure to account for the length of postoperative follow-up; (3) muddying the analysis by inclusion of multiple surgical techniques and combining two or more series with differing definitions and standards; (4) combining patients with and without thymoma in a composite analysis; (5) including re-operations in the primary analysis when most patients at the time of the re-operation had severe symptoms of long duration and may have failed earlier thymectomy for undetermined reasons; (6) use of meta-analysis based on mixed and uncontrolled data; (7) failure to report relapses; and (8) failure to consider the rate of spontaneous remissions.

6. THE RESULTS OF THYMECTOMY

6.1. Introduction

There are no well-designed, well-controlled prospective studies that compare medical management with thymectomy or compare the various resectional techniques. Such studies are necessary to determine whether thymectomy is truly beneficial and to determine the relative merits of the various resectional techniques *(100)*. This, of course, makes comparative analysis of these issues, based on existing published information, extremely difficult if not completely unreliable.

In addition to prospective studies, the use of uniform definitions and classifications is required. Toward this end, we firmly support the recommendations of the Task Force of the Myasthenia Gravis Foundation of America (MGFA) which was formed to address the need for standards of evaluation *(92, 101)* (Table 1). The recommendations include the establishment of uniform clinical classifications, quantitative methods of patient evaluation, and classifications of therapy status, thymectomy type and extent, post-intervention status, morbidity and mortality information, and recommendations for clinical trials and outcomes analysis.

Until such time as controlled prospective studies have been performed, however, physicians and patients in making therapeutic decision have no alternative but to rely on the *best available* information available, *properly evaluated*. This requires selecting only those studies that report at least some significant data and have employed reliable statistical methods of analysis; it is also necessary to ignore those studies that do not fulfill some minimally acceptable criteria *(1)*.

Table 1
MGFA Thymectomy Classification (*modified*) (101)

T-1 Transcervical thymectomy
a-Basic
b-Extended
c-*Extended with partial sternal split*
d-*Extended with videoscopic technology*
T-2 Videoscopic thymectomy
a-Classic VATS *(unilateral)*
b-VATET *(bilateral with neck dissection)*
c-*Videoscopic with robotic technology (unilateral)*
d-*Videoscopic with robotic technology (bilateral)*
T-3 Transsternal thymectomy
a-Standard
b-Extended
T-4 Transcervical and transsternal thymectomy

Modifications to the MGFA classification appear necessary since each change in operative technique potentially changes the amount of thymus removed and the results. *The modifications are recorded in italics.* Presumably additional definitions will be required as additional minimally invasive modifications are developed.

We believe the following discussion represents such an analysis and that it suggests a thymect-omy benefit in the treatment of non-thymomatous MG and supports the concept that the less thymus left behind the better the results.

6.2. Thymectomy Versus Medical Management

As the surgical community debates and analyzes the different methods and techniques of thymectomy, this debate must be viewed in the context that conclusive evidence of the efficacy of thymectomy is itself still lacking many years after it was introduced as a therapeutic modality.

A retrospective review in 1976 by Buckingham et al. from the Mayo Clinic used computer-assisted matching to study patients with non-thymomatous MG (ages 17 to >60) and demon-strated considerably better results following thymectomy than with medical management alone *(102)*. Eighty of 104 surgical patients were matched "satisfactorily" from the files of 459 medically treated patients. Using crude data corrected for length of follow-up, complete remission was experienced in 35% of the surgical patients (at an average 19.5 years follow-up) compared to 7.5% of the medical group (at an average 23 years follow-up), even though only an apparently limited "standard" transsternal thymectomy (T-3a) was being performed at that time. Remission was defined as "no medication and no MG symptoms except in rare instances when a mild inconstant deficit was present".

An evidence-based review of thymectomy in non-thymomatous MG between 1953 and 1998 has been performed *(103)*. The analysis included 21 controlled but non-randomized studies in which crude or corrected crude remission rates were analyzed. The authors found positive associations in most studies between thymectomy and MG remission (*we note this positive association even though the thymectomy techniques in this review were in many instances not defined or consisted of the limited basic transcervical thymectomies (T-1a) or the limited standard trans-sternal thymectomies (T-3a), indicating presumably only partial resections*). After thymectomy, patients were twice as likely to attain medication-free remission status than un-operated patients. However, there were significant confounding differences in baseline characteristics of prognostic importance between the thymectomy and non-thymectomy patient groups. Importantly, the author's final conclusion was that the benefit of thymectomy in non-thymomatous autoimmune MG has not been established conclusively and that prospective studies were indicated.

To help define the role of thymectomy in non-thymomatous MG, a clinical trial "A Multi-Center, Single-Blind, Randomized Study Comparing Thymectomy to No Thymectomy in Non-thymoma-tous Myasthenia Gravis (MG) Patients Receiving Prednisone" established by John Newsom-Davis and supported by NIH-NINDS is underway *(104)*: As the title indicates, the trial is designed to establish the role of thymectomy in the management of MG patients receiving prednisone. This type of evaluation was undertaken, recognizing the difficulty of developing a direct comparison of thymectomy with no thymectomy, since corticosteroids are now being used increasingly either as the sole treatment or in conjunction with thymectomy. An extensive form of the *extended transster-nal thymectomy* is the surgical procedure being employed. This thymectomy technique was selected to ensure complete removal of the mediastinal thymus and to avoid exposing the recurrent laryngeal nerves to injury. Accordingly, the cervical extensions are removed from below without a formal neck dissection. This well-designed and well-supervised study should go a long way to help resolve the role of thymectomy in MG.

6.3. Comparative Results Following Thymectomy

The available evidence continues to suggest that the less thymus left behind, including extra-lobar lobules of thymus and microscopic thymus in the peri-thymic fat, the better the long-term results. However, not only has this observation not been confirmed by prospective studies but there are also no well-designed prospective studies that conclusively indicate which resectional technique is the thymectomy of choice.

Table 2
Remission Rates (life table analysis) Following 11 Thymectomies for Non-thymomatous MG

Thym(type)	N	Class (% of patients)				Preop(durat)	Post-OP FU (% CSR)				
		Ocu	Mild	Mod	Sev	Mean years	2-3 years	5 years	6 years	7.5 years	10 years
Sponraneous											
None[1]	149	9	16	73	2	–	10	15	–	18	20
Transcervical Thymectomy											
T-1a[2]	651	–	–	–	–	–	14	23	–	33	40
T-1b[3]	78	15	32	39	14	–	31	43	–	–	–
T-1c[4]	300	0	27	56	17	2.5	15	33	–	–	–
T-1d[5]	120	11	32	37	20	0.8	–	30	38	55	91
Videoscopic Thymectomy											
T-2a[6]	36	22	41	14	23	3.0	–	13	20	20	75
T-2b[7]	159	12	31	45	12	2.6	33	51	51	–	–
T-2c[8]	92	7	34	43	16	2.2	–	20	32	32	–
Transsternal Thymectomy											
T-3b[9]	73	–	–	–	–	5.5	30	40	–	50	52
T-3b[10]	47	0	30	59	11	2.0	20	44	49	–	–
T-3b[11]	98	18	41	23	18	2.3	–	30	–	41	45
Transcervical-Transsternal Thymectomy											
T-3[12]	72	0	18	52	30	2.8	30	50	–	81	–

This analysis compares Kaplan-Meier life table analysis of 11 thymectomy reports and one report of spontaneous remissions. For an interpretation of this table see Section 6.3. Comparative Results Following Thymectomy.

In the interpretation of Table 2 we would like first to make the following comments:

(1) We believe that the most reliable *available data* for this type of analysis are reports that employ Kaplan–Meier life table analysis. We were able to find 12 such reports (Table 2). We recognize, however, that even though these studies have employed Kaplan–Meier analysis as we believe important, they are not truly comparable because of significant variations in the severity and duration of the preoperative illness, variable use of corticosteroids and other immunosuppressive therapies, various definitions of remission, varying methods of postoperative evaluation including the frequent use of evaluation by telephone, and a total lack of quantitative (QMG) evaluation. However, this represents what we consider to be the *best available evidence* at this time.

(2) Since we did not find for comparison *spontaneous* life table analysis remission rates for adults, we included life table analysis remission rates in children *(105)*. However, not included in this figure is a study of standard trans*sternal* thymectomies limited to children under the age of 17 years. It had a Kaplan–Meier remission rate at 7.5 years of 44% *(105)*.

(3) We have limited the review to the analysis of *remissions* since, as previously stated, they are the most reliable determinant. Without controlled prospective studies and quantitative evaluation, the comparative evaluation of levels of *improvement* has virtually no significance and therefore has also not been included in this analysis.

(4) In analyzing this uncontrolled retrospective data, even though it consists of remission rates determined by Kaplan–Meier analysis, the known *favorable* and *unfavorable* preexisting clinical status of the patients in each series must be considered in comparing the results. As of now, although not confirmed by well-controlled prospective studies, the less severe the preoperative MG manifestations and shorter the preoperative duration of the illness *(25, 69, 106, 107, 108)* the better the results.

(5) We have omitted a presentation of *crude data and adjusted crude data* because, as previously discussed, the use of this form of analysis is not a valid statistical method to compare the results of the various thymectomy techniques *(109)*. It is of note that the oft-repeated statement that "results are equal" has usually been based on the comparative analysis of *crude data*.

(6) Until a total thymectomy is proven not to be the goal of surgery, we continue to believe that the combined transcervical–transsternal "maximal" thymectomy (T-4) should remain the *benchmark*

against which other thymectomy techniques are measured (whether or not it is the surgical technique of choice) because it routinely and predictably removes all the *anatomic variations* of the thymus that have been generally accepted as the surgical anatomy of the thymus (Fig. 1). It is of note that 50 consecutive such T-4 thymectomies were employed in defining these anatomical findings *(31)*.

(7) Since this analysis consists of uncontrolled retrospective studies, there remain a number of unknown factors that make the analysis only speculative. With this as a background, we submit our speculative interpretation of the data.

If the information in Table 2 is viewed at face value, the 75 and 91% remission rates at 10 years suggest that these somewhat novel approaches are clearly the thymectomy of choice at this time. Of note, however, is that in the former series (#6, *T-1d, an extended transcervical thymectomy in which a transcervical thoracoscope was employed*) *(71)*, 63% of the patients had purely ocular or mild preoperative MG manifestations and in the later series (#5, *the T-2a, unilateral videoscopic VATS*) *(110)*, 43% of the patients had ocular or mild MG manifestations *and* their mean duration of preoperative MG manifestations was only 10 months (0.8 years). Thus, the high remission rates, though admirable, may in fact represent a very favorable cohort of patients rather than the result of the surgical technique employed per se. If this interpretation is correct, from the therapeutic standpoint it is still necessary to determine which of the two events, or what other event, was responsible. The jump from 20 to 75% and from 55 to 91% within 2.5 years seems odd but in any case is unexplained.

Therefore, we believe it is more helpful to review the 5-year follow-up data because both a post-thymectomy evaluation at 5 years is a good starting point and all 12 of the reports have data at this follow-up interval. The VATET (#7 – *combined cervical and bilateral videoscopic thymectomy*) *(111)* and the maximal thymectomy (#12 – *combined transcervical–transsternal thymectomy*) *(25)* have 51 and 50%, respectively, remission rates at that interval. However, 43% of the VATET patients had ocular or mild MG manifestations compared to 18% of the T-4 patients suggesting that the VATET cohort were a more favorable cohort. Similarly, both the extended trans*cervical* thymectomy (T-1b) *(67)* and two of the *extended transsternal thymectomies* (#9 and #10) *(107, 111)*, with remission rates in the low 40% range, also had a high percentage of patients with ocular or mild preoperative MG manifestations. Thus, this evaluation suggests that at 5 years the T-4 maximal thymectomy has the highest remission rate, with the extended transsternal thymectomy close behind.

Also of note is the striking difference at 5 years between the most extensive resection (the "maximal" T-4 thymectomy) and what is clearly a very limited resection (the "basic" T-1a thymectomy). The former has a remission rate of 50% compared to 23% for the latter, and this difference is seen to persist. This finding strongly supports the concept that the less thymus left behind the better the results.

Whether or not the above *analysis* is valid, and certainly this analysis is speculative, we believe this data makes it clear that the various resectional techniques do not produce equal results and that probably the thesis that the less thymus left behind the better the results is valid. *Most importantly, however, it is clear that resolution of this debate will require well-designed and well-executed prospective studies.*

7. INDICATIONS FOR THYMECTOMY

In view of the lack of controlled prospective studies, probable limited thymic resections in a number of instances, and some invalid analyses of the available retrospective data, it is not surprising that neurologists hold differing views as to when, and even if, a thymectomy is indicated in the course of treatment of non-thymomatous MG *(112)*. A better understanding of the indications for surgery awaits well-designed prospective studies.

Keesey's algorithm in reference to moderate generalized MG *(113)* represents a centrist position in the treatment of adult patients: "Thymectomy is recommended for those relatively healthy patients whose myasthenic symptoms interfere with their lives enough for them to consider undergoing major thoracic surgery. While it is expensive and invasive, thymectomy is the only treatment available that offers a chance of an eventual drug-free remission. This chance seems to be better earlier in the disease than later. The potential benefit of thymectomy also decreases as the adult patient gets older and as the thymus naturally involutes with time. Further, the risk of surgery increases with age. The age at which the risks outweigh the potential benefits must be individualized for every patient". Keesey's group routinely performed an extended transsternal thymectomy *(46)*.

Although we believe the preponderance of evidence indicates that patients with less severe illness and operated upon early in their illness do better *(69)*, it is our opinion at this time that a thymectomy should be considered for most adult patients with more than very mild generalized non-thymomatous MG, *regardless* of the duration of symptoms or age of the patient, and probably virtually all patients with respiratory or oropharyngeal symptoms, including those with mild such symptoms (MGFA Class IIa) *(101, 114)*, unless a specific contraindication exists. The patient's medical condition has to be satisfactory for the planned surgery and patient selection over the age of 65–70 years has to be selective.

Although the role of thymectomy for patients with purely ocular symptoms is even less well defined and some have not recommended elective thymectomy *(115)*, it is probably indicated for patients in whom the ocular MG interferes with life style or work, and when immunosuppressive or other therapy is contraindicated or not effective *(116, 117)*. Thymectomy within the first year of the illness in adults, "early" thymectomy, has been recommended and appears to make sense *(108, 118, 119, 120)*, but lacks prospective documentation. The indication for thymectomy in children is a separate issue addressed by others *(105, 121, 122, 123, 124)*.

8. SELECTING THE THYMECTOMY TECHNIQUE

In the absence of controlled prospective studies comparing the various thymectomy techniques, it is not possible to state with certainty which technique should be the procedure of choice. Accordingly, it is necessary for the referring neurologist to have an understanding of the debate herein presented in determining what procedure to recommend. However, regardless of the surgical approach, surgical expertise and experience are required. The surgeon should be convinced of the importance of complete removal of the thymus and be willing to commit the necessary time and care to achieve this goal *safely*. Since the need for complete removal of all gross and microscopic thymus has not been definitively confirmed, it is generally considered preferable to leave behind small amounts of suspected thymus, or even likely thymus, than injure the recurrent laryngeal, the left vagus, or the phrenic nerves. Injury to these nerves can be devastating to a patient with MG.

Based on an analysis of the *available data*, it appears that the more thorough the removal of all tissue that may contain thymus, the better the long-term results. Therefore, *again conceptually*, the combined transcervical–transsternal thymectomy best fulfills these criteria; from the practical standpoint, however, this procedure should only be undertaken by thoracic surgeons experienced in neck surgery or with the assistance of such a surgeon. Alternatively, we support the extended transsternal thymectomy since in the hands of experienced surgeons it predictably removes all but a possible small amount of the thymic variations in the neck, has less risk of injuring the recurrent laryngeal nerves, has produced good results, and at this time is probably the most commonly performed form of thymectomy.

Clearly, however, if one of the *minimally invasive* techniques can be demonstrated to be as effective in the treatment of MG as the procedures that employ a median sternotomy, they would be our choice for most patients. Accordingly, *in the hands of surgeons experienced with these techniques*, the extended trans*cervical* thymectomy (*if* performed as defined by Cooper *(65)*) and the VATET thymectomy (bilateral thoracoscopic thymectomy combined with a formal neck exploration) may offer acceptable alternatives. Although there are a number of enthusiastic proponents of the unilateral VATS procedure, and experienced and respected surgeons are performing it, there is insufficient data available at this time to unequivocally determine its role. Based on the anatomy of the thymus, the specimens removed, and the available results, the basic trans*cervical* (T-1a) and the standard trans*sternal* (T-3a) resections appear to represent incomplete resections and to give less good results.

As stated, the *minimally invasive* surgical techniques are clearly very desirable, *assuming* the eventual results are equivalent to the *more invasive* procedures. However, the thymectomy technique should probably be selected in most instances on the basis of the long-term results and the selection should ordinarily not compromise the therapeutic goals. Accordingly, if it is determined by controlled prospective studies that the more traditional approaches produce significantly better results compared to the less invasive procedures (including more remissions with fewer subsequent hospitalizations and/or complications of prolonged immunosuppression), the initial disadvantages of the traditional thymectomies, including the reticence of young men and women to undergo "elective" surgery, will be outweighed by their subsequent advantages. If this is not the case and the less invasive procedures prove as effective or more effective, or perhaps nearly as effective, then they would become the thymectomy of choice in experienced hands.

The review and recommendations noted above are supported by many of the experts in the field, although clearly not all. In view of the lack of unambiguous prospective studies, however, it is not possible to be dogmatic. Hopefully, these recommendations will, at the least, be seriously considered in the decision-making process. In addition to the Newsom-Davis prospective randomized study underway *(104)*, properly controlled prospective studies, comparing the results of the benchmark procedure with the "less invasive" ones, are required to resolve the many issues discussed.

9. RE-OPERATION

There is a need to recognize that a re-operation, repeat thymectomy, may be appropriate and highly desirable for patients who have an unsatisfactory result after one of the more limited thymic resections *(125)*. This recommendation appears to be straightforward when applied to a patient who still has, or progresses to, an incapacitating and poorly controlled illness, especially if repeated hospitalizations and ICU stays have been required 3–5 or more years following a basic trans*cervical* (T-1a) or a standard trans*sternal* thymectomy (T-3a).

It is more difficult to recommend re-operation for less severe symptoms or after more extensive original operations, but it probably is appropriate in selected instances. Unfortunately at this time, it is not possible to determine the location, *or even presence*, of even moderate amounts of residual thymus prior to surgery or *even by gross inspection at the time of surgery*. Negative CT scans or MRI examinations do not exclude the presence of residual thymus and antibody studies are usually not helpful *(27)*. However, a review of the operative note and pathological report of the original surgery, which is mandatory in any case, usually makes it clear that an incomplete resection had been performed. Hopefully the day will come when techniques are developed that can determine the location of even small amounts of thymic tissue. The timing of re-operations is also a difficult decision. Although a 3- to 5-year wait frequently seems prudent, earlier re-operation may be more appropriate and many years later should not disqualify a patient from re-operation.

If re-operation is undertaken for persistent or recurrent symptoms, it should be as extensive as the combined transcervical–transsternal T-4 resections in both the neck and mediastinum, regardless of the thymectomy technique employed in the initial operation. This recommendation is supported by the findings following re-operations using the "maximal" technique (Table 3). Obviously, however, a re-operation is more difficult and time consuming than a primary resection, and the risk to the nerves and the thoracic duct is greater. Therefore, wide exposure in the neck and mediastinum, with the use of a "T" incision rather than separate neck and sternal

Table 3
FifteenRe-operations for Non-thymomatous MG Following Previous "Basic" Transcervical and "Standard" Transsternal Thymectomies

Patient characteristics (Duration 1–30 years)		
Moderately Severe	3	
Severe	7	
Crisis	5	
Residual thymus (2–23 gms)		
Previous transcervical (9)		In Neck (1 of 4)
		In Med. (9 of 9)
Previous transsternal (6)		In Neck (6 of 6)
		In Med. (4 of 6)
Results (Mean F.U. 3.5 years)		
Remission	1	
Pharmacologic remission	5	
Minimal manifestations	3	
Improved	4	
Unchanged	2	

Employing the "maximal" thymectomy technique, residual thymus was found in all cases. Thymus was found in the previously "unopened" compartment (neck or mediastinum) in 100% of the cases. It was also found in the neck in one-quarter of the previous transcervical cases and in the mediastinum in two-thirds of the previous transsternal cases.

incisions, appears highly desirable *(31)*. If the surgeon is not experienced in neck surgery, the assistance of a surgeon experienced in this area is suggested.

We believe it is illogical to limit the re-operation procedure, as some have done, to a repeat transcervical approach *(11, 12, 106)*, a limited transsternal approach *(11, 24)*, or even an extended transsternal resection without a formal neck dissection *(21)*. Thoracoscopic surgery may also be inappropriate for re-operations. Incomplete resections at re-operations presumably explain the occasional failure to find thymus or for a lack of benefit *(12)*.

10. TRACHEOSTOMY: TIMING AND TECHNIQUE

Although tracheostomy is no longer necessary on a routine basis in the peri-operative management of these patients, there are instances when it is either mandatory or at least very desirable (to make the care easier and safer). It is important, however, that unless there be several weeks between the performance of the tracheostomy and the mediastinal dissection, the tracheostomy wound should be biologically isolated from the mediastinum; this is to prevent bacterial contamination of the mediastinum and the development of mediastinitis.

If the patient has a tracheostomy in place at the time of any of the transsternal resections, anesthesia is given via an oral endotracheal tube and the sternal wound is sealed from the tracheostomy site. To complete the cervical portion of a combined transcervical–transsternal thymectomy, if that operation is the choice, a formal neck dissection must be performed *after* the tracheostomy tube has been removed and the wound is well healed. If a tracheostomy is required within the first 2–3 weeks after a transcervical or transsternal resection, it must be a two-stage procedure, the stages separated by at least 4–5 days. If the need for a tracheostomy postoperatively is anticipated at the time of the thymectomy, the first stage can be performed at the time of the initial surgery.

11. SURGICAL MANAGEMENT OF THYMOMA

Because approximately 10% of patients with MG will have a thymoma *(126)*, a chest CT scan is indicated prior to thymectomy in all these patients. It should identify thymomas in most instances *(127)*. The presence of antibodies to striated muscle antigens also predicts the existence of a thymoma *(15)*. However, small thymomas may first be discovered at surgery. In the resection of thymic neoplasms (regardless of size, lack of evidence, or suspicion of invasion), we believe it is inappropriate to use any approach that circumvents a complete sternotomy *(128, 129)*. To avoid tumor seeding and late recurrences, the well-encapsulated, as well as the invasive, tumors require wide local resection with as good tumor margins as practical, including the removal of adherent pericardium or wedges of adherent lung if necessary. Although a phrenic nerve may also appear to be involved, although always a difficult and individual decision, in most instances it can and perhaps should be preserved in patients with MG. In addition, a total thymectomy should be performed unless a specific contraindication exists.

12. PERI-OPERATIVE PATIENT MANAGEMENT

Patients undergoing thymectomy in the treatment of MG require the care of a team of neurologists, pulmonologists, respiratory therapists, intensive care specialists, anesthesiologists, and surgeons who have had experience in the care of these patients. Such teamwork has been demonstrated to improve outcome following thymectomy for MG *(130)*, as has operative volume in other cardio-thoracic operations *(131, 132, 133)*. And regardless of the surgical technique employed, the surgeons should be totally conversant with the special problems of patients with MG and committed to the frequently difficult postoperative care. To accomplish these goals, these operations should be performed at centers where such teams exist.

It is, of course, important that the diagnosis of MG be unequivocally established prior to thymectomy and, as emphasized by Kaminski *(120)*, the patient and the patient's family should have a thorough understanding of the illness, non-surgical and surgical treatment options, anticipated postoperative course and morbidity, and of course, the potential results, or lack thereof, of the surgery.

The special problems associated with the peri-operative care of these patients are directly related to the severity of the MG manifestations at the time of surgery. The major risk is the presence of oropharyngeal and respiratory muscle weakness with the potential for aspiration of oral secretions, inability to cough effectively postoperatively, and respiratory failure. However, if the oropharyngeal and respiratory weakness is eliminated preoperatively, *by other than anti-cholinesterase medication*, the effect of which may not persist postoperatively, the risks are those of any patient without MG. Accordingly, the preoperative preparation of these patients is central to the success of surgery *(134)*. The patient should be as symptom-free as possible at the time of surgery, especially free of signs and symptoms of oropharyngeal and respiratory weakness, using plasmapheresis and immunosuppression when necessary, although its role is less well defined *(33, 49, 135, 136, 137)*. The concern about wound healing and infection associated with steroids is valid but probably overemphasized and much less of a concern than performing an operation on an inadequately prepared patient. On the other hand, cholinesterase inhibitors in the preoperative preparation of these patients should not be relied upon to achieve control of oropharyngeal and respiratory weakness because these inhibitors only temporarily mask the symptoms, which may then flare up in the early postoperative period complicating the care. Such patients can be expected to have a high rate of postoperative respiratory complications *(138, 139)*.

The preoperative evaluation should include a detailed evaluation of the pulmonary function status. Vital capacity (VC) and respiratory muscle force measurements are recommended, both before and after cholinergic inhibitors (1.5–2 mg neostigmine i.m. for adults in association with

0.4 mg atropine), if the patient is receiving this medication and can tolerate its withdrawal for the 6–8 h interval necessary for this type of testing. The dual "before" and "after" measurements gives an indication of the deficits that may be masked by the cholinesterase medication and the potential need for preoperative plasmapheresis or immunosuppression. The measurement of Maximum Expiratory Force (MEF) is easy to perform, is an excellent measure of cough effectiveness (an important determination), and is more sensitive and reliable than the VC in the evaluation of these patients, both preoperatively and in the early postoperative period *(140)*. An MEF of less than 40–50 cm H_2O indicates a potential for postoperative respiratory complications and respiratory failure. These determinations are also helpful in assessing the timing of extubation in those patients that require postoperative ventilatory support *(134, 140)*.

During surgery, many believe that muscle relaxants should be avoided if possible. However, it is preferable to face the potential need for prolonged postoperative ventilation, perhaps as a result of a muscle relaxant, than have the patient suffer the consequences of hypoxia. Accordingly, should intubation be difficult, whether during anesthesia induction or at anytime intubation is required, immediate muscle paralysis may be mandatory to achieve intubation rapidly and successfully. Curare can be employed, using 10% of the standard dose. The use of succinylcholine is somewhat controversial and can usually be avoided; however in emergencies, ordinary doses can also be used.

Postoperatively, these patients should be managed in an intensive care setting *(141, 142)* where they can be closely observed by experienced personnel. The use of epidural anesthesia for control of pain following a median sternotomy is extremely helpful. An institutional protocol for the management of the postoperative MG patient is helpful and can include details of the ventilatory support management, defining the role of MEF measurements in deciding when to extubate, the role of physical therapy and bronchoscopy in maintaining a clear airway, pain control techniques, immunosuppression if indicated, the role of cholinergic medication, and timing and technique of a two-staged tracheostomy if necessary.

If the patient is well prepared preoperatively and the preoperative MEF off cholinergic medication is satisfactory, regardless of the surgical technique employed and including trans*sternal* incisions, extubation may be acceptable immediately postoperatively. However, emergency re-intubation and respiratory support should be instituted immediately at any time for *early* signs of fatigue, progressive weakness, or impending respiratory failure. The use of cholinergic medication at such a time is usually ineffective and may delay needed intubation. The need for intubation may even occur after discharge from a special care unit. Under these circumstances the temptation is to await the availability of a special care bed. To delay may, however, be catastrophic; re-intubation may have to be performed as a true emergency before an ideal accommodation can be arranged, or even considered.

13. HOSPITAL MORBIDITY AND MORTALITY

If the MG patient is well prepared preoperatively, thymectomy should be a safe procedure and the mortality should be that of a similar operative procedure in a patient without MG. The recent operative mortality for thymectomy has been reported in the 0–2% range. In the 250 "maximal" thymectomies for MG performed at the Columbia Presbyterian Medical Center in NYC, there were no nerve injuries and one death in a 65-year-old severely ill patient (0.04%).

Clearly, the trans*cervical* and thoracoscopic approaches avoid the postoperative pain and unpleasant scar of the sternotomy. However, the sternotomy pain is now usually well controlled and the intercostal pain of the thoracoscopic procedures may persist for some time. In the evaluation of therapeutic options, in addition to the determination of the remission-improvement status, postoperative morbidity, complications directly related to each form of therapy, quality of life and cost–benefit assessment *(92)*, and the number and duration of subsequent hospitalizations and ICU stays, is required *(101)*.

If it is determined by prospective studies that the combined cervical-mediastinal or aggressive transsternal operations produce significantly improved results, especially complete stable remissions, and if they avoid multiple subsequent hospitalizations and the need for prolonged immunosuppressive therapy with its inherent risks, the initial disadvantages of the combined and aggressive transsternal procedures may be outweighed by their subsequent advantages.

14. OUTCOMES RESEARCH

Well-designed and well-controlled prospective studies are required to begin to resolve the many conflicting statements and unanswered questions that exist concerning thymectomy in the treatment of MG. This goal must be achieved if patient protocols and operative techniques are to be properly evaluated. In this era of evidenced-based therapy, these steps are not only desirable but essentially mandatory.

The ideal method of such evaluation is to undertake a prospective randomized clinical trial, Class I evidence in the American Academy of Neurology (AAN) nomenclature *(143)*. However, since the development of such a study comparing thymectomy techniques is not only unlikely but probably unnecessary, a prospective risk-adjusted outcome analysis of non-randomly assigned treatment *(144)*, Class II evidence in the AAN nomenclature, is an acceptable and achievable method of study that if properly controlled and carefully monitored should resolve many of the unresolved issues. To eliminate bias, the use of two or more surgical centers, comparing their respective preferences, is preferable to a single team performing and comparing two types of thymectomies, *We strongly recommended that one or more such Class II studies be undertaken.*

In addition to a prospective study, the use of clinical research standards, which include definitions of clinical classification, *quantitative* assessment of disease severity, grading systems of post-intervention status, and approved methods of analysis are required. The Data Bank concept, appropriately developed and rigorously monitored *(101)*, should be particularly useful and practical for multiple institutions to compare the relative value of the various thymectomy techniques.

The primary focus of comparative analysis of thymectomy for MG should remain complete stable remission. Remissions are not only the most reliable measure of success but the most desirable result from the patient standpoint. "Survival" instruments, which are used in the analysis of remissions, are the most reliable determinant. The Kaplan–Meier life table analysis is the technique of choice. Different levels of clinical improvement should also be *quantitatively* evaluated *(101, 145)*. The use of uncorrected crude data has no place in these analyses.

Qualify of life (QoL) instruments should also be employed because therapy for MG is usually not innocuous and frequently does not produce a completely stable remission. QoL measures evaluate the impact of intermediate levels of clinical improvement and morbidity of therapy and complement information provided by survival and clinical improvement analysis. *They should not, however, replace "survival" instrument evaluation.* Although a functional status instrument assessing activities of daily living has been developed for MG *(146)*, there are no disease-specific QoL instruments for MG at this time. They are needed and it is recommended that they be developed *(147)*.

Experts in the field of biostatistics and outcomes analysis should be *included* in the design of all studies, in the collection of the information, and in the evaluation of the data. An outcomes analysis guideline *(92)* and the accompanying diagram defining their inter-relationships (Fig. 4) give background information on the available analytical techniques.

Fig. 4. The Inter-relationship Between Outcomes Instruments. The inter-relationship between Survival-Improvements instruments, the Quality of Life instruments, and the Quality adjusted Survival-Cost Effective instruments is demonstrated. Remissions are a survival function. (Reprinted with permission from *Neurology* **55** (1):16–30, 2000. <http://www.neurology.org> Lippincott Williams & Wilkins).

15. RECOMMENDATIONS

We believe that at this time the *available evidence* supports the role of thymectomy in the treatment of MG. This recommendation and practice are bolstered by the modern-day ability to perform the procedure with a very low morbidity and mortality, thus fulfilling the basic surgical tenant of risk versus benefit. Given that recommendation and practice, we clearly understand the limits of the data to date and support the randomized trial underway comparing thymectomy to no thymectomy in non-thymomatous patients with MG receiving prednisone.

In terms of surgical approach, our bias at this time is toward some type of maximal or extended thymectomy. We believe this can be accomplished best by sternotomy or by the VATET technique. However, this practice paradigm must be viewed with the understanding that the published results to date do not clearly support any one particular approach and the use of transcervical thymectomies or unilateral thoracoscopic (VATS) thymectomies is performed by a number of accomplished thoracic surgeons.

In the final analysis, the onus is on the neurologic and thoracic surgical community to definitively prove the benefit of thymectomy in myasthenia gravis and determine the most appropriate thymectomy technique. Accordingly, we support the impending trial of thymectomy versus no thymectomy in patients receiving corticosteroids. In addition, we urge the thoracic surgical and neurologic community to undertake well-designed and well-supervised prospective studies as outlined above and as soon as practical, comparing the various thymectomy techniques.

16. EPILOGUE

Although arguments can undoubtedly be submitted to refute some of the statements herein, we hope that this presentation will lead to a better understanding of the thymectomy controversies and improved results following thymic resection for MG. We also hope, and expect, the day will come when some form of targeted immunosuppression or other non-surgical therapy, with no significant side effects, produces long-term remissions in patients with autoimmune non-thymomatous myasthenia gravis. At that time, thymectomy in any form, especially the transsternal procedures, will be considered barbaric. Until such time, however, a thymectomy, *properly performed*, should be considered an integral part of the therapy of MG.

Twenty-six years ago, Buckingham et al. made basically the same statement: "It seems likely that future discoveries will permit the management of MG without surgical intervention but, for the present, the beneficial role of thymectomy for these unfortunate patients has been clearly demonstrated" *(102)*. We agree.

REFERENCES

1. Jaretzki A, III. Thymectomy for myasthenia gravis: an analysis of the controversies regarding technique and results. Neurology 1997;48(Suppl. 5)(April):S52–S63.
2. Blalock A, Harvey AM, Ford FF, Lilienthal J, Jr. The treatment of myasthenia gravis by removal of the thymus gland. JAMA 1941;117(18):1529–1533.
3. Keynes G. The surgery of the thymus gland. Brit J Surg 1946;32:201–214.
4. Kirschner PA, Osserman KE, Kark AE. Studies in myasthenia gravis: transcervical total thymectomy. JAMA 1969;209(6):906–10.
5. Kark AE, Kirschner PA. Total thymectomy by the transcervical approach. Br J Surg 1971;56:321–326.
6. Faulkner SL, Ehyai AS, Fisher RD, Fenichel GM, Bender HW. Contemporary management of myasthenia gravis: the clinical role of thymectomy. Ann Thorac Surg 1977;23:348–352.
7. Olanow CW, Wechsler AS, Roses AD. A prospective study of thymectomy and serum acetylcholine receptor antibodies in myasthenia gravis. Ann Surg 1982;196:113–121.
8. Vincent A, Newsom-Davis J, Newton P, Beck N. Acetylcholine receptor antibody and clinical response to thymectomy in myasthenia gravis. Neurology 1983;33:1276–1282.
9. Hankins JR, Mayer RF, Satterfield JR, Turney SZ, Attar A, Sequeira AJ, et al. Thymectomy for myasthenia gravis: 14 year experience. Ann Surg 1985;201:618–625.
10. Otto TJ, Strugalska H. Surgical treatment for myasthenia gravis. Thorax 1987;42:199–204.
11. Cooper J, Jaretzki A, III, Mulder DG, Papatestas AS. Symposium: thymectomy for myasthenia gravis. Contemp Surg 1989;34(June):65–86.
12. Papatestas AE, Cooper J, Jaretzki A, III, Mulder DG. Symposium: thymectomy for myasthenia gravis. Contemp Surg 1989;34(June):65–85.
13. Verma P, Oger J. Treatment of acquired autoimmune myasthenia gravis: a topic review. Can J Neurol Sci 1992;19:360–375.
14. Drachman DB. Myasthenia gravis. N Engl J Med 1994;330(25):1797–1810.
15. Penn AS. Thymectomy for myasthenia gravis. In: Handbook of Myasthenia Gravis and Myasthenic Syndromes. RP Lisak, ed. 1994;Marcel Dekker, Inc, New York:321–339.
16. Mack MJ, Landreneau RD, Yim AP, Hazelrigg SR, Scruggs GR. Results of video-assisted thymectomy in patients with myasthenia gravis. J Thor Cardiovasc Surg 1996;112(5):1352–1360.
17. Penn AS, Jaretzki A, III, Wolff M, Chang HW, Tennyson V. Thymic abnormalities: antigen or antibody? Response to thymectomy in myasthenia gravis. Ann NY Acad Sci 1981;377:786–803.
18. Wekerle H, Muller-Hermelink HK. The thymus in myasthenia gravis. Curr Top Pathol 1986;75:179–206.
19. Younger DS, Worrall BB, Penn AS. Myasthenia gravis: historical perspective and overview. Neurol Suppl 1997;48(Suppl 5):S1–S17.
20. Penn AS, Lovelace RE, Lange DJ, et al. Experimental myasthenia gravis in neonatally thymectomized rabbits. Neurology 1977;27(April):365.
21. Masaoka A, Monden Y, Seike Y, Tanioka T, Kagotani K. Reoperation after transcervical thymectomy for myasthenia gravis. Neurology 1982;32:83–85.
22. Rosenberg M, Jauregui WO, Vega MED, Herrera MR, Roncoroni AJ. Recurrence of thymic hyperplasia after thymectomy in myasthenia gravis: its importance as a cause of failure of surgical treatment. Am J Med 1983;74:78–82.
23. Henze A, Biberfeld P, Christensson B, Matell G, Pirskanen R. Failing transcervical thymectomy in myasthenia gravis, an evaluation of transsternal re-exploration. Scand J Thor Cardiovasc Surg 1984;18:235–238.
24. Rosenberg M, Jauregui WO, Herrera MR, Roncoroni A, Rojas OR, Olmedo GSM. Recurrence of thymic hyperplasia after trans-sternal thymectomy in myasthenia gravis. Chest 1986;89(6):888–891.
25. Jaretzki A, III, Penn AS, Younger DS, Wolff M, Olarte MR, Lovelace RE, et al. "Maximal" thymectomy for myasthenia gravis: results. J Thorac Cardiovasc Surg 1988;95:747–757.
26. Miller RG, Filler-Katz A, Kiprov D, Roan R. Repeat thymectomy in chronic refractory myasthenia gravis. Neurology 1991;41:923–924.
27. Kornfeld P, Merav A, Fox S, Maier K. How reliable are imaging procedures in detecting residual thymus after previous thymectomy. Ann NY Acad Sci 1993;681:575–576.
28. Monden MAY. Comparison of the results of transsternal simple, transcervical simple and extended thymectomy. Ann NY Acad Sci 1981;377:755–764.
29. Matell G, Lebram G, Osterman PO, Pirskanen R. Follow up comparison of suprasternal vs transsternal method for thymectomy in myasthenia gravis. Ann NY Acad Sci 1981;377:844–845.
30. Masaoka A, Nagaoka Y, Kotake Y. Distribution of thymic tissue at the anterior mediastinum: current procedures in thymectomy. J Thorac Cardiovasc Surg 1975;70(4):747–754.
31. Jaretzki A, III, Wolff M. "Maximal" thymectomy for myasthenia gravis: surgical anatomy and operative technique. J Thorac Cardiovasc Surg 1988;96:711–716.

32. Fukai I, Funato Y, Mizuno Y, Hashimoto T, Masaoka A. Distribution of thymic tissue in the mediastinal adipose tissue. J Thorac Cardiovasc Surg 1991;101:1099–1102.

33. Bulkley GB, Bass KN, Stephenson R, Diener-West M, George S, Reilly PA, et al. Extended cervicomediastinal thymectomy in the integrated management of myasthenia gravis. Ann Surg 1997;226(3), Sept:324–335.

34. Clark RE, Marbarger JP, West PN, Sprat JA, Florence JM, Roper CL, et al. Thymectomy for myasthenia gravis in the young adult. J Thorac Cardiovasc 1980;80:696–701.

35. Jaretzki A, III. Transcervical-transsternal "maximal" thymectomy for myasathenia gravis (Video). Society of Thoracic Surgeons' Film Library 1996;P.O. Box 809285, Chicago, IL 60680–9285.

36. Lennquist S, Andaker L, Lindvall B, Smeds S. Combined cervicothoracic approach in thymectomy for myasthenia gravis. Acta Chir Scand 1990;156:53–61.

37. Andersen M, Jorgensen J, Vejlsted H, Clausen PP. Transcervical thymectomy in patients with non-thymomatous myasthenia gravis. Scand J Thorac Cardiovasc Surg 1986;20:233–235.

38. Kagotani K, Monden Y, Nakaha K, Fujii Y, Seike Y, Kitamura S, et al. Anti-acetylcholine receptor antibody titer with extended thymectomy in myasthenia gravis. J Thorac Cardiovasc Surg 1985;90:7–12.

39. Monden Y, Nakahara K, Fujii Y, Hashimoto J, Ohno K, Masaoka A, et al. Myasthenia gravis in elderly patients. Ann Thor Surg 1985;39:433–436.

40. Mulder DG, Graves M, Hermann C, Jr. Thymectomy for myasthenia gravis: recent observations and comparison with past experience. Ann Thorac Surg 1989;48:551–555.

41. Fujii N, Itoyama Y, Machi M, Goto I. Analysis of prognostic factors in thymectomized patients with myasthenia gravis: correlation between thymic lymphoid cell subsets and postoperative clinical course. J Neurol Sci 1991;105(2):143–149.

42. Nussbaum MS, Rosenthal GJ, Samaha FJ, Grinvalsky HT, Quinlan JG, Schmerler M, et al. Management of myasthenia gravis by extended thymectomy with anterior mediastinal dissection. Surgery 1992;112:681–688.

43. Frist WH, Thirumalai S, Doehring CB, Merrill WH, Stewart JR, Fenichel GM, et al. Thymectomy for the myasthenia gravis patient: factors influencing outcome. Ann Thor Surg 1994;57:334–338.

44. Masaoka A, Yamakawa Y, Niwa H, Fukai I, Condo S, Kobayashi M, et al. Extended thymectomy for myasthenia gravis: a 20 year review. Ann Thor Surg 1996;62(Sept):853–859.

45. Detterbeck FC, Scott WW, Howard JF, Egan TM, Keagy BA, Starck PJK, et al. One hundred consecutive thymectomies for myasthenia gravis. Ann Thorac Surg 1996;62:242–245.

46. Mulder DG. Extended transsternal thymectomy. Chest Surg Clin N Am 1996;6(1):95–105.

47. Fischer JE, Grinvalski HT, Nussbaum MS, Sayers HK, Cole RE, Samaha FJ. Aggressive surgical approach for drug-free remission from myasthenia gravis. Ann Surg 1987;205:496–503.

48. Hatton P, Diehl J, Daly BDT, Rheinlander HF, Johnson H, Shrader JB, et al. Transsternal radical thymectomy for myasthenia gravis: a 15-year review. Ann Thorac Surg 1989;47:838–840.

49. Mulder DM. Extended Transsternal Thymectomy. General Thoracic Surgery. TW Shields ed.. 5th edition. 2000; Lippincott Williams & Wilkins. Philadelphia, PA (Chapter 171):2233–2237.

50. Steinglass KM. Extended Transsternal Thymectomy for Myasthenia Gravis (Videocassette). The Society of Thoracic Surgeons Chicago, IL. 2002.

51. Blalock A, Mason MF, Morgan HJ, Riven SS. Myasthenia gravis and tumors of the thymus region: report of a case in which the tumor was removed. Am Surg 1939;110(4):544–561.

52. Clagett OT, Eaton LM. Surgical treatment of myasthenia gravis. J Thorac Surg 1947;16:62–80.

53. Blalock A. Thymectomy in the treatment of myasthenia gravis: report of 20 cases. J Thorac Surg 1944;13:1291–1308.

54. Olanow CW. The surgical management in myasthenia gravis. In: Surgery of the Chest. 6th edition. Sabiston DC Jr, Spencer FC, eds. 1995;WB Saunders Co, Philadelphia, PA.

55. LoCicero J, III. The combined cervical and partial sternotomy approach for thymectomy. Chest Surg Clin N Am 1996;6(1):85–93.

56. Trastek VF. Thymectomy. Mastery of Cardiothoracic Surgery. LR Kaiser, IL Krin, TL Spray eds. 1998;Lippincott-Raven, Philadelphia, PA:105–111.

57. deCampos JRM, Filomeno LTB, Marchiori PE, Jatene FB. Partial sternotomy approach to the thymus. In Minimally Access Cardiothoracic Surgery. Yim, AP, Hazelrigg, SR, Izzat, B, Landreneau, RJ, Mack, MJ, Naunheim, KS, eds. 1999;WB Saunders Co, Philadelphia, PA (Chapter 27):205–208.

58. Granone P, Margaritora S, Cesario A, Galetta D. Thymectomy (Transsternal) in myasthenia gravis via video-assisted infra-mammary cosmetic incision. Eur J Cardiothorac Surg 1999;15(6):861–863.

59. Olanow CW, Wechsler AS. The surgical management of myasthenia gravis. In: Gibbon's Surgery of the Chest. 4th edition. DC Sabiston & FC Spencer, ed. 1983;WB Saunders Co, Philadelphis:849–869.

60. Olanow CW, Wechsler AS. The surgical management in myasthenia gravis. In: Textbook of Surgery. 14th edition. D.C. Sabiston Jr., ed. 1991;WB Saunders Co, Philadelphia, PA:1801–1814.

61. Trastek VF, Pairolero PC. Standard thymectomy. In: Mediastinal Surgery. TW Shields, ed. 1991;Lea & Febiger, Philadelphia (Chapter 39):365–368.

62. Trastek VF, Pairolero PC. Surgery of the thymus gland. In: General Thoracic Surgery, 4th Ed. TW Shields, ed. 1994; William & Wilkins Baltimore:1770–1781.

63. Olanow CW, Wechsler AS, Sirotkin-Roses M, Stajich J, Roses AD. Thymectomy as primary therapy in myasthenia gravis. Ann NY Acad Sci 1987;505:595–606.

64. Wilkins EW, Jr. Thymectomy. Modern Techn in Surg 1981;38:1–13.

65. Cooper J, Al-Jilaihawa A, Pearson F, Humphrey J, Humphrey HE. An improved technique to facilitate transcervical thymectomy for myasthenia gravis. Ann Thorac Surg 1988;45:242–247.

66. Bril V, Kojic S, Ilse W, Cooper J. Long-term clinical outcome after transcervical thymectomy for myasthenia gravis. Ann Thorac Surg 1998;65:1520–1522.

67. Shrager JB, Deeb ME, Mick R, Brinster CJ, Childers HE, Marshall MB, et al. Transcervical thymectomy for myasthenia gravis achieves results comparable to thymectomy by sternotomy. Ann Thorac Surg 2002;74:320–327.

68. Klingen G, Johansson L, Westerholm CJ, Sundstroom C. Transcervical thymectomy with the aid of mediastinoscopy for myasthenia gravis: eight years, experience. Ann Thorac Surg 1977;23:342–347.

69. Durelli L, Maggi G, Casadio C, Ferri R, Rendine S, Bergamini L. Actuarial analysis of the occurrence of remission following thymectomy for myasthenia gravis in 400 patients. J Neurol Neurosurg Psychiatry 1991;54(5):406–411.

70. Maggi G, Cristofori RC, Marzio PD, Pernazza F, Turello D, Ruffini E. Transcervical thymectomy with partial sternal split. Osp Ital Chir 2004;10(1):53–60.

71. dePerrot M, Bril V, McRae K, Keshavjee S. Impact of minimally invasive trans-cervical thymectomy on outcome in patients with myasthenia gravis. Eur J Cardiothoracic Surg 2003;24(5):677–683.

72. Bramis J, Diamantis T, Tsigris C, Pikoulis E, Papaconstaninou I, Nikolaou A, et al. Video-assisted transcervical thymectomy. Surg Endosc 2004;18(10):1535–1538.

73. Zielinski M, Kuzdzal J, Szlubowski A, Soja J. Transcervical-subxiphoid-videothoracoscopic "Maximal" thymectomy – Operative Technique and early results. Ann Thorac Surg 2004;78(1):399–403.

74. Kirby TJ, Ginsberg RJ. Transcervical thymectomy. In General Thoracic Surgery, 6th edition. Shields T. MD, LoCicero, J. MD, Ponn, R. MD ed. 2005;Lippincott Williams & Wilkins. Philadelphia, PA. 2(Section XXIX, Chapter 177):2634–2637.

75. Kark AE, Papatestas AE. Some anatomic features of the transcervical approach for thymectomy. Mt Sinai J Med 1971;38:580–584.

76. Papatestas AE, Genkins G, Kornfeld P, Horowitz S, Kark AE. Transcervical thymectomy in myasthenia gravis. Surg Gynecol Obst 1975;140(April):535–540.

77. Sugarbaker DJ. Thoracoscopy in the management of anterior mediastinal masses. Ann Thor Surg 1993;56(3):653–656.

78. Roviaro G, Rebuffat C, Varoli F, Vergani C, Maciocco M, Scalambra SM. Videothoracoscopic excision of mediastinal masses: indications and technique. Ann Thorac Surg 1994;58:1679–1684.

79. Sabbagh MN, Garza JJS, Patten B. Thoracoscopic thymectomy in patients with myasthenia gravis. Muscle and Nerve 1995;18(12):1475–1477.

80. Yim APC, Kay RLC, Ho JKS. Video-assisted thoracoscopic thymectomy for myasthenia gravis. Chest 1995;108(5):1440–1443.

81. Mack MJ, Scruggs GR. Video-assisted thymectomy. In General thoracic surgery. 5th edition. Shields T. MD, LoCicero, J. MD, Ponn, R. MD ed. 2000;Lippincott Williams & Wilkins, Philadelphia, PA 2 (Section XXVIII, Chapter 173):2243–2249.

82. Mineo TC, Pompeo E, Ambrogi V, Sabato AF, Bernardi G, Casciani CU. Adjuvant pneumomediastinum in thoracoscopic thymectomy for myasthenia gravis. Ann Thor Surg 1996;62:1210–1212.

83. Novellino L, Longoni M, Spinelli L, Andretta M, Cozzi M, Faillace G, et al. "Extended" thymectomy, without sternotomy, performed by cervicotomy and thoracoscopic technique in the treatment of myasthenia gravis. Int Surg 1994;79(4):378–381.

84. Scelsi R, Ferro MT, Scelsi L, Novellino L, Mantegazza R, Cornelio F, et al. Detection and morphology of thymic remnants after video-assisted thoracoscopic thymectomy (VATET) in patients with myasthenia gravis. Int Surg 1996;81:14–17.

85. Shigemura N, Shiono H, Inour M, Minami M, Ochta M, Okumura M, et al. Inclusion of the transcervical approach in video-assited thoracoscopic extended thymectomy (VATET) for myasthenia gravis: a prospective trial. Surg Endosc 2006;June 22, prepub.

86. Morgan JA, Ginsburg ME, Sonett JR, Morales DLS, Kohmoto T, Gorenstrin LA, et al. Advanced thoracoscopic procedures are facilitated by computer aided robotic technology. Eur J Cardiothoracic Surg 2003;23:883–887.

87. Ashton RC, McGinnis KM, Connery CP, Swistel DS, Ewing DR, DeRose JJ. Total endoscopic robotic thymectomy for myasthenia gravis (case report). Ann Thorac Surg 2003;75:569–571.

88. Kernstine KH. Robotic Thymectomy for Myasthenia Gravis. Personal Communication 2003.

89. Rea F, Marulli G, Bortolotti L, Feltracco P, Zuin A, Sartori F. Experience with the "Da Vinci" robotic system for thymectomy in patients with myasthenia gravis: report of 33 cases. Ann Thorac Surg 2006;81:455–459.

90. Hsu CP, Chuang CY, Hsu NY, Chen CY. Comparison between the right side and subxiphoid bilateral approaches in performing video-assisted thoracoscopic extended thymectomy for myasthenia gravis. Surg Endosc 2004;18(5):821–824.

91. Uchiyama A, Shimizu S, Murai H, Kuroki S, Okido M, Tanaka M. Infrasternal mediastinoscopy thymectomy in myasthenia gravis: surgical results in 23 patients. Ann Thor Surg 2001;72:1902–1905.

92. Weinberg A, Gelijns A, Moskowitz A, Jaretzki A. Myasthenia gravis: outcomes analysis. <<www.myasthenia.org/research/Clinical_Research_Standards.htm> (Appendix – Outcome Analysis.doc) Reprints: Myasthenia Gravis Foundation of America, Inc. 1821 University Avenue West (Suite S256), St. Paul, MN-55104–2897. 2000.

93. Lawless JF. Life tables, graphs, and related procedures. In: Statistical Models and Methods for Life Table Data. By J F Lawless, ed. John Wiley & Son Inc, New York 1982:52–94.

94. Olak J, Chiu RCJ. A surgeon's guide to biostatistical inferences. Part II. ACS Bull 1993;78(2):27–31.

95. Blackstone EH. Outcome analysis using hazard function methodology. Ann Thorac Surg 1996;61:S2–S7.

96. McQuillen MP, Leone MG. A treatment carol: thymectomy revisited. Neurology 1977;12(77):1103–1106.

97. Viet HR, Schwab RS. Thymectomy for myasthenia gravis. In: Records of Experience of Mass Gen Hosp. Veit HR, ed. 1960; Charles C Thomas, Springfield, IL:597–607.

98. Oosterhuis HJ. Observations of the natural history of myasthenia gravis and the effect of thymectomy. Ann NY Acad Sci 1981;377:678–689.

99. Sanders DB, Kaminski HJ, Jaretzki A, III, Phillips L, H II. Thymectomy for myasthenia gravis in older patients (Letter to Editor). J Am Coll Surg 2001;193(3):340–341.

100. Sonett JR. Thymectomy for Myasthenia Gravis. Difficult Decisions in Thoracic Surgery – An Evidence-Based Approach. MK Freguson, ed. 2007;London, Springer:469–473.

101. TaskForce M, Jaretzki A, III,Chair, Barohn RB, Ernstoff RM, Kaminski HJ, Keesey JC, et al. Myasthenia gravis: recommendations for clinical research standards. Neurology 2000;55:16–23.

102. Buckingham JM, Howard FM, Bernatz PE. The value of thymectomy in myasthenia gravis: a computer-assisted matched study. Ann Surg 1976;184:453–458.

103. Gronseth S, Barohn RJ. Practice parameters: thymectomy for autoimmune myasthenia gravis (an evidence base review). Neurology 2000;55:7–15.

104. Wolfe GI, Kaminski HJ, Jaretzki A, III, Swan A, Newsom-Davis J. Development of a thymectomy trial in nonthymomatous myasthenia gravis patients receiving immunosuppressive therapy. Ann NY Acad Sci 2003;998(Aug):473–480.

105. Rodriguez M, Gomez MR, Howard F, Taylor WF. Myasthenia gravis in children: long term follow-up. Ann Neurol 1983;13:504–510.

106. Papatestas AE, Genkins G, Kornfield P, Eisenkraft JB, Fagerstrom RP, Posner J, et al. Effects of thymectomy in myasthenia gravis. Ann Surg 1987;206(1):79–88.

107. Lindberg C, Andersen O, Larsson S, Oden A. Remission rate after thymectomy in myasthenia gravis when the bias of immunosuppressive therapy is eliminated. Acta Neurol Scand 1992;86(3):323–328.

108. Rubin JW, Ellison RG, Moore HV, Pai GP. Factors affecting response to thymectomy for myasthenia gravis. J Thorac Cardiovasc Surg 1981;82(5):720–728.

109. Jaretzki A, III, Steinglass KM, Sonett JR. Thymectomy in the management of myasthenia gravis. Semin Neurol 2004;24(1):121–134.

110. Manlula A, Lee TW, Wan I, Law CY, Chang C, Garzon JC, et al. Video-assisted thoracic surgery thymectomy for nonthymomatous myasthenia gravis. Chest 2005;128(5):3454–3460.

111. Novellino L, Spinelli L, Albani AP, Cirelli B, Mancin A, Bettonalgi M, et al. Thymectomy by cervicotomy and bilateral thoracoscopy to treat non-thymomatous myasthenia. Osp Ital Chir 2004;10:61–74.

112. Lanska D. Indications for thymectomy in myasthenia gravis. Neurology 1990;40:1828–1829.

113. Keesey JC. A treatment algorithm for autoimmune myasthenia in adults. Ann NY Acad Sci 1998;841:753–768.

114. TaskForce MG. Recommendations for clinical research standards. www.myasthenia.org/clinical/research/standards2002; www.myasthenia.org/research/MGFA_Standards.pdf.

115. Daroff RB. Ocular myasthenia. In Myasthenia Gravis and Related Disorders. Kaminski HJ, ed. 2003;Humana Press, Totowa, NJ:115–128.

116. Schumm F, Wietholter H, Fateh-Mogbadan A, Dichgans J. Thymectomy in myasthenia with pure ocular symptoms. J Neurol Neurosurg Psych 1985;48(4):332–337.

117. Evoli A, Batocchi AP, Minisci C, DiShino C, Tonali P. Therapeutic options in ocular myasthenia gravis. Neuromuscul Disord 2001;11(2):208–216.

118. Monden Y, Nakahara K, Kagotani K, Fujii Y, Nanjo S, Masaoka A, et al. Effects of preoperative duration of symptoms on patients with myasthenia gravis. Ann Thorac Surg 1984;38:287–291.

119. DeFilippi VJ, Richman DP, Ferguson MK. Transcervical thymectomy for myasthenia gravis. Ann Thor Surg 1994;57:194–197.

120. Kaminski HJ. Treatment of myasthenia gravis. In Myasthenia Gravis and Related Disorders. Kaminski HJ, ed. 2003; Humana Press, Totowa, NJ:197–221.

121. Adams C, Theodorescu D, Murphy E, Shandling B. Thymectomy in juvenile myasthenia gravis. J Child Neurol 1990;5:215–218.

122. Andrews PI, Massey JM, Howard JF, Sanders DB. Race, sex, and puberty influence onset, severity, and outcome in juvenile myasthenia gravis. Neurology 1994;44:1208–1214.

123. Andrews PI. A treatment algorithm for autoimmune myasthenia gravis in children. Ann NY Acad Sci 1998;841:789–802.

124. Seybold ME. Thymectomy in childhood myasthenia gravis. Ann NY Acad Sci 1998;841:731–741.

125. Kirschner PA. Reoperation on the thymus. A critique. Chest Surg Clin N Am 2001;11(2):439–445.

126. Wekerle H, Muller-Hermelink H. The thymus in myasthenia gravis. Current Topics in Pathology. The Human Thymus. Hispphysiology and Pathology. Muller-Hermelink, HK, ed. 1986; Springer-Verlag Co, New York:75.

127. Ellis K, Austin JHM, Jaretzki A, III. Radiologic detection of thymoma in patients with myasthenia gravis. Am J Radiol 1988;151:873–881.

128. Lovelace RE, Younger DS. Myasthenia gravis with thymoma. Neurology 1997;48(Suppl. 5):S76–S81.

129. Shields TW. Thymic Tumors. General Thoracic Surgery. 5th Edition. 2000;Lippincott Williams & Wilkins, Philadelphia, PA; 2:2181–2205.

130. Mussi A, Lucchi M, Murri L, Ricciardi R, Luchini L, Angeletti CA. Extended thymectomy in myasthenia gravis: a team-work of neurologist, thoracic surgeon, and anaesthetist may improve the outcome. Eur J Cardio-thorac Surg 2001;19:570–575.

131. Houghton A. Variation in outcome of surgical procedures. Br J Surg 1994;81:653–660.

132. Crawford FA, Anderson RP, Clark RE, Grover FL, Kouchoukos NT, Waldhausen JA, et al. Volume requirements for cardiac surgery credentialing: a critical examination. Ann Thorac Surg 1996;61:12–16.

133. Silvestri GA, Handy J, Lackland D, Corley E, Reed CE. Specialists achieve better outcomes than generalists for lung cancer surgery. Chest 1998;114:675–680.

134. Krucylak PE, Naunheim KS. Preoperative preparation and anesthetic management of patients with myasthenia gravis. Sem in Thorac and Cardiovasc Surg 1999;11:47–53.

135. d'Empaire G, Hoaglin DC, Perlo VP, Pontoppidan H. Effect of prethymectomy plasma exchange on postoperative respiratory function in myasthenia gravis. J Thorac Cardiovasc Surg 1985;89:592–596.

136. Cumming WJK, Hudgson P. The role of plasmapheresis in preparing patients with myasthenia for thymectomy. Muscle Nerve 1986;9:S155–S158.

137. Gotti P, Spinelli A, Marconi G, Duranti R, Gigliotti F, Pizzi A, et al. Comparative effects of plasma exchange and pyridostigmine on expiratory muscle strength and breathing pattern in patients with myasthenia gravis. Thorax 1995;50:1080–1086.

138. Kas J, Kiss D, Simon V, Svastics E, Major L, Szobor A. Decade-long experience with surgical therapy of myasthenia gravis: early complications of 324 transsternal thymectomies. Ann Thorac Surg 2001;72:1691–1697.

139. Jaretzki A, III, Phillips L, II, Aarli J, Kaminski HJ, Sanders DB. Preoperative preparation of patients with myasthenia gravis forstalls postoperative respiratory complications following thymectomy – Letter to Editor. Ann Thor Surg 2003;75:1068–1069.

140. Younger DS, Braun NMT, Jaretzki A, III, Penn AS, Lovelace AE. Myasthenia gravis: determinants for independent ventilation after transsternal thymectomy. Neurology 1984;34:336–340.

141. Mayer SA. Intensive care of the myasthenic patient. Neurology 1997;48(Suppl. 5):S70–S81.

142. Juel VC. Myasthenia gravis: management of myasthenic crisis and perioperative care. Semin Neurol 2004;24:75–81.

143. Miller RG, Rosenberg JA, Force APPT. Care of patients with ALS: report of the Quality Standards Subcommittee of the AAN. Neurol 1999;52:1311–1323.

144. Kirklin JW, Blackstone EH. Clinical studies with non-randomly assigned treatment. In Cardiac Surgery. 2nd edition. Kirklin, JW, Barrett-Boyes, BG eds. 1993;Churchill-Livingstone, New York (Chapter 6):269–270.

145. Heagerty PJ, Zeger SL. Marginal regression models for clustered ordinal measurements. J Am Stat Assoc 1996;91(435):1024–1036.

146. Wolfe GI, Herbelin L, Nations SP, Foster B, Bryan WW, Barohn RJ. Myasthenia gravis activities of daily living profile. Neurology 1999;52:1487–1489.

147. Jaeschke R, Guyatt GH. How to develop and validate a new quality of life instrument. Quality of Life Assessments in Clinical Trials, Spilker B, ed. 1990;Raven Press, Ltd. New York (Chapter 5):47–57.

148. Park IK, Choio SS, Lee JG, Kim DJ, Chung KY. Complete stable remission after extended transsternal thymectomy in myasthenia gravis. Eur J Cardiothoracic Surg 2006;30:525–528.

13

Lambert–Eaton Syndrome

Charles M. Harper and Vanda A. Lennon

1. INTRODUCTION

The Lambert–Eaton syndrome (LES) is an autoimmune presynaptic disorder of peripheral cholinergic neurotransmission characterized by proximal weakness and mild autonomic dysfunction. Approximately 60% of cases are associated with small-cell lung carcinoma (SCLC), which can appear before or after the clinical manifestations of LES. The P/Q-type voltage-gated calcium channel in motor nerve terminals, classified by structure and function as $Ca_v2.1$ *(1)*, is the presumptive target of antibodies that cause LES. Antigenically similar calcium channels are expressed in the surface membranes of SCLC tumor cells. When neuronal $Ca_v2.1$ channels are functionally impaired, nerve-stimulated quantal release of the neurotransmitter acetylcholine (ACh) is reduced. This lowers the safety margin for synaptic transmission at the neuromuscular junction and in certain autonomic nerve terminals, leading to weakness at rest that is temporarily improved by exertion, as well as xerostomia, impotence and hypohidrosis.

2. HISTORY

The first reported association of a myasthenic syndrome with lung cancer is commonly attributed to Anderson, Churchill-Davidson and Richardson in 1953 *(2)*. This trio of English clinicians described a patient with bronchial carcinoma who presented with generalized proximally predominant weakness, diplopia and dysphagia. No electrophysiological abnormality was found. A neurological diagnosis of myasthenia was made on the basis of a positive edrophonium test and therapeutic benefit of neostigmine. The patient may indeed have had myasthenia gravis *(3, 4, 5)*, which has been reported, albeit rarely, with lung carcinomas of small-cell and non-small-cell types *(4)*. Lambert was the first to describe the presynaptic myasthenic syndrome that later bore his name. Beginning in 1956, he and colleagues at the Mayo Clinic in Minnesota described in a series of papers the clinical and electrodiagnostic features of a unique "myasthenic syndrome associated with malignant tumors" *(6, 7, 8, 9, 10)*. The characteristic findings were facilitation of both muscle strength and amplitude of the compound muscle action potential (CMAP) after exercise or high-frequency electrical stimulation. Elmqvist and Lambert *(11, 12)* demonstrated the presynaptic origin of LES by using microelectrodes to study biopsied intercostal nerve-muscle preparations. They documented a severe reduction in endplate potential (EPP) amplitude and quantal content, with preservation of the miniature endplate potential (MEPP) amplitude.

Progress in understanding the pathogenesis and in the treatment of LES has continued over the half-century since its discovery. Studies linking LES with a high frequency of autoimmune disorders and autoantibodies were first published in the 1970s and early 1980s *(13, 14)*. In 1982, the Mayo Clinic team of Fukunaga, Engel, Osame and Lambert demonstrated that the active zones for vesicle release in presynaptic nerve terminals were disorganized in LES patients

Dedicated to the memory of Edward H. Lambert, MD, PhD, 1915–2003

From: *Current Clinical Neurology*
Myasthenia Gravis and Related Disorders
Edited by: H.J. Kaminski, DOI 10.1007/978-1-59745-156-7_13, © Humana Press, New York, NY

(15). This observation coincided with recognition that those transmembrane particles were voltage-gated calcium channels, which play a critical role in ACh release *(16)*. The autoimmune nature of LES was reinforced by Lang, Newsom-Davis and colleagues at Oxford University, who were the first group to document transfer of the microelectrode findings of LES to mice using IgG from serum of LES patients *(17)*. This observation was confirmed by Kim *(18)* and extended by the Oxford and Mayo groups who demonstrated that mice receiving LES-IgG had a loss of presynaptic membrane-active zones *(19)* and developed muscle weakness and characteristic EMG findings of LES *(20)*. By demonstrating the interaction of LES serum antibodies with calcium channels expressed in cultured lung cancer cells, the Oxford group were able to link SCLC pathogenically to LES *(21)*. Development of Ca_v channel subtype-specific ligands and sensitive radioimmunoassays advanced understanding of the basic pathophysiology of LES by enabling its serological diagnosis *(22, 23, 24, 25, 26)*. Treatment of LES was improved by (1) advances in imaging and treatment of small-cell lung carcinoma *(27)*, (2) immunomodulatory therapy *(28, 29, 30, 31)* and (3) drugs that promote the quantal release of ACh *(32, 33, 34, 35)*.

3. PATHOGENESIS

3.1. *Physiology of Neuromuscular Transmission*

Normal neuromuscular transmission ensures the generation of a single muscle fiber action potential in response to each motor nerve action potential firing at rates of 50 Hz or greater. The safety margin of neuromuscular transmission is the "excess" voltage of the EPP above the level required to reach the threshold for muscle action potential activation. Postsynaptic factors that affect the size of the EPP (and therefore the size of the safety margin) are the number of functioning nicotinic acetylcholine receptors (AChRs) located on the crests of the folds of the muscle's postsynaptic membrane, the presence of acetylcholinesterase in the basal lamina and the three-dimensional configuration of the synapse *(36)*. Presynaptic factors that affect the size of the EPP are the number of ACh-containing vesicles in the active zones of the nerve terminal and the number of functional $Ca_v2.1$ channels in the nerve terminal membrane. In normal as well as diseased endplates, the EPP amplitude falls with slow rates of repetitive nerve stimulation (2–5 Hz) due to depletion of immediately available stores of ACh *(37)*. When the EPP amplitude drops below threshold, the muscle fiber action potential is blocked. The safety margin of neuromuscular transmission is large enough to prevent blocking at normal endplates *(37, 38, 39, 40)*. The characteristic abnormality observed in neuromuscular junction disease of either presynaptic or postsynaptic origin is blocking of neurotransmission at multiple endplates. This produces fatigable weakness and correlates with increased jitter and blocking on single fiber EMG studies, motor unit potential (MUP) amplitude variation on standard concentric needle EMG and a decrement of the compound muscle action potential (CMAP) observed at slow rates of repetitive stimulation on nerve conduction studies *(38, 39, 40)*.

During exercise or rapid rates of repetitive nerve stimulation, the mobilization of ACh stores and influx of Ca^{2+} into the nerve terminal through the $Ca_v2.1$ channel temporarily improve the safety margin of neuromuscular transmission (post-activation facilitation). This is followed by several minutes of reduced safety margin below baseline levels (post-activation exhaustion). A decrement of the EPP or CMAP observed at slow rates of repetitive nerve stimulation is often repaired by brief exercise or by increasing the rate of repetitive stimulation above 10 Hz, but depending on severity of the underlying disorder, the decrement invariably returns and increases for several minutes. In patients with a severe disorder of neuromuscular transmission, the EPP fails to reach threshold in a majority of fibers even at rest or after a single nerve stimulus. This results in objective weakness and a reduction of the CMAP amplitude below the lower limit of normal at rest. In these cases brief exercise or repetitive stimulation at a rapid rate produces a transient increase in the EPP above threshold and temporary facilitation of the CMAP.

The mechanism by which Ca^{2+} facilitates the release of ACh from the motor nerve terminal has been partially elucidated *(16, 37, 38)*. The $Ca_v2.1$ channels are arranged within the terminal membrane in double parallel rows that lie immediately adjacent to the crests of the postsynaptic folds. This brings the ACh vesicle release point in close proximity to the area of postsynaptic membrane where AChR concentration is the greatest. Entry of Ca^{2+} through the $Ca_v2.1$ channel is triggered by nerve terminal depolarization. Prolongation of the time constant of depolarization (e.g., by blocking voltage-dependent potassium channels using 3,4-diaminopyridine) increases the amount of Ca^{2+} influx. The abrupt local rise in Ca^{2+} concentration in the nerve terminal is "sensed" by active zone proteins exemplified by synaptotagmin, a vesicular membrane soluble *N*-ethylmaleimide-sensitive factor attachment protein receptor (v-SNARE) protein. The binding of synaptotagmin to the "synaptic core complex" (formed by association of vesicular synaptobrevin and nerve terminal proteins syntaxin-1 and SNAP-25) destabilizes the vesicular and nerve terminal membranes creating a "fusion pore" leading to exocytosis of ACh. The $Ca_v2.1$ channel itself is a nerve terminal target membrane (t-SNARE) protein interacting noncovalently with the "synaptic core complex" and with soluble cytoplasmic factors, including NSF and α-SNAP *(37, 38)*. Following exocytosis, vesicle membranes are recycled from the plasma membrane by clathrin-dependent and independent processes. Synaptotagmin plays a role in the former, which involves sequential interaction of dynamin and amphiphysin with other highly conserved cytoplasmic proteins *(16, 37, 38)*.

Neuronal $Ca_v2.1$ channels consist of five subunits *(41, 42)*. The α1 subunit is largest and contains the voltage sensor and cation pore. It has four nearly identical domains, each with six transmembrane segments. The M4 segment contains the voltage sensor, and loops between M5 and M6 segments in each domain form the ion channel and determine Ca^{2+}-selective permeability. The α1 subunit also bears high-affinity-binding sites for channel antagonists (including ω-conopeptides) and for regulatory G-proteins and SNARE proteins. Its sequence determines the family and subfamily of the Ca_v channel complex (Table 1). The auxiliary subunits, β, α2–δ and γ, influence the insertion and stability of α1 in the plasma membrane and modify the conductance and kinetics of the channel.

3.2. Pathophysiology of Neuromuscular Transmission in LES

LES was recognized initially as a disorder of neuromuscular transmission because of its clinical similarities to myasthenia gravis, including fatigable weakness that is improved with cholinesterase inhibitors. The characteristic electrophysiological findings described by Lambert and colleagues were established by standard electrodiagnostic studies and by microelectrode studies *(6, 9, 12)*. The microelectrode findings are normal amplitude and frequency of the miniature endplate potential (MEPP), with reduction of the EPP amplitude secondary to a low quantal content of the nerve action potential *(12)*. The quantal content and the size of the EPP increase during high-frequency nerve stimulation and with addition of extracellular Ca^{2+}. Biochemical studies of nerve-muscle preparations from LES patients have revealed a reduction in the amount of ACh released and normal function of acetylcholinesterase *(43)*.

The fully developed defect of neuromuscular transmission in LES is typically severe, reducing the EPP amplitude below threshold in the majority of fibers in resting muscle. The patient is weak at rest and CMAP responses to initial supramaximal nerve stimuli are low in amplitude (Fig. 1). Brief exercise produces transient clinical improvements in strength and tendon reflex responses and electromyographic facilitation of the CMAP amplitude. If exercise is impractical for the patient, repetitive stimulation at 20–50 Hz can be used to elicit facilitation of the CMAP amplitude. At this stage, standard concentric needle EMG shows MUP amplitude variation and stimulated single fiber EMG shows increased jitter and blocking, both of which improve at higher rates of stimulation *(44)*.

Table 1
Classification of Voltage-Gated Calcium (Ca$_v$) Channels*

Class	Type	α1 subunit	Antagonists	Effector function	Predominant cell expression
Ca$_v$1.1	L	S	Dihydropyridines, phenylalkylamines, benzothiazepines	Contraction	Skeletal muscle
Ca$_v$1.2		C		Contraction	Heart
Ca$_v$1.3		D		Secretion; gene transcription	Neuron soma; pancreatic islet; kidney; ovary; cochlea
Ca$_v$1.4		F		Gene transcription	Retina
Ca$_v$2.1	P/Q	A	ω-conotoxin M$_{VIIC}$, ω-agatoxin $_{IIIA}$ and $_{IVA}$	Neurotransmitter release	Central, autonomic and peripheral motor neurons
Ca$_v$2.2	N	B	ω-conotoxin G$_{VIA}$ and M$_{VIIA}$	Neurotransmitter release	Central sensory and motor, and autonomic neurons
Ca$_v$2.3	R	E	SNX-482, ω-agatoxin $_{IIIA}$	Neurotransmitter release	Central neurons, cochlea, retina, heart, pituitary
Ca$_v$3.1	T	G	ω-agatoxin $_{IIIA}$, succinimides	Pacemaker activity	Central and peripheral neuron soma
Ca$_v$3.2		H		Neurotransmitter release	Central neurons, heart, kidney, liver
Ca$_v$3.3		I		Neurotransmitter release	Central neurons

*Channels of Ca$_v$1 and Ca$_v$2 classes are activated by high voltage; class Ca$_v$3 (T-type) channels are activated by low voltage.

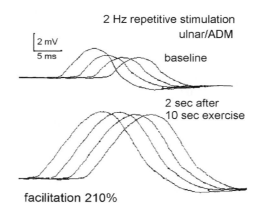

Fig. 1. Repetitive stimulation study in a patient with moderate–severe weakness caused by LES. *x*-axis, time (ms); *y*-axis, CMAP amplitude (mV); ADM, abductor digiti minimi; sec, seconds.

The patient with early LES may have a relatively mild defect of neuromuscular transmission in which the safety margin is not reduced sufficiently to drop the EPP below threshold at rest. This ensures relatively preserved strength at rest and a normal CMAP amplitude (Fig. 2). During repetitive stimulation at slow rates (2–5 Hz), there is decrement of the CMAP, which repairs with exercise or higher rates of repetitive stimulation. Thus, when LES is mild, the results of electrophysiological tests of neuromuscular transmission are very similar to those of myasthenia gravis

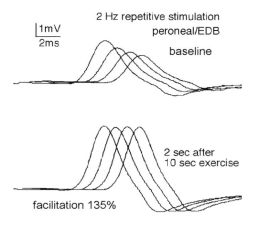

2 Hz repetitive stimulation
peroneal/EDB

baseline

2 sec after
10 sec exercise

facilitation 135%

Fig. 2. Repetitive stimulation study in a patient with mild weakness caused by LES. x-axis, time (ms); y-axis, CMAP amplitude (mV); EDB, extensor digitorum brevis; sec, seconds.

or other postsynaptic disorders of neuromuscular transmission *(44)*. The abnormalities revealed by standard clinical electrophysiological testing (nerve conduction studies, repetitive stimulation and needle EMG) reflect the severity more than the synaptic element responsible for the defect of neuromuscular transmission.

3.3. Immunopathophysiology of LES

LES may occur as a paraneoplastic syndrome or as an idiopathic organ-specific autoimmune disease *(14, 31, 45)*. Thyrogastric autoimmune diseases, which include thyroiditis, pernicious anemia and type 1 diabetes mellitus and their serological markers are frequent in LES patients, particularly in those without evidence of lung cancer *(13, 14, 45, 46)*. The P/Q-type ($Ca_v2.1$) calcium channel in the presynaptic membrane at the neuromuscular junction is strongly impli-cated but still not proven definitively to be the target of pathogenic autoantibodies. It is also implicated in autonomic manifestations of LES *(47)*. The finding of P/Q-type calcium channel antibody in the serum of more than 90% of non-immunosuppressed LES patients *(25)* is compelling but circumstantial evidence for it being the effector of neuromuscular transmission impairment, as is the capacity of polyclonal IgG from serum of LES patients to transfer to mice clinical, electrophysiological and ultrastructural nerve terminal lesions characteristic of LES *(18, 19, 20)*. There have been no reports of a monoclonal or affinity-purified antibody of P/Q-type calcium channel specificity transferring the neuromuscular or autonomic transmission defect of LES to laboratory animals. This long-awaited evidence will definitively prove the pathogenicity of P/Q-type calcium channel antibodies. A Japanese report of electrophysiologi-cal abnormalities consistent with LES being induced in rats immunized with a peptide corre-sponding to an extracellular segment of the α_{1A} subunit of the $Ca_v2.1$ channel complex awaits confirmation *(48)*.

Small-cell lung carcinoma (SCLC) is the most common neoplasm associated with paraneo-plastic LES and is usually occult when neurological manifestations appear. The $Ca_v2.1$ channel at the neuromuscular junction is related antigenically to the Ca_v channels in SCLC and other neuroendocrine cells *(21, 22, 49)*. Reports that the prognosis of SCLC is better when associated with LES *(50)* and that successful treatment of SCLC leads to remission of LES in 70% of patients *(27, 31)* support the hypothesis that paraneoplastic LES arises as an immune response initiated by antigens expressed in the patient's neoplasm. The immunogen that initiates and perpetuates Ca_v channel autoimmunity in the idiopathic form of LES has not been identified.

The pathophysiology of LES is entirely reproducible in mice using patient's serum IgG as the effector. As we reviewed above, LES patients' IgG confers electrophysiological and morphological characteristics of LES at the motor nerve terminal (17, 18, 19, 20). Mice receiving IgG from individual selected patients become profoundly weak, exhibit characteristic EMG findings of LES and die from respiratory failure (20). Patients with exceptionally potent IgG presumably produce IgG with extremely high affinity for $Ca_V2.1$ channel ectodomain epitopes common to human and mouse. Impairment of Ca_V channel function is independent of complement (20), and is attributed to antigenic modulation, which is an IgG-mediated acceleration of Ca_V channel degradation through bivalent cross-linking of adjacent channels (51, 52). LES IgG reduces Ca^{2+} influx through multiple Ca_V channel subtypes in cultured SCLC cells and motor neurons (21, 53, 54, 55, 56, 57) and reduces neurotransmitter release at the mouse neuromuscular junction (20, 47, 57). The benefit of immunotherapy in LES patients is explained most plausibly by clearance of circulating antibodies, effected by plasmapheresis or intravenous immune globulin therapy (58) or by reduced production of antibodies (immunosuppressive medications). Effector T lymphocytes have not been implicated in the pathophysiology of LES, but the production of antigen-specific IgG by plasma cell progeny of activated B lymphocytes requires continuing helper T-lymphocyte activity (59).

4. EPIDEMIOLOGY

The incidence and prevalence of LES are significantly lower than those of myasthenia gravis. In the Netherlands the annual incidence rate of LES was 14 times lower (0.48×10^{-6}) than that of MG (6.48×10^{-6}) (60). LES was 46 times less prevalent than MG in the same study. We estimate that LES accounts for approximately 10% of the 4-6 per million patients newly diagnosed annually with MG which is estimated to have a prevalence of 4060 per million of the US population. Females predominate in idiopathic LES (31), which has an onset age ranging from the first decade to old age. Patients with idiopathic LES are frequently non-smokers and often have a personal or family history of organ-specific autoimmune diseases and a higher prevalence of certain HLA gene products than patients with paraneoplastic LES (61, 62). The frequency of LES in patients with newly diagnosed SCLC is estimated to be 3% (63). In its earliest descriptions, paraneoplastic LES predominantly affected men. This sex bias was attributed to cigarette smoking habits. The epidemiology of lung cancer has paralleled the sociology of smoking, with a two-decade lag. Thus, in 1987, lung cancer became the leading cause of cancer death in North American women (64). Nevertheless, men still outnumber women in lung cancer prevalence. Among cases of SCLC-related LES (not complicated by other autoimmune neurological disorders) identified serologically in Mayo Clinic's Neuroimmunology Laboratory since 1987, men marginally outnumber women (authors' unpublished observation; $n > 150$ patients). By contrast, in that same period women have accounted for 66% of patients diagnosed with SCLC-related paraneoplastic neurological disorders defined serologically by the type 1 anti-neuronal nuclear autoantibody (ANNA-1, or anti-Hu) (65). Unlike LES, neurological disorders associated with the ANNA-1 autoantibody are characterized by inflammatory lesions affecting the peripheral or central nervous system or both. The cytokine milieu at the level of helper T-lymphocyte activation determines whether or not the immunological outcome of an immune response is inflammatory. It is conceivable that the outcome is influenced both by sex hormones and by onconeural antigen dominance among proteins released to lymph nodes by necrosing tumor cells. A pro-inflammatory response would yield cytotoxic T-cell effectors, for which nuclear and cytoplasmic-directed antibodies serve as a surrogate marker (66), while a non-inflammatory response may favor production of plasma membrane ion channel antibodies.

Cancer is diagnosed in about 45% of contemporary LES cases, and SCLC accounts for 90% of paraneoplastic cases (45). Overall, 18% of patients who have SCLC without clinical evidence of

LES are seropositive for P/Q-type calcium channel antibodies *(25)*. The diagnosis of LES may be overlooked when multiple paraneoplastic disorders coexist, particularly in the setting of severe peripheral neuropathy or subacute cerebellar ataxia *(67)*. Other cancers reported with LES include thymoma, lymphoma, reproductive tract tumors and renal cell cancer *(45, 68, 69, 70, 71, 72, 73)*, but there is a high likelihood that the patient has a coexisting occult SCLC *(22, 65, 66)*. Anticipation of SCLC in these cases is supported by observations that the diagnosis of LES can antedate the detection of SCLC by 5–8 years *(45)*; and unpublished observations of V.A. Lennon and E.H. Lambert). Primary small-cell carcinoma in an extrapulmonary site, such as skin (Merkel cell carcinoma), tongue, larynx, pancreas, breast, cervix or prostate *(74)*, should also be considered in patients presenting with LES, particularly in the absence of known risk factors for lung cancer.

5. CLINICAL PRESENTATION

5.1. Symptoms and Signs

Subacute progressive fatigue, weakness and unanticipated injurious falls are the most common presenting manifestations of LES. The weakness increases with exertion and improves transiently with rest. Sometimes a history of transient clinical facilitation of strength following brief exercise is elicited, but more commonly there are complaints of increased weakness with exercise. Aching pain in the hip and posterior thigh region is a characteristic of paraneoplastic LES, but other sensory symptoms are rare unless LES is accompanied by paraneoplastic manifestations of sensory neuropathy or radiculoplexopathy. Muscle atrophy is rare, and fasciculations and cramps are not observed. The distribution of weakness is characteristic, with symptoms beginning in hip flexors and other proximal lower limb muscles. Patients commonly complain of difficulty arising from low chairs or a squatted position or trouble climbing stairs and inclines. In one series of 50 LES patients, weakness began in the lower limbs in 65% and was generalized in 12% of patients *(45)*. Even after weakness is more generalized, the hip girdle muscles are involved disproportionately in most patients. Proximal muscles of the upper limb and neck and interossei muscles of the hand also are preferentially affected by LES. Cranial muscle is involved at some point in approximately 25% of patients. Ptosis, facial weakness, dysphagia, dysarthria and difficulty chewing are milder than in myasthenia gravis and usually occur after the onset of limb weakness *(35, 45, 75)*. Respiratory involvement is usually mild with a restrictive functional pattern, unless complicated by emphysema related to smoking, in which case respiratory failure can occur terminally. In rare cases respiratory failure is the presenting manifestation of LES *(76)*. Deep tendon reflexes are characteristically reduced or absent in LES, but it is important to note that reflexes may be preserved early in the course of the illness. Facilitation of the reflex after brief exercise is an important sign supporting the diagnosis of LES and is easier to demonstrate at the bedside than facilitation of muscle strength.

Symptoms of autonomic dysfunction occur in approximately 80% of LES patients and may be the presenting symptoms in 6% of patients *(45)*. Common autonomic symptoms are impotence (men) and xerostomia (both sexes). Other common abnormalities of autonomic testing involve sweating, cardiovagal reflexes and salivation *(46)*. Slow pupillary reflexes, gastrointestinal dysmotility, orthostatic hypotension and urinary retention are sometimes noted, but these signs usually indicate a coexisting paraneoplastic autonomic neuropathy in the setting of SCLC and are often accompanied by one or more marker autoantibodies specific for neuronal nuclear or cytoplasmic antigens, particularly CRMP-5-IgG, ANNA-1, ANNA-2, ANNA-3, PCA-2 or amphiphysin autoantibody (*(66, 67, 77, 78)* and V.A. Lennon, unpublished observation) or the anti-glial/neuronal nuclear antibody AGNA-1/ANNA-4 *(79, 80)*.

The sensory system is spared in LES, but sensory symptoms and signs may be present if there is an associated paraneoplastic sensory neuronopathy, peripheral neuropathy, myelopathy or encephalopathy. Constitutional symptoms of anorexia and weight loss are usually attributed to an underlying malignancy and may indicate autoimmune gastroparesis for which autoantibodies of ANNA-1, CRMP-5, voltage-gated N-type calcium channel and potassium channel and ganglionic AChR specificities are valuable serological markers *(81, 82, 83, 84)*. Symptoms of lung cancer itself, such as hemoptysis and chest pain, are very uncommon. On the other hand, manifestations of coexisting organ-specific autoimmune disorders (e.g., pernicious anemia, hyperthyroidism or hypothyroidism) are very common in patients with idiopathic LES. The authors have personally encountered two patients with idiopathic LES in whom weight loss was attributed to concurrent but initially unrecognized Graves' disease.

5.2. Natural History

The course of LES is more variable among patients with lung cancer or at high risk for it than among those without lung cancer risk. LES frequently presents several months or years prior to the diagnosis of cancer. If sputum cytology, chest X-ray and CT are negative in LES patients at risk for SCLC, the diagnostic yield is increased by performing MRI imaging or PET scanning of the chest, bronchoscopy and transbronchial biopsy, transesophageal ultrasound or thoracoscopy if imaging studies are suspicious for a neoplasm. A search for extrapulmonary small-cell carcinoma is warranted when chest studies are negative. These have been recorded in skin, tongue, larynx, pancreas, cervix, prostate, breast and ovary *((74)* and V.A. Lennon, unpublished observations). The clinical course of LES, with and without cancer, is generally progressive in the first year, with less fluctuation and a lower spontaneous remission rate than autoimmune myasthenia gravis. Strength often improves following effective treatment of the underlying malignancy. It has been reported that SCLC in the context of LES may have a less aggressive natural history and a better response to therapy *(50)*. It is not clear whether this relates to early diagnosis due to appearance of the paraneoplastic syndrome or reflects a more effective immune response to the tumor. In the absence of cancer, weakness and fatigue due to LES may produce long-term disability *(31)*. Most patients benefit from continued therapy with an oral cholinesterase inhibitor and 3-4 diaminopyridine, an agent that increases the release of ACh induced by nerve stimulus *(85, 86)*. Guanidine is also beneficial *(34)* in this regard, but its use is limited by renal and bone marrow toxicity. Immunosuppressive therapy also is beneficial *(31)*, but symptomatic response to immunosuppressive therapy tends to be less in LES than in myasthenia gravis.

6. DIAGNOSIS

6.1. Clinical Manifestations

LES should be suspected in a patient who complains of generalized fatigue or weakness that worsens with exertion and should be considered in the differential diagnosis of prolonged postoperative apnea involving a neuromuscular blocking drug *(5, 87)*. Factors favoring the diagnosis of LES rather than myasthenia gravis include early and prominent involvement of hip flexor muscles, xerostomia, impotence and other manifestations of autonomic dysfunction, reduced or absent reflexes (in lower limbs initially and more diffuse in the established disease), facilitation of tendon reflexes or strength on clinical examination, a personal history of smoking, asbestos or tobacco smoke exposure or lung cancer or a family history of lung cancer. A personal or family history of autoimmunity is common in both idiopathic and paraneoplastic LES as well as in myasthenia gravis.

The diagnosis of LES should also be suspected when a patient who presents with paraneoplastic sensorimotor or sensory neuropathy related to lung cancer has prominent weakness, sicca symptoms, recent onset erectile dysfunction or is seropositive for P/Q-type calcium channel antibody, or when a patient has respiratory failure of undetermined cause or is referred with a diagnosis of seronegative myasthenia gravis.

6.2. Electrodiagnostic Studies

For unambiguous electrodiagnosis of LES it is critically important that the patient be well rested and warm, that medications affecting the neuromuscular junction be withheld for several hours before testing and that clinically weak muscles be tested. When weakness is moderate to severe, electrodiagnostic studies reveal a pathognomonic pattern (Table 2) *(44)*. The baseline amplitude of the CMAP is reduced. A decrement is observed with slow rates of repetitive nerve stimulation, and facilitation of more than 200% (doubling of the baseline amplitude) is observed following brief exercise or high-frequency repetitive nerve stimulation. Needle examination shows low-amplitude varying MUP. Fibrillation potentials are usually absent but may occur in severe cases. Single fiber EMG shows increased jitter and blocking that transiently improve at higher firing rates. The electrodiagnostic findings are usually demonstrated most readily in small intrinsic hand muscles.

When weakness is mild, as in the early course of LES (Table 2), the CMAP amplitude may be reduced only slightly or may be in the lower range of normal. In this setting the CMAP decrement is usually modest and facilitation is often less than 200%. Needle EMG is normal or shows only mild MUP amplitude variation. Mild cases of LES are difficult to differentiate from myasthenia gravis *(44)*. The distribution of clinical symptoms, the profile of serum autoantibodies and the result of stimulated single fiber EMG testing can help, but in some cases the correct diagnosis requires continued observation and retesting at a later date after the disease progresses.

6.3. Serological Tests

P/Q-type ($Ca_v2.1$) calcium channel antibodies are detectable in more than 90% of non-immunosuppressed patients with LES (Table 3) with or without cancer of any type *(25)*. It remains to be determined whether or not other motor nerve terminal antibodies may be defined as causative of LES. It was recently reported from Japan that serum IgG from two "seronegative" LES patients each transferred to mice microelectrophysiological findings consistent with LES within 48 h of intravenous injection *(88)*. However, as reported for myasthenia gravis *(89)*, we personally have encountered apparent seronegativity at disease onset with seroconversion when reevaluated at 12 months. N-type ($Ca_v2.2$) calcium channel antibodies are detectable in 75% of LES patients with lung cancer and in 40% without cancer risk *(25)*, but less frequently in LES patients who have other types of cancer *(22)*. Eighteen percent of SCLC patients without clinical evidence of LES

Table 2
Electrodiagnostic Characteristics of Lambert–Eaton Syndrome

Severity of weakness	Nerve conduction studies CMAP* amplitude	Needle examination
Moderate–severe	(1) Baseline (initial stimulus): low	(1) Insertional activity: normal
	(2) During RS[†] at 2–3 Hz: decrement	(2) Motor unit potentials: small with amplitude variation
	(3) After brief exercise or after RS at 20–50 Hz: facilitation > 200%	(3) Single fiber EMG: increased jitter and blocking
Mild	(1) Baseline (initial stimulus): normal	(1) Insertional activity: normal
	(2) During RS at 2–3 Hz: decrement	(2) Motor unit potentials: normal to small with amplitude variation
	(3) After brief exercise or after RS at 20–50 Hz: facilitation < 200%	(3) Single fiber EMG: increased jitter and blocking

*CMAP, compound muscle action potential elicited by supramaximal stimulation of nerve in clinically weak muscle, with patient well rested, warm and with neuromuscular-active medication withheld.
[†]RS, repetitive stimulation.

Table 3
Serum Profiles of Organ-Specific Autoantibodies in Non-immunosuppressed Adult Patients with Lambert–Eaton Syndrome, Generalized Myasthenia Gravis and controls*

Autoantibody	Frequency (% seropositive)				
	LES without cancer	LES with lung cancer	MG without thymoma	MG with thymoma	Non-autoimmune neurological diseases
Calcium channel, P/Q	91	99	< 2	< 2	< 2
Calcium channel, N	40	75	< 2	< 2	< 5
Muscle AChR	7	7	90	100	< 5
Striational	5	5	30	80	< 2
Thyrogastric[†]	55	20	50[‡]	30	20
Ganglionic AChR	10	10	< 2	8	< 2
CRMP-5	< 2	0[††]	< 5[¥]	18	< 2
Anti-glial nuclear (AGNA-1)	0	43	< 2	< 2	< 2
One or more of above	96	100	95	100	20

*Based on Refs. *(26, 78, 111)* and authors' unpublished data.

[†]Thyroperoxidase, thyroglobulin, gastric parietal cell, intrinsic factor blocking or glutamic acid decarboxylase (GAD65) antibody.

[‡]Prevalence highest in ocular myasthenia.

[††]Positive only if coexisting paraneoplastic accompaniments.

[¥]Positive only in patients with thymoma-predictive autoantibody profile *(26)*, but negative chest imaging (usually elder-onset MG).

have P/Q-type calcium channel antibodies and 22% have N-type calcium channel antibodies *(25)*. Nicotinic AChR antibodies, of muscle type or ganglionic type, or striational antibodies are found in ∼13% of patients with LES, with and without cancer *(26, 90, 91)*. Calcium channel antibodies of N-type and P/Q-type also serve as serological markers of paraneoplastic neurological disorders affecting the central and peripheral nervous system (i.e., other than LES) in the context of small-cell lung carcinoma, ovarian carcinoma and breast carcinoma *(25, 92)*. Calcium channel antibody values in those patients tend to be lower than in untreated LES patients. The anti-glial/neuronal nuclear antibody (AGNA-1/ANNA-4) is an IgG marker of neurological autoimmunity related to small-cell lung carcinoma and has been reported positive in 43–64% of patients with paraneoplastic LES *(79, 80)*. Its autoantigen is the astrocytic/neuronal nuclear transcription factor, SOX1, which is expressed as an onconeural protein in small-cell lung carcinomas *(93, 94)*. Other IgG neuronal nuclear and cytoplasmic autoantibody markers of small-cell lung carcinoma (e.g., ANNA-1, CRMP-5, amphiphysin, PCA-2 and ANNA-3) *(66, 77, 78)* are rarely found in patients with paraneoplastic LES in the absence of a coexisting encephalomyeloradiculopathy or neuropathy.

6.4. Differential Diagnosis

The differential diagnosis of LES includes autoimmune myasthenia gravis and congenital myasthenic syndromes and defects of neuromuscular transmission induced by drugs or toxins. A congenital myasthenic syndrome that resembles LES is severe at birth *(95, 96)*. Autoimmune myasthenia gravis can usually be distinguished from LES by differences in clinical manifestations as well as electrodiagnostic tests and autoimmune serological tests *(26)*. Cases of mild LES or severe myasthenia gravis can yield quite similar electrophysiological findings with baseline CMAP amplitudes in the low normal range, a decrement at low rates of repetitive nerve stimulation and facilitation of less than 200% following exercise or high rates of repetitive stimulation *(44)*. The distribution of weakness and the autoantibody profile *(26)* usually differentiate the disorders. Sometimes additional time for observation and retesting is needed, as LES typically progresses in severity if not treated. Generally, if a patient complains of severe weakness, negative electromyographic findings are more common when the cause is myasthenia gravis than when it is LES. Although cases of an overlap or combination of LES and myasthenia gravis have been reported, no instance of a dual syndrome has been documented by in vitro microelectrode studies performed on biopsied intercostal or anconeus nerve-muscle preparations. These reports may reflect misinterpretation of autoantibody profiles *(97)*.

Botulism usually presents as an acute and severe weakness of cranial, axial and limb muscles, requiring respiratory insufficiency and emergent hospitalization. LES usually presents more gradually and rarely presents with respiratory failure. Electrodiagnostic studies in botulism reveal CMAPs of low amplitude with less prominent decrement and facilitation than LES. On needle EMG, fibrillation potentials are typically widespread in botulism but are rare in LES. The diagnosis of botulism is usually made on clinical grounds and confirmed by electrodiagnostic studies and demonstration of the toxin in the stool. Toxins delivered through the bite of venomous snakes and spiders may produce findings similar to botulism. These as well as acute intoxication with magnesium, neuromuscular blocking drugs and organophosphates are unlikely to be mistaken for the diagnosis of LES.

7. TREATMENT

7.1. Symptomatic Treatment

7.1.1. Cholinesterase Inhibitors

Because the quantal content is usually severely reduced in LES, the beneficial effect of AChE inhibition is usually modest unless it is combined with medications that simultaneously increase the release of ACh (e.g., 3,4-diaminopyridine or guanidine).

7.1.2. Guanidine

The beneficial effect of guanidine was initially described by Lambert and colleagues *(8, 10)*. Guanidine increases intracellular calcium within the nerve terminal by inhibiting calcium uptake by organelles *(98)*. Its clinical utility is limited by significant hemopoietic and renal toxicity *(99, 100)*, but guanidine is reported to be effective with less risk of toxicity when the dose is strictly limited and the drug is used in combination with pyridostigmine *(34, 101)*.

7.1.3. 3,4-Diaminopyridine

The effectiveness of 3,4-diaminopyridine (DAP) in LES has been documented in placebo-controlled prospective trials and in long-term follow-up studies *(30, 102)*. This agent blocks K_v channels in the nerve terminal, thus prolonging the action potential and enhancing Ca^{2+} entry. Objective measures of strength are improved, including respiratory muscle function. Electromyographically there is an increase in amplitude of the CMAP, decreased CMAP decrement and SFEMG reveals less jitter and block *(103, 104)*. The peak beneficial effect of a single oral dose of 3,4-DAP occurs in about 1 h and lasts for 2–5 h *(103)*. The usual effective dose is 10–20 mg 4–5 times daily (85, 86, 102). Adverse effects include perioral and acral paresthesiae, lightheadedness, headache, insomnia and seizures. Seizures are rare if the baseline EEG is normal and the daily dose does not exceed 100 mg. Cardiac arrhythmia was described in a patient given an intentional overdose of 3,4-DAP *(105)*. There is a theoretical risk of cardiac arrhythmia in patients with the prolonged QT syndrome, but 3,4-DAP toxicity has not been reported in that setting. Because 3,4-DAP is not approved for routine use in the United States, its use requires application as an investigational drug on a compassionate use basis. The drug is contraindicated if preliminary screening by electrocardiogram and electroencephalogram reveals a prolonged QT interval or epileptiform discharges. It is Mayo Clinic practice to monitor complete blood count, liver function and serum creatinine 3 times per year, but the authors have not yet encountered bone marrow, liver or kidney toxicity, attributable to 3,4-DAP therapy.

7.2. Treatment of the Associated Neoplasm

When cancer is found, its effective treatment may be followed by amelioration of LES manifestations, or even remission *(50)*. This has been documented particularly for SCLC treated by chemotherapy and radiation. Because neurological improvement is not universal, patients should be cautioned that their neuromuscular disorder may need continued symptomatic treatment or immunotherapy even if their underlying cancer is cured.

7.3. Immunotherapy

The general approach for LES is as for myasthenia gravis and other autoimmune neurological disorders. Immunotherapy is reserved for patients who fail to respond adequately to symptomatic therapy or, in the case of paraneoplastic LES, when tumor therapy is not accompanied by neurological improvement. Alleviation of symptoms in LES is generally less dramatic than in myasthenia gravis *(31)*. Nevertheless, benefit has been documented with plasmapheresis *(106)*, intravenous immune globulin (IVIG) *(107, 108, 109)* and corticosteroids *(31)*. In patients who are considered at high risk for cancer, immunotherapy is customarily reserved for those who are severely disabled by weakness and failing to respond to symptomatic therapy. Reluctance to treat stems from fear of compromising the patient's survival by promoting evasion of the suspected tumor from an effective immune response. Parenthetically, however, we do not recall having observed noteworthy emergence of metastatic cancer in two decades of using prednisone and azathioprine to treat numerous LES patients at risk for lung cancer. This anecdotal observation implies that an inherently effective cancer immune response is not impaired by the standard immunosuppressant protocols used to treat neurological autoimmune disorders.

The type of immunotherapy chosen is tailored to the individual patient's need, depending on the magnitude and duration of benefit. Plasmapheresis is beneficial in LES, but because the duration of benefit is limited to several weeks, it is generally reserved for severely weak patients, particularly in the setting of respiratory insufficiency. Occasionally plasmapheresis is used as maintenance therapy at a frequency of 1–2 exchanges every 3–4 months. This usually requires placement of a central line or arterio-venous shunt, which increases the risk of complications from infection, coagulation disorders and fluid/electrolyte imbalance.

Evidence for the benefit of IVIG therapy is less convincing than for plasmapheresis. A small controlled trial did show statistical improvement in muscle strength and decreased concentration of calcium channel antibodies in the IVIG-treated LES patients *(109, 110)*. The effect peaked at 2–4 weeks and disappeared by 8 weeks. The high cost of IVIG and relatively short duration of action have limited its use as a maintenance therapy in LES. Treatment protocols have utilized 400–500 mg/kg daily for 2–5 consecutive days or intermittent infusions at a frequency of 1–2 every 2–8 weeks.

Oral corticosteroid therapy benefits most LES patients, but improvement of strength is less dramatic than observed in myasthenia gravis. In one reported study, long-term treatment with prednisilone at 20–30 mg daily or on alternate days produced modest symptomatic benefit in most patients *(31)*. Azathioprine is also effective either as a steroid-sparing agent or as a stand-alone immunosuppressant *(31)*. Insufficient data are available concerning the efficacy of cyclosporine, cyclophosphamide, mycophenolate or Rituximab as alternative immunosuppressants in LES. In principle, experimental treatment strategies that are being investigated for myasthenia gravis should be applicable to the treatment of LES, including the deletion of antigen-specific B cells or T cells and interference with costimulatory molecules that perpetuate the autoimmunization process.

REFERENCES

1. Ertel EA, Campbell KP, Harpold MM, et al. Nomenclature of voltage-gated calcium channels. Letter to the Editor. Neuron 2000; 25:533–535.
2. Anderson HJ, Churchill-Davidson HC, Richardson AT. Bronchial neoplasm with myasthenia: prolonged apnoea after administration of succinylcholine. Lancet 1953; 2: 1291–1293.
3. Sciamanna MA, Griesmann GE, Williams CL, Lennon VA. Nicotinic acetylcholine receptors of muscle and neuronal (α7) types coexpressed in a small cell lung carcinoma. J Neurochem 1997; 69: 2302–2311.
4. Griesmann GE, Harper CM, Lennon VA. Paraneoplastic myasthenia gravis and lung carcinoma: distinction from Lambert-Eaton myasthenic syndrome and hypothesis of aberrant muscle acetylcholine receptor expression. Muscle Nerve 1998 7(Suppl. 7): S122–S32.
5. Churchill-Davidson HC. Muscle relaxants. In: Langton Hewer C, ed. Recent Advances in Anesthesia and Analgesia, 9th ed., Chapter 3. London: J and A Churchill, Ltd., 1963: 79–110.
6. Lambert EH, Eaton LM, Rooke ED. Defect of neuromuscular transmission associated with malignant neoplasm. Am J Physiol 1956; 187: 612.
7. Eaton LM, Lambert EH. Electromyography and electrical stimulation of nerves in diseases of the motor unit: observations on myasthenic syndrome associated with malignant tumors. JAMA 1957; 163: 1117–1124.
8. Rooke ED, Lambert ED, Thomas JE. A myasthenic syndrome closely related to several malignant intrathoracic tumors. Dtsch Med Wochenschr 1961;86:1660–1664.
9. Lambert EH, Rooke ED, Eaton LM, Hodgson CH. Myasthenic syndrome occasionally associated with bronchial neoplasm: neurophysiologic studies. In: Viets HR, ed. Myasthenia Gravis. Springfield: Charles C. Thomas, 1961: 362–410.
10. Lambert EH, Rooke ED. Myasthenic state and lung cancer. In: Brain WR, Norris FH, Eds. The Remote Effects of Cancer on the Nervous System (Contemporary Neurology Symposia), vol. 1. New York: Grune & Stratton, 1965: 67–80.
11. Elmqvist D, Lambert EH. Detailed analysis of neuromuscular transmission in a patient with the myasthenic syndrome sometimes associated with bronchogenic carcinoma. Mayo Clin Proc 1968; 43: 689–713.
12. Lambert EH, Elmqvist D. Quantal components of end-plate potentials in the myasthenic syndrome. Ann NY Acad Sci 1971; 183: 183–199.
13. Gutmann L, Crosby TW, Takamori M, Martin JD. The Eaton-Lambert syndrome and autoimmune disorders. Am J Med 1972; 53: 354–356.
14. Lennon VA, Lambert EH, Whittingham S, Fairbanks V. Autoimmunity in the Lambert-Eaton myasthenic syndrome. Muscle Nerve 1982; 5: S21–S25.
15. Fukunaga H, Engel AG, Osame M, Lambert EH. Paucity and disorganization of presynaptic membrane active zones in the Lambert-Eaton myasthenic syndrome. Muscle Nerve 1982; 5: 686–697.
16. Engel AG. Anatomy and molecular architecture of the neuromuscular junction. In: Engel AG, ed. Myasthenia Gravis and Myasthenic Syndromes. New York: Oxford Press, 1999: 3–40.
17. Lang B, Newsom-Davis J, Wray D, Vincent A, Murray N. Autoimmune etiology for myasthenic (Eaton-Lambert) syndrome. Lancet 1981; 2: 224–226.
18. Kim YI. Passively transferred Lambert-Eaton syndrome in mice receiving purified IgG. Muscle Nerve 1986; 9: 523–530.
19. Fukunaga H, Engel AG, Lang B, Newsom- Davis J, Vincent A. Passive transfer of Lambert-Eaton myasthenic syndrome with IgG from man to mouse depletes the presynaptic membrane active zones. Proc Natl Acad Sci 1983; 80: 7636–7640.
20. Lambert EH, Lennon VA. Selected IgG rapidly induces Lambert-Eaton myasthenic syndrome in mice: Complement independence and EMG abnormalities. Muscle Nerve 1988; 11: 1133–1145.
21. Roberts A, Perera S, Lang B, Vincent A, Newsom-Davis J. Paraneoplastic myasthenic syndrome IgG inhibits ^{45}Ca flux in human small cell carcinoma line. Nature 1985; 317: 737–739.
22. Lennon VA, Lambert EH. Autoantibodies bind solubilized calcium channel-omega-conotoxin complexes from small cell lung carcinoma: a diagnostic aid for Lambert-Eaton myasthenic syndrome. Mayo Clin Proc 1989; 64: 1498–1504.
23. Sher E, Canal N, Piccolo G, Gotti C. Scoppetta C, Evoli A, Clementi F. Specificity of calcium channel autoantibodies in Lambert-Eaton myasthenic syndrome. Lancet 1989; ii: 640–643.

24. Leys K, Lang B, Johnston I, Newsom-Davis J. Calcium channel autoantibodies in the Lambert-Eaton myasthenic syndrome. Ann Neurol 1991; 29: 307–314.

25. Lennon VA, Kryzer TJ, Griesmann GE, O'Suilleabhain PE, Windebank AJ, Woppmann A, Miljanich GP, Lambert EH. Calcium-channel antibodies in the Lambert-Eaton syndrome and other paraneoplastic syndromes. N Engl J Med 1995; 332: 1467–1471.

26. Lennon VA. Serologic profile of myasthenia gravis and distinction from Lambert-Eaton myasthenic syndrome. Neurology 1997; 48 (Suppl. 5): S23–S27.

27. Chalk CH, Murray NM, Newsom-Davis J, O'Neill JH, Spiro SG. Response of the Lambert-Eaton myasthenic syndrome to treatment of associated small-cell lung carcinoma. Neurology 1990; 40: 1552–1556.

28. Streib EW, Rothner AD. Lambert-Eaton myasthenic syndrome: long-term treatment of three patients with prednisone. Ann Neurol 1981; 10: 448–453.

29. Newsom-Davis J, Murray NM. Plasma exchange and immunosuppressive drug treatment in the Lambert-Eaton myasthenic syndrome. Neurology 1984; 34: 480–485.

30. Lundh H, Nilsson O, Rosen I. Current therapy of the Lambert-Eaton myasthenic syndrome. Prog Brain Res 1990; 84:163–170.

31. Maddison P, Lang B, Mills K, Newsom-Davis J. Long-term outcome in Lambert-Eaton myasthenic syndrome without lung cancer. J Neurol Neurosurg Psychiatry 2001; 70: 212–217.

32. Lundh H, Nilsson O, Rosen I. Treatment of Lambert-Eaton syndrome: 3,4-diaminopyridine and pyridostigmine. Neurology 1984; 34:1324–1330.

33. McEvoy KM, Windebank AJ, Daube JR, Low PA. 3,4-Diaminopyridine in the treatment of Lambert-Eaton myasthenic syndrome. N Engl J Med. 1989; 321: 1567–1571.

34. Oh SJ, Kim DS, Head TC, Claussen GC. Low dose guanidine and pyridostigmine: relatively safe and effective long-term symptomatic therapy in Lambert-Eaton myasthenic syndrome. Muscle Nerve 1997; 20: 1146–1152.

35. Tim RW. Massey JM. Sanders DB. Lambert-Eaton myasthenic syndrome: electrodiagnostic findings and response to treatment. Neurology. 2000; 54: 2176–2178.

36. Lambert EH, Okihiro M, Rooke ED. Clinical physiology of the neuromuscular junction. In: Paul WM, Daniel EE, Kay CM, Monckton G, Eds. Muscle. Oxford: Pergamon Press 1965; 487–497.

37. Boonyapisit K, Kaminski HJ, Ruff RL. The molecular basis of neuromuscular transmission disorders. Am J Med 1999; 160: 97–113.

38. Hughes BW, Kusner LL, Kaminski H J. Molecular architecture of the neuromuscular junction. Muscle Nerve 2006; 33: 445–461.

39. Anderson CR, Stevens CF. Voltage clamp analysis of acetylcholine produced end-plate current fluctuation at frog neuromuscular junction. J Physiol (London) 1973; 235: 655–691.

40. Jablecki CK. Electrodiagnostic evaluation of patients with myasthenia gravis and related disorders. Neurol Clin 1985; 3: 557–572.

41. Jones SW. Overview of voltage-dependent calcium channels. J Bioenerg Biomembr 1998; 30: 299–312.

42. Black JL, Lennon VA. Identification and cloning of human neuronal high voltage-gated calcium channel γ-2 and γ-3 subunits: neurological implications. Mayo Clin Proc 1999; 74: 357–361.

43. Molenaar PC, Newsom-Davis J, Polak RL, Vincent A. Eaton-Lambert syndrome: acetylcholine and choline acetyltransferase in skeletal muscle. Neurology 1982; 32: 1061–1065.

44. Harper CM. Electrodiagnosis of endplate disease. In: Engel AG, ed. Myasthenia Gravis and Myasthenic Syndromes. New York: Oxford Press, 1999: 65–86.

45. O'Neill JH, Murray NM, Newsom-Davis J. The Lambert-Eaton myasthenic syndrome. A review of 50 cases. Brain 1988; 111: 577–596.

46. O'Suilleabhain P, Low PA, Lennon VA. Autonomic dysfunction in the Lambert-Eaton myasthenic syndrome: serological and clinical correlates. Neurology 1998; 50: 88–93.

47. Waterman SA, Lang B, Newsom-Davis J. Effect of Lambert-Eaton myasthenic syndrome antibodies on autonomic neurons in the mouse. Ann Neurol 1997; 42: 147–156.

48. Komai K, Jwasa K, Takamori M. Calcium channel peptide can cause an autoimmune-mediated model of Lambert-Eaton myasthenic syndrome in rats. J Neurol Sci 1999; 166: 126–130.

49. Oguro-Okano M, Griesmann GE, Wieben ED, Slaymaker S, Snutch TP, Lennon VA. Molecular diversity of neuronal type calcium channels identified in small cell lung carcinoma. Mayo Clin Proc 1992; 67: 1150–1159.

50. Maddison P, Newsom-Davis J, Mills K R, Souhami RL. Favorable prognosis in Lambert-Eaton myasthenic syndrome and small-cell lung carcinoma. Lancet 1999; 353: 117–118.

51. Prior C, Lang B, Wray D, Newsom-Davis J. Action of Lambert-Eaton myasthenic syndrome IgG at mouse motor nerve terminals. Ann Neurol 1985; 17: 587–592.

52. Fukuoka T, Engel AG, Lang B, Newsom-Davis J, Prior C, Wray DW. Lambert-Eaton myasthenic syndrome: I. Early morphological effects of IgG on the presynaptic membrane active zones. Ann Neurol 1987; 22: 193–199.

53. De Aizpurua HJ, Lambert EH, Griesmann GE, Olivera BM, Lennon VA. Antagonism of voltage-gated calcium channels in small cell carcinomas of patients with and without Lambert-Eaton myasthenic syndrome by autoantibodies, ω-conotoxin and adenosine. Cancer Res 1988; 48: 4719–4724.

54. Lang B, Vincent A, Murray NM, Newsom-Davis J. Lambert-Eaton myasthenic syndrome: immunoglobulin G inhibition of Ca2 + influx in tumor cells correlates with disease severity. Ann Neurol 1989; 25: 265–271.

55. Meriney SD, Hulsizer SC, Lennon VA, Grinnell AD. Lambert-Eaton myasthenic syndrome immunoglobulins react with multiple types of calcium channels in small cell lung carcinoma. Ann Neurol 1996; 40: 739–749.

56. Garcia KD, Beam KG. Reduction of calcium currents by Lambert-Eaton syndrome sera: Motoneurons are preferentially affected, and L-type currents are spared. J Neurosci 1996; 16: 4903–4913.

57. Lang B, Newsom-Davis J, Peers C, Prior C, Wray DW. The effect of myasthenic syndrome antibody on presynaptic calcium channels in the mouse. J Physiol (London) 1987; 390: 257–270.

58. Yu Z, Lennon VA. Mechanism of intravenous immune globulin therapy in antibody-mediated autoimmune diseases. N Engl J Med 1999; 340: 227–228.

59. Giuntoli RL, Lu J, Kobayashi H, Kennedy R, Celis E. Direct costimulation of tumor-reactive CTL by helper T cells potentiate their proliferation, survival and effector function. Clin Cancer Res 2002; 8: 922–931.

60. Wirtz PW, Nijnuis MG, Sotodeh M, Willems LNA, Brahim JJ, Putter H, Wintzen AR, Verschuuren JJ. J Neurology 2003; 250: 698–701.

61. Willcox N, Demaine AG, Newsom-Davis J, Welsh KI, Robb SA, Spiro SG. Increased frequency of IgG heavy chain marker Glm(2) and of HLA-B8 in Lambert-Eaton myasthenic syndrome with and without associated lung carcinoma. Hum Immunol 1985; 14: 29–36.

62. Parsons KT, Kwok WW, Gaur LK, Nepom GT. Increased frequency of HLA class II alleles DRB1*0301 and DQB1*0201 in Lambert-Eaton myasthenic syndrome without associated cancer. Hum Immunol 2000; 61: 828–833.

63. Elrington GM, Murray NME, Spiro SG, Newsom-Davis J. Neurological paraneoplastic syndromes in patients with small cell lung cancer: a prospective survey of 150 patients. J Neurol Neurosurg Psychiatry 1991; 54: 764–767.

64. Seltzer V. Cancer in women: prevention and early detection. J Womens Health Gender Based Med 2000; 9: 483–488.

65. Lucchinetti CF, Kimmel DW, Lennon VA. Paraneoplastic and oncological profiles of patients seropositive for type 1 anti-neuronal nuclear autoantibodies. Neurology 1998; 50: 652–657.

66. Yu Z, Kryzer TJ, Griesmann GE, Kim K-K, Benarroch E, Lennon VA. CRMP-5 neuronal autoantibody: marker of lung cancer and thymoma-related autoimmunity. Ann Neurol 2001; 49: 146–154.

67. Pittock SJ, Kryzer TJ, Lennon VA. Paraneoplastic antibodies coexist and predict cancer, not neurological syndrome. Ann Neurol 2004; 56: 715–719.

68. Gutmann L, Phillips LH, Gutmann L. Trends in the association of Lambert-Eaton myasthenic syndrome with carcinoma. Neurology 1992; 42: 848–850.

69. Argov Z, Shapira Y, Averbuch-Heller L, Wirguin I. Lambert-Eaton myasthenic syndrome (LES) in association with lymphoproliferative disorders. Muscle Nerve 1995; 18: 715–719.

70. Burns TM, Juel VC, Sanders DB, Phillips LH. Neuroendocrine lung tumors and disorders of the neuromuscular junction. Neurology 1999; 52: 1490–1491.

71. Collins DR, Connolly S, Burns M, Offiah L, Grainger R, Walsh JB. Lambert-Eaton myasthenic syndrome in association with transitional cell carcinoma: a previously unrecognized association. Urology 1999; 54: 162.

72. Dalmau J, Gultekin HS, Posner JB. Paraneoplastic neurologic syndromes: pathogenesis and physiopathology. Brain Pathol 1999; 9: 275–284.

73. Oyaizu T, Okada Y, Sagawa, Yamakawa K, Kuroda H, Fujihara K, Itoyama Y, Tanita T, Motomura M, Kondo T. Lambert-Eaton myasthenic syndrome associated with an anterior mediastinal small cell carcinoma. J Thoracic Cardiovasc Surg 2001; 121: 1005–1006.

74. Kennelly KD, Dodick DW, Pascuzzi RM, Albain KS, Lennon VA. Neuronal autoantibodies and paraneoplastic neurological syndromes associated with extrapulmonary small cell carcinoma. Neurology 1997; 48 (Suppl. 2): A31.

75. Burns TM, Russell JA, Lachance DH, Jones, HR. Oculobulbar involvement is typical with Lambert-Eaton Myasthenic Syndrome. Ann Neurol 2003; 53: 270–273.

76. Barr CW, Claussen G, Thomas D, Fesenmeier JT, Pearlman RL, Oh SJ. Primary respiratory failure as the presenting symptom in Lambert-Eaton myasthenic syndrome. Muscle Nerve 1993; 6: 712–715.

77. Vernino S, Lennon VA. New Purkinje cell antibody (PCA-2): marker of lung cancer-related neurological autoimmunity. Ann Neurol 2000; 47: 297–305.

78. Chan KH, Vernino S, Lennon VA. ANNA-3 anti-neuronal nuclear antibody: Marker of lung cancer-related autoimmunity. Ann Neurol 2001; 50: 301–311.

79. Graus F, Vincent A, Pozo-Rosich P, Sabater L, Saiz A, Lang B, Dalmau J. Anti-glial nuclear antibody: marker of lung cancer-related paraneoplastic neurological syndromes. J Neuroimmunol 2005; 165: 166–171.

80. Lachance D, Pittock, SJ, Kryzer TJ, Lennon V, Chan KH. Anti-neuronal nuclear antibody type 4 (ANNA-4), a novel paraneoplastic marker of small cell lung carcinoma (SCLC). Neurology 2006; 66 (Suppl. 2): A340.

81. Lee H-R, Lennon VA, Camilleri M, Prather CM. Paraneoplastic gastrointestinal motor dysfunction: clinical and laboratory characteristics. Am J Gastroenterol 2001; 96: 373–379.

82. Vernino S, Low PA, Fealey RD, Stewart JD, Farrugia G, Lennon VA. Autoantibodies to ganglionic acetylcholine receptors in autoimmune autonomic neuropathies. N Engl J Med 2000; 343: 847–855.

83. Pasha SF, Lunsford TN, Lennon VA. Autoimmune gastrointestinal dysmotility treated successfully with pyridostigmine – A Case Report. Gastroenterology 2006; 131: 1592–1596.

84. Knowles CH, Lang B, Clover L, Scott SM, Gotti C, Vincent A, Martin JE. A role for autoantibodies in some cases of acquired non-paraneoplastic gut dysmotility. Scand J Gastroenterol 2002; 37: 166–170.

85. Lundh H, Nilsson O, Rosen I, Johansson S. Practical aspects of 3,4-diaminopyridine treatment of the Lambert-Eaton myasthenic syndrome. Acta Neurologica Scand 1993; 88: 136–140.

86. Sanders DB. 3,4-Diaminopyridine (DAP) in the treatment of Lambert-Eaton myasthenic syndrome (LES). Ann NY Acad Sci 1998; 841: 811–816.

87. Small S, Ali HH, Lennon VA, Brown RH, Carr DB, de Armendi A. Anesthesia for unsuspected Lambert-Eaton myasthenic syndrome with autoantibodies and occult small cell lung carcinoma. Anesthesiology 1992; 76: 142–145.

88. Nakao YK, Motomura M, Fukudome T, Fukuda T, Shiraishi H, Yoshimura T, Tsujihata M, Eguchi K. Seronegative Lambert-Eaton myasthenic syndrome: study of 110 Japanese patients. Neurology 2002; 59: 1773–1775.

89. Chan KH, Lachance DH, Harper CM, Lennon VA. Frequency of seronegativity in adult-acquired generalized myasthenia gravis. Muscle Nerve 2006: 36: 651–658.

90. Lennon VA. Serological diagnosis of myasthenia gravis and the Lambert-Eaton myasthenic syndrome. In: Lisak R, ed. Handbook of Myasthenia Gravis, Chapter 7. New York: Marcel Dekker, 1994; 149–164.

91. Vernino S, Adamski J, Kryzer TJ. Lennon VA. Neuronal nicotinic AChR antibody in subacute autonomic neuropathy and cancer-related syndromes. Neurology 1998; 50: 1806–1813.

92. Lennon VA, Kryzer TJ. Correspondence: Neuronal calcium channel autoantibodies coexisting with type 1 Purkinje cell cytoplasmic autoantibodies (PCA-1 or "anti-Yo"). Neurology 1998; 51: 327–328.

93. Sabater L, Titulaer M, Saiz A, Verschurren J, Gure AO, Graus F. SOX1 antibodies are markers of paraneoplastic Lambert-Eaton myasthenic syndrome. Neurology 2008; 70: 924–928.

94. Gure AO, Stockert E, Scanlan MJ, et al. Serological identification of embryonic neural proteins as highly immunogenic tumor antigens in small cell lung cancer. Proc Natl Acad Sci USA; 2000; 97: 4198–4203.

95. Albers JW, Faulkner JA, Dorovini-Zis K. Abnormal neuromuscular transmission in an infantile myasthenic syndrome. Ann Neurol 1984, 16: 28–34.

96. Bady B, Chauplannaz G, Carrier H, Congenital Lambert-Eaton myasthenic syndrome. J Neurol Neurosurg Psychiatr 1987; 50: 476–478.

97. Katz JS, Wolfe GI, Bryan WW, Tintner R, Barohn RJ. Acetylcholine receptor antibodies in the Lambert-Eaton myasthenic syndrome. Neurology 1998; 50: 470–475.

98. Kamenskaya MA, Elmqvist D, Thesleff S. Guanidine and neuromuscular transmission. IL Effect on transmitter release in response to repetitive nerve stimulation. Arch Neurol 1975; 32: 510–518.

99. Cherington M. Guanidine and germine in Eaton-Lambert syndrome. Neurology. 1976; 26: 944–946.

100. Blumhardt LD, Joekes AM, Marshall J, Philalithis PE. Guanidine treatment and impaired renal function in the Eaton-Lambert syndrome. Br Med J 1977; 1:946–947.

101. Silbert PL, Hankey GJ, Barr AL. Successful alternate day guanidine therapy following guanidine-induced neutropenia in the Lambert-Eaton myasthenic syndrome. Muscle Nerve 1990; 13: 360–361.

102. McEvoy KM, Windebank AJ, Daube JR, Low PA. 3,4-Diaminopyridine in the treatment of Lambert-Eaton myasthenic syndrome. N Engl J Med 1989; 321: 1567–1571.

103. Harper CM, McEvoy KM, Windebank AJ, Daube JR. Effect of 3,4-Diaminopyridine on neuromuscular transmission in patients with Lambert-Eaton myasthenic syndrome. Electroencephalogr Clin Neurophysiol 1990; 75: 557.

104. Kim DS, Claussen GC, Oh SJ. Single-fiber electromyography improvement with 3,4-diaminopyridine in Lambert-Eaton myasthenic syndrome. Muscle Nerve 1998; 21: 1107–1108.

105. Boerma CE, Rommes JH, van Leeuwen RB, Bakker J. Cardiac arrest following an iatrogenic 3,4-diaminopyridine intoxication in a patient with Lambert-Eaton myasthenic syndrome. Clin Toxicol 1995; 33: 249–251.

106. Dau PC, Denys EH. Plasmapheresis and immunosuppressive drug therapy in the Eaton-Lambert syndrome. Ann Neurol 1982; 11: 570–575.

107. Bird SJ. Clinical and electrophysiologic improvement in Lambert-Eaton syndrome with intravenous immunoglobulin therapy. Neurology 1992; 42: 1422–1423.

108. Motomura M, Tsujihata M, Takeo G, Matsuo H, Kinoshita I, Nakamura T, Yoshimura T, Nagataki S. The effect of high-dose gamma-globulin in Lambert-Eaton myasthenic syndrome. Neurol Ther 1994; 11: 377–383.

109. Takano H, Tanaka M, Koike R, Nagai H, Arakawa M, Tsuji S. Effect of intravenous immunoglobulin in Lambert-Eaton myasthenic syndrome with small-cell lung cancer: correlation with the titer of anti-voltage-gated calcium channel antibody. Muscle Nerve 1994; 17: 1074–1075.

110. Bain PG, Motomura M, Newsom-Davis J, Misbah SA, Chapel HM, Lee ML, Vincent A, Lang, B. Effects of intravenous immunoglobulin (IVIg) treatment on muscle weakness and calcium channel autoantibodies in the Lambert-Eaton myasthenic syndrome. Neurology 1996; 47: 678–683.
111. Vernino S, Lennon VA. Autoantibody profiles and neurological correlations of thymoma. Clinical Cancer Research 2004; 10: 7270–7275.

14

Acquired Neuromyotonia

Angela Vincent and Ian Hart

1. INTRODUCTION

Neuromyotonia (NMT; Isaacs' syndrome) is a syndrome of nerve hyperexcitability that usually presents with clinical and electrodiagnostic features of continuous skeletal muscle overactivity. Although the syndrome can be caused by several different pathogenic mechanisms, here we concentrate on the acquired autoimmune form. We review the clinical features, including CNS features, immune and tumor associations, and the role of antibodies to voltage-gated ion channels.

2. CLINICAL FEATURES OF NMT AND RELATED SYNDROMES

2.1. Clinical Manifestations

The typical presenting features are myokymia (muscle twitching) and cramps (1, 2, 3, 4, 5), which can be associated with muscle stiffness, pseudomyotonia (delayed muscle relaxation after contraction), pseudotetany (for example, Chvostek's and Trousseau's signs) and weakness. The limb and trunk muscles are most commonly affected, although facial, bulbar and respiratory muscles can also be involved. There is often increased sweating. Muscle contraction often provokes or exacerbates the symptoms. Reflexes are usually normal but can be reduced or absent when there is an accompanying neuropathy. Chronic cases may develop muscle hypertrophy.

Some patients with NMT have sensory symptoms such as paraesthesia and numbness without evidence of peripheral neuropathy, suggesting that sensory as well as motor nerves may be hyperexcitable (6). Perhaps more interestingly, in view of the antibody-mediated pathology, central nervous system (CNS) symptoms ranging from personality change and insomnia to a psychotic-like state with delusions and hallucinations can occur. This association is often called chorée fibrillaire or Morvan's syndrome (7, 8), which can also include autonomic dysfunction (9, 10). Although the full-blown Morvan's syndrome is apparently rare, there is an emerging spectrum of related CNS disorders with mild or even absent peripheral features (see below).

Neuromyotonia can occur at any age, and since the severity of the clinical features ranges from the inconvenient to the incapacitating, it is not at all clear how common the disorder is. The course is variably progressive, although spontaneous remissions can occur and some forms may be monophasic. These have been related to established infections in a few cases (11, 12) or to a preceding anaphylactic shock (13). No HLA association has been detected in 12 patients examined (Hart unpublished data).

A review of the clinical features of NMT showed that very similar features are found in patients with cramp-fasciculation syndrome (CFS, see (14)), a proportion of whom also have VGKC antibodies (Table 1, (14)). This, together with other evidence for autoimmunity in some CFS patients (14), strongly indicates that CFS and NMT should be considered as parts of a spectrum of autoimmune peripheral nerve hyperexcitability. This may also apply to myokymia/cramp syndrome (15) and other conditions associated with muscle hyperexcitability and hyperhidrosis (16, 17). Gutmann et al. (18) have discussed the relationship between electrical neuromyotonic

From: *Current Clinical Neurology*
Myasthenia Gravis and Related Disorders
Edited by: H.J. Kaminski, DOI 10.1007/978-1-59745-156-7_14, © Humana Press, New York, NY

Table 1
Similarities Between Neuromyotonia and the Cramp-Fasciculation Syndrome

Feature	Neuromyotonia (42 patients) % positive	Cramp-fasciculation syndrome (18 patients) % positive
Cramps	88	100
Muscle twitching	100	94
Pseudomyotonia	36	22
Sensory symptoms	33	39
CNS symptoms	29	22
Other autoimmune diseases	59	28
Thymoma	19	11
Lung tumor	10	6
Other autoantibodies	50	17
VGKC antibodies	38	28

Data from reference *(14)*.

and myokymic discharges and the clinical disorder and come to a similar conclusion. Here we will mainly discuss NMT because most of the experimental studies have been performed in patients who fulfill this diagnosis.

2.2. Electrophysiological Features

The classical electrophysiological feature of NMT is the spontaneous firing of single motor units as doublet, triplet or multiplet discharges (Fig. 1). These neuromyotonic discharges occur at irregular intervals of 1–30 s and have an intraburst frequency of 40–400 Hz. They may be present in the absence of visible muscle twitching. In addition, voluntary or electrical activation may trigger prolonged after-discharges at similar frequencies. Fasciculation and fibrillation potentials can also occur. The presence of these different electrophysiological phenomena in NMT has recently been summarized *(19)* (Table 2).

The spontaneous and repetitive muscle activity arises in the peripheral motor nerve, rather than in the muscle itself *(4, 20)*. Isaacs *(20)* showed that the discharges were abolished when

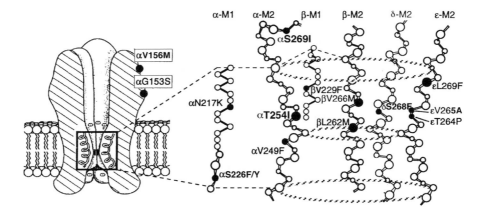

Fig. 1. Needle electromyography recording from the right extensor digitorum communis muscle of a neuromyotonia patient. There is a multiplet discharge with an intraburst frequency of 150 Hz followed at irregular intervals by several doublets and single discharges (fasciculations), all from a single motor unit.

Table 2
Frequency of Different Electrophysiological Findings in NMT

Electrophysiological finding	Frequency (%) in 11 patients
Spontaneous neuromyotonia	11 (100)
Stimulus-induced neuromyotonia	7 (65)
Fibrillation potentials	2 (18)
Fasciculations	11 (100)
Neuromyotonic discharges	
Doublets	11 (100)
Triplet	10 (90)
Multiplets	8 (73)
Maximum intraburst frequency (Hz)	50–220

Data taken from Maddison et al. *(19)*.

curare was used to block neuromuscular transmission, but persisted after either proximal or distal nerve block. A macro EMG study also suggested that spontaneous discharges are generated distally *(21)*. These observations, and recent studies using botulinum toxin, suggested that the activity was generated in the terminal arborization of the motor axons *(21, 22)*. However, other studies indicate a more proximal initiation of the hyperactivity with reduction by proximal nerve block *(23, 24, 25)* or by epidural anesthesia *(26)*. The simplest explanation for these data is that the activity can originate at different sites including the lower motor neuron cell body in some cases. Some patients also with stiff person syndrome present with muscle hyperactivity and cramps. Sleep or general anesthesia abolishes muscle activity in stiff person syndrome but not in NMT, and this is a clinically important distinction between the two conditions.

Although it seems likely that NMT is usually a functional disorder of the motor nerve, rather than a secondary phenomenon caused by structural nerve or myelin damage, about 5–10% of patients have, in addition to NMT, electrical evidence of a subclinical idiopathic axonal or demyelinating polyneuropathy.

3. SERUM AND CEREBROSPINAL FLUID TESTS

Although most patients with NMT reported so far do not have marked CNS features, it is common to find a raised total IgG and oligoclonal bands in the CSF *(1)*. Curiously, VGKC antibodies (see below) are often not found, perhaps because the serum levels are relatively low and the corresponding CSF level undetectable.

4. PHYSIOLOGICAL BASIS OF EXCITABILITY OF MOTOR AXONS IN NMT PATIENTS

It is possible to measure the response of subthreshold stimuli and the stimulus intensity (threshold) that is required to excite the motor nerve *(27, 28)*. One of these measures is the strength–duration time constant (SDTC), which is a property of the nodal membrane determined by the passive membrane time constant and the ion conductances acting at the nerve threshold. The SDTC was prolonged in the motor but not in the sensory axons in some NMT patients in one study *(29)*. However, a subsequent study, using more sensitive measures of both motor and sensory axonal excitability in the median nerve, found that all excitability indices were normal in all eight NMT patients examined *(30)*. Thus, a general alteration of axonal membrane permeability is probably not present in these patients, as predicted by the previous studies, which indicated a distal pathology in most patients (see above).

4.1. Clinical and Experimental Evidence for an Autoimmune Basis

4.1.1. Associated Autoimmune Syndromes

As in disorders of the neuromuscular junction, one of the first indications that NMT had an autoimmune basis was the recognition of associated autoimmune diseases. These conditions include myasthenia gravis (MG) with or without thymoma, rheumatoid arthritis, systemic lupus erythematosus and diabetes (Table 1). Individual patients with NMT and systemic sclerosis or idiopathic thrombocytopenia, or NMT occurring after bone marrow transplantation, are also reported *(31, 32, 33, 34, 35, 36, 37, 38, 39)*. Some patients may have an inflammatory neuropathy, although in these cases one might argue that the neuropathy is the primary event *(40)*.

The strongest tumor association is with thymoma. This occurs in patients with or without associated MG, and particularly in the former can precede the neuromyotonia by several years. NMT developing after treatment of thymoma is often a sign of tumor recurrence. Small-cell lung cancers are also described in a few patients (Table 1, *(25)*), but in these cases the neuromyotonia usually precedes the cancer with latencies between onset of NMT and diagnosis of SCLC of up to 4 years *(1)*. There are isolated reports of neuromyotonia with Hodgkin's lymphoma *(41)* or with a plasmacytoma and paraproteinemia *(42)*.

4.1.2. Response to Immunotherapies

The first clinical experiment that suggested an autoimmune etiology for NMT was in a patient in whom plasma exchange resulted in marked reduction in the frequency of spontaneous motor unit potentials *(43)*. Although there have been no formal trials of immunotherapies, there are many anecdotal reports of improvement following plasma exchange, intravenous immunoglobulins or corticosteroids *(1)*. In most patients, clinical improvement begins within 2–8 days of plasma exchange and lasts for about 1 month.

4.1.3. Passive Transfer of Immunoglobulins to Mice

Passive transfer of NMT IgG to mice provided further evidence for the autoimmune basis of the disorder. After injection of purified IgG intraperitoneally for 10–12 days, there was an enhancement of neuromuscular transmission in mouse hemidiaphragm preparations, as demonstrated by a resistance to bath-applied curare *(43)*. Mice injected with NMT IgG released significantly more packets of acetylcholine from their motor nerve endings, consistent with the increased efficiency of neuromuscular transmission *(44)*. However, there were no spontaneous motor unit potentials or double responses to single stimuli, as observed in the human disease *(43)*. Dorsal root ganglion cell cultures incubated with NMT IgG, however, showed repetitive action potentials *(44)*. Together these studies suggested that an IgG autoantibody is responsible for at least some of the features of neuromyotonia. Moreover, the striking similarity between the changes seen in the mouse preparations and those found after small doses of the VGKC antagonists, 3,4-diaminopyridine or 4-aminopyridine, makes it likely that the target of the antibodies is a VGKC *(44)*. The main function of neuronal VGKC is to restore the membrane potential after each action potential, preventing re-opening of voltage-calcium channels. Thus, a reduction in VGKC numbers or function would lead to prolonged depolarization of the motor nerve with enhanced transmitter release. Indeed dendrotoxins from species of *Dendroaspis mambas (45)*, that block the function of some of the VGKCs of the Kv1 family, produce repetitive or spontaneous activity in the motor nerve terminal not dissimilar from that in NMT.

VGKCs function as delayed rectifiers or type A rapid inactivators and are involved in nerve repolarization *(45)*. A variety of studies suggest that they are mainly restricted to the paranodal and juxtanodal regions and nerve terminals of the peripheral motor nerves. In addition, they are also widely expressed in the central nervous system and peripheral sensory nerves. There are at least nine VGKC genes, termed Kv1.1–1.9 *(46)*. In the nodes, both Kv1.1 and 1.2 are expressed and co-localize with binding of NMT IgG *(47)*. The identity of the VGKCs at the motor nerve terminal is not so clear, but both Kv1.2 and 1.6 may be present as well as other types of potassium channel (K. Kleopa, personal communication).

5. ANTIBODIES TO VGKC

In order to look for antibodies to VGKCs, the same approach was used as in MG (Chapter 8). VGKCs were extracted from human or mammalian brain and allowed to bind ^{125}I-α-dendrotoxin *(44, 48)*. The labeled complex was then used in radioimmunoprecipitation assays. The results show significant levels of antibodies to VGKCs in around 40% of NMT patients (Fig. 2). A more sensitive but more difficult technique is to measure antibody binding to *Xenopus* oocytes that have been made to express various VGKC genes. Using a sensitive molecular-immunohistochemical assay, Hart et al. showed antibody binding to the different VGKC Kv1 subunits in 80–90% of patients [*(48)*]) and unpublished results]. At present, however, this approach is not widely available.

Functional VGKCs consist of four α subunits, each of which has six transmembrane regions (S1–S6) *(46)*. Together, the α subunits' S5 and S6 regions form the central ion pore. VGKC Kv1s are expressed as gene products that need to combine as homomultimeric or heteromultimeric tetramers and to associate with intracellular β subunits *(49a)*. In the immunoprecipitation assay only those complexes that include subtypes that bind dendrotoxin (Kv1.1, 1.2 and 1.6) are labeled. This means that antibodies to other subtypes that are associated with dendrotoxin-binding subunits in a heteromultimer will be detected, but antibodies to those that do not associate with dendrotoxin-binding subunits will not be detected. In practice, the assay is also limited by the concentrations of the various channel subtypes in the brain extracts currently employed. For instance, in rabbit brain extracts, the Kv1.2 subtype seems to predominate and many NMT autoantibodies appear to bind to it (Watanabe, Clover and Vincent, unpublished findings), but there may be other subtypes or combinations of subtypes that are not present in the extract and are potential targets for the antibodies. Attempts to improve the sensitivity of the assay by employing different neurotoxins such as charybdotoxin, apamin or margatoxin have not been successful. In fact, margatoxin, which is thought to bind to the Kv1.3 subtype as well as Kv1.2, produced results that were essentially identical to those with dendrotoxin (Clover et al. unpublished observations).

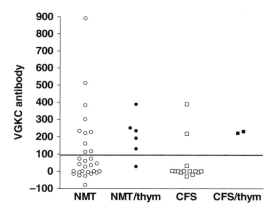

Fig. 2. VGKC antibodies in neuromyotonia patients without or with a thymoma. Data taken from Hart et al. *(14)*.

Immunohistochemical assays show that different sera contain antibodies to different combinations of VGKC subunits, indicating heterogeneity of the antibody response *(48)*, similar to that found for antibodies to acetylcholine receptors in MG. However, the studies need to be extended to all known VGKC subtypes in a more systematic manner. One way to do this would be to express each VGKC gene individually in a mammalian cell line and to use immunofluorescence to measure binding of the antibodies to the expressed VGKCs. Similar approaches are now being used to improve the sensitivity of antibody detection in myasthenia gravis (Leite et al. 2008) *(49b)*. This method has the advantage that one would be measuring only antibodies to the extracellular determinants, which are likely to be the pathogenic antibodies, and eventually the approach could be modified to provide a precise and quantitative assay. However, the efficiency of cell surface expression of individual VGKC Kv1 subunits is variable and this method has not yet been found helpful in the detection of antibodies in NMT.

In most NMT patients, the anti-VGKC antibodies are IgG. However, one patient with a plasmacytoma and IgM monoclonal gammopathy developed NMT *(42)* and IgM purified from another NMT patient bound to an α-dendrotoxin-labeled protein by Western blotting and showed immunoreactivity with both cultured PC12 cells and human intramuscular nerve axons *(50, 51)*. These findings suggest that in some patients NMT is IgM autoantibody-mediated.

The mechanism of action of NMT autoantibodies is probably to modulate the number of functional VGKCs, rather than to cause a complement-mediated attack on the motor nerve

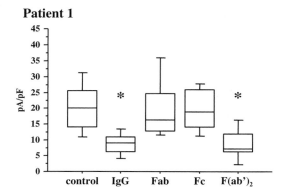

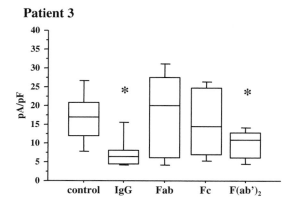

Fig. 3. Effects of IgG subfractions purified from a patient with neuromyotonia on VGKC currents in NB-1 cells. Whole IgG and F(ab')2 reduced currents substantially, but monovalent Fab and Fc did not. Taken from Tomimitsu et al. *(53)* with permission.

terminal or to directly block the ion channel. Patch clamp studies of cultured PC12 and NB-1 cell lines, and also of non-neuronal cell lines transfected with VGKC Kv1.1 or 1.6, demonstrated that NMT IgG suppresses voltage-dependent potassium currents *(51, 52, 53)*. This effect did not occur within a few hours but was found after overnight incubation. Moreover, the role of complement-mediated damage was excluded, since the effects are dependent on divalent antibodies or F(ab)2 and not found with monovalent Fabs or Fc *(53)* (Fig. 3). The lack of complement-mediated attack in both the Lambert–Eaton myasthenic syndrome and the NMT is interesting, because previous studies in Miller–Fisher syndrome, in which there are IgG antibodies directed against the ganglioside GQ1b on the motor nerve terminal, indicate that complement can effectively damage the motor endplate *(54)*. One explanation could be that the antibodies in LEMS and NMT are of the IgG2 or IgG4 subclasses, which do not bind complement efficiently. Indeed, most VGKC antibodies appear to be of IgG4 subclass predominantly, although extensive studies have not yet been done (Clover, Vincent unpublished observations).

A direct block of function should be seen within 1 or 2 h. When NMT IgG is applied to dorsal root ganglion cells, repetitive firing was observed after 24 h but not in the initial 2 h *(44)*. Similarly, in the patch clamp studies of PC12 cells, application of the NMT IgG for short periods did not have any effect, and positive results were only found after one or more days *(51, 52, 53)*. Thus, autoantibodies appear to be able to reduce VGKC function by lowering the available numbers of channels rather than by directly affecting their activity.

5.1. Other Possible Autoantibodies in NMT

A variety of neurotoxins from species as diverse as dinoflagellates, scorpions, snails and sea anemones, alter the closing of voltage-gated sodium channels to prolong sodium currents and produce repetitive firing of axons. Thus antibodies to sodium channels, if they also prolonged activations, could cause NMT. However, a patch clamp study of NB-1 cells did not detect any significant effect of NMT IgG on sodium currents *(52)*.

Other possible antigens targeted by the autoimmune process are the modulatory receptors that are present on the motor nerve terminal. These bind a variety of substances released by the motor nerve, such as acetylcholine, adenosine triphosphate, adenosine and calcitonin gene-related peptide, or circulating substances such as epinephrine. These substances are thought to exist in order to regulate acetylcholine release and prevent overexcitation of the motor endplate by excessive acetylcholine. Therefore, any reduction in their expression could lead to muscle overactivity and manifestations of NMT.

A radioimmunoassay using ^{125}I-labeled epibatidine (a specific agonist drug) identified autoantibodies to neuronal (rather than muscle) nicotinic acetylcholine receptors in sera from three of the six NMT patients studied *(55)*. However, it is not clear what effect these antibodies would have on the activity of the motor nerve terminal as most studies in mammalian or amphibian muscles suggest that the functions of these presynaptic receptors are subtle and can only be demonstrated under relatively extreme conditions.

In some patients, autoantibodies may be directed against myelin antigens. The electromyographic features of NMT have been demonstrated in mice with targeted deletions in the genes for P0 or Pmp22 and in the Trembler mouse, which has a point mutation in Pmp22 *(56. 57)*. The similarities with NMT are striking (discussed in *(58)*) and suggest that manifestations of neuromyotonia could also be caused by antibodies to myelin proteins in humans.

6. VGKC ANTIBODIES IN OTHER MUSCLE DISEASES

A study of 42 patients with NMT and 18 patients with cramp-fasciculation syndrome (CFS) suggests considerable overlap between the two syndromes, as summarized in Table 1. Indeed, immunoglobulins from two CFS patients suppressed outward potassium currents of patch-clamped NB-1 cells in culture in a manner similar to NMT Ig. In addition, VGKC antibodies were found in a patient with "neuromuscular hyperexcitability", very similar to a previous case of rippling muscle disease (59). Both these patients had coexisting MG and are therefore assumed to have an autoimmune disease.

NMT can occur in patients with GBS (40), an acute inflammatory demyelinating polyneuropathy, and at least one patient had VGKC antibodies (Ebers and Vincent unpublished observations). Defects in myelination by themselves can induce nerve hyperexcitability, as mentioned above, presumably by secondary effects on the regional expression of ion channels along the length of the axon. Purified Ig from two GBS patients reduced VGKC currents in NB-1 cells (52), but there is no evidence that these antibodies can immunoprecipitate VGKCs and their significance is uncertain.

7. CENTRAL NERVOUS SYSTEM ABNORMALITIES IN NMT

7.1. Morvan's Syndrome

The clinical association of neuromyotonia with hallucinations, delusions, insomnia and personality change was first recognized by Morvan, who called it chorée fibrillare (7, 8). Although he described the case fully, the central symptoms were only recorded in one of the ten patients with symptoms and signs of NMT that he described. If they are sought specifically, the incidence of central nervous system manifestations in NMT is probably around 20% (Table 1, (14)).

The question arises as to whether autoantibodies enter the CNS by diffusion, by leakage at sites where the blood–brain barrier is known to be weak or whether there is intrathecal synthesis of autoantibody. Oligoclonal bands or an increase in CSF total IgG are commonly present in NMT patients, suggesting that there is some intrathecal IgG synthesis (1). Preliminary immunohistochemical studies demonstrated binding of both CSF and serum IgG from anti-VGKC antibody-positive NMT patients to neurons in the dentate nucleus of the human cerebellum (60, 61), and the pattern was similar to that of a polyclonal rabbit antiserum raised against a peptide of the Kv1.1 subtype. However, in general, the VGKC antibody levels in the CSF of patients are very low, even when they have clear CNS manifestations (Vincent and Clover, unpublished finding). Thus, the relationship between VGKC antibodies and CNS abnormalities is not understood.

This is underscored by study of a patient with classical Morvan's syndrome (10) (Fig. 4). His onset was at 76 years of age and no neoplasm was identified antemortem, although a small adenocarcinoma of the lung was found postmortem. He developed severe insomnia, memory loss, confusion and hallucinations over a few months, associated with extreme sweating, salivation and severe constipation. An electroencephalogram showed complete lack of non-REM sleep and a tendency to oscillate between a state of sub-wakefulness and REM sleep. Polysomnographic studies showed florid neuromyotonic discharges and cardiac arrhythmias. Strikingly, the peripheral, autonomic and central features improved substantially after plasma exchange. However, in addition to the VGKC antibodies there were marked changes in circadian rhythms of serum levels of various hormones under CNS control, such as cortisol and melatonin. These changes could have been secondary to the effects of the autoantibodies on the CNS, or conversely, the effects of the autoantibodies on the periphery might have altered hormone levels with secondary effects on CNS functions (10).

There have been increasing numbers of case reports or short studies of Morvan's syndrome over the last few years, with variable presentations including peripheral, autonomic and central manifestations. Neurophysiology and clinical features in two recent cases are described (62). VGKC antibodies appear to be present in only around 50% of the patients with these histories (A. Vincent, unpublished results) and other antibodies may be present in clinically similar cases (63). About half the patients (but not only the VGKC antibody-positive ones) have a thymoma. The

Morvan patient's serum   Control serum

Fig. 4. Evidence for VGKC antibodies capable of binding CNS neurones in a patient with neuromyotonia, autonomic dysfunction and CNS disorder (Morvan's syndrome; see reference *(10)*). The patient's serum IgG antibodies (*left*, pink stain) bind to thalamic neurones and to different layers of the hippocampus. In general white matter is spared. The distribution is similar to that of antibodies to Kv1.1 and 1.2 subtypes. Control serum IgG (*right*) does not bind appreciably. Courtesy of Ms Linda Clover.

full clinical spectrum and range of antibodies in these patients requires further study but, despite the growing recognition of the condition, it remains a rare disease.

7.2. Limbic Symptoms Without Overt Neuromyotonia

Two other patients were reported who contrast with Morvan's syndrome *(64)*. Both were diagnosed as having a limbic encephalitis on the basis of memory loss, confusion and anxiety with abnormalities of the hippocampus on MRI. Neither was reported to have neuromyotonia or sleep disorders at the time or subsequently, but both had excessive secretions – sweating in one and salivation in the other. In a retrospective study, both patients had high levels of VGKC antibodies at the time of their limbic manifestations, and both improved when VGKC antibody levels fell – in one case after plasma exchange and in the other spontaneously. Interestingly, one of these patients was a woman with a long history of MG with thymoma. The limbic manifestations arose following a recurrence of the thymoma, 13 years after the tumor had been removed. At the time, the MG was poorly controlled, but there was a striking dissociation between acetylcholine receptor autoantibody levels that fluctuated quite widely and the single peak of VGKC antibodies that coincided with the limbic abnormalities. Plasma exchange was effective in reducing the CNS signs *(64)*.

These observations stimulated us to look for antibodies to VGKC in a variety of patients with cognitive, psychiatric or sleep disorders, similar to those previously associated with paraneoplastic limbic encephalitis. Studies on VGKC antibody-positive cases *(65, 66)* indicate a growing number of patients with non-paraneoplastic limbic encephalitis that respond well to immunosuppressive treatment regimens. The antibody levels in these patients are typically high (often >1000 pM) and fall dramatically following adequate immunosuppression. In many cases, the antibodies do not recur even when immunotherapies are stopped. Around 200 cases have been identified in the UK over the last five 5 years (Vincent, Clover, unpublished observations) and a detailed study is in progress. This condition and other "autoimmune channelopathies" are reviewed in more detail elsewhere *(67)*.

8. TREATMENT OF NMT

Anti-epileptic drugs, such as carbamazepine and phenytoin, probably act by reducing sodium conductance and thus stabilize the nerve membrane making it less excitable. Many patients require only these drugs, either separately or together. In general, baclofen, benzodiazepines, carbonic anhydrase inhibitors and quinine are unhelpful.

Plasma exchange often produces short-term relief *(1, 43)* and a trial of plasma exchange can help establish the diagnosis in the difficult situation of a patient with suspected autoimmune

NMT but with no VGKC antibodies. There is no evidence for a differential response to plasma exchange among patients with and without detectable VGKC antibodies. Nevertheless, a positive response to plasma exchange may be helpful in predicting those patients who are likely to benefit from long-term immunosuppression. Intravenous immunoglobulin has also been used, although reports suggest that some patients do not do well *(68)*, or do better with plasma exchange *(69)*. Combinations of prednisolone with azathioprine or methotrexate have helped some patients *(1)*. A recent review from Europe summarizes the main approaches to treatment of NMT *(70)* but randomized control trials are needed to evaluate these therapies more fully.

REFERENCES

1. Newsom-Davis J, Mills KR. Immunological associations of acquired neuromyotonia (Isaacs' syndrome). Report of five cases and literature review. Brain 1993;116:453–469.
2. Denny-Brown D, Foley DM. Myokymia and the benign fasciculation of muscular cramps. Trans Assoc Am Physicians 1948;61:88–96.
3. Gamstorp I, Wohlfart G. A syndrome characterized by myokymia, myotonia, muscular wasting and increased perspiration. Acta Psychiatr Scand 1959; 34:181–194.
4. Isaacs H. A syndrome of continuous muscle-fibre activity. J Neurol Neurosurg Psychiatry 1961;24: 319–325.
5. Mertens HG, Zschoke S. Neuromyotonie. Klin Wocehenscher 1965;43:917–925.
6. Lance JW, Burke D, Pollard J. Hyperexcitability of motor and sensory neurons in neuromyotonia. Ann Neurol 1979;5:523–532.
7. Morvan A, De la chorée fibrillaire. Gaz Hebdon de Med Chirurg 1890;27:173–200.
8. Serratrice G, Azulay JP. Que reste-t-il de la choreé fibrillaire de Morvan? Rev Neurol 1994;150:257–265.
9. Halbach M, Homberg V, Freund HJ. Neuromuscular, autonomic and central cholinergic hyperactivity associated with thymoma and acetylcholine receptor–binding antibody. J Neurol (Berlin) 1987;234:433–436.
10. Liguori R, Vincent A, Clover L, et al. Morvan's syndrome: peripheral and central nervous system and cardiac involvement with antibodies to voltage-gated potassium channels. Brain 2001;124:2417–2426.
11. Maddison P, Lawn N, Mills KR, Vincent A, Donaghy M. Acquired neuromyotonia in a patient with spinal epidural abscess. Muscle Nerve 1998;21:672–674.
12. Tahmouch AJ, Alonso RJ, Tahmouch GP, et al. Cramp-fasciculation syndrome: a treatable hyperexcitable peripheral nerve disorder. Neurology 1991;41:1021–1024.
13. Turner MR, Madkhana A, Ebers GC, Clover L, Vincent A, McGavin G, Sarrigiannis P, Kennett R, Warrell DA. Wasp sting induced autoimmune neuromyotonia J Neurol Neurosurg Psychiatry. 2006;77:704–705.
14. Hart IK, Maddison P, Vincent A, Mills K, Newsom-Davis J. Phenotypic variants of autoimmune peripheral nerve hyperexcitability. Brain 2002; 125:1887–1895.
15. Smith KKE, Claussen GY, Fesenmeier JT, Oh SJ. Myokymia-cramp syndrome: evidence of hyperexcitable peripheral nerve. Muscle Nerve 1994;17:1065–1067.
16. Wakayama Y, Ohbu S, Machida H. Myasthenia gravis, muscle twitch, hyperhidrosis, and limb pain associated with thymoma: proposal of a possible new myasthenic syndrome. Tohoku J Exp Med 1991;164:285–291.
17. Ho WKH, Wilson JD. Hypothermia, hyperhidrosis, myokymia and increased urinary excretion of catecholamines associated with a thymoma. Med J Aust 1993;158:787–788.
18. Gutmann L, Libell D, Gutmann L. When is myokymia neuromyotonia? Muscle Nerve 2001;24:151–153.
19. Maddison P, Mills KR, Newsom-Davis J. Clinical, electrophysiological characterization of the acquired neuromyotonia phenotype of autoimmune peripheral nerve hyperexcitability. Muscle Nerve 2006;33:801–808.
20. Isaacs H. Continuous muscle fibre activity in an Indian male with additional evidence of terminal motor fibre abnormality. J Neurol Neurosurg Psychiatry 1967;30:126–133.
21. Arimura K, Arimura Y, Ng A, et al. The origin of spontaneous discharges in acquired neuromyotonia. A macro EMG study. Muscle Nerve 2005;116:1835–1839.
22. Deymeer F, Oge AE, Serdaroglu P, et al. The use of botulinum toxin in localizing neuromyotonia to the terminal branches of the peripheral nerve. Muscle Nerve 1998;21:643–646.
23. García-Merino A, Cabella A, Mora JS, Liaño H. Continuous muscle fiber activity, peripheral neuropathy, and thymoma. Ann Neurol 1991;29:215–218.
24. Irani PF, Purohit AV, Wadia HH. The syndrome of continuous muscle fiber activity. Acta Neurol Scand 1977;55:273–288.
25. Partanen VSJ, Soininen H, Saksa M, Riekkinen P. Electromyographic and nerve conduction findings in a patient with neuromyotonia, normocalcemic tetany and small-cell lung cancer. Acta Neurol Scand 1980;61:216–226.
26. Hosokawa S, Shinoda H, Sakai T, Kato M, Kuroiwa Y. Electrophysiological study on limb myokymia in three women. J Neurol Neurosurg Psychiatry 1987;50:877–881.

27. Bostock H, Cikurel K, Burke D. Threshold tracking techniques in the study of human peripheral nerve. Muscle Nerve 1998;21:137–158.
28. Burke D. Excitability of motor axons in neuromyotonia. Muscle Nerve 1999;22:797–799.
29. Maddison P, Newsom-Davis J, Mills KR. Strength-duration properties of peripheral nerve in acquired neuromyotonia. Muscle Nerve 1999;22: 823–830.
30. Kiernan MC, Hart IK, Bostock H. Excitability of motor axons in patients with spontaneous motor unit activity. J Neurol Neurosurg Psychiatry 2001;70:56–64.
31. Harman JB, Richardson AT. Generalized myokymia in thyrotoxicosis: report of a case. Lancet 1954;2:473–474.
32. Reeback J, Benton S, Swash M, Schwartz MS. Penicillamine-induced neuromyotonia. Brit Med J 1979;279:1464–1465.
33. Vilchez JJ, Cabello A, Benedito J, Villarroya T. Hyperkalaemic paralysis, neuropathy and persistent motor unit discharges at rest in Addison's disease. J Neurol Neurosurg Psychiatry 1980;43:818–822.
34. Benito-Leon J, Martin E, Vincent A, Fernandez-Lorente J, de Blas G. Neuromyotonia in association with essential thrombocythemia. J Neurol Sci 2000:173:78–79.
35. Gutmann L, Gutmann L, Schochet SS. Neuromyotonia and type 1 myofiber predominance in amyloidosis. Muscle Nerve 1996;19:1338–1341.
36. Hadjivassiliou M, Chattopadhyay AK, Davies-Jones GA, et al. Neuromuscular disorder as a presenting feature of coeliac disease. J Neurol Neurosurg Psychiatry 1997;63:770–775.
37. Le Gars L, Clerc D, Cariou D, et al. Systemic juvenile rheumatoid arthritis and associated Isaacs' syndrome. J Rheumatol 1997;24:178–180.
38. Benito-Leon J, Miguelez R, Vincent A, et al. Neuromyotonia in association with systemic sclerosis. J Neurol 1999;246:976–977.
39. Liguori R, Vincent A, Avoni P, et al. Acquired neuromyotonia after bone marrow transplantation. Neurology 2000;54:1390–1931
40. Vasilescu C, Alexianu M, Dan A. Muscle hypertrophy and a syndrome of continuous motor unit activity in prednisone-responsive Guillain–Barré polyneuropathy. J Neurol (Berlin) 1984;231:276–279.
41. Caress JB, Abend WK, Preston DC, Logigian EL. A case of Hodgkin's lymphoma producing neuromyotonia. Neurology, 1997;49:258–259.
42. Zifko U, Drlicek M, Machacek E, et al. Syndrome of continuous muscle fiber activity and plasmacytoma with IgM paraproteinemia. Neurology 1994;44:560–561.
43. Sinha S, Newsom-Davis J, Mills K, et al. Autoimmune aetiology for acquired neuromyotonia (Isaacs' syndrome). Lancet 1991;338: 75–77.
44. Shillito P, Molenaar PC, Vincent A, et al. Acquired neuromyotonia: Evidence for autoantibodies directed against K$^+$ channels of peripheral nerves. Ann Neurol 1995;38:714–722.
45. Harvey AL, Rowan EG, Vatanpour H, Fatehi M, Castaneda O, Karlsson E. Potassium channel toxins and transmitter release. Ann NY Acad Sci 1994;710:1–10.
46. Ramaswami R, Gautam, M, Kamb A, et al. Human potassium channel genes: molecular cloning and functional expression. Mol Cell Neurosci 1990;1:214–223.
47. Kleopa KA, Elman LB, Lang B, Vincent A, Scherer SS. Neuromyotonia and limbic encephalitis sera target mature Shaker-type K+ channels: subunit specificity correlates with clinical manifestations. Brain. 2006 Jun;129 (Part 6):1570–1584. Epub 2006 Apr 13.
48. Hart IK, Waters C, Vincent A, et al. Autoantibodies detected to expressed K$^+$ channels are implicated in neuromyotonia. Ann Neurol 1997;41:238–46
49a. Rettig J, Heinemann SH, Wunder F, et al. Inactivation properties of voltage-gated potassium channels altered by presence of β-subunit. Nature 1994;369:289–294.
49b. Leite MI, Jacob S, Viegas S, Cossins J, Clover L, Morgan BP, Beeson D, Willcox N, Vincent A. IgG1 antibodies to acetylcholine receptors in "seronegative" MG. Brain 2008;131:1940–1952.
50. Arimura K, Watanabe O, Kitajima I, et al. Antibodies to potassium channels of PC12 in serum of Isaacs' syndrome: western blot and immunohistochemical studies. Muscle Nerve 1997;20:299–305
51. Sonoda Y, Arimura K, Kurono A, et al. Serum of Isaacs' syndrome suppresses potassium channels in PC-12 cell lines. Muscle Nerve 1996;19:1439–1446.
52. Nagado T, Arimura K, Sonoda Y, et al. Potassium current suppression in patients with peripheral nerve hyperexcitability. Brain 1999;122:2057–2066.
53. Tomimitsu H, Arimura K, Nagado T, et al. Mechanism of action of voltage-gated K + channel antibodies in acquired neuromyotonia. Annals Neurol 2004;56:440–444.
54. O'Hanlon GM, Plomp JJ, Chakrabarti M, et al. Anti-GQ1b ganglioside antibodies mediate complement-dependent destruction of the motor nerve terminal. Brain 2001;124:893–906
55. Verino S, Adamski J, Kryzer TJ, et al. Neuronal nicotinic ACh receptor antibody in subacute autonomic neuropathy and cancer-related syndromes. Neurology 1998;50:1806–1813.

56. Toyka KV, Zielasek J, Ricker K, et al. Hereditary neuromyotonia: a mouse model associated with deficiency or increased gene dosage of the PMP22 gene. J Neurol Neurosurg Psychiatry 1997;63:812–813.
57. Zielasek J, Martini R, Suter U, Toyka KV. Neuromyotonia in mice with hereditary myelinopathies. Muscle Nerve 2000;23:696–701.
58. Vincent A. Understanding neuromyotonia. Muscle Nerve 2000;23:655–657.
59. Verino S, Auger RG, Emslie-Smith AM, et al. Myasthenia, thymoma, presynaptic antibodies, and a continuum of neuromuscular hyperexcitability. Neurology 1999;53:1233–1239.
60. Hart IK, Leys K, Vincent A, et al. Autoantibodies to voltage-gated potassium channels in acquired neuromyotonia. Neuromusc Disord 1994;4:535(Abstr.).
61. Hart IK, Waters C, Newsom-Davis J. Cerebrospinal fluid and serum from acquired neuromyotonia patients seropositive for anti-potassium channel antibodies label dentate nucleus neurones. Ann Neurol 1996;40:554–555.
62. Josephs KA, Silber MH, Fealey RD, et al. Neurophysiologic studies in Morvan 2004;21:440–445.
63. Batocchi AP, Marca GD, Mirabella, M, et al. Relapsing-remitting autoimmune agrypnia. Ann Neurol 2001;50:668–671.
64. Buckley C, Oger J, Clover L, et al. Potassium channel antibodies in two patients with reversible limbic encephalitis. Ann Neurol 2001;50:74–79
65. Vincent A, Buckley C, Schott JM, Baker I, Dewar BK, Detert N, Clover L, Parkinson A, Bien CG, Omer S, Lang B, Rossor MN, Palace J. Potassium channel antibody-associated encephalopathy: a potentially immunotherapy-responsive form of limbic encephalitis. Brain. 2004 Mar;127(Part 3):701–12. Epub 2004 Feb 11.
66. Thieben MJ, Lennon VA, Boeve BF, Aksamit AJ, Keegan M, Vernino S. Potentially reversible autoimmune limbic encephalitis with neuronal potassium channel antibody. Neurology 2004 Apr 13;62(7):1177–1182.
67. Vincent A, Lang B, Kleopa KA.Autoimmune channelopathies and related neurological disorders. Neuron 2006 Oct 5;52(1):123–138.
68. Ishii A, Hayashi A, Ohkoshi N, et al. Clinical evaluation of plasma exchange and high dose intravenous immunoglobulin in a patient with Isaacs' syndrome. J Neurol Neurosurg Psychiatry 1994;57:840–842.
69. Van den Berg JS, van Engelen BG, Boerman RH, de Baets MH. Acquired neuromyotonia: superiority of plasma exchange over high-dose intravenous human immunoglobulin. J Neurol 1999;246:623–625.
70. Skeie GO, Apostolski S, Evoli A, Gilhus NE, Hart IK, Harms L, Hilton-Jones D, Melms A, Verschuuren J, Horge HW. Guidelines for the treatment of autoimmune neuromuscular transmission disorders. Eur J Neurol 2006 Jul;13(7):691–699.

15

Congenital Myasthenic Syndromes

David Beeson

1. INTRODUCTION

Congenital myasthenic syndromes (CMS) are rare inherited disorders of neuromuscular transmission characterized by fatigable muscle weakness *(1)*. Their overall prevalence is uncertain but is thought to be in the order of 1 in 200 000 of the population in the UK. They are genetically determined (usually autosomal recessive – so a history of consanguinity is common) non-autoimmune disorders. Major clinical features include onset in infancy, fatigable weakness, a decremental response to repetitive nerve stimulation and absence of autoantibodies to the muscle acetylcholine receptor (AChR) or muscle-specific tyrosine kinase (MuSK). Remarkable differences in severity occur even within families harboring the same mutation. Although impairment of neuromuscular transmission may often give rise to similar clinical presentation, detailed analysis of intact biopsied muscle fibers from patients using electrophysiology, microscopy and biochemical techniques demonstrates distinct molecular and cellular mechanisms. The syndromes may be classified on the basis of the site of the defect of neuromuscular transmission, but this may only be available at a few centers and is not always certain. A revised classification is likely to evolve as the molecular basis of each is identified. Diagnosis depends upon electrophysiological tests, morphological studies of the endplate region in muscle biopsy specimens and increasingly on identification of the specific genetic defect *(2)*.

2. MOLECULAR GENETIC CLASSIFICATION

The CMS can be broadly grouped according to the location of the defective neuromuscular junction protein and thus are often classed as presynaptic, synaptic and postsynaptic disorders *(3)* (Table 1). Within these headings different genes may be defective. Thus far, genetic diagnosis has identified mutations that cause CMS in 11 different genes *(4)*, but it is clear that there are many further subgroups of patients where the underlying defect has not yet been identified.

CMS with episodic apnea is the only neuromuscular junction presynaptic disorder in which the genetic origin has been defined and is due to mutations in the enzyme choline acetyltransferase *(5, 6)*. Other presynaptic disorders have been identified that have a paucity of synaptic vesicles, reduced quantal release or resemble the autoimmune Lambert–Eaton myasthenic syndrome; however, the genetic defects have proved extremely difficult to pinpoint *(7)*. Acetylcholinesterase (AChE) deficiency is the only CMS directly affecting proteins in the synaptic space and results from mutations in *COLQ*, which encodes an AChE-associated collagen tail that both anchors and concentrates AChE in the synaptic cleft *(8, 9)*. By contrast, a series of different genes have been found to cause postsynaptic disorders. Mutations in the genes encoding the AChR subunits were the first to be identified. These may give rise to kinetic abnormalities of AChR function or AChR deficiency or a combination of altered AChR kinetics and AChR deficiency *(9, 10, 11, 12, 13)*. However, it is becoming evident that mutations in other proteins located on the postsynaptic side of the NMJ that may be involved in synaptic function, in AChR clustering or in the maturation of the synaptic structure, can

From: *Current Clinical Neurology*
Myasthenia Gravis and Related Disorders
Edited by: H.J. Kaminski, DOI 10.1007/978-1-59745-156-7_15, © Humana Press, New York, NY

Table 1
Classification of Congenital Myasthenic Syndromes (CMS) and Their Genetic Loci*

Syndrome	Gene
Presynaptic CMS	
CMS with episodic apnea	*CHAT*
Synaptic CMS	
Congenital endplate acetylcholinesterase deficiency	*COLQ*
Postsynaptic CMS	
AChR deficiency syndromes	*CHRNA, CHRNB, CHRND* and *CHRNE*
Multiple pterygium syndromes due to AChR γ-subunit mutations	*CHRNG*
AChR deficiency syndromes due to mutations in rapsyn	*RAPSN*
Slow-channel CMS	*CHRNA, CHRNB, CHRND* and *CHRNE*
Fast-channel CMS	*CHRNA, CHRND* and *CHRNE*
CMS due to voltage-gated sodium channel mutations	*SCN4A*
CMS due to mutations in MuSK	*MUSK*
CMS due to mutations in Dok-7	*DOK7*

*The genetic origin of many presynaptic and postsynaptic CMS are yet to be defined.
AChR, muscle acetylcholine receptor; MuSK, muscle-specific tyrosine kinase; Dok-7, downstream of kinase-7.

commonly underlie CMS. Mutations in *MUSK (14)* and *SCN4A (15)* have proved to be rare, but mutations of *RAPSN (16)* or *DOK7 (17)* are common causes of CMS.

3. DIAGNOSTIC METHODS

CMS should be considered in any person presenting with fatigable muscle weakness during infancy or early childhood. Clinical features may help to pinpoint which of the diverse range of genetic loci is involved and provide clues about the underlying molecular pathogenesis. Thus electromyography (EMG), the clinical phenotype and in some cases muscle biopsies may all provide important diagnostic pointers.

Standard EMG can often detect impaired neuromuscular transmission, especially if the muscle under test is weak. Decrement in the compound muscle action potential (CMAP) elicited by repetitive nerve stimulation at low frequency (2–3 Hz) is suggestive of impaired neuromuscular transmission but is generally considered less sensitive than single-fiber electromyography (SFEMG) and can occur in other disorders, whereas SFEMG revealing abnormal jitter and block usually indicates defective neuromuscular transmission.

Phenotypic clues may be gleaned for many of the CMS and greatly facilitate targeted genetic screens. More detailed descriptions will be given later in this chapter, but illustrative examples are that a repetitive CMAP in response to a single nerve stimulus is frequently seen in syndromes involving overexcitation such as the slow-channel congenital myasthenic syndromes or that AChE deficiency syndromes and mild arthrogryposis multiplex congenita are strong pointers to AChR deficiency due to RAPSN mutations, although it can also occur rarely with mutations in other genes.

In specialist centers, muscle biopsy is essential for characterizing disorders where the underlying genetic cause remains unknown. Electrophysiology of endplates can determine quantal content, the amplitude of miniature endplate potentials and currents (MEPPs and MEPCs) and endplate potentials and currents (EPPs and EPCs). Their size and decay times may suggest abnormalities in the number and kinetic properties of the endplate AChR. Electron microscopy can define the ultrastructure of the pre- and postsynaptic apparatus of the neuromuscular junction, providing clues to whether the defect is pre- or postsynaptic. Binding of iodinated or fluorescence-labeled α-neurotoxins, such as α-bungarotoxin (α-BuTx), can be used to determine the localization and number of the AChR. Similarly, histochemistry may be used to establish the presence or absence of AChE at the endplates *(3)*.

Response to treatments may provide additional clues about the disorder, although the tensilon test, where the short-term response to intravenous edrophonium is measured, may give misleading results. In all CMS, autoimmune myasthenia gravis should be excluded through testing for antibodies to AChR or to MuSK. Parental consanguinity or a positive family history is suggestive of hereditary rather than autoimmune myasthenia, and the onset of myasthenia gravis at less than 1 year is very rare. Although most CMS first present in infancy or early childhood and show recessive inheritance, a significant exception to this generalization is the slow-channel myasthenic syndrome, which may present in infancy or adult life and is usually inherited as an autosomal-dominant trait. In addition, late-onset cases of CMS associated with mutations in *RAPSN* and *DOK7* have been reported *(18, 19)*.

4. PRESYNAPTIC CONGENITAL MYASTHENIC SYNDROMES

These are the least well characterized of the myasthenic disorders. Electrophysiology and ultrastructure studies of the endplate regions in muscle biopsies have identified apparent defects in the presynaptic apparatus in various cases of CMS, but in many the identification of the genetic abnormality has remained elusive. These include cases where the electrophysiology shows similarities to that seen for the Lambert–Eaton myasthenic syndrome and others where there is a paucity of synaptic vesicles in the presynaptic bouton and reduced quantal release *(1)*. The presynaptic disorder where the molecular mechanism has been elucidated is associated with impaired resynthesis of acetylcholine (ACh) *(5)*.

4.1. Congenital Myasthenic Syndrome with Episodic Apnea

4.1.1. Clinical Features

This is an autosomal-recessive disorder that was previously called familial infantile myasthenic syndrome. However, the new terminology is not ideal since similar severe episodic apneic attacks are now known to occur in other forms of CMS. Clinical symptoms typically manifest at birth and consist of hypotonia, bulbar and respiratory muscle weakness, ptosis and mild extraocular muscle weakness. Respiratory insufficiency with recurrent apnea is a hallmark of this disorder. The episodic crises, which may be life-threatening in early life, are frequently induced by infections and fever, stress or overexertion, but become less frequent with age. Patients usually respond well to anticholinesterase medication, which may be taken prophylactically in anticipation of a crisis.

Diagnosis may be helped by the characteristic EMG profile of these patients. In rested muscle, the CMAP elicited by repetitive nerve stimulation at 3 Hz and SFEMG may be normal. However, after exercise or repeated nerve stimulation at 10 Hz for up to 5 min, a decremental response and abnormal jitter can be seen. In vitro studies similarly show normal EPP and MEPP amplitudes in rested muscle that decrease after continuous 10 Hz stimulation *(3)*. Endplate AChR shows normal number and distribution.

4.1.2. Molecular Basis

Mutations in the gene encoding choline acetyltransferase (*ChAT*) have been shown to underlie this disorder. ChAT catalyses the reversible synthesis of ACh from acetyl coenzyme A and choline at the synapse. The mutations identified are mostly missense and cause either a reduction in enzyme expression or in catalytic activity or a combination of the two *(5, 6, 20)*. Impairment in the ability to resynthesize ACh sufficiently fast, and thus reduction of ACh within the presynaptic vesicles, is responsible for the activity-dependent weakness, evident as decrement and reduced MEPP amplitude following continuous stimulation at 10 Hz (Fig. 1).

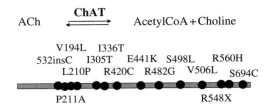

Fig. 1. Linear diagram of mutations identified in ChAT illustrating that mutations are located along the length of the coding region.

5. SYNAPTIC CONGENITAL MYASTHENIC SYNDROME

5.1. Endplate Acetylcholinesterase Deficiency

5.1.1. Clinical Features

Endplate acetylcholinesterase deficiency (EAD) is an autosomal-recessive disorder with onset at birth or in early childhood. Weakness is often severe affecting facial, cervical axial and limb muscles that may result in lordosis and kyphoscoliosis. Difficulty feeding, respiratory distress and delayed motor milestones are common. In some cases delayed papillary light reflexes are evident and may be used to distinguish EAD from other CMS. Symptoms are refractory to anticholinesterase medication, and patients do not respond well to AChR open channel blockers, such as quinidine sulfate or fluoxetine, that have been used in cases of the slow-channel syndrome. Ephedrine has been reported to be beneficial in a few cases *(21)*.

EMG shows features associated with defective neuromuscular transmission, with CMAP decrement on repetitive stimulation and jitter and block evident with SFEMG. Single nerve stimulation elicits a repetitive CMAP, which is another diagnostic hallmark and is only seen in EAD or the slow-channel syndrome. On muscle biopsy, microelectrode studies show prolonged decay of the MEPPs and EPPs indicating prolonged activation of the AChR ion channels owing to the absence of AChE, and histochemical staining of the neuromuscular junctions shows this absence or severe reduction of AChE at the endplates. Further analysis of the endplate region often demonstrates a compensatory reduction in nerve terminal size and extension of Schwann cells into the synaptic cleft.

5.1.2. Molecular Basis

The *COLQ* gene encodes the collagen-like tail (ColQ) that attaches the asymmetric form of AChE to the basal lamina at the neuromuscular junction. Mutations in ColQ rather than AChE itself have been found to cause endplate EAD. The ColQ protein contains an N-terminal proline-rich domain (PRAD) that binds AChE tetramers, a collagen domain with 63 Gxy repeats and a C-terminal domain responsible for anchoring in the basal lamina and for initiating the collagen triple helix structure. Many mutations have been identified in *COLQ (9, 21, 22, 23)*. They occur in all the putative functional domains, although there is some evidence that mutations in the C-terminal domain that affect the anchoring to the basal lamina are less severe *(24)*. The result of the mutations is the loss of the asymmetric form of AChE from the synaptic cleft and consequently the increase in the time that ACh is available to bind the AChR. Prolonged exposure to ACh will cause desensitization of AChR and the persistent depolarization of the endplate will inactivate the voltage-gated sodium channels in the depths of the postsynaptic folds and thus block signal transmission. In addition, the prolonged stimulation of the endplates may lead to overload of calcium ions at the endplate and an 'endplate myopathy'.

6. POSTSYNAPTIC CONGENITAL MYASTHENIC SYNDROMES

The majority of CMS are caused by mutations in genes that encode postsynaptic proteins. Initial studies identified mutations in the genes encoding the AChR subunits that impair ion channel gating or reduce the number of endplate receptors or a combination of the two, giving rise to 'slow channel', 'fast channel' or AChR deficiency syndromes *(25, 11,12)*. However, CMS may also arise from mutations in proteins involved in the formation and maintenance of the neuro-muscular junction such as rapsyn *(16, 18, 26)*, MuSK *(14)* or Dok-7 *(17)*.

6.1. Genetic Disorders of the AChR

Muscle AChRs are glycosylated transmembrane molecules that mediate synaptic transmission. They are allosteric (existing in several different conformations) and the binding of two ACh to each receptor is thought to favor a conformational change that results in a brief activation of the channel and the influx of cations. Each AChR is made up of five subunits arranged in a pentameric structure around the central ion pore. In mammalian muscle there are two types of AChR: a form found in fetal muscle that consists of $\alpha_2\beta\gamma\delta$ and an adult muscle form $\alpha_2\beta\delta\varepsilon$. The subunits are homologous, vary in size from 437 to 495 amino acids and are encoded by separate genes of between 10 and 12 exons. In embryonic muscle, before innervation, fetal AChR are distributed along the length of the muscle fibers. During innervation, the AChR are clustered on the postsynaptic membrane and are lost from extrasynaptic sites. At the same time expression of the γ-subunit (fetal) mRNA is repressed and is replaced by ε-subunit (adult) mRNA transcribed from subsynaptic nuclei *(27)*. In humans the γ-subunit is readily detectable in neuromuscular junctions of fetal muscle up to around 31 weeks gestation *(28)* and continues to be expressed at very low levels in adult muscle *(29)*.

6.2. AChR Deficiency due to Mutations in the AChR Subunits

6.2.1. Clinical Features

Hereditary AChR deficiency is the most common CMS *(4)*. It is an autosomal-recessive disorder in which mutations in the AChR subunit genes cause a primary deficiency of AChR at the endplate. The phenotype may vary from mild to severe. Weakness is usually evident at birth or within the first year of life and is characterized by feeding difficulties, ptosis, impaired eye movements and delayed motor milestones. Patients sometimes show improvement in adolescence and in general the disease course is not progressive. In general, patients improve with anticholinesterase medication or 3,4-diaminopyridine. Electromyography typically shows decrement of CMAPs at 3 Hz stimulation and single-fiber EMG shows increased jitter and block. Intracellular microelectrode recordings from the endplate region of biopsied muscle show that MEPPs and MEPCs may be decreased to around 8–26% and 20–42% of normal values. Consistent with MEPP and MEPC findings, staining with ^{125}I-α-BuTx or Alexa Fluor-conjugated α-BuTx shows reduced numbers of AChR that are often distributed abnormally along the muscle fiber *(30)*. In addition, electron microscopy shows a severe reduction of the postsynaptic folds (Table 2).

Table 2
Distinguishing Clinical Characteristics of AChR Deficiency Due to Mutations in CHRNE and RAPSN

Clinical feature	Early-onset rapsyn mutations	AChR deficiency ε-subunit mutations
Arthrogryposis	Common	Absent
Episodic crises	Common	Rare
Ophthalmoplegia	Absent	Common
Spontaneous improvement	Common	Rare

6.2.2. Molecular Basis

Mutations in each of the AChR subunits, *CHRNA, CHRNB, CHRND* and *CHRNE*, may underlie AChR deficiency syndromes *(4)*. However, the overwhelming majority are in *CHRNE*. At least 70 different mutations, located along the length of the ε-subunit gene, have been identified and may cause premature termination of translation, affect the promoter *(31, 32, 33)* or the signal peptide or affect assembly of the AChR pentamer *(4)*. Many are null mutations *(34)*. It is thought that in these cases residual expression of the γ- or fetal subunit is incorporated into the endplate receptors which accumulate at low levels at the endplate, are able to mediate synaptic transmission and partially compensate for loss of adult AChR *(13)*. Evidence to support this hypothesis includes recordings from endplates which demonstrate that the receptors have functional properties of fetal AChR and that animal models in which the adult AChR is replaced by fetal AChR mimic the human condition *(35)*. The partial compensation through γ-subunit expression explains why the majority of AChR mutations are located in the ε-subunit gene. Most *CHRNE* mutations are limited to a few families, although some evidence for founder effects is evident within European populations. In particular, the mutation ε1267delG *(36, 37)*, which tends to generate a relatively mild phenotype, is common in southeastern Europe, particularly in the Romany gypsies *(38)*. Mutations in *CHRNA, CHRNB* and *CHRND* underlying AChR deficiency tend to cause a severe phenotype probably because they cannot be compensated by expression of an alternative subunit. In these cases the mutations usually affect assembly of the AChR pentamer *(39)*.

7. KINETIC ABNORMALITIES OF THE ACHR

Mutations that underlie AChR deficiency may often also change the kinetics or functional properties of the AChR. In these cases the primary pathogenic defect is loss of endplate AChR and the altered channel kinetics has a secondary effect. However, if the numbers of endplate AChR are not severely reduced, the phenotype is determined by the altered ion channel gating.

7.1. Slow-Channel Congenital Myasthenic Syndrome

7.1.1. Clinical Features

This syndrome was first described by Andrew Engel and colleagues *(40)*. It is an autosomal-dominant disorder, with an age of onset of weakness that may occur neonatally or may not arise until adolescence, adulthood or during pregnancy. There is also a wide range of severity and in some cases variable penetrance *(41)*. Weakness and associated wasting of cervical and scapular muscles and of finger extensors is often an early feature. Frequently there is only mild ptosis and extraocular muscle involvement. Unlike other CMS, this disorder is commonly slowly progressive, involving the respiratory, limb and bulbar muscles.

On EMG, decrement of CMAPs in response to 3 Hz stimulation may only be seen in affected muscles. However, as for endplate AChE deficiency, a characteristic repetitive CMAP response to a single nerve stimulus is often, but not always, present. In vitro microelectrode studies show prolongation of the EPPs and EPCs and of the MEPPs. Single-channel recordings directly from endplate regions show AChR with abnormally prolonged activations that account for the abnormally long decay of the EPPs and EPCs. Ultrastructural studies show an endplate myopathy in which there is a widening of the postsynaptic cleft, areas of degenerating junctional folds, degenerating and swollen mitochondria, apoptotic subsynaptic nuclei, calcium deposits and vacuole formation *(39)*.

7.1.2. Molecular Basis

At least 18 different mutations have been identified that give rise to slow-channel congenital myasthenic syndromes *(4)*. They occur not only in each of the AChR subunits *(42, 43)* but also in different functional domains within each subunit *(44)*, although mutations within the M2 channel pore region are most common *(45)*. They are all single amino acid changes that result in a pathogenic gain of function for the AChR explaining the dominant inheritance.

The AChR is thought to adopt multiple conformations, and at least three interconvertible functional states have been recognized: a resting state in the absence of ACh in which the probability of opening is small; an active state in the presence of ACh in which the probability of opening is high and a closed state that results from prolonged exposure to high ACh concentrations in which the AChR is 'desensitized'. A simplistic but illustrative mechanism for the activation of the AChR (derived from recordings of single-channel currents) is given in Fig. 2.

This scheme does not take into account channel openings of unliganded or monoliganded AChRs or various desensitized states and assumes that the binding of the two ACh to a single receptor is equivalent. In neuromuscular transmission there is a transient saturating concentration of ACh, leading to rapid binding of two ACh (A_2R, where the receptor is bound but closed), and the rapid open rate β ensures fast opening of the channel (A_2R^*). The hydrolysis of ACh by AChE rapidly clears ACh from the synaptic cleft, lowering the ACh concentration so that there is no rebinding once ACh dissociates from the receptor. Because the opening rate β and the dissociation rate k_{-2} are roughly similar, the receptor will oscillate between an open and a closed state before the ACh finally dissociates from the receptor. The activations of wild-type channels depend largely upon β (the opening rate), α (the closing rate) and k_{-2} (the dissociation rate). Mutations of the AChR that affect channel kinetics are likely to alter one or more of these rates *(44)*.

Single-channel recordings of AChR harboring slow-channel mutations show prolonged ion channel activations both from mutant AChR expressed in HEK 293 cells or *Xenopus* oocytes and in recordings direct from muscle biopsies. For mutation αG153S *(10, 44)*, which is located close to the predicted ACh-binding site, kinetic analysis shows that prolonged activations arise primarily through a reduction in the rate of dissociation of ACh from the AChR (k_{-2}), thereby increasing the number of channel openings during ACh occupancy. Thus, the primary effect of the αG153S mutation is to alter the affinity of AChR for ACh. However, the majority of slow-channel syndrome mutations are in the M2 transmembrane domains. Mutations in the M2 domain, such as ϵL264P *(25)*, primarily slow the rate of channel closure (α) so that within activations the duration of individual opens are increased. Although severity of disease is variable, in general, patients with mutations in the M2 region tend to be more severely affected. The prolonged ion channel activations explain the extended decay phase for the EPPs and MEPPs observed in slow-channel syndrome patients.

The evidence supports the theory that prolonged channel activations lead to excess entry of calcium or calcium overload, which in turn activates a variety of enzymatic pathways leading to the degenerative changes on the postsynaptic side of the synapse. The endplate myopathy can lead to defective neuromuscular transmission through reducing the number of endplate AChR and reducing efficiency of transmission through widening of the synaptic cleft and decay of the postsynaptic folds. Neuromuscular transmission may also be compromised by an increased propensity of the mutant channels to desensitize, and by depolarization block of the voltage-gated sodium channels due to summation of the endplate potentials at physiological rates of stimulation *(46)*.

$$A + R \underset{k_{-1}}{\overset{k_{+1}}{\rightleftharpoons}} A + AR \underset{k_{-2}}{\overset{k_{+2}}{\rightleftharpoons}} A_2R \underset{\alpha}{\overset{\beta}{\rightleftharpoons}} A_2R^*$$

Fig. 2. Basic kinetic framework for analysis of the activation of the AChR. A, ACh; R, AChR; A_2R^* indicates the ion channel in the open state.

Both quinidine and fluoxetine that, among other actions, block the AChR channel when it is open have been partially successful in treating patients and have improved symptoms *(47, 48, 49)*. However, both these drugs can have potentially serious side effects and so patients should be monitored.

7.2. Fast-Channel Congenital Myasthenic Syndrome

7.2.1. Clinical Features

The phenotypes of fast-channel congenital myasthenic syndromes and AChR deficiency syndromes share many features. Fast-channel syndromes show recessive inheritance *(11)*, except in one reported case *(50)*, with the fast-channel mutation usually found in combination with a low expressor or null allele. Weakness is usually evident at birth or in the first months and is characterized by feeding difficulties, ptosis, impaired eye movements and delayed motor milestones. Patients with fast-channel syndromes tend to be more severely affected than AChR deficiency syndromes due to ε-subunit mutations, and in one case joint contractures at birth were reported *(51)*. EMG typically shows decrement at 3 Hz stimulation and single-fiber EMG reveals an increase in jitter and block. Intracellular microelectrode recordings from the endplate of biopsied muscles show small MEPPs and MEPCs. However, biochemical and morphological analysis helps to differentiate the syndromes; endplates do not show a severe loss of receptors, postsynaptic folds or other morphological changes associated with AChR deficiency. Fast-channel patients show a beneficial response to cholinesterase inhibitors, to 3,4-diaminopyridine or to a combination of the two.

7.2.2. Molecular Basis

A series of different mutations have been identified that alter the AChR channel properties causing abnormally brief channel activations, in direct contrast to the slow-channel mutations. The mutations have been identified in genes encoding each of the AChR subunits *(11, 50, 51, 52)*. They cause a loss of response to ACh and thus show recessive inheritance. When a fast-channel mutation segregates with a null mutation or a second fast-channel mutation, the fast-channel phenotypic footprint is uncovered. Mutations have been identified that reduce AChR affinity for ACh or alternatively affect the channel-gating properties. For instance, εP121L has been identified in combination with εS143L, εG-8R or Y15H. In each case, the low-expresser second allele unmasks the phenotypic effects of εP121L that are generated by channel activations that are fewer and shorter than normal. εP121L slows the rate of channel opening (β) but has little effect on dissociation (k_{-2}). Since channel opening depends upon β/k_{-2}, εP121L will result in reduced channel reopening when ACh is bound and consequently shorter activations result in a reduction in signal transmission *(11)*.

8. MUTATIONS AFFECTING ACHR CLUSTERING AND SYNAPTIC STRUCTURE

Efficient synaptic transmission depends upon the apposition of nerve terminal and postsynaptic apparatus and the correct localization of all the key functional components *(53)*. Just as studies of mutations underlying the fast- and slow-channel syndromes provide insights into AChR function, so too studies of recently identified CMS provide novel insights into synaptogenesis and the maintenance of the neuromuscular synapse.

8.1. AChR Deficiency due to RAPSN Mutations

8.1.1. Clinical Features

Mutations in the AChR-clustering protein rapsyn also cause endplate AChR deficiency *(16)*. Onset of symptoms is usually at birth, 'early onset', although occasional 'late-onset' cases presenting from early adulthood to middle age have been reported *(18)*. Early-onset cases are frequently associated with hypotonia, marked bulbar dysfunction often necessitating nasogastric feeding and may require assisted ventilation. Joint contractures (arthrogryposis multiplex congenita) of hands and ankles are common. In childhood the course of disease is associated with severe exacerbations often presenting with life-threatening respiratory failure. Patients tend to improve over time; severe apneic episodes are rarer over the age of 6, and in many cases in adulthood disability is minimal. Late-onset cases may be mistaken for 'seronegative' immune-mediated myasthenia gravis. Bulbar, speech and respiratory problems were not observed. Weakness of ankle dorsiflexion, which is uncommon in myasthenia gravis, may provide a clue that *RAPSN* mutations underlie the condition *(18)*. Both early- and late-onset cases show abnormal decrement on EMG and jitter on single-fiber EMG, although it is not always easy to detect. Patients with *RAPSN* mutations respond well to anticholinesterase medication, although some may gain further benefit from the addition of 3,4-diaminopyridine.

On muscle biopsy, endplates from patients with AChR deficiency due to ε-subunit mutations or AChR deficiency due to rapsyn mutations appear similar; however, the two conditions differ in distinctive clinical features that may enable a targeted genetic screen *(54)*. The underlying cause of these differential features is unclear. However, whereas patients with ε-subunit null mutations most likely survive through maintained low-level expression of the fetal (γ-) subunit, patients with rapsyn mutations express low levels of the ε-subunit. Thus, one clear difference between the rapsyn group and the ε-subunit group is the type of AChR that mediated synaptic transmission *(55)*.

8.1.2. Molecular Basis

In about a third of AChR deficiency cases mutations are not detected in the AChR subunits. Many of these cases are due to the recessive inheritance of mutations within the AChR-clustering protein rapsyn *(55, 56, 57)*. Various functional domains have been proposed for rapsyn including an N-terminal myristoylation signal involved in membrane association, a string of tetratricopeptide repeats involved in rapsyn self-association, a zinc finger/coiled-coil domain implicated in the interaction of rapsyn with the AChR and a RING-H2 domain thought to be involved in binding to scaffold proteins. Mutations are observed along the length of the rapsyn protein. To date, more than 30 different mutations have been identified, but there are no clear phenotypic associations with their positions within rapsyn. However, the overwhelming majority of patients harbor the missense mutation N88K on at least one allele, suggesting an original founder mutation *(58)*. The observations suggest that rapsyn-N88K retains at least partial function, whereas many of the others are null mutations. This is supported by cell culture experiments in which several mutations were found to drastically inhibit rapsyn function and AChR/rapsyn association, whereas rapsyn-N88K was able to mediate agrin-induced AChR clusters, but these clusters were found to be less stable than clusters formed with wild-type rapsyn *(59)*. Thus, it may be that in patients with the N88K mutation AChR deficiency is due to instability of the endplate rapsyn-N88K/AChR clusters.

Not all patients with AChR deficiency harbor N88K. Rare cases have been reported where patients have other mutations in the coding region that result in partially functional rapsyn. In addition, a number of cases have been identified with mutations in the promoter region of the *RAPSN* gene. Some of these promoter mutations severely reduce rapsyn mRNA transcription, but one, −38A>G, has a less-drastic effect on transcription and has been found homozygous in patients of Iranian Jewish origin who commonly show facial malformation *(60)*. The facial malformation, high-arched palate and joint contractures observed in patients with rapsyn mutations are all thought to result from akinesis in the womb, presumably due to failed neuromuscular transmission at critical stages in fetal development. Thus the rapsyn mutations must affect clustering of both the adult and the fetal AChR subtypes.

A novel mutation recently identified in the AChR δ-subunit gene, δE381K, does not affect AChR function but rather impairs rapsyn-induced AChR clustering, presumably through impaired interaction with rapsyn *(61)*. The patient phenotype bears all the hallmarks of a 'rapsyn deficiency' rather than an AChR ε-subunit deficiency. This mutation may shed light on the molecular basis for AChR–rapsyn interaction, which remains unresolved despite many years of study.

8.2. Congenital Myasthenic Syndrome with Proximal Weakness due to Mutations in DOK7

8.2.1. Clinical Features

Different muscle groups can be differentially affected in the CMS. A group of patients have been identified in whom proximal muscles are more affected than distal muscle groups *(17)*. These have been termed 'limb girdle' congenital myasthenia *(62)*. However, this term may lead to confusion with other non-myasthenic limb girdle muscle disorders, and they may be better classified as CMS with proximal muscle weakness.

Clinical onset of disease is generally characterized by difficulty in walking after initially achieving normal walking milestones. Patients occasionally have earlier symptoms of ptosis, floppy tone and bulbar and respiratory problems, but even in these, walking onset is not delayed. The difficulty in walking or running tends to worsen in childhood and is often accompanied by upper limb weakness and loss of ambulation in some cases. A waddling and lordotic gait is often seen associated with proximal lower limb and truncal weakness.

Ptosis is often present from an early age, though it may develop and progress in childhood. Eye movements are usually normal, while facial, jaw and neck weakness is common and tongue wasting has been observed in around 50% of cases. Bulbar problems typically develop later in the clinical course than limb weakness. Features seen in patients with rapsyn mutations such as congenital joint deformity and weakness of ankle dorsiflexion, which are associated with reduced fetal movement in the womb, have not been observed. However, fluctuations in symptoms were common. In many cases studied, diagnosis as a myasthenic disorder was delayed and disorders such as muscular dystrophy and congenital myopathy were suggested *(63, 64)*.

8.2.2. Molecular Basis

The classic view of the development of the neuromuscular synapse derives from a series of experiments in which components of the neuromuscular junction were 'knocked out' *(65, 66, 67, 68)*. They highlighted a pathway in which agrin, released from the motor nerve terminal, activates MuSK, a receptor tyrosine kinase, which in turn activates a kinase pathway in which the AChR β-subunit is phosphorylated, and rapsyn clusters and stabilizes the AChR on the post-synaptic membrane *(69)* (Fig. 3).

However, AChR clusters can form in the absence of neural agrin and it had been postulated that a muscle-derived factor(s) can also interact with MuSK and play a role in the AChR-clustering pathway. Recent findings in the laboratory of Yuji Yamanashi appear to verify this postulate *(70)*. The 'Dok' family of cytoplasmic proteins bind specific receptor phosphotyrosine kinases and play a role in enhancing the signaling. Dok-7 was found to bind specifically to MuSK and when expressed in cultured C2C12 cell-line myotubes was able to induce large AChR clusters in the absence of neural agrin. When the *DOK7* gene was knocked out in mice, neuromuscular junctions failed to form and offspring failed to survive past birth *(70)*. The three-dimensional structure has yet to be determined but the primary structure of Dok-7 has 504 amino acids, with an N-terminal pleckstrin homology domain, a phosphotyrosine-binding domain and a large C-terminal region containing several tyrosine residues.

Splice site, missense and frameshift mutations have been identified in *DOK7* (Fig. 4) and recessive inheritance of these results in a myasthenic syndrome with the characteristic proximal or limb girdle-type weakness described above *(17)*. The majority of the mutations are located in the

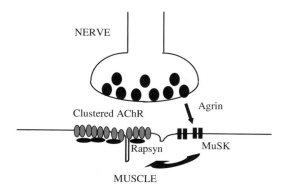

Fig.3. Illustrative diagram highlighting the classical pathway believed to be responsible for formation and maintenance of the neuromuscular junction.

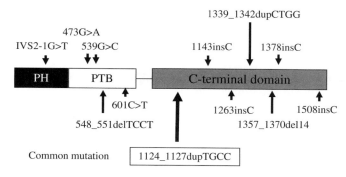

Fig. 4. Diagrammatic representation of Dok-7 and the position of identified mutations. PH, pleckstrin homology domain; PTB, phosphotyrosine-binding domain.

large 3′ exon of the gene that encodes the C-terminal region of Dok-7. Mutation 1124_1127dupTGCC is common, occurring in at least one allele in 20/24 kinships reported in a recent study *(63)*. It is thought that Dok-7 activates MuSK through the Dok-7 phosphotyrosine-binding domain interacting with the juxtamembrane phosphotyrosine-binding motif of MuSK. The binding can occur even when the C-terminal region is truncated. In accordance, all patients to date have been found to have at least one allele with a mutation in the 3′ exon of *DOK7*, and by contrast, patients have yet to be identified with heteroallelic mutations confined to the PH and PTB domains, implying that this combination may be lethal *(17, 63, 64)*. As stated, this disorder shows recessive inheritance; however, it is reported that a second mutation within the coding region of *DOK7* has not been found in a number of patients *(63, 64)*. It is likely that mutations in the non-coding region of this gene will be identified.

Studies of the motor endplates from patients harboring *DOK7* mutations found that components of the neuromuscular junction were present at normal density and showed normal function but that the size of the pre- and postsynaptic structures was reduced *(71)*. These observations, in combination with functional studies on the action of mutant Dok-7 on the AChR-clustering pathway, suggest that mutations of *DOK7* result in impaired maturation and maintenance of neuromuscular junction structure *(17)*.

Not all congenital myasthenic syndromes with limb girdle patterns of muscle weakness are due to *DOK7* mutations. *DOK7* mutations have not been identified in a subgroup of patients with a limb girdle phenotype that is associated with 'tubular aggregates' in muscle biopsies *(63, 64)*. Whereas patients with *DOK7* mutations usually fail to respond to long-term treatment with anticholinesterase medications, the tubular aggregates subgroup show a good response. Interestingly, study of a motor endplate biopsy from a limb girdle patient with tubular aggregates indicated that the extent of the postsynaptic folds and the endplate AChR number and density were reduced *(71)*. The observations suggest these patients have a form of endplate AChR deficiency, and this would explain their positive response to anticholinesterase medication. Similar to patients with *DOK7* mutations, eye movements in the subgroup of patients with tubular aggregates are seldom affected. However, their pattern of weakness usually shows a greater focus on truncal muscle groups. It is likely that the limb girdle congenital myasthenic syndromes will be due to mutations in many different genes of which *DOK7* is the first to be identified.

8.3. Prenatal Hereditary Myasthenia due to Mutations in CHRNG

Neuromuscular transmission at nearly all normal adult muscle endplates is mediated by AChR consisting of $\alpha_2\beta\delta\epsilon$-subunits. However, for crucial periods of fetal development, in utero transmission is mediated through the fetal form ($\alpha_2\beta\delta\gamma$) of the AChR *(72)*. Loss of fetal movement during these periods can lead to a series of developmental abnormalities.

8.3.1. Clinical features

Multiple pterygium syndromes or Escobar's syndrome is an autosomal-recessive condition that manifests in orthopedic and cranial abnormalities. Characteristically there is short stature, arthrogryposis multiplex congenita, pterygia of the neck and anomalies of the head including low-set ears, ptosis, a pointed and receding chin and high-arched palate *(73)*. Intrauterine death and stillbirths are common.

8.3.2. Molecular Basis

Mutations of the AChR γ-subunit gene *CHRNG* have been found to underlie many cases of Escobar's syndrome *(74, 75)*. At least 18 different mutations have been identified to date. They may be splice site, short duplications, missense or nonsense mutations that result in either truncation or low expression levels of the γ-subunit. The loss of fetal AChR function associated with *CHRNG* mutations is thought to result in fetal akinesia, which in turn causes the associated multiple developmental abnormalities. Surprisingly, some patients who harbor γ-subunit null alleles can survive, suggesting that early expression of the ε-subunit that partially compensates for loss of the γ-subunit might occur. Similarly, the severity of the condition varies in patients with the same mutations. Following birth, neuromuscular transmission is mediated by adult AChR and patients show little or no progression of their condition. Since the disorder results from lack of neuromuscular transmission at crucial developmental phases, it might be expected that recessive inheritance of loss of function mutations in other essential components of the neuromuscular junction, such as rapsyn, MuSK or *DOK-7* would also result in fetal akinesia.

8.4. Rare Postsynaptic Congenital Myasthenic Syndromes

Single cases have been reported of heteroallelic mutations in *MuSK (14)* and heteroallelic mutations in the postsynaptic voltage-gated sodium channel, *SCN4A (15)*. The patient with *MuSK* mutations had a fluctuating course of disease with episodes of severe respiratory distress in early childhood. Muscle biopsy revealed reduced endplate AChR levels; however, the patient failed to show a beneficial response to anticholinesterase medication on its own but did respond to combined pyridostigmine and 3,4-diaminopyridine. The precise underlying molecular mechanism is not clear but the mutation V790M appeared to reduce expression and stability of MuSK.

More recently, it has been hypothesized that this mutation alters MuSK conformation, thereby altering the interaction of MuSK with Dok-7 and thus MuSK/Dok-7 signaling *(70)*.

In a severely affected patient with mutations in *SCN4A*, endplate potentials of normal amplitude failed to activate the postsynaptic voltage-gated sodium channels. This is due to the rapid inactivation of the $Na_v1.4$ channels resulting from the presence of mutation V1442E. This mutation may show dominant inheritance, although it was identified in a patient who also harbored a clinically silent mutation S246L that causes small but detectable biophysical changes *(15)*.

REFERENCES

1. Engel AG, Sine SM. Current understanding of congenital myasthenic syndromes. Curr Opin Pharmacol 2005;5:308–321.
2. Engel AG, Ohno K, Sine SM. Sleuthing molecular targets for neurological diseases at the neuromuscular junction. Nat Rev Neurosci 2003;4:339–352.
3. Beeson D, Hantai D, Lochmuller H, Engel AG. 126th International Workshop: Congenital Myasthenic Syndromes, 24–26 September 2004, Naaden, the Netheralnds. Neuromuscul Disord 2005;15:498–512.
4. Ohno K, Engel AG. Congenital myasthenic syndromes: gene mutations. Neuromuscul Disord. 2004;14:117–22.
5. Ohno K, Tsujino A, Brengman JM, et al. Choline acetyltransferase mutations cause myasthenic syndrome associated with episodic apnea in humans. Proc Natl Acad Sci U S A. 2001;98:2017–22.
6. Maselli RA, Chen D, Mo D, Bowe C, Fenton G, Wollmann RL. Choline acetyltransferase mutations in myasthenic syndrome due to deficient acetylcholine resynthesis. Muscle Nerve. 2003;27:180–187.
7. Milone M, Fukuda T, Shen XM, Tsujino A, Brengman J, Engel AG. Novel congenital myasthenic syndromes associated with defects in quantal release. Neurology 2006;66:1223–1229.
8. Donger C, Krejci E, Serradell AP, et al. Mutation in the human acetylcholinesterase-associated collagen gene, COLQ, is responsible for congenital myasthenic syndrome with endplate acetylcholinesterase deficiency (Type 1c). Am J Hum Genet 1998;63:967–975.
9. Ohno K, Brengman J, Tsujino, A, Engel, AG. Human endplate acetylcholinesterase deficiency caused by mutations in the collagen-like tail subunit (ColQ) of the asymmetric enzyme. Proc Natl Acad Sci USA 1998;95:9654–9659.
10. Sine SM, Ohno K, Bouzat C, et al. Mutation of the acetylcholine receptor alpha subunit causes a slow-channel myasthenic syndrome by enhancing agonist binding affinity. Neuron.1995;15:229–239.
11. Ohno K, Wang HL, Milone M, et al. Congenital myasthenic syndrome caused by decreased agonist binding affinity due to a mutation in the acetylcholine receptor epsilon subunit. Neuron. 1996;17:157–170.
12. Engel AG, Ohno K, Bouzat C, Sine SM, Griggs RC. End-plate acetylcholine receptor deficiency due to nonsense mutations in the epsilon subunit. Ann Neurol. 1996;40:810–817.
13. Ohno K, Quiram PA, Milone M, et al. Congenital myasthenic syndromes due to heteroallelic nonsense/missense mutations in the acetylcholine receptor epsilon subunit gene: identification and functional characterization of six new mutations. Hum Mol Genet.1997;6:753–766.
14. Chevessier F, Faraut B, Ravel-Chapuis A, et al. MuSK, a new target for mutations causing congenital myasthenic syndrome. Hum Mol Genet 2004;13:3229–3240.
15. Tsujino A, Maertens C, Ohno K, et al. Myasthenic syndrome caused by mutation of the SCN4A sodium channel. Proc Natl Acad Sci USA 2003;100:7377–7382.
16. Ohno K, Engel AG, Shen XM, et al. Rapsyn mutations in humans cause endplate acetylcholine-receptor deficiency and myasthenic syndrome. Am J Hum Genet 2002;70:875–885.
17. Beeson D, Higuchi O, Palace J, et al. Dok-7 mutations underlie a neuromuscular junction synaptopathy. Science 2006;313:1975–1978.
18. Burke G, Cossins J, Maxwell S, et al. Rapsyn mutations in hereditary myasthenia; distinct early- and late-onset phenotypes. Neurology 2003;61:826–828.
19. Muller JS, Herczegfalvi A, Vilchez J, et al. Phenotypical spectrum of DOK7 mutations in congenital myasthenic syndromes. Brain 2007;130:1497–1506.
20. Barisic N, Muller JS, Paucic-Kirincic E, et al. Clinical variability of CMS-EA (congenital myasthenic syndrome with episodic apnea) due to identical CHAT mutations in two infants. Eur J Paediatr Neurol 2005;9:7–12.
21. Bestue-Cardiel M, Saenz de Cabezon-Alvarez A, et al. Congenital endplate acetylcholinesterase deficiency responsive to ephedrine. Neurology 2005;65:144–146.
22. Ohno K, Engel AG, Brengman JM, et al. The spectrum of mutations causing end-plate acetylcholinesterase deficiency. Ann Neurol 2000;47:162–170.
23. Muller JS, Petrova S, Kiefer R, et al. Synaptic congenital myasthenic syndrome in three patients due to a novel missense mutation (T441A) of the COLQ gene. Neuropediatrics 2004;35:183–189.

24. Kimbell LM, Ohno K, Engel AG, Rotundo RL. C-terminal and heparin-binding domains of collagenic tail subunit are both essential for anchoring acetylcholinesterase at the synapse. J Biol Chem 2004;279:10997–11005.

25. Ohno K, Hutchinson DO, Milone M, et al. Congenital myasthenic syndrome caused by prolonged acetylcholine receptor channel openings due to a mutation in the M2 domain of the epsilon subunit. Proc Natl Acad Sci U S A 1995; 92:758–762.

26. Muller JS, Mildner G, Muller-Felber W, et al. Rapsyn N88K is a frequent cause of congenital myasthenic syndromes in European patients. Neurology 2003;60:1805–1810.

27. Croxen R, Young C, Slater C, et al. End-plate gamma- and epsilon-subunit mRNA levels in AChR deficiency syndrome due to epsilon-subunit null mutations. Brain 2001;124:1362–1372.

28. Hesselmans LF, Jennekens FG, Van den Oord CJ, Veldman H, Vincent A. Development of innervation of skeletal muscle fibers in man: relation to acetylcholine receptors. Anat Rec 1993;236:553–562.

29. MacLennan C, Beeson D, Buijs AM, Vincent A, Newsom-Davis J. Acetylcholine receptor expression in human extraocular muscles and their susceptibility to myasthenia gravis. Ann Neurol 1997;41:423–431.

30. Vincent A, Cull-Candy SG, Newsom-Davis J, Trautmann A, Molenaar PC, Polak RL. Congenital myasthenia: end-plate acetylcholine receptors and electrophysiology in five cases. Muscle Nerve 1981;4:306–318.

31. Nichols P, Croxen R, Vincent A, et al. Mutation of the acetylcholine receptor ε-subunit promoter in congenital myasthenic syndrome. Ann Neurol 1999;45:439–443.

32. Ohno K, Anlar B, Engel AG. Congenital myasthenic syndrome caused by a mutation in the Ets-binding site of the promoter region of the acetylcholine receptor epsilon subunit gene. Neuromuscul Disord 1999;9:131–135.

33. Abicht A, Stucka R, Schmidt C, et al. A newly identified chromosomal microdeletion and an N-box mutation of the AChR epsilon gene cause a congenital myasthenic syndrome. Brain 2002;125:1005–1013.

34. Cossins J, Webster R, Maxwell S, Burke G, Vincent A, Beeson D. A mouse model of AChR deficiency syndrome with a phenotype reflecting the human condition. Hum Mol Genet. 2004;13:2947–2957.

35. Ealing J, Webster R, Brownlow S, et al. Mutations in congenital myasthenic syndromes reveal an ε-subunit C-terminal cysteine, C470, crucial for maturation and surface expression of adult AChR. Hum Mol Genet 2002;11:3087–3096.

36. Croxen R, Newland C, Betty M, Vincent A, Newsom-Davis J, Beeson D. Novel functional epsilon-subunit polypeptide generated by a single nucleotide deletion in acetylcholine receptor deficiency congenital myasthenic syndrome. Ann Neurol 1999;46:639–647.

37. Abicht A, Stucka R, Karcagi V, et al. A common mutation (epsilon1267delG) in congenital myasthenic patients of Gypsy ethnic origin. Neurology 1999;53:1564–1569.

38. Morar B, Gresham D, Angelicheva D, et al. Mutation history of the roma gypsies. Am J Hum Genet 2004;75:596–609.

39. Quiram PA, Ohno K, Milone M, et al. Mutation causing congenital myasthenia reveals acetylcholine receptor beta/delta subunit interaction essential for assembly. J Clin Invest 1999;104:1403–1410.

40. Engel AG, Lambert EH, Mulder DM, et al. A newly recognized congenital myasthenic syndrome attributed to a prolonged open time of the acetylcholine-induced ion channel. Ann Neurol 1982;11:553–569.

41. Croxen R, Hatton C, Shelley C, et al. Recessive inheritance and variable penetrance of slow-channel congenital myasthenic syndromes. Neurology 2002;59:162–168.

42. Gomez CM, Maselli R, Gammack J, et al. A beta-subunit mutation in the acetylcholine receptor channel gate causes severe slow-channel syndrome. Ann Neurol. 1996;39:712–723.

43. Gomez CM, Maselli RA, Vohra BP, et al. Novel delta subunit mutation in slow-channel syndrome causes severe weakness by novel mechanisms. Ann Neurol. 2002;51:102–112.

44. Hatton CJ, Shelley C, Brydson M, Beeson D, Colquhoun D. Properties of the human muscle nicotinic receptor, and of the slow-channel myasthenic syndrome mutant epsilonL221F, inferred from maximum likelihood fits. J Physiol 2003;547:729–760.

45. Croxen R, Newland C, Beeson D, et al. Mutations in different functional domains of the human muscle acetylcholine receptor alpha subunit in patients with the slow-channel congenital myasthenic syndrome. Hum Mol Genet 1997;6:767–774.

46. Engel AG, Ohno K, Milone M, et al. New mutations in acetylcholine receptor subunit genes reveal heterogeneity in the slow channel congenital myasthenic syndrome. Hum Mol Genet 1996;5:1217–1227.

47. Harper CM, Engel AG. Quinidine sulfate therapy for the slow-channel congenital myasthenic syndrome. Ann Neurol 1998;43:480–484.

48. Harper CM, Fukodome T, Engel AG. Treatment of slow-channel congenital myasthenic syndrome with fluoxetine. Neurology 2003;60:1710–1713.

49. Colomer J, Muller JS, Vernet A, et al. Long-term improvement of slow-channel congenital myasthenic syndrome with fluoxetine. Neuromuscul Disord 2006;16:329–333.

50. Webster R, Brydson M, Croxen R, Newsom-Davis J, Vincent A, Beeson D. Mutation in the AChR ion channel gate underlies a fast channel congenital myasthenic syndrome. Neurology 2004;62:1090–1096.

51. Wang HL, Milone M, Ohno K, et al. Acetylcholine receptor M3 domain: stereochemical and volume contributions to channel gating. Nat Neurosci 1999;2:226–233.

52. Brownlow S, Webster R, Croxen R, et al. Acetylcholine receptor delta subunit mutations underlie a fast-channel myasthenic syndrome and arthrogryposis multiplex congenita. J Clin Invest 2001;108:125–130.

53. Sanes JR, Lichtman JW. Induction, assembly, maturation and maintenance of a postsynaptic apparatus. Nat Rev Neurosci 2001;2:791–805.

54. Burke G, Cossins J, Maxwell S, et al. Distinct phenotypes of congenital acetylcholine receptor deficiency. Neuromuscul Disord 2004;14:356–364.

55. Muller JS, Mildner G, Muller-Felber W, et al. Rapsyn N88K is a frequent cause of congenital myasthenic syndromes in European patients. Neurology 2003;60:1805–1810.

56. Dunne V, Maselli RA. Identification of pathogenic mutations in the human rapsyn gene. J Hum Genet 2003;48:204–207.

57. Maselli RA, Dunne V, Pascual-Pascual SI, Rapsyn mutations in myasthenic syndrome due to impaired receptor clustering. Muscle Nerve 2003;28:293–301.

58. Muller JS, Abicht A, Burke G, et al. The congenital myasthenic syndrome mutation RAPSN N88K derives from an ancient Indo-European founder J Med Genet 2004;41:e104.

59. Cossins J, Burke G, Maxwell S, et al. Diverse molecular mechanisms involved in AChR deficiency due to rapsyn mutations. Brain 2006;129:2773–2783.

60. Ohno K, Sadeh M, Blatt I, Brengman JM, Engel AG. E-box mutations in the RAPSN promoter region in eight cases with congenital myasthenic syndrome. Hum Mol Genet 2003;12:739–748.

61. Muller JS, Baumeister SK, Schara U, et al. CHRND mutation causes a congenital myasthenic syndrome by impairing co-clustering of the acetylcholine receptor with rapsyn. Brain 2006;129:2784–2793.

62. McQuillen MP. Familial limb-girdle myasthenia. Brain 1966;89:121–132.

63. Palace J, Lashley D, Newsom-Davis J, et al. Clinical features of the DOK7 neuromuscular junction synaptopathy. Brain 2007;130:1507–1515.

64. Muller JS, Herczegfalvi A, Vilchez JJ, et al. Phenotypical spectrum of DOK7 mutations in congenital myasthenic syndromes. Brain. 2007;130:1497–1506.

65. Missias AC, Mudd J, Cunningham JM, Steinbach JH, Merlie JP, Sanes JR. Deficient development and maintenance of postsynaptic specializations in mutant mice lacking an 'adult' acetylcholine receptor subunit. Development 1997;124:5075–5086.

66. Gautam M, Noakes PG, Mudd J, et al. Failure of postsynaptic specialization to develop at neuromuscular junctions of rapsyn-deficient mice. Nature 1995;377:232–236.

67. Gautam M, Noakes PG, Moscoso L, et al. Defective neuromuscular synaptogenesis in agrin-deficient mutant mice. Cell 1996;85:525–535.

68. DeChiara TM, Bowen DC, Valenzuela DM, et al. The receptor tyrosine kinase MuSK is required for neuromuscular junction formation in vivo. Cell 1996;85:501–512.

69. Willmann R, Fuhrer C. Neuromuscular synaptogenesis: clustering of acetylcholine receptors revisited. Cell Mol Life Sci 2002;59:1296–1316.

70. Okada K, Inoue A, Okada A, et al. The muscle protein Dok-7 is essential for neuromuscular synaptogenesis. Science 2006;312:1802–1805.

71. Slater CR, Fawcett PR, Walls TJ, et al. Pre- and post-synaptic abnormalities associated with impaired neuromuscular transmission in a group of patients with 'limb-girdle' myasthenia. Brain 2006;129:2061–2076.

72. Mishina M, Takai T, Imoto K, et al. Molecular distinction between fetal and adult forms of muscle acetylcholine receptor. Nature 1986;321:406–411.

73. Escobar V, Bixler D, Gleiser S, Weaver DD, Gibbs T. Multiple pterygium syndrome. Am J Dis Child 1978;132:609–611.

74. Morgan NV, Brueton LA, Cox P, et al. Mutations in the embryonal subunit of the acetylcholine receptor (CHRNG) cause lethal and Escobar variants of multiple pterygium syndrome. Am J Hum Genet 2006;79:390–395.

75. Hoffmann K, Muller JS, Stricker S, et al. Escobar syndrome is a prenatal myasthenia caused by disruption of the acetylcholine receptor fetal gamma subunit. Am J Hum Genet 2006;79:303–312.

16

Toxic Neuromuscular Transmission Disorders

James F. Howard, Jr.

1. INTRODUCTION

The neuromuscular junction (NMJ) is uniquely sensitive to the effects of neurotoxins. Unlike the blood–brain barrier that protects the brain and spinal cord and the blood–nerve barrier that protects peripheral nerve, there are no barriers to protect the NMJ from the deleterious effects of these agents. Several forms of neurotoxins are directed against the NMJ. Many occur as natural substances of plants or animals, other result from the actions of widely prescribed pharmaceutical compounds and still others are environmental hazards. In nearly all instances of NMJ neurotoxicity, there is a reduction in the safety factor of neuromuscular transmission by one of several mechanisms. These neurotoxins may affect either the pre-synaptic or the post-synaptic elements of the NMJ. The clinical features of these neurotoxins are quite varied as many have associated toxicity of other parts of the central, peripheral or autonomic nervous systems. Many will have other systemic effects as well. While feared as the purveyor of morbidity and mortality, many of these neurotoxins have led to significant advances in our understanding of the molecular mechanisms of pharmacology and physiology and their associated diseases. For example, was it not for the recognition that α-bungarotoxin binds to the acetylcholine receptor (AChR), our advances in the diagnosis and treatment of myasthenia gravis (MG) would have been delayed *(1)*. Worldwide, the most common neurotoxicity of the neuromuscular junction results from envenomation. More concern to the clinical neurologist are those situations that result from the direct effects of various pharmacologic agents routinely used in the practice of medicine that produce significant aberrations of neuromuscular transmission in susceptible individuals. The potential for environmental intoxication has been limited by the stringent regulation of federal and international regulatory agencies.

The neurotoxins of interest may be broadly classified into three major categories: biological, environmental and pharmacological. This chapter focuses on the direct effects of neurotoxins affecting neuromuscular transmission of man but is not meant to be a treatise on the broad topic of neuromuscular neurotoxicology. It is not possible to elaborate on all of the pharmacological and physiological effects of particular toxins beyond the scope of the NMJ nor will it discuss in detail the neuromuscular blocking effects of these neurotoxins in animals or experimental preparations.

All forms of NMJ neurotoxicity are characterized by progressive, typically symmetrical, muscle weakness. Muscles of eye movement or the eyelid are most often involved as well as the muscles of neck flexion and the pectoral and pelvic girdles. There may be involvement of bulbar and respiratory musculature depending upon the toxin involved, the dose acquired and the duration of toxin exposure. Cognition and sensation are spared, unless other elements of the central nervous system are simultaneously involved. Muscle stretch reflexes are often preserved or only minimally diminished particularly during the early phases of the illness but may be lost if the weakness is severe.

From: *Current Clinical Neurology*
Myasthenia Gravis and Related Disorders
Edited by: H.J. Kaminski, DOI 10.1007/978-1-59745-156-7_16, © Humana Press, Newyork, NY, 2009

2. PHARMACOLOGICAL NEUROTOXICITY

The adverse reaction of drugs on synaptic transmission may be classified as acting either *pre-synaptically*, with a reduction in acetylcholine (ACh) release secondary to local anesthetic-like activity on the nerve terminal, alteration or impairment of calcium flux into the nerve terminal or a hemicholinium effect; *post-synaptically*, with antibody blockade of ACh receptors, curare-like effects or potentiation of depolarizing or non-depolarizing neuromuscular blocking agents; or, in varying degrees, *both*. Each of these pharmacological interactions may result in any of the clinical situations described above. Since the publication of the summaries of Barrons, Howard and Kaeser, describing disorders of neuromuscular transmission occurring as the result of adverse drug reactions, many more reports have surfaced adding to the list of potentially dangerous drugs *(2, 3, 4, 3, 4, 5, 6)*. An up-to-date list of these potential drug–disorder interactions is maintained on the website of the Myasthenia Gravis Foundation of America (http://www.myasthenia.org/docs/MGFA_MedicationsandMG.pdf). Unfortunately, much of the literature is anecdotal and there are only a few comprehensive in vitro studies of drug effects on neuromuscular transmission in animal or human nerve-muscle preparations. The potential adverse effects of these medications must be taken into consideration when deciding which drugs to use in treating patients who have disorders of synaptic transmission.

With the possible exceptions of d-penicillamine, α-interferon and botulinum toxin, there are no drugs that are absolutely contraindicated in patients with MG and Lambert–Eaton syndrome (LES). There are, however, numerous drugs that interfere with neuromuscular transmission and will make the weakness of these patients worse or prolong the duration of neuromuscular block in patients receiving muscle relaxants. Drug-induced disturbances of synaptic transmission resemble MG with varying degrees of ptosis, ocular, facial, bulbar, respiratory and generalized muscle weakness. Treatment includes discontinuation of the offending drug and when necessary reversing the neuromuscular block with intravenous infusions of calcium, potassium or cholinesterase inhibitors. In rare instances, these drugs may induce an autoimmune form of MG (d-penicillamine and interferon alpha). In these situations, the treating physician must utilize therapies that are typically used for other forms of autoimmune MG.

While it is most desirable to avoid drugs that may adversely affect neuromuscular transmission, in certain instances they must be used for the management of other illnesses. In such situations a thorough knowledge of the deleterious side effects can minimize their potential danger. If at all possible it is wise to use the drug within a class of drugs that has been shown to have the least effect on neuromuscular transmission. Unfortunately, studies, which allow such comparisons, are quite few.

The most frequently encountered problems are the effects of antibiotics (aminoglycoside and macrolides) and β-adrenergic blocking agents acutely worsening the strength of patients with MG. Less commonly encountered is prolonged muscle weakness and respiratory embarrassment post-operatively in patients with disorders of neuromuscular transmission.

2.1 Antibiotics

The aminoglycoside antibiotics may produce neuromuscular weakness irrespective of their route of administration *(7)*. The weakness is related to serum levels of the drug and is reversible in part by cholinesterase inhibitors, calcium infusion and the aminopyridines *(8)*. These drugs have pre- and post-synaptic actions; many have elements of both. Neuromuscular toxicity data exist for several of the antibiotics including amikacin, gentamicin, kanamycin, neomycin, netilmicin, streptomycin and tobramycin *(9)*. Of the group neomycin is the most toxic; tobramycin the least. Clinically, gentamicin, kanamycin, neomycin, tobramycin and streptomycin have been implicated in producing muscle weakness in non-myasthenic patients *(4)*. Neuromuscular blockade is not limited to the aminoglycoside antibiotics. Myasthenic patients given the macrolides, erythromycin or azithromycin will report a mild exacerbation of their weakness *(10, 11)*. The newly recognized ketolide, telithromycin, has produced abrupt and severe worsening in MG or unmasked previously undiagnosed MG [Howard, JF, personal observation] *(12)*. The polypeptide and monobasic amino-acid antibiotics, penicillins, sulfonamides, tetracyclines and fluroquinolones cause transient worsening of myasthenic weakness, potentiate the weakness of neuromuscular blocking agents or have theoretical reasons for blocking synaptic transmission *(13)*. Lincomycin and clindamycin can cause neuromuscular blocking which is not readily reversible with cholinesterase inhibitors *(14, 15)*. Polymyxin B, colistimethate and colistin are also reported to produce neuromuscular weakness particularly in patients with renal disease or when used in combination with other antibiotics or neuromuscular blocking agents *(7, 16)*. These drugs, ampicillin, the tetracycline analogs, oxytetracycline and rolitetracycline and, more recently, ciprofloxacin *(17)* exacerbate MG, although the mechanisms for each medication is not fully understood *(18, 19)*.

2.2 Cardiovascular Drugs

Many cardiovascular drugs are implicated in adversely affecting the strength of patients with MG and LES, and they (along with the antibiotics) account for the majority of adverse drug reactions in patients with neuromuscular disorders. Beta-adrenergic blockers may cause exacerbation of MG, or their use may coincide with the onset of myasthenic manifestations *(20, 21)*. Even those drugs instilled topically on the cornea are capable of producing such weakness *(22, 23)*. Atenolol, labetolol, metoprolol, nadolol, propranolol and timolol cause a dose-dependent reduction in the efficacy of neuromuscular transmission in normal rat skeletal muscle and human myasthenic intercostal muscle biopsies *(21)*. Different β-blockers have reproducibly different pre- and post-synaptic effects on neuromuscular transmission. Of the group, propranolol is most effective in blocking neuromuscular transmission, and atenolol the least. The effects of calcium channel blockers on skeletal muscle are not understood, and studies have provided conflicting information. Some demonstrated neuromuscular blockade with post-synaptic curare-like effects, pre-synaptic inhibition of ACh release and both pre- and post-synaptic effects *(24, 25, 26)*. The oral administration of calcium channel blockers to cardiac patients without neuromuscular disease does not produce altered neuromuscular transmission by single-fiber electromyography (SF-EMG) measures *(27)*. Acute respiratory failure was temporally associated with administration of oral dose of verapamil in a patient with LES and small cell carcinoma of the lung *(28)*. One patient with moderately severe, generalized MG developed acute respiratory failure following verapamil initiation (author's observations). Low doses of verapamil and its timed-release preparation have been used successfully for the treatment of hypertension in patients with MG receiving cyclosporin (author's observations).

Procainamide may produce acute worsening of strength in patients with MG *(29)*. The rapid onset of neuromuscular block and the rapid resolution of symptoms following discontinuation of the drug suggest the drug has a direct toxic effect on synaptic transmission, rather than the induction of an autoimmune response against the neuromuscular junction. The postulated mechanism of action is primarily at the pre-synaptic membrane with impaired formation of ACh or its release, although it is known to have post-synaptic blocking effects as well. Two case reports suggest that the anti-arrhythmic P-glycoprotein inhibitor, propafenone, may cause acute exacerbations of myasthenic weakness *(30, 31)*. Like the effects of procainamide, the rapid onset of worsening and resolution following the discontinuation of the drug implicates a direct toxic effect on neuromuscular transmission.

The earliest report of quinidine (the steroisomer of quinine) administration aggravating MG was by Weisman *(32)*. There are several reports of the unmasking of previously unrecognized MG following treatment with quinidine *(33, 34)*. The neuromuscular block is both pre-synaptic, impairing either the formation or release of ACh or, in larger doses, post-synaptic with a curare-like action *(35)*. It has been claimed the ingestion of small amounts of quinine, for example in a gin and tonic, may acutely worsen weakness in a myasthenic patient, although this cannot be substantiated with objective reports.

2.3 Cholesterol-Lowering Agents

Several published works suggest that the statin cholesterol-lowering agents may be causal in the exacerbation of myasthenic weakness *(36, 37, 38, 39, 40)*. The mechanism for this worsening is not clear but several postulates have been proposed. It is well recognized that HMG-CoA reductase therapy may produce a myopathy *(41, 42, 43, 44)*. Is it possible the myasthenic worsening could be due to a co-existing disorder of the muscle membrane? Statins have immunomodulatory properties, with the ability to induce production of the Th2 cytokines interleukin (IL)-4, IL-5 and IL-10 *(45)*. Animal and human studies suggest that these Th2 cytokines play a role in the development of MG *(46)*. Therefore, it is possible that upregulation of Th2 cytokine production could lead to worsening MG. Statins have been postulated to cause mitochondrial dysfunction by depleting endogenous coenzyme Q10 *(47)*. Statin-induced mitochondrial failure in the nerve terminal has been proposed as a mechanism to impair neuromuscular transmission given the high content of mitochondria in the nerve terminal *(36)*. There is no evidence to suggest that HMG-CoA reductase directly interferes with neuromuscular transmission *(48)*.

2.4 Magnesium

Hypermagnesemia is an uncommon clinical complication from the use of magnesium-containing drugs *(49)*. Renal failure predisposes to hypermagnesemia, and is reason to avoid magnesium-containing antacids and laxatives *(50, 51)*. Excessive use of enemas containing Mg^{++} may produce hypermagnesemia, but this is usually in patients with underlying gastrointestinal (GI) tract disease *(52, 53)*. Hypermagnesemia commonly results from administration of high doses of parenteral $MgSO_4$ for treatment of eclampsia, at times resulting in serious side effects to the mother or the newborn *(54, 55, 56)*. The clinical manifestations of hypermagnesemia correlate with serum Mg^{++} levels *(57, 53, 58)*. Muscle stretch reflexes become reduced when the serum Mg^{++} level exceeds 5 mEq/L; levels of 9–10 mEq/L are associated with absent reflexes and muscle weakness. During treatment of pre-eclampsia, muscle stretch reflexes are monitored and Mg^{++} administration is discontinued if the reflexes disappear *(56)*. Fatal respiratory failure can occur at levels greater than 10 mEq/L *(54, 56)*. Serum levels greater than 14 mEq/L can induce acute cardiac arrhythmia, including heart block and arrest. Symptoms of autonomic dysfunction in hypermagnesemia include dry mouth, dilated pupils, urinary retention, hypotension and flushing, and are thought to result from pre-synaptic blockade at autonomic ganglia *(59)*. The muscles of ocular motility tend to be spared and the clinical findings of hypermagnesemia resemble those of LES rather than MG *(60)*. Magnesium inhibits release of ACh by competitively blocking calcium entry at the motor nerve terminal *(61)*. There may also be a milder post-synaptic affect. Magnesium also potentiates the action of neuromuscular blocking agents, and this must be considered in women undergoing cesarean section after receiving Mg^{++} for pre-eclampsia *(62, 63)*. Patients with underlying junctional disorders are more sensitive to Mg^{++}-induced weakness. Patients with MG and LES may become weaker after receiving Mg^{++} even when serum Mg^{++} levels are normal or only slightly elevated *(64, 65, 66, 67)*. Reports exist of the uncovering of previously unrecognized MG following treatment of pre-eclampsia with magnesium salts (Howard JF, unpublished observation) *(68)*. Increased MG symptoms most often occur when Mg^{++} is administered parenterally, but on occasion is seen with oral use *(67)*. Therefore, parenteral Mg^{++} administration should be avoided and oral Mg^{++} preparations should be used with caution in patients with known disorders of synaptic transmission, such as

MG, LES and botulism. Patients with MG or LES respond poorly to calcium and may respond better to cholinesterase inhibitors *(64)*.

2.5 Recreational Drugs

Several cases of MG exacerbation following recreational use of cocaine have been reported *(69, 70, 71)*. These attacks frequently include respiratory insufficiency requiring ventilatory support *(70, 71)* and improve with therapeutic apheresis *(71)* (D.B. Sanders – unpublished observations) or high-dose intravenous immunoglobulin *(71)*. Myasthenic exacerbations have been associated with elevated serum creatine kinase in some patients *(70)*. Cocaine reduces the skeletal muscle response to repetitive nerve stimulation in mice by decreasing the muscle and nerve excitability, but without apparent effect on NMT *(71)*. Cocaine inhibits nicotinic AChR in cultured muscle cells *(72)*.

2.6 Rheumatologic Drugs

D-Penicillamine (D-P) is used in the treatment of rheumatoid arthritis (RA), Wilson's disease and cystinuria. A number of autoimmune diseases occur in patients receiving D-P of which MG is most frequent *(73, 74, 75)*. The MG induced by D-P is usually mild and may be restricted to the ocular muscles. In many patients the symptoms are not recognized, and it may be difficult to demonstrate mild weakness of the limbs in the presence of severe arthritis. It is unlikely that D-P has a direct effect on neuromuscular transmission as MG begins after prolonged D-P therapy in most patients and has a relatively low incidence in patients receiving D-P for Wilson's disease compared with those receiving it for RA *(76)*. It is more likely that D-P induces MG by stimulating or enhancing an immunological reaction against the neuromuscular junction. When MG begins while the patient is receiving D-P, it remits in 70% of patients within 1 year after the discontinuation of the drug *(77)*. In a few patients the MG persists after D-P is discontinued, implying that a subclinical myasthenic state existed prior to the initiation of the D-P (author's observations).

Chloroquine is used primarily as an antimalarial drug, but in higher doses it is also used in the treatment of several collagen vascular disorders including RA, discoid lupus erythematosus and porphyria cutanea. It may produce a number of neurological side effects among which are disorders of neuromuscular transmission. Reported mechanisms of action for this have been both pre- and post-synaptic, but chloroquine may also alter immune regulation producing a clinical syndrome of MG similar to that reported with D-P. One patient with RA and another with systemic lupus erythematosus developed the typical clinical, physiological and pharmacological picture of MG following prolonged treatment with chloroquine. Antibodies to the AChR were identified and subsequently slowly disappeared, as did the clinical and electrophysiological abnormalities, with discontinuation of the drug *(38, 39)*. A patient is described with a transient post-synaptic disorder of neuromuscular transmission following 1 week of chloroquine therapy that was thought to be due to a direct toxic effect on the neuromuscular junction rather than a derangement of immune function *(78)*.

2.7 Other

2.7.1 Interferon Alpha

Generalized MG may occur after starting interferon alpha therapy for leukemia, during interferon alpha-2b treatment for malignancy and during treatment for chronic active hepatitis C *(79, 80, 81, 82, 83)*. Myasthenic crisis may even develop with interferon alpha therapy *(84)*. The mechanism of interferon-induced MG is not known. Expression of interferon gamma at motor endplates of transgenic mice results in generalized weakness and abnormal NMJ function, which improves with cholinesterase inhibitors. Immunoprecipitation studies identified an 87-kD target antigen recognized by sera from these transgenic mice and from human MG patients. Such studies suggest that the expression of interferon gamma at the motor endplates provoked an autoimmune humoral response, similar to that which occurs in human MG *(85)*.

2.7.2 Botulinum Neurotoxin

Botulinum neurotoxin, when used therapeutically for focal dystonia, has unmasked subclinical LES *(86)*. Worsening of weakness or crisis has also been reported in MG following injections of botulinum toxin *(87, 88, 89, 90)*. A known defect of NMT is considered to be a relative contra-indication to the use of botulinum toxin although others have reported its successful use in patients with MG *(91, 92)*.

Numerous other drugs may interfere with neuromuscular transmission. Many local anesthetics, certain anticonvulsants, magnesium, iodinated contrast dyes and, of course, the neuromuscular blocking agents used by anesthesiology during surgery are included in this list. Newly reported adverse drug-disorder interactions are reported frequently. It is beyond the scope of this chapter to discuss them all in detail and the reader is referred to a more comprehensive review of the topic *(20)* or to the web address of MG Foundation of America.

3. BIOLOGICAL NEUROTOXINS

3.1 Botulism

Botulism is caused by a clostridial neurotoxin that blocks the release of ACh from the motor nerve terminal[*] *(93)*. The result is a long-lasting, severe muscle paralysis. Botulism may be classified clinically according to presentation as noted in Table 1. Of eight different types of botulinum toxins, types A and B cause most cases of botulism in the United States. Type E is transmitted in seafood and Type A is thought to produce the most severe manifestations. The classic form of the disease usually follows ingestion of foods that were contaminated by inadequate sterilization. Not all persons ingesting contaminated food become symptomatic. Nausea and vomiting are the first symptoms and the neuromuscular features begin 12–36 hours after exposure. The clinical findings are stereotypical. Ptosis, blurred vision, dysphagia and dysarthria are the presenting features. The pupils may be dilated and poorly reactive to light. Descending weakness progresses for 4–5 days and then plateaus. Respiratory paralysis may occur rapidly.

Table 1
Clinical Classification of Botulism

Classic form
Infantile form
Wound botulism
Traumatic or surgical
Drug abuse
Intranasal
Intravenous
Hidden form

[*]While tetanus toxin also binds to the neuromuscular junction, its mechanism of action is distinctly different. This toxin is translocated into the nerve terminal and then moves in a retro-axonal fashion to the synaptic space between the alpha motoneuron and inhibitory neurons. There it inhibits exocytosis resulting in paresis. Because it does not directly involve the motor nerve terminal it will not be discussed further.

Most patients have autonomic dysfunction, such as dry mouth, constipation, urinary retention and cardiovascular instability.

Infantile botulism is caused by the chronic absorption of toxin from *Clostridium botulinum* growing in the infant gastrointestinal tract *(94)*. Honey is a common source of contamination. The onset of symptoms, constipation, lethargy and poor suck, usually occur between 1 week and 12 months of age, most commonly between 2 and 8 months of age. The majority of reported cases occur with Type A or B toxin. A descending pattern of weakness occurs and may produce widespread cranial nerve and limb muscle involvement. The pupils are poorly reactive and the tendon reflexes are hypoactive. Most babies will require ventilatory support.

Wound botulism occurs due to the contamination of a wound with *Clostridium botulinum*. Its rarity may be related to the difficulty of spore germination in a wound environment. The clinical presentation of wound botulism is similar to the classical form of the disease as noted above. It may occur as a complication of intranasal and parenteral use of cocaine *(95)*.

Hidden botulism refers to those cases in which no source of botulism can be identified and in whom no apparent source or exposure is known *(96, 97)*. Some argue that these cases represent adult forms of infantile botulism *(98)*. This is suggested by the high prevalence of colonic diverting procedures, achlorhydria, Crohn's disease or recent antibiotic treatment among these patients *(99)*.

Response to antitoxin treatment is generally poor probably because once toxin binds to the nerve terminal it is no longer accessible to the antitoxin. Antibiotic therapy is not effective unless the botulism is the infantile or hidden form of the disease. Otherwise, treatment is supportive with respiratory assistance when necessary. Cholinesterase inhibitors are not usually beneficial; guanidine or 3,4-diaminopyridine (3,4-DAP) may improve strength but not respiratory function. Recovery takes many months but is usually complete.

EMG abnormalities evolve as the disease progresses and may not be present at onset of symptoms. CMAP amplitude is decreased in affected muscles, but motor and sensory nerve conduction is normal. Some patients demonstrate a decremental pattern and most have post-tetanic facilitation in some muscles of 30–100% at some time to low-frequency stimulation. These findings are similar to LES but have a more restricted distribution. SF-EMG shows markedly increased jitter and blocking. The organism can be recovered from the stool of infected infants or in the hidden form of the disease.

3.2 Envenomation

Most biological toxins of animal origin affect the cholinergic system and either facilitate the release of neurotransmitter from the pre-synaptic nerve terminal or block the AChR. In general, bites from snakes, scorpions and ticks are more common during summer months when they are inadvertently encountered. In contrast, exposure to marine toxins may occur at any time as they are acquired through ingestion and less rarely by injection or penetration. Specific geographic loci can be demonstrated for each of these vectors. For example, tick envenomation predominates in states west of the Rocky Mountains, the western provinces of Canada and in Australia. The geography of snake envenomation is species specific. The cobras are found in Asia and Africa, the kraits in Southeast Asia, the mambas in Africa, the coral snake in North America and the sea snakes in the waters of the Pacific near Australia and New Guinea.

3.2.1 Arthropods

The venoms of the phylum Arthropod are used to incapacitate prey for feeding or as a defense against predators *(100)*, an observation made in antiquity. Few of the arthropods, however, produce toxicity at the NMJ, but when produced, it is by three mechanisms (Table 2). In the first there is an initial augmentation of ACh release followed by pre-synaptic depletion of neurotransmitter. The second leads to a facilitation of ACh release without a subsequent pre-synaptic

Table 2
Mechanisms of Arthropod Blockade of Synaptic Transmission

Facilitation of ACh release with subsequent exhaustion of neurotransmitter
Facilitation of ACh release without subsequent exhaustion of neurotransmitter
Depletion of ACh release with subsequent exhaustion of neurotransmitter

depletion of neurotransmitter. The third mechanism causes a depletion of ACh release without a subsequent pre-synaptic depletion of neurotransmitter.

3.2.2 Spider Bites

Only a few spider venoms affect the neuromuscular junction. The funnel web spider and the redback spider of Australia are the most dangerous spiders in this group. In North America, only the bite of the black widow spider is of concern. The usual victim of a black widow spider bite is a small boy, perhaps the result of their inquisitiveness of nooks and crannies.

Lathrotoxins found in the venoms from the spider genus *Latrodectus* (black widow spider) cause systemic lathrodectism. These toxins produce a marked facilitation in neurotransmitter release by depolarization of the pre-synaptic nerve terminal and increasing Ca^{++} influx into the nerve terminal at all neurosecretory synapses including the neuromuscular junction (101, 102, 103). There is a subsequent depletion of neurotransmitter from the nerve terminal resulting in a blockade of synaptic transmission. This toxin exerts its effects on the pre-synaptic nerve terminal by several mechanisms. The toxin binds to neurexin and thereby activates the pre-synaptic protein complex of neurexin, syntaxin, synaptotagmin and the N-type calcium channel to massively facilitate ACh release (104). Neurotransmitter release in nerve-muscle preparations, as measured by MEPP frequency, increases several hundred fold within a few minutes (105). There is a subsequent depletion of synaptic vesicles, disruption of the highly organized active zone region of the pre-synaptic nerve terminal thus inhibiting the docking of synaptic vesicles to the terminal membrane and effective recycling of vesicular membrane (106, 107, 108, 109, 110, 110, 111).

Signs of a black widow spider bite begin within minutes of the bite and reflect the massive release of neurotransmitter from peripheral, autonomic and central synapses (112). Severe muscle rigidity and cramps precedes generalized muscle weakness due to the depolarizing neuromuscular blockade. The black widow spider bite is rarely fatal but cardiovascular collapse may occur in the elderly or in young children. Treatment is primarily supportive. The administration of calcium gluconate may be helpful in alleviating muscle cramps and rigidity (113). Magnesium salts may be beneficial by reducing neurotransmitter release (112). The administration of horse-serum antivenom is effective and rapidly reverses the neurotoxic effects (114).

3.2.3 Tick Paralysis

Tick paralysis, a worldwide disorder, was first described at the turn of the 20th century in North America and Australia although there is vague reference to an earlier case in the early 1800s (115, 116, 117, 118). It is one of several kinds of neuromuscular disorders that result from tick venom exposure. Tick paralysis results from the introduction of a neurotoxin from one of more than 60 tick species (119, 120). In North America, the *Dermacentor andersoni, D. variabilis, D. occidentalis, Amblyomma americanum and A. maculatum* species are toxic. The vectors in Europe and the Pacific are *Ixodes ricinus* and *I. cornuatus* and in Australia it is *I. holocyclus*. Geographically, tick paralysis is more common in states west of the Rocky Mountains and in British Columbia and Alberta (121).

The symptoms are stereotypical. Within 5–6 days of attachment, there is a prodrome of paresthesia, headache, malaise, nausea and vomiting. The prodromal period parallels the feeding pattern of the tick. Over the next 24–48 hours an ascending paralysis occurs. It begins symmetrically in the lower extremities and progresses to involve the trunk and arms. In most instances when a tick is found, it is fully engorged. In contrast to the vectors found most commonly in North America (*Dermacentor* and *Amblyomma species*), the weakness of the Australian tick is more severe and much slower to resolve. In these patients there is often a worsening of clinical signs 24–48 hours following the removal of the tick *(122)*. Sensation is preserved but muscle stretch reflexes are often diminished or not present suggesting the Landry–Guillain–Barré syndrome (Table 3), a common misdiagnosis *(123)*. There is no demonstrable response to cholinesterase inhibitors *(124, 125)*. There is some indication of an association between the proximity of the site of attachment to the brain and the severity of the disease. Antitoxin may be of benefit in some situations, but the high frequency of acute allergic reactions makes its widespread use less useful *(126)*. Resolution of manifestations is dependent in part on how quickly the tick is removed, suggesting that the amount of muscle weakness is a dose-dependent process. Often, improvement begins within hours of removing the tick. Paralyses due to the *Dermacentor* species may continue over several days. However, prolonged weakness is reported *(127)*. Death may occur due to respiratory failure from severe bulbar and respiratory muscle weakness and the clinical picture may be clouded by the presences of central nervous system manifestations *(128, 121)*.

Where once these envenomations carried a 12–25% mortality, the improvement in critical care over the latter half of the 20th century makes these intoxications rarely fatal *(129, 115)*. Children are more prone to the disorder than adults. This may be due in part to their play habits and their lower body mass relative to the amount of toxin acquired. The head and neck is the most common site for tick attachment, although any part of the body may be bitten. Some studies suggest that girls are more often affected because their on average longer hair than boys allows the tick to remain hidden for longer periods and therefore allows prolonged feeding *(123, 130)*. The identification of a tick bite is often delayed resulting in misdiagnosis. Tick paralysis may be confused with Landry–Guillain–Barré syndrome, myasthenia gravis, spinal cord disease, periodic paralysis, diphtheria, heavy-metal intoxication, insecticide poisoning, porphyria and hysteria *(123, 131)*. In many instances the tick is located by the nurse, the house officer, the mortician or autopsy personnel *(132, 129, 123)* Careful, systematic inspection of the scalp, neck and perineum, often with a fine-toothed comb, is necessary to locate the tick.

The mechanisms of paralysis following tick envenomation remain controversial. The most potent toxin is from the Australian tick, *Ixodus holocyclus*. Holocyclotoxin, isolated from the salivary glands of female ticks, causes a temperature-dependent blockade of neuronally evoked release of ACh *(133)*. Others have suggested a post-synaptic block of neuromuscular transmission *(134)*. The tick paralysis of the *Dermacentor* species is understood less clearly, and no direct abnormality of synaptic transmission may occur. Rather, the abnormality may be due to impaired depolarization of the nerve terminal with the consequence of decreased ACh release *(135, 136)*.

Table 3
Comparative Features of Ascending Paralysis

Clinical and laboratory features	Tick paralysis	Landry–Guillain–Barré syndrome
Rate of progression	Hours to days	Days to 1–2 weeks
Sensory loss	Absent	Mild
Muscle stretch reflexes	Diminished or absent	Diminished or absent
Time to recovery	< 24 hours after tick removal	Weeks to months
CSF WBC count	< 10 per mm^2	< 10 per mm^2
CSF protein	Normal	Elevated

CSF: Cerebrospinal fluid; mm^2: per square millimeter; <: less than.

Prolonged distal motor latencies, slowed nerve conduction velocities and reduced compound muscle action potential (CMAP) amplitudes are described *(120, 137, 138, 139, 140)*.

3.2.4 Scorpion Bites

The peptides contained in Scorpion neurotoxins may cause a variety of neurological effects, the most significant of which are those that modulate Na^+ and K^+ channel function. Some, however, affect the neuromuscular junction and produce an enhanced pre-synaptic depolarization resulting in neurotransmitter release *(141)*. Increased excretion of catecholamines is demonstrated after scorpion sting and may relate to the primary effect of the venom or to a secondary sympathetic adrenergic surge. Treatment is non-specific and focuses on maintaining respiratory, cardiac and coagulation function. Antivenom appears not to be efficacious *(142, 143)*.

3.2.5 Snakebites

Envenomation by snakebite occurs from four major groups; *Viperidae* (true vipers) *Crotalidae* (rattlesnakes and pit vipers) *Elapidae* (American coral snake, cobras, kraits, mambas), *Hydrophiodae* (sea snakes). Neuromuscular blockade occurs primarily from the *Elapidae* and *Hydrophiodae* species *(144, 145, 146)*. One *Crotalidae* species, *Croatulus durissus terrificus*, a South American rattlesnake has a very potent neuromuscular blocking venom. Other rattlesnakes and pit vipers act through hematological and cardiovascular mechanisms. Venom is produced and stored in salivary glands and inoculation occurs through fangs or modified premaxillary teeth *(144)*.

Snake toxins may act by pre-synaptic or post-synaptic mechanisms. Pre-synaptic toxins, β-neurotoxins (β-bungarotoxin, notexin and taipoxin), act to inhibit the normal release of acetylcholine from the pre-synaptic cell of the neuromuscular junction. Often, there is an initial augmentation of acetylcholine release followed by pre-synaptic depletion of neurotransmitter. They tend to be more potent than post-synaptic toxins. Post-synaptic neurotoxins, α-neurotoxins, produce a curare-mimetic, non-depolarizing neuromuscular block and vary in the degree of the reversibility of the block in experimental preparations.

Most venoms are a mixture of the two types of neurotoxin, although one type may predominate in a given venom. For example, the venom of the Thai cobra is composed primarily of a single post-synaptic neurotoxin *(147)*. In contrast, the venom of *Bungarus multicintus* contains β-bungarotoxin, four other pre-synaptic toxins, α-bungarotoxin and two other post-synaptic toxins *(148)*. The venoms of *Hydrophiodae* species is more toxic than land snakes although the amount of toxin injected by sea snakes is smaller than that of land-based snakes *(149, 150)*. The α-neurotoxins (post-synaptic), like curare, bind to the muscle nicotinic AChR. They have a slower onset of action, a longer duration of effect and are 15–40 times more potent than d-tubocurarine *(151)*. There are numerous subforms of β-neurotoxins (pre-synaptic). Most have a phospholipase component that is essential for the pre-synaptic effects of the toxin. All suppress the release of ACh from the nerve terminal although there is some variability in the precise mechanism by which this occurs. In experimental preparations, toxins from different species potentiate each other suggesting different binding sites at the neuromuscular junction *(152)*. Taipoxin from the Australian and Papua New Guinean taipan snake is unique. In addition to its potent pre-synaptic blockade of synaptic transmission, it also has a direct myotoxic component. This produces rapid muscle necrosis and degeneration. There is species variation in the susceptibility to toxin exposure. The venom of the Australian mulga snake is fatal in man, produces ptosis in monkeys and does not produce a neuromuscular block in the rabbit *(153, 154)*.

The clinical course of snake envenomation follows a specific pattern. After the bite of a pit viper or cobra, there is local pain, which is often absent following the bite of other *Elapidae* (mambas, kraits, coral snakes) and *Hydrophiodae*. Swelling typically follows within an hour of the bites from *Viperidae, Crotalidae* or the cobra but is not seen following bites from other *Elapidae* and *Hydrophiodae*. A preparalytic stage develops with headache, vomiting, loss of consciousness, paresthesias, hematuria or hemoptysis *(155)*. These manifestations are not common after envenomation by cobras or mambas. The time between snakebite and paralysis may vary from 0.5 to 19 hours *(156)*. The first signs of neuromuscular toxicity are usually ptosis and ophthalmoparesis though these are absent following the bite of the South American rattlesnake. Facial and bulbar weakness develops over hours following the ocular signs *(157)*. For 2–3 days, limb, diaphragmatic and intercostal weakness follows and may continue to evolve *(144, 158)*, and without appropriate treatment, cardiovascular collapse, seizures and coma ensue. There is no sensory abnormality other than that from the bite itself. Other systemic effects of neurologic importance relate to coagulation deficits. Cerebral and subarachnoid hemorrhage may occur after bites from many species and is the leading cause of death following viper bites in several regions of the world *(159, 160)*.

Treatment consists of antivenom which are most effective in bites that do not contain significant amounts of phospholipase, a component of pre-synaptic neurotoxins *(161, 157, 162)*. If the type of snake is known, a high-titer monovalent type is administered, but more often snake variety is not known necessitating the use of polyvalent antivenom. The goal of antivenom is to shorten the duration of weakness, and frequently the addition of respiratory, cardiovascular and hematological support is required. Supportive measures are the mainstay of care for most victims of coral snake bite. Intensive care treatment and airway maintenance is similar to patients with myasthenia gravis. Some authors recommend treatment with cholinesterase inhibitors in cases that are predominantly caused by a post-synaptic abnormality and suggest that electrodiagnostic testing may be useful in determining their effectiveness *(163, 164)*.

3.2.6 Marine Toxins

The rapid rise in marine pollution has spurred a renewed interest in marine toxins. Previously they were only of interest to the physiologist and pharmacologist who use them in the investigation of biological systems. Examples of marine neurotoxicology are scattered throughout the literature dating to biblical times (Exodus 7:20–21). The reader is referred to Southcott's paper for an excellent review of the subject *(165)*. Marine neurotoxins affecting the neuromuscular junction are rare and occur primarily from poisonous fish, a few mollusks and perhaps dinoflagellates. Unlike the poisoning from arthropods and snakes, the majority of marine intoxications occur as the result of ingestion. Unique to some marine toxins is the increase in concentration of toxin through successive predatory transvection up the food chain.

Dinoflagellates are single-celled, biflagellated, algae-like organisms. Diatoms, similar to dinoflagellates, are not flagellated and are encased by a silica shell. The toxins produced by these organisms cause a variety of systemic and neurological effects, but neuromuscular junction effects are rare and indirect. Paralytic shellfish poisoning results from neurotoxins produced by less than 1% of the 2,000–3,000 species of known dinoflagellates and diatoms *(166)*. The toxin is rapidly absorbed through the gastrointestinal tract and symptoms begin within 30 min of ingestion. Characteristically, there is an initial burning or paresthesiae of the face and mouth, spreading quickly to involve the neck and limbs. Slowly, the sensations abate and are replaced with numbness, some ataxia and, in severe cases, progressive generalized weakness and respiratory failure. Overall, the mortality approaches 10%. Most neurotoxins from dinoflagellates and diatoms are sodium channel blockers (e.g., saxitoxin and tetrodotoxin). Brevetoxin, a milder neurotoxin that cause the non-lethal neurotoxic shellfish poisoning, depolarizes cholinergic systems, by opening sodium channels and resulting in neuromuscular transmission alterations indirectly. Ciguatoxin, from *Gambierdiscus toxicus*, is commonly found in the Caribbean and South Pacific regions. This heat-stable lipid enhances the release of ACh from the neuromuscular junction by prolonged sodium channel opening. Onset of symptoms is within hours of ingestion and respiratory muscle paralysis may occur quickly. Recovery usually occurs within a few weeks.

Conontoxins are a diverse group of toxins from predatory cone snails that inject their venom through a small harpoon-like dart *(167)*. It is only the fish predatory species (*Conus geographus, C. textile, C. marmoreus and C. omaria*) of this mollusk that appear dangerous to humans *(168, 169, 170)*. The effects of these toxins are variable among species and within a single species, and several have direct effects on the neuromuscular junction. α-Conotoxins block the binding of ACh to the ligand-binding site *(171, 172, 173)*. These venoms function similarly to the snake α-neurotoxins described earlier. The ω-conotoxins block the voltage-gated calcium channel of the pre-synaptic nerve terminal *(174)*. The latter toxin has played an important role in understanding of the LES and serves as the basis for the currently used antibody assay *(175, 176)*. Following the injection of toxin there is an intense local pain quickly followed by malaise, headache and within 30 min progressive generalized weakness. Respiratory failure often occurs within 1–2 hours. Most cone shell bites are preventable. These shells should be handled carefully with forceps and thick gloves. The proboscis protrudes from the small end of the shell but it is flexible and long enough to sting the holder at the other end. The live shells should never be placed in a pocket as the dart may penetrate cloth *(165)*. Treatment is directed toward respiratory and cardiovascular support. There is no available antivenom. There is no literature discussing the potential efficacy of cholinesterase inhibitors. More than 60% of stings are fatal *(177, 170)*.

The most venomous fish is the stonefish, *Synanceja horrida, S. traachynis, S. verrucosa* found in the Indo-Pacific oceans and Red Sea as well as the genus *Inimicus* found off the coast of Japan *(178)*. The toxin, stonustoxin, is inflicted by injection through the 13 dorsal spines when the victim steps on the small fish that is buried in the sand. Neuromuscular blockade results from induced neurotransmitter release with depletion of ACh stores, similar to that of other pre-synaptic toxins *(179, 180)*. Envenomation results in immediate, excruciating pain that may last for 1–2 days. Severe edema occurs due to the actions of hyaluronidase that promotes the rapid spread of venom through the tissue and tissue necrosis may occur *(165)*. In addition to gastrointestinal, autonomic and cognitive effects, the victim may experience generalized muscle weakness due to the mechanism noted above. Death occurs from cardiotoxicity. Treatment is supportive and in some patients a specific antitoxin may be administered.

3.2.7 Plant Toxins

Rarely, plant neurotoxins affect the human NMJ, but more toxic effects are observed in animals. Neurotoxicity is dependent upon the potency, concentration and interaction with other toxins or substrates in the victim. Many are alkaloids. Coniine, the neurotoxin from the herb, *Conium maculatum* (poison hemlock), produces a rapidly ascending paralysis often resulting in death. Sensory abnormalities are common and prominent *(181)*. The death of Socrates is attributed to hemlock *(182)*. The mechanisms of action of this piperidine alkaloid neurotoxin are not completely understood. There is evidence that the toxin acts as a curare-mimetic *(183)*.

4. OCCUPATIONAL NEUROTOXINS

4.1 Heavy Metals

Numerous polyvalent cations affect neuromuscular transmission and are often used to study basic mechanisms of synaptic transmission. These include barium, erbium, cadmium, cobalt, gadolinium, lanthanum, manganese, nickel, praseodymium, triethyltin, zinc *(184, 185, 186, 187, 188, 189, 190, 191, 192, 193, 194, 195, 196)*. Nearly all of these intoxicants have multiple effects on synaptic transmission, but they predominantly block the release of ACh as well as facilitating spontaneous quantal release of neurotransmitter. They exert their effects by the block of the flux of Ca^{++} through voltage-gated calcium channels and disrupt intracellular stores of Ca^{++} *(197)*.

Heavy-metal intoxication is a rare cause of clinical NMJ toxicity. Interest in this topic arose from the 1971 contamination of grain in Iraq with a methylmercury fungicide. Despite appropriate warnings, the grain was fed to animals, ground for flour and used for making bread *(198)*. Symptoms began within 1 month of consumption ultimately affecting more than 6500 people and killing nearly 8% *(199)*. Patients experienced ataxia, fatigue, generalized muscle weakness and occasionally optic atrophy. While one of the expected abnormalities following mercurial poisoning is a peripheral neuropathy (based on the Minamata experience), extensive electrodiagnostic examinations of the affected population did not demonstrate this *(200, 201)*. Repetitive nerve stimulation studies demonstrated a decremental response that was partially reversible with cholinesterase inhibitors *(202)*. Similar abnormalities were demonstrated in experimental animals *(203)*.

4.2 Organophosphate and Carbamate Poisoning

The earliest use of a cholinesterase inhibitor as a neurotoxin is attributed to tribesman in Africa who used the Calabar bean as a right of passage or an "ordeal poison" *(17)*. Organophosphates (OP) are a class of more than 20,000 compounds that irreversibly inhibit cholinesterases including acetylcholinesterase (AChE) *(204)*. They are widely used in the agricultural, manufacturing and pharmaceutical industries as well as a weapon of mass destruction *(205, 206)*. Exposure to OP compounds occurs in the workplace, in food, drinking water and in the environment. OP intoxication is infrequent in the United States because OP-containing insecticides are not readily available. However, these are used commonly in many other countries, where intoxication most commonly results from attempted suicide by ingestion of insecticides, indiscriminate handling and storage by poorly informed workers *(207, 208, 209, 210)*.

The physiochemical properties of these compounds vary. They may be solid, liquid or gaseous and soluble in various media. Some are highly corrosive, others are not; some highly volatile, others not. Dermal contact, respiratory inhalation and gastrointestinal absorption may lead to OP absorption. These various physiochemical properties lend themselves to the wide range of applications noted above as well as to the inherent danger of their use *(211)*.

Four neuromuscular toxicological syndromes occur from OP poisoning; an acute cholinergic crisis (type 1 and 2), an intermediate syndrome, a myopathy and a delayed induce neuropathy (Table 4) *(212)*. Only the type 2 cholinergic crisis and the intermediate syndrome are the result of NMJ toxicity. OP compounds exert their NMJ toxicity by the irreversible inhibition of AChE. This results in the excessive accumulation of ACh at the NMJ as well as other cholinergic synapses of the central, peripheral and autonomic nervous systems *(213)*. The excessive accumulation of ACh produces a depolarizing neuromuscular block at the NMJ that is followed by desensitization of the AChR *(214, 215, 216)*. Electrodiagnostic studies demonstrate normal nerve conduction studies, reduced CMAP amplitudes, a decremental response to repetitive nerve stimulation and CMAP afterdischarges to a single nerve stimulus *(209, 217)*.

Carbamate salts and esters, which are primarily used as pesticides, are synthetic analogues of the alkaloid physostigmine (eserine). They may directly or indirectly affect the NMJ. Like the OP compounds, carbamates also inhibit the action of AChE at cholinergic synapses. They are easily absorbed into the central nervous system because of their lipid solubility characteristics. Unlike OP compounds, the effects of carbamate agents are reversible. However, the manifestations of

Table 4
Neuromuscular Syndromes
of Organophosphate Poisoning

Acute cholinergic crisis
Intermediate syndrome
Myopathy
Delayed toxin-induced neuropathy

carbamate poisoning are indistinguishable from OP poisoning. Neurotoxicity occurs rapidly following significant exposure to both classes of compounds. Mortality rates are high with death usually occurring from respiratory paralysis, which develops in 40% of poison victims *(218)*.

4.2.1 Pesticides

OP chemistry had its origin around 1820 when Lassainge synthesized triethyl phosphate, but not until the turn of the century did they become commonly used for their insecticidal properties. The OP insecticides are all derivatives of phosphoric acid. There are many subclasses within this group of compounds and their various moieties (e.g., sulfur, amides) confer variation in overall toxicity. Despite recognition of their toxicity, their use continues to rise, particularly in developing countries where demand was predicted to more than double in the 1990s *(219)*. Most fatal intoxications result from suicidal ingestion *(207, 208, 209, 210, 220, 221, 222, 223)*. Reports of carbamate NMJ toxicity are few *(224)*. The largest episode of carbamate poisoning occurred in 1985 when aldicarb was illegally used as an insecticide on watermelons *(225)*. Seventy-seven percent of 1376 exposed individuals were poisoned; each exhibiting a dose-related spectrum of nicotinic and muscarinic cholinergic receptor toxicity. Fatalities are rare and only occur at high exposure levels *(226, 227, 228, 229)*. Symptoms appear rapidly, often within an hour, peaking in 2–3 hours with full recovery within 72 hours *(230)*.

4.2.2 Agents of War and Terrorism

The highly dangerous toxicity of OP compounds was recognized in 1932 leading to the development of a series of G-agents (G = German). These compounds, GA (ethyl *N,N*-dimethyl-phosphoramidocyanidate; tabun) in 1936, GB (*O*-isopropyl methylphosphonofluoridate; sarin) in 1938, GD (*O*-pinacolyl methylphosphonofluoridate; soman) in 1944 and CF (cyclohexyl methylphosphonofluoridate; cyclosarin) were developed specifically as agents of war *(231)*. The V-series (V = venomous) followed and included four compounds, VE (*S*-(diethylamino)ethyl *O*-ethyl ethylphosphonothioate), VG (*O,O*-diethyl-*S*-[2-(diethylamino)ethyl] phosphorothioate; also called Amiton or Tetram), VM (phosphonothioic acid, methyl-, *S*-(2-(diethylamino)ethyl) *O*-ethyl ester) and VX (*O*-ethyl-*S*-[2(diisopropylamino)ethyl] methylphosphonothioate; venom X) in 1952. Little is known about a Russian agent, coded VR-55 *(232)*. Great Britain ceased nerve gas weapons research in 1959 and the US transiently discontinued their efforts between 1969 and 1981 *(233)*. Other countries continue to develop their weapons programs. It is speculated that GA was used in the 1980s during the Iraq–Iran conflict causing innumerable deaths *(232)*. G-agents share a number of similar characteristic properties. They are highly volatile in room air. As a result their toxicity may be from either inhalation or by contact. These compounds are soluble in fat and water. This allows ready absorption through the skin eyes and mucous membranes. Vapor agents are absorbed initially through the eyes producing local irritation and then through the respiratory tree. Liquid agents penetrate through the skin at the point of contact and are able to produce more severe and generalized symptoms.

The V-series toxins are the most highly toxic chemical warfare agents. These agents of war are termed persistent agents as they remain active on skin, clothes and other surfaces for prolonged periods of time. VX was synthesized by the British in 1957. Other agents in the series (Ve, Vg, Vm and V-gas) are less well known as information about them is very limited.

Local effects (sweating, mucosal irritation) occur within seconds and paralysis, and apnea may occur as quickly as 1–2 min of exposure. Comparative inhalation toxicities are summarized in Table 5. Absorption may be through inhalation, ingestion or cutaneous contact. VX is the most potent and GA the least. Terrorist attacks in Japan resulted in the exposure of civilians to GB and VX *(234, 235, 236, 237, 238)*.

Table 5
Comparative Human LCt_{50} and LD $_{50}$ to Nerve Agents

Nerve agent	Aerosolized (LCt_{50})	Percutaneous (LD$_{50}$)
VX vapor	10 mg·min/m^3	6–10 mg
Soman vapor	50 mg·min/m^3	350 mg
Sarin vapor	100 mg·min/m^3	1,700 mg
Tabun vapor	400 mg·min/m^3	1,000 mg

Abbreviations: **LCt_{50}** – the dose of vapor necessary to cause death in 50% of the exposed population where *C* is concentration and *t* is time; **LD$_{50}$** – the dose of cutaneous exposure necessary to cause death in 50% of the exposed population; mg – milligram; min/m^3 – minutes per meter squared.

4.2.3 Pathophysiology

The G- and V-series nerve agents inhibit the hydrolysis of acetylcholine by acetylcholinesterase by binding to and phosphorylating the active site of AChE. The resulting depolarizing neuro-muscular block leads to rapid and profound muscle weakness. Death occurs by respiratory failure. The inhibition of AChE by these agents becomes irreversible, a phenomenon known as "aging". The aging half-time is variable; a short as 2 min for GD and as long as 48 hours for VX *(239)*. Aging is an irreversible phenomenon. Prior to this reaction, the enzyme can be reactivated with the use of oximes which remove the neurotoxin from the AChE molecule. Oximes dissociate the toxic phosphate moiety from the esteratic site on AChE, thus reactivating esterase and restoring normal NMT *(240)*. Following attachment of the nerve agent to AChE, a portion of the nerve agent, called the leaving group, is cleaved from the bound molecule. This is followed by a second reaction during which an alkyl group leaves the nerve agent. This results in an aged complex for which oximes have no effect *(241)*.

4.2.4 Treatment

Treatment is directed toward the prevention of chemical exposure by appropriate clothing and the decontamination of exposed victims *(242)*. Aggressive cardiopulmonary support is necessary. Atropine is an effective antidote to block excessive cholinergic activity at muscarinic receptors in both organophosphate and carbamate intoxications. AChE reactivators (oximes, e.g., 2-PAM) and anticholinergics are often helpful for acute OP intoxications but appear to have little effect to reverse weakness caused by the intermediate syndrome. Oximes dissociate the toxic phosphate moiety from the esteratic site on AChE, thus reactivating esterase and restoring normal neuro-muscular transmission *(243)*. Soman is the least responsive of the nerve agents to oxime therapy because the agent–enzyme complex rapidly undergoes "aging". This refers to the compound undergoing a time-dependent conformational change that is no longer responsive to reactivators *(244)*. In contrast to OP intoxications, oxime reactivators are contraindicated in carbamate poisoning as these compounds enhance the effects of the carbamate and promote further junctional toxicity *(245)*. In cases where the offending compound is not known, it is possible to assay the reactivation of AChE activity in vivo and possibly differentiate between OP and carbamate poisoning *(246)*. Neuromuscular blockade with curariform drugs block repetitive discharges although it is not clear whether there is any clinical benefit to their use *(247)*. The carbamate, pyridostigmine was used as a "pre-treatment" for organophosphorous poisoning during the Gulf War but their use is controversial. While studies showed that a 40% inhibition in AChE activity by physostigmine protected experimental animals from acute cholinergic toxicity following exposure to soman, such findings have not been conclusively demonstrated for pyridostigmine *(248)*.

REFERENCES

1. Fambrough DM, Drachman DB, Satymurti S. Neuromuscular function in myasthenia gravis: decreased acetycholine receptors. Science 1973;182:293–295.
2. Barrons RW. Drug-induced neuromuscular blockade and myasthenia gravis. Pharmacotherapy 1977;17:1220–1232.
3. Howard JF. Adverse drug effects on neuromuscular transmission. Semin Neurol 1990;10:89–102.
4. Kaeser HE. Drug-induced myasthenic syndromes. Acta Neurologica Scandinavica 1984;70:39–37.
5. Swift TR. Disorders of neuromuscular transmission other than myasthenia gravis. Muscle & Nerve 1981;4:334–353.
6. Argov Z, Mastaglia FL. Disorders of neuromuscular transmission caused by drugs. N Engl J Med 1979;301:409–413.
7. Pittinger C, Adamson R. Antibiotic blockade of neuromuscular function. Annu Rev Pharmacol 1972;12:109–184.
8. Singh YN, Marshall IG, Harvey AL. Reversal of antibiotic-induced muscle paralysis by 3, 4- diaminopyridine. J Pharm Pharmacol 1978;30:249.
9. Caputy AJ, Kim YI, Sanders DB. The neuromuscular blocking effects of therapeutic concentrations of various antibiotics on normal rat skeletal muscle: a quantitative comparison. J Pharmacol Exp Therapeut 1981;217:369–378.
10. Snavely SR, Hodges GR. The neurotoxicity of antibacterial agents. [Review] [293 refs]. Ann Intern Med 1984;101:92–104.
11. Cadisch R, Streit E, Hartmann K. [Exacerbation of pseudoparalytic myasthenia gravis following azithromycin (Zithromax)]. [German]. Schweizerische Medizinische Wochenschrift 1996;Journal Suisse de Medecine. 126:308–310.
12. Perrot X, Bernard N, Vial C, Antoine JC, Laurent H, Vial T, Confavreux C, Vukusic S. Myasthenia gravis exacerbation or unmasking associated with telithromycin treatment. Neurology 2006;67:2256–2258.
13. Roquer J, Cano A, Seoane JL, Pou Serradell A. Myasthenia gravis and ciprofloxacin [letter]. Acta Neurologica Scandinavica 1996;94:419–420.
14. Samuelson RJ, Giesecke AH Jr, Kallus FT, Stanley VF. Lincomycin-curare interaction. Anesth Analg 1975;54:103–105.
15. Fogdall RP, Miller RD. Prolongation of a pancuronium-induced neuromuscular blockade by clindamycin. Anesthesiology 1974;41:407–408.
16. McQuillen MP, Engbaek L. Mechanism of colistin-induced neuromuscular depression. Arch Neurol 1975;32:235–238.
17. Moore B, Safani M, Keesey J. Possible exacerbation of myasthenia gravis by ciprofloxacin [letter]. Lancet 1988;1:882.
18. Decker DA, Fincham RW. Respiratory arrest in myasthenia gravis with colistimethate therapy. Arch Neurol 1971;25:141–144.
19. Argov Z, Brenner T, Abramsky O. Ampicillin may aggravate clinical and experimental myasthenia gravis. Arch Neurol 1986;43:255–256.
20. Howard JF Jr. Adverse drug effects on neuromuscular transmission. [Review] [202 refs]. Semin Neurol 1990;10:89–102.
21. Howard JF, Johnson BR, Quint SR. The effects of beta-adrenergic antagonists on neuromuscular transmission in rat skeletal muscle. Society for Neuroscience abstracts 1987;13:147.
22. Coppeto JR. Timolol-associated myasthenia gravis. Am J Ophthalmol 1984;98:244–245.
23. Verkijk, A. Worsening of myasthenia gravis with timolol maleate eyedrops. Ann Neurol 1985;17:211–212.
24. Bikhazi GB, Leung I, Foldes FF. Interaction of neuromuscular blocking agents with calcium channel blockers. Anesthesiology 1982;57:A268.
25. Van der Kloot W, Kita H. The effects of verapamil on muscle action potentials in the frog and crayfish and on neuromuscular transmission in the crayfish. Comp Biochem Physiolo 1975;50C:121–125.
26. Ribera AB, Nastuk WL. The actions of verapamil at the neuromuscular junction. Comp Biochem Physiol C: Comp Pharmacol Toxicol 1989;93C:137–141.
27. Adams RJ, Rivner MH, Salazar J, Swift TR. Effects of oral calcium antagonists on neuromuscular transmission. Neurology 1984;34:132–133.
28. Krendel DA, Hopkins LC. Adverse effect of verapamil in a patient with the Lambert-Eaton syndrome. Muscle & Nerve 1986;9:519–522.
29. Kornfeld P, Horowitz SH, Genkins G, Papatestas AE. Myasthenia gravis unmasked by antiarrhythmic agents. Mt Sinai J Med 1976;43:10–14.
30. Lecky BR, Weir D, Chong E. Exacerbation of myasthenia by propafenone [letter]. J Neurol, Neurosurg Psychiatry 1991;54:377
31. Fierro B Castiglione MG, Salemi G, Savettieri G. Myasthenia-like syndrome induced by cardiovascular agents. Report of a case. Ital J Neurol Sci 1987;8:167–169.
32. Weisman SJ. Masked myasthenia gravis. JAMA 1949;141:917–918.
33. Shy ME, Lange DJ, Howard JF, Gold AP, Lovelace RE, Penn AS. Quinidine exacerbating myasthenia gravis: a case report and intracellular recordings. Ann Neurol 1985;18:120.
34. Stoffer SS, Chandler JH. Quinidine-induced exacerbation of myasthenia gravis in patient with myasthenia gravis. Arch Intern Med 1980;140:283–284.

35. Miller RD, Way WL, Katzung BG. The neuromuscular effects of quinidine. Proc Soc Exp Biol Med 1968;129:215–218.
36. Cartwright MS, Jeffery DR, Nuss GR, Donofrio PD. Statin-associated exacerbation of myasthenia gravis. Neurology 2004;63:2188.
37. Engel WK. Reversible ocular myasthenia gravis or mitochondrial myopathy from statins? Lancet 2003;361:85–86.
38. O'Riordan J, Javed M, Doherty C, Hutchinson M. Worsening of myasthenia gravis on treatment with imipenem/cilastatin [letter]. J Neurol Neurosurg Psychiatry 1994;57:383
39. Parmar B, Francis PJ, Ragge NK. Statins, fibrates, and ocular myasthenia. Lancet 2002;360:717
40. Purvin V, Kawasaki A, Smith KH, Kesler A. Statin-associated myasthenia gravis: report of 4 cases and review of the literature. Medicine (Baltimore) 2006;85:82–85.
41. Law M, Rudnicka AR. Statin safety: a systematic review. Am J Cardiol 2006;97:S52–S60.
42. Jacobson TA. Statin safety: lessons from new drug applications for marketed statins. Am J Cardiol 2006;97:S44–S51.
43. Huynh T, Cordato D, Yang F, Choy T, Johnstone K, Bagnall F, Hitchens N, Dunn R. HMG CoA reductase-inhibitor-related myopathy and the influence of drug interactions. Intern Med J 2002;32:486–490.
44. Evans M, Rees A. Effects of HMG-CoA reductase inhibitors on skeletal muscle: are all statins the same? Drug Saf 2002;25:649–663.
45. Youssef S, Stuve O, Patarroyo JC, Ruiz PJ, Radosevich JL, Hur EM, Bravo M, Mitchell DJ, Sobel RA, Steinman L, Zamvil SS. The HMG-CoA reductase inhibitor, atorvastatin, promotes a Th2 bias and reverses paralysis in central nervous system autoimmune disease. Nature 2002;420:78–84.
46. Milani M, Ostlie N, Wang W, Conti-Fine BM. T cells and cytokines in the pathogenesis of acquired myasthenia gravis. Ann NY Acad Sci 2003;998:284–307.
47. Hargreaves IP, Heales S. Statins and myopathy. Lancet 2002;359:711–712.
48. Wierzbicki AS, Poston R, Ferro A. The lipid and non-lipid effects of statins. Pharmacol Ther 2003;99:95–112.
49. Krendel DA. Hypermagnesemia and neuromuscular transmission. Semin Neurol 1990;10:42–45.
50. Castlebaum AR, Donofrio PD, Walker FO, Troost BT. Laxative abuse causing hypermagnesemia quadriparesis and neuromuscular junction defect. Neurology 1989;39:746–747.
51. Randall RE, Cohen MD, Spray CC, Rossmeise EC. Hypermagnesemia in renal failure. Ann Intern Med 1964;61:73–88.
52. Collins EN, Russell PW. Fatal magnesium poisoning. Cleve Clin Q 2002;16:162–166.
53. Mordes JP, Wacker WEC. Excess magnesium. Pharmacol Rev 1978;29:273–300.
54. Flowers CJ. Magnesium in obstetrics. Am J Obstet Gynecol 1965;91:763–776.
55. Lipsitz PJ. The clinical and biochemical effects of excess magnesium in the newborn. Pediatrics 1971;47:501–509.
56. Pritchard JA. The use of magnesium sulfate in preeclampsia/eclampsia. J Reprod Med 1979;23:107–114.
57. Fishman RA. Neurological aspects of magnesium metabolism. Arch Neurol 1965;12:562–596.
58. Somjen G, Hilmy M, Stephen CR. Failure to anesthetize human subjects by intravenous administration of magnesium sulfate. J Pharmacol Exp Ther 1966;154:652–659.
59. Hutter OF, Kostial K. Effect of magnesium ions upon the release of acetylcholine. J Physiol 1953;120:53P
60. Swift TR. Weakness from magnesium containing cathartics. Muscle Nerve 1979;2:295–298.
61. Del Castillo J, Engback L. The nature of the neuromuscular block produced by magnesium. J Physiol 1954;124:370–384.
62. De Silva AJC. Magnesium intoxication: an uncommon cause of prolonged curarization. Br J Anaesth 1973;45:1228–1229.
63. Ghoneim MM, Long JP. The interaction between magnesium and other neuromuscular blocking agents. Anesthesiology 1970;32:23–27.
64. Cohen BA, London RS, Goldstein PJ. Myasthenia gravis and preeclampsia. Obstet Gynecol 1976;48:35S–37S.
65. George WK, Han CL. Calcium and magnesium administration in myasthenia gravis. Lancet 1962;ii:561.
66. Gutmann L, Takamori M. Effect of Mg^{++} on neuromuscular transmission in the Eaton-Lambert syndrome. Neurology 1973;23:977–980.
67. Strieb EW. Adverse effects of magnesium salt cathartics in a patient with the myasthenic syndrome. Ann Neurol 1973;2:175–176.
68. Bashuk RG, Krendel DA. Myasthenia gravis presenting as weakness after magnesium administration. Muscle Nerve 1990;13:708–712.
69. Berciano J, Oterino A, Rebollo M, Pascual J. Myasthenia gravis unmasked by cocaine abuse. N Engl J Med 1991;325:892.
70. Daras M, Samkoff LM, Koppel BS. Exacerbation of myasthenia gravis associated with cocaine use. Neurology 1996;46:271–272.
71. Venkatesh S, Rao A, Gupta R. Exacerbation of myasthenia gravis with cocaine use [letter]. Muscle Nerve 1996;19:1364
72. Krivoshein AV, Hess GP. Mechanism-based approach to the successful prevention of cocaine inhibition of the neuronal (alpha3beta4) nicotinic acetylcholine receptor. Biochemistry 2004;43:481–489.

73. Balint G, Szobor A, Temesvari P, Zahumenszky Z, Bozsoky S. Myasthenia gravis developed under d-penicillamine treatment. Scan J Rheumatol 1975;12–21.

74. Bucknall RC, Dixon A, Glick EN, Woodland J, Zutshi DW. Myasthenia gravis associated with penicillamine treatment for rheumatoid arthritis. Br Med J 1975;1:600–602.

75. Czlonskowska A. Myasthenia syndrome during penicillamine treatment. Br Med J 1975;2:726–727.

76. Masters CL, Dawkins RL, Zilko PJ, Simpson JA, Leedman RJ Penicillamine-associated myasthenia gravis, anti-acetylcholine receptor and antistriational antibodies. Am J Med 1977;63:689–694.

77. Albers JW, Beals CA, Levine SP. Neuromuscular transmission in rheumatoid arthritis, with and without penicillamine treatment. Neurology 1981;31:1562–1564.

78. Robberecht W, Bednarik J, Bourgeois P, Van Hees J, Carton H. Myasthenic syndrome caused by direct effect of chloroquine on neuromuscular junction. Arch Neurol 1989;46:464–468.

79. Perez A, Perella M, Pastor E, Cano M, Escudero J. Myasthenia gravis induced by alpha-interferon therapy. Am J Hematol 1995;49:365–366.

80. Batocchi AP, Evoli A, Servidei S, Palmisani MT, Apollo F, Tonali P. Myasthenia gravis during interferon alfa therapy. Neurology 1995;45:382–383.

81. Lensch E, Faust J, Nix WA, Wandel E. Myasthenia gravis after interferon alpha treatment. Muscle Nerve 1996;19:927–928.

82. Piccolo G, Franciotta D, Versino M, Alfonsi E, Lombardi M, Poma G. Myasthenia gravis in a patient with chronic active hepatitis C during interferon-alpha treatment [letter]. J Neurol, Neurosurg Psychiatry 1996;60:348

83. Mase G, Zorzon M, Biasutti E, Vitrani B, Cazzato G, Urban F, Frezza M. Development of myasthenia gravis during interferon-alpha treatment for anti-HCV positive chronic hepatitis [letter]. J Neurol Neurosurg Psychiatry 1996;60:348–349.

84. Konishi, T. [A case of myasthenia gravis which developed myasthenic crisis after alpha-interferon therapy for chronic hepatitis C]. [Review] [14 refs] [Japanese]. Rinsho Shinkeigaku – Clinical Neurology 1996;36:980–985.

85. Gu D, Wogensen L, Calcutt N, Zhu S, Merlie JP, Fox HS, Lindstrom JM, Powell HC, Sarvetnick N. Myasthenia gravis-like syndrome induced by expression of interferon in the neuromuscular junction. J Exp Med 1995;18:547–557.

86. Erbguth F, Claus D, Engelhardt A, Dressler D. Systemic effect of local botulinum toxin injections unmasks subclinical Lambert-Eaton myasthenic syndrome. J Neurol Neurosurg Psychiatry 1993;56:1235–1236.

87. Borodic G. Myasthenic crisis after botulinum toxin. Lancet 1998;352:1832

88. Emmerson J. Botulinum toxin for spasmodic torticollis in a patient with myasthenia gravis. Mov Disord 1994;9:367

89. Tarsy D, Bhattacharyya N, Borodic G. Myasthenia gravis after botulinum toxin A for Meige syndrome. Mov Disord 2000;15:736–738.

90. Martinez-Matos JA, Gascon J, Calopa M, Montero J. [Myastenia gravis unmasked by botulinum toxin]. Neurologia 2003;18:234–235.

91. Goncalves MR, Barbosa ER, Zambon AA, Marchiori PE. Treatment of cervical dystonia with botulinum toxin in a patient with myasthenia gravis. Arq Neuropsiquiatr 1999;57:683–685.

92. Fasano A, Bentivoglio AR, Ialongo T, Soleti F, Evoli A. Treatment with botulinum toxin in a patient with myasthenia gravis and cervical dystonia. Neurology 2005;64:2155–2156.

93. Cherington M. Clinical spectrum of botulism. Muscle Nerve 1998;21:701–710.

94. Pickett J, Berg B, Chaplin E, Brunstetter-Shafer, M. Syndrome of botulism in infancy: clinical and electrophysiologic study. N Engl J Med 1976;295:770–792.

95. MacDonald KL, Rutherford SM, Friedman SM, Dieter JA, Kaye BR, McKinley GF, Tenney JH, Cohen ML. Botulism and botulism-like illness in chronic drug users. Ann Intern Med 1985;102:616–618.

96. Chia JK, Clark JB, Ryan CA, Pollack M. Botulism in an adult associated with food-borne intestinal infection with Clostridium botulinum. N Engl J Med 1986;315:239–241.

97. Dowell VR Jr, McCroskey LM, Hatheway CL, Lombard GL, Hughes JM, Merson MH. Coproexamination for botulinal toxin and clostridium botulinum. A new procedure for laboratory diagnosis of botulism. JAMA 1977;238:1829–1832.

98. McCroskey LM, Hatheway CL, Woodruff BA, Greenberg JA, Jurgenson P. Type F botulism due to neurotoxigenic *Clostridium baratii* from an unknown source in an adult. J Clin Microbiol 1991;29:2618–2620.

99. Griffin PM, Hatheway CL, Rosenbaum RB, Sokolow R. Endogenous antibody production to botulinum toxin in an adult with intestinal colonization botulism and underlying Crohn's disease. J Infect Dis 1997;175:633–637.

100. Maretic, Z. Venoms of Theridiidae, genus Latrodectus. B. Epidemiology of envenomation, symptomatology, pathology and treatment. In Arthropod Venoms. Handbuch der experimentellen Pharmakologie. Volume 48 (S. Bettini, ed.) Springer Verlan, Berlin. 1978 pp. 185–207.

101. Rosenthal L. Alpah-lathrotoxin and related toxins. Pharmacolo Ther 1989;42:115–134.

102. Hurlbut WP, Iezzi N, Fesce R, Ceccarelli B. Correlation between quantal secretion and vesicle loss at the frog neuromuscular junction. J Physiol 1990;424:501–526.

103. Henkel AW, Sankaranarayanan S. Mechanisms of α-latrotoxin action. Cess Tissue Res 1999;296:229–233.

104. Ushkaryov YA, Petrenko AG, Geppert M, Sudhof TC. Neurexins: Synaptic cell surface proteins related to the α-lathrotoxin receptor and laminin. Science 1992;257:50–56.

105. Longenecker HE, Hurlbut WP, Mauro A, Clark AW. Effects of black widow spider venom on the frog neuromuscular junction. Effects on end-plate potential, miniature end-plate potential and nerve terminal spike. Nature 1970;225:701–703.

106. Clark AW, Hurlbut WP, Mauro A. Changes in the fine structure of the neuromuscular junction of the frog caused by black widow spider venom. J Cell Biol 1972;52:1–14.

107. Clark AW, Mauro A, Longenecker HE, Hurlbut WP. Effects of black widow spider venom on the frog neuromuscular junction. Effects on the fine structure of the frog neuromuscular junction. Nature 1970;225:703–705.

108. Ceccarelli B, Grohovaz F, Hurlbut WP. Freeze-fracture studies of frog neuromuscular junctions during intense release of neurotransmitter. I. Effects of black widow spider venom and Ca2 + -free solutions on the structure of the active zone. J Cell Biol 1979;81:163–177.

109. Pumplin DW, Reese TS. Action of brown widow spider venom and botulinum toxin on the frog neuromuscular junction examined with the freeze-fracture technique. J Physiol 1977;273:443–457.

110. Gorio A, Rubin LL, Mauro A. Double mode of action of black widow spider venom on frog neuromuscular junction. J Neurocytol 1978;7:193–202.

111. Howard BD. Effects and mechanisms of polypeptide neurotoxins that act presynaptically. Ann Rev Pharmacol Toxicol 1980;20:307–336.

112. Gilbert WW, Stewart CM. Effective treatment of arachiodism by calcium salts. Am J Med Sci 1935;189:532–536.

113. Miller TA. Bite of the black widow spider. Am Fam Physician 1992;45:181–187.

114. D'Amour EF, Becker FE, Van Riper W. The black widow spider. Q Rev Biol 1936;11:123–160.

115. Temple IU. Acute ascending paralysis, or tick paralysis. Med Sentinel 1912;20:507–514.

116. Todd JL. Tick bite in British Columbia. CMAJ 1912;2:1118–1119.

117. Cleland JB. Injuries and diseases of man in Australia attributable to animals (except insects). Australas Med Gaz 1912;32:295–299.

118. Gregson, JD. Tick paralysis: an appraisal of natural and experimental data. Monograph Canadian Department of Agriculture 973;9:4–109.

119. Tick paralysis – Washington, 1995. From the Centers for Disease Control and Prevention. JAMA 1996;275:1470

120. Gothe R, Kunze K, Hoogstraal H. The mechanisms of pathogenicity in the tick paralyses. J Med Entomol 1979;16:357–369.

121. Weingart JL. Tick paralysis. Minn Med 1967;50:383–386.

122. Brown AF, Hamilton DL. Tick bite anaphylaxis in Australia. J Accid Emerg Med 1998;15:111–113.

123. Felz MW, Smith CD, Swift TR. A six-year-old girl with tick paralysis [see comments]. N Engl J Med 2000;342:90–94.

124. Cherington M, Synder RD. Tick paralysis. Neurophysiologic studies. N Engl J Med 1968;278:95–97.

125. Swift TR, Ignacio OJ. Tick paralysis: electrophysiologic studies. Neurology 1975;25:1130–1133.

126. Grattan-Smith PJ, Morris JG, Johnston HM, Yiannikas C, Malik R, Russell R, Ouvrier RA. Clinical and neurophysiological features of tick paralysis. Brain 1997;120:1975–1987.

127. Donat JR, Donat JF. Tick paralysis with persistent weakness and electromyographic abnormalities. Arch Neurol 1981;38:59–61.

128. Lagos JC, Thies RE. Tick paralysis without muscle weakness. Arch Neurol 1969;21:471–474.

129. Rose I. A review of tick paralysis. Can Med Assoc J 1954;70:175–176.

130. Dworkin MS, Shoemaker PC, Anderson DE. Tick paralysis: 33 human cases in Washington State, 1946–1996. Clin Infect Dis 1999;29:1435–1439.

131. Jones HR, Jr. Guillain-Barre syndrome in children. Curr Opin Pediatr 1995;7:663–668.

132. Stanbury JB, Huyck JH. Tick paralysis: critical review. Medicine 1945;24:219–242.

133. Stone BF, Aylward JH. Holocyclotoxin – the paralysing toxin of the austrlian paralsys tick Ixodus holocyclus; chemical and immunological characterization. Toxicon 1992;30:552–553.

134. Rose I, Gregson JD. Evidence of neuromuscular block in tick paralysis. Nature 1959;178:95–96.

135. Gothe R, Neitz AWH. Tick paralyses: Pathogensis and etiology. Adv Dis Vector Res 1991;8:177–204.

136. Stone BF. Tick paralysis, particularly involving Ixodes holocyclus and other *Ixodes* species. Adv Dis Vector Res 1988;5:61–85.

137. Esplin DW, Phillip CB, Hughes LE. Impairment of muscle stretch reflexes in tick paralysis. Science 1960;132:958–959.

138. Cherington M, Snyder RD. Tick paralysis: neurophysiological studies. N Engl J Med 1968;278:95–97.

139. DeBusk FL, O'Connor S. Tick toxicosis. Pediatrics 1972;50:329–329.

140. Haller JS, Fabara JA. Tick paralysis. Case report with emphasis on neurological toxicity. Am J Dis Child 1972;124:915–917.

141. Warnick JE, Albuquerque EX, Diniz CR. Electrophysiological observations on the action of the purified scorpion venom, Tityus-toxin, on nerve and skeletal muscle of the rat. J Pharmacol Exp Ther 1976;198:155–167.
142. Belghith M, Boussarsar M, Haguiga H, Besbes L, Elatrous S, Touzi N, Boujdaria R, Bchir A, Nouira S, Bouchoucha S, Abroug F. Efficacy of serotherapy in scorpion sting: a matched-pair study. J Toxicol Clin Toxicol 1999;37:51–57.
143. Sofer S, Shahak E, Gueron M. Scorpion envenomation and antivenom therapy. Pediatrics 1994;124:973–978.
144. Campbell CH. The effects of snake venoms and their neurotoxins on the nervous system of man and animals. Contemp Neurol Ser 1975;12:259–293.
145. Vital Brazil O. Venoms: their inhibitory action on neuromuscular transmission. In: Cheymol, J., Editor, Neuromuscular Blocking and Stimulating Agents, Pergamon Press, New York, 1972 pp. 145–167.
146. Lee CY. Elapid neurotoxins and their mode of action. Clin Toxicol 1970;3:457–472.
147. Karlsson E, Arnberg H, Eaker D. Isolation of the principal neurotoxin of tow *Naja naja* subspecies. Eur J Biochem 1971;21:1–16.
148. Lee CY, Chang SL, Kau ST, Luh SH. Chromatographic separation of the venon of *Bungarus multicinctus* and characteristics of its components. J Chromatogr 1972;72:71–82.
149. Tu AT, Tu T. Sea snakes from southeast Asia and the Far East and their venoms. In: Halstead, B.W., Editor, Poisonous and Venomous Marine Animals of the World Vol. 3, U.S. Government Printing Office, Washington, DC, 1970 pp. 885–903..
150. Barme M. Venomous sea snakes of Viet Nam and their venoms. In Venomous and Poisonous Animals and Noxious Plants of the Pacific Region, ed. by H. L. Keegan and W. V. McFarlane. The Macmillan Co., New York, 1963 pp. 373–378.
151. Karlsson E. Chemistry of protein toxins in snake venoms. In: Lee, C.Y., Editor, Snake Venoms, Handbook of Experimental Pharmacology vol. 52, Springer, Berlin, 1979 pp. 159–212.
152. Chang CC, Su MJ. Mutual potentiation a nerve terminals, between toxins from snake venoms that contain phospholipase A activity: β–bungarotoxin, crotoxin, taipoxin. Toxicon 1980;18:641–648.
153. Rowlands JB, Mastaglia FL, Kakulas BA, Hainsworth D. Clinical and pathological aspects of a fatal case of mulga (Pseudechis australis) snakebite. Med J Aust 1969;1:226–230.
154. Kellaway CH. The peripheral action of the Australian snake venoms. 2. The curari-like action in mammals. Aust J Exp Biol Med Sci 1932;10:181–194.
155. Bouquier JJ, Guibert J, Dupont C, Umdenstock R. Les piqures de vipere chez l'enfant. Archives Francaises de Pediatrie 1974;31:285–296.
156. Mitrakul C, Dhamkrong-At A, Futrakul P, Thisyakorn C, Vongsrisart K, Varavithya C, Phancharoen S. Clinical features of neurotoxic snake bite and response to antivenom in 47 children. Am J Trop Med Hyg 1984;33:1258–1266.
157. Reid HA. Cobra-bites. Br Med J 1964;2:540–545.
158. Warrell DA, Barnes HJ, Piburn MF. Neurotoxic effects of bites by the Egyptian cobra (Naja haje) in Nigeria. Trans R Soc Trop Med Hyg 1976;70:78–79.
159. Kerrigan KR. Venomous snake bites in Eastern Ecuador. Am J Trop Med Hyg 1991;44:93–99.
160. Ouyang C, Teng C-M, Huang T-F. Characterization of snake venom components acting on blood coagulation and platelet function. Toxicon 1992;30:945–966.
161. Reid HA. Antivenom in sea-snake bite poisoning. Lancet 1975;i:622–623.
162. Theakston RDG, Phillips RE, Warrell DA, Galagedera Y, Abeysekera DTDJ, Dissanayaka P, De Silva A, Aloysius DJ. Envenoming by the common krait (*Bungarus caeruleus*) and Sri Lankan cobra (*Naja naja naja*): efficacy and complications with Haffkine antivenom. Trans R Soc Trop Med Hyg 1990;84:301–308.
163. Kumar S, Usgaonkar RS. Myasthema gravis like picture resulting from snake bite. J Indian Med Assoc 1968;50:428–429.
164. Pettigrew LC, Glass JP. Neurologic complications of coral snake bite. Neurology 1985;35:589–592.
165. Southcott RV. The neurologic effects of noxious marine creatures. Contemp Neurol Ser 1975;165–258.
166. Steidinger KA, Steinfield HJ. Toxic marine dinoflagellates. In: Spector DL, ed. Dinoflagellates. New York, Academic. 1984;201–206.
167. Olivera BM, Gray WR, Zeikus R, McIntosh JM, Varga J, Rivier J, de Santos V, Cruz LJ. Peptide neruotoxins from fish-hunting cone snails. Science 1985;230:1338–1343.
168. Kohn AJ. Venomous marine snails of the genus Conus. In Venomous and Poisonous Animals and Noxious Plants of the Pacific Region, ed. by H. L. Keegan and W. V. McFarlane. Permagon Press, The Macmillan Co., New York, 1963 pp. 83.
169. Kohn AJ. Recent cases of human injury due to venomous marine snails of the genus *Conus*. Hawaii Med J Inter-Island Nurses' Bulletin 1958;17:528–532.
170. Cruz LJ, White J. Clinical toxicology of Conus snail stings. In: Clinical Toxicology of Animal Venoms, ed. J. Meier, and J. White, Boca Raton FL: CRC Press, 1995 pp. 117–127.
171. Gray WR, Luque A, Olivera BM, Barrett J, Cruz LJ. Peptide toxins from *Conus geographus* venom. J Biol Chem 1981;256:4734–4740.

172. Hopkins C, Grilley M, Miller C, Shon KJ, Cruz LJ, Gray WR, Dykert J, Rivier J, Yoshikami D, Olivera BM. A new family of Conus peptides targeted tothe nicotinic acetylcholine receptor. J Biol Chem 1995;270:22361–22367.

173. McIntosh M, Cruz LJ, Hunkapiller MW, Gray WR, Olivera BM. Isolation and structure of a peptide toxin from the marine snail *Conus magnus*. Arch Biochem Biophys 1982;218:329–334.

174. McCleskey EW, Fox AP, Feldman D, Cruz LJ, Olivera BM, Tsien RW, Yoshikami D. Calcium channel blockade by a peptide from *Conus*: specificity and mechanism. Proc Natl Acad Sci U S A 1987;84:4327–4331.

175. Adams DJ, Alewood PF, Craik DJ, Drinkwater RD, Lewis RJ. Conotoxins and their potential pharmaceutical applications. Drug Dev Res 1999;46:219–234.

176. De Aizpurua HJ, Lambert EH, Griesmann GE, Olivera M, Lennon VA. Antagonism of voltage-gated calcium channels in small cell carcinomas of patients with and without Lambert-Eaton myasthenic syndrome by autoantibodies omega-conotoxin and adenosine. Cancer Res 1988;48:4719–4724.

177. Yoshiba S. [An estimation of the most dangerous species of cone shell, Conus (Gastridium) geographus Linne, 1758, venom's lethal dose in humans]. Japanese J Hyg 1984;39:565–572.

178. Halstead BW. Poisonous and venomous marine animals (U.S.Govt.Print.Office, Wash.DC, vol 2, 1967, vol 3 1970.

179. Gwee MC, Gopalakrishnakone P, Yuen R, Khoo HE, Low KS. A review of **stonefish** venoms and toxins. Pharmacol Ther 1994;64:509–528.

180. Kreger AS, Molgo J, Comella JX, Hansson B, Thesleff S. Effects of stonefish (*Synanceia trachynis*) venom on murine and frog neruomuscular junctions. Toxicon 1993;31:307–317.

181. Hardin JW, Arena JM. Human Poisoning from Native and Cultivated Plants. 2nd edition, Duke University Press, Durham, NC. 1974.

182. Davies ML, Davies TA. Hemlock: murder before the lord. Med Sci Law 1994;34:331–333.

183. Panter K E, Keeler RF. Piperidine alkaloids of poison hemlock (Conium maculatum). In Cheeke, PR., ed. Toxicants of plant origin. Vol. I. Alkaloids. CRC Press, Inc., Boca Raton, 1989. pp. 109–132.

184. Silinsky EM. On the role of barium in supporting the asynchronous release of acetylcholine quanta by moter nerve impulses. J Physiol (Lond) 1978;274:157–171.

185. Silinsky EM. Can barium support the release of acetylcholine by nerve impulses? Br J Pharmacol 1977;59:215–217.

186. Metral S, Bonneton C, Hort-Legrand C, Reynes J. Dual action of erbium on transmitter release at the from neuromuscular synapse. Nature 1978;271:773–775.

187. Cooper GP, Manalis RS. Cadmium: effects on transmitter release a the frog neuromuscular junction. Eur J Pharmacol 1984;99:251–256.

188. Forshaw PJ. The inhibitory effect of cadmium on neuromuscular transmission in the rat. Eur J Pharmacol 1977;42:371–377.

189. Weakly JN. the action of cobalt ions on neruomuscular transmission in the frog. J Physiol (Lond) 1973;234:597–612.

190. Molgo J, del Pozo E, Banos JE, Angaut-Petit D. Changes in quantal transmitter release caused by gadolinium ions at the frog neuromuscular junction. Br J Pharmacol 1991;104:133–138.

191. Kajimoto N, Kirpekar SM. Effects of manganese and lanthanum on spontaneous release of acetylcholine at frog motor nerve terminals. Nature 1972;235:29–30.

192. Balnave RJ, Gage PW. The inhibitory effect of manganese on transmitter release at the neuromuscular junction of the toad. Br J Pharmacol 1973;47:339–352.

193. Kita H, Van der Kloot W. Action of Co and Ni at the frog neuromuscular junction. Nature 1973;245:52–53.

194. Alnaes E, Rahaminoff R. Dual action of praseodymium (Pr3 +) on transmitter release at the frog neuromuscular synapse. Nature 1975;247:479

195. Allen JE, Gage PW, Leaver DD, Leow ACT. Triethyltin decreases evoked transmitter release at the mouse neuromuscular junction. Chem Biol Interact 1980;31:227–231.

196. Benoit PR, Mambrini J. Modification of transmitter release by ions which prolong the presynaptic action potential. J Physiol (Lond) 1970;210:681–695.

197. Cooper GP, Manalis RS. Influence of heavy metals on synaptic transmission. Neurotoxicology 2001;4:69–84.

198. Rustam H, Hamdi T. Methylmercury poisoning n Iraq; a neurological study. Brain 1974;97:499–510.

199. Bakir F, Damluji SF, Amin-Saki L, Murthada M, Khalida A, Al-Rawi NY, Tikriti S, Dhahir HJ, Clarkson T, Smith JC, Doherty RA. Methylmercury poisoning in Iraq. Science 1973;181:230–241.

200. Igata A. Neurological aspects of methylmercury poisoning in Minamata. In: Tsubaki T, Takahashi H (eds) Recent advances in Minamata disease studies. Kodansha, Tokyo, 1986 pp 41–57.

201. LeQuense P, Damluji SF, Berlin M. Electrophysiological studies of peripheral nerves in patients with organic mrcury poisoning. J Neurol Neurosurg Psychiatry 1974;37:333–339.

202. Rustam H, von Burg R, Amin-Saki L, Elhassani S. Evidence of a neuromuscular disorder in methymercury poisoning. Arch Environ Health 1975;30:190–195.

203. Atchinson WD, Narahashi T. Methylmercury induced depression of neuromuscular transmission in the rat. Neurotoxicology 1982;3:37–50.

204. Taylor P. Anticholinesterase agents. In: Gilman, AG, Goodman, LS, Rall, T.W. and Murad, F., Editors, The Pharmacological Basis of Therapeutics, Macmillan, New York, 1985 pp. 110–129.
205. Edmundson RS. Dictionary of organphosphorous compounds [electronic resource]. 1988
206. Gunderson CH, Lehmann CR, Sidell FR, Jabari B. Nerve effects: a review. Neurology 1992;42:946–950.
207. Fernando, R. Pesticides in Sri Lanka, Documentation of selected literature and legal aspects. Friedrich-Ebert Press-Colombo. 1989 pp. 27–34.
208. Besser R, Gutmann L, Dilimann U, Weilemann LS, Hopf HC. End plate dysfunction in acute organophosphate intoxication. Neurology 1989;39:561–567.
209. Gutmann L, Besser R. Organophosphate intoxication: pharmacologic, neurophysiologic, clinical, and therapeutic considerations. Semin Neurol 1990;10:46–51.
210. Jeyaratnam J. Acute pesticide poisoning: a major health problem. World Health Stat Q 1990;43:139–145.
211. Aldridge WN, Reiner E. Enzyme Inhibitors as Substrates. Interactions of Esterases with Esters of Organophosphorus and Carbamic Acids, North Holland, Amsterdam 1972 pp. 1–328.
212. Marrs TC. Organophosphate poisoning. Pharmacol Ther 1993;58:51–66.
213. Namba T, Nolte CT, Jackrel J, Grob D. Poisoning due to organophosphorous insecticides. Am J Med 2001;50:475–492.
214. Karalliedde L, Senanayake N. Organophosphorous insecticide poisoning. Br J Anaesth 1989;63:736–750.
215. De Wilde V, Vogblaers D, Colarddyn F, Vanderstraeten G, Van Den Neucker K, De Bleecker J, De Reuck J, Van den Neede M. Postsynaptic neuromuscular dysfunction in orgaophosphate induced intermediate syndrome. Klinische Wochenschrift 1991;69:177–183.
216. Good JL, Khurana RK, Mayer RF, Cintra WM, Albuquerque EX. Pathophysiological studies of neuromuscular function in subacute organophosphate poisoning induced by phosmet. J Neurol, Neurosurg Psychiatry 1993;56:290–294.
217. Maselli RA, Soliven BC. Analysis of the organophosphate-induced electromyographic response to repetitive nerve stimulation: paradoxical response to edrophonium and D-tubocurarine. Muscle Nerve 1991;14:1182–1188.
218. Tsao TC, Juang Y, Lan R, Shieh W, Lee C. Respiratory failure of acute organophosphate and carbamate poisoning. Chest 1990;98:631–636.
219. Anonymous. Public Health Impact of Pesticides Used in Agriculture, World Health Organization/UNEP, 1990 pp. 1–128.
220. Haddad LM. Organophosphate poisoning – Intermediate syndrome. J Toxicol:Clin Toxicol 1992;30:331–332.
221. De Bleecker J, Willems J, Van Den Neucker K, De Reuck J, Vogelaers D. Prolonged toxicity with intermediate syndrome after combined parathion and methyl parathion poisoning. Clin Toxicol 1992;30:333–345.
222. Güler K, Taşçioglu C, Özbey N. Organophosphate poisoning. Isr J Med Sci 1996;32:791–792.
223. Chaudhry R, Lall SB, Mishra B, Dhawan B. Lesson of the week – a foodborne outbreak of organophosphate poisoning. Br Med J 1998;317:268–269.
224. Cranmer MF. Carbaryl. A toxicological review and risk analysis. Neurotoxicology 1986;7:247–328.
225. Goldman LR, Smith DF, Neutra RR, Saunders LD, Pond EM, Stratton J, Waller K, Jackson J, Kizer KW. Pesticide food poisoning from contaminated watermelons in California. Arch Environ Health 1990;45:229–236.
226. Freslew KE, Hagardorn AN, McCormick WF. Poisoning from oral ingestion of carbofuran (Furandan 4F), a cholinesterase-inhibiting carbamate insecticide, and its effects on cholinesterase activity in various biological fluids. J Forensic Sci 1992;37:337–344.
227. Jenis EH, Payne RJ, Goldbaum LR. Acute meprobamate poisoning: a fatal case following a lucid interval. J Am Med Assoc 1969;207:361–362.
228. Klys M, Kosún J, Pach J, Kamenczak A. Carbofuran poisoning of pregnant women and fetus per injestion. J Forensic Sci 1989;34:1413–1416.
229. Maddock RK, Bloomer HA. Meprobamate overdosage. Evaluation of its severity and methods of treatment. J Am Med Assoc 1967;201:123–127.
230. Ecobichon DJ. Carbamate Insecticides. Ch. 52. In Handbook of Pesticide Toxicology, R. Krieger (Ed.), Academic Press, 2001 pp. 1087–1106.
231. Maynard RL, Ballantyne B, Marrs T, Turner P. Toxicology of chemical warfare agents Stockton Press, New York, NY (USA). 1993.
232. Spencer PS, Wilson BW, Albuquerque EX. Sarin other nerve agents, and their antidotes. In Experimental and Clinical Neurotoxicology, 2nd Ed., P.S. Spencer, and H.H. Schaumburg, eds. New York: Oxford University Press. 2000 pp. 1073–1093.
233. Meselson M, Perry Robinson J. Chemical warfare and disarmament. Sci Am 1980;242:34–47.
234. Nozaki H, Aikawa N, Fujishima S, Suzuki M, Shinozawa Y, Hori S, Nogawa S. A case of VX poisoning and the difference from sarin. Lancet 1995;346:698–699.
235. Nozaki H, Aikawa N, Shinozawa Y, Hori S, Fujishima S, Takuma K, Sagoh M. Sarin poisoning in Tokyo subway. Lancet 1995;345:980–981.

236. Morita H, Yanagisawa N, Nakajima T, Shimizu M, Hirabayashi H, Okudera H, Nohara M, Midorikawa Y, Mimura S. Sarin poisoning in Matsumoto, Japan. Lancet 1995;346:290–293.

237. Nakajima T, Saro S, Morita H, Yanagisawa N. Sarin poisoning of a rescue team in the Matsumoto sarin incident in Japan. Occup Environ Med 1997;54:697–701.

238. Woodall J. Tokyo subway gas attack. Lancet 1997;350:296.

239. Dunn MA, Hackley BE, Sidell FR. Pretreatment for nerve agent exposure. In Sidell FR, Takafuji ET, Franz DR, eds. Textbook of Military Medicine: Medical Aspects of Chemical and Biological Warfare. Washington, DC: Borden Institute, Walter Reed Army Medical Center; 1997:pp. 181–196.

240. Dawson RM. Review of oximes available for treatment of nerve agent poisoning. J Appl Toxicol 1994;14:317–331.

241. Worek F, Szinicz L, Eyer P, Thiermann H. Evaluation of oxime efficacy in nerve agent poisoning: development of a kinetic-based dynamic model. Toxicol Appl Pharmacol 2005;209:193–202.

242. Holstege CP, Kirk M, Sidell FR. Chemical warfare. Nerve agent poisoning. Criti Care Med 1997;13:923–942.

243. Dawson RM. Review of oximes available for treatment of nerve agent poisoning. J Appl Toxicol 1994;14:317–331.

244. Becker G, Kawan A, Szinicz L. Direct reaction of oximes with sarin, soman or tabun in vitro. Arch Toxicol 1997;71:714–718.

245. Ecobichon DJ. Carbamic acid ester insecticides. In: D.J. Ecobichon and R.M. Joy, Editors, Pesticides and Neurobiological Diseases (2nd Edition ed.),, CRC, Boca Raton, FL 1994 pp. 313–351.

246. Rotenberg M, Shefi M, Dany S, Dore I, Tirosh M, Almog S. Differentiation between organophosphate and carbamate poisoning. Clinica Chimica Acta 1995;234:11–21.

247. Besser R, Vogt T, Gutmann L. High pancuronium sensitivity of axonal nicotinic-acetylcholine receptors in humans during organophosphate intoxication. Muscle Nerve 1991;14:1197–1201.

248. Miller SA, Blick DW, Kerenyi SZ, Murphy MR. Efficacy of physostigmine as a pretreatment for organophosphate poisoning. Pharmacol Biochem Behav 1993;44:343–347.

17

The Impact of Myasthenia Gravis on Mood, Cognitive Function, and Quality of Life

Robert H. Paul, Larry L. Mullins and James M. Gilchrist

1. INTRODUCTION

In the previous edition of this chapter we noted that for the most part clinical and basic science research in the field of myasthenia gravis (MG) had largely focused on understanding the etiology, clinical sequelae, and treatment of the disease with little attention directed toward the psychological aspects of the condition. Unfortunately, this situation has not improved. Despite evidence that the prevalence of MG has increased in the past decade *(1)*, few studies have explored the psychological status of patients with the disease. As we noted previously, the significance of this oversight is underscored by the absence of a cure for MG and the reality that patients are therefore required to live with a chronic medical condition. We have limited data obtained from cohorts of MG patients, but there is a large literature from patients with other medical diseases that living with a chronic medical condition places patients and their caregivers and loved ones at risk for experiencing significant psychological distress and reductions in their self-reported quality of life. Since a patient's functional response to a disease and overall quality of life represent an important aspect of clinical care, it is important to focus on the factors that both challenge and support the psychological well-being of patients.

Broadly, the psychological aspects of MG can be categorized into two areas: (1) the effect of a patient's psychological health on the expression of their disease and (2) the effect of disease on the psychological health of the patient. In this chapter we examine the research in each of these areas, beginning with the more controversial issue of whether psychological function directly influences disease onset and progression followed by a review of psychological outcome among patients with MG. We then discuss the controversial issue of cognitive abnormalities caused by the disease and the relationship between perceived fatigue and memory function. Finally, we review factors that support psychological health in chronic disease, with a particular emphasis on MG.

2. THE EFFECTS OF PSYCHOLOGICAL HEALTH ON DISEASE

Concern that psychological health may directly affect disease expression is predicated on the theoretical perspective that psychological distress can lead to immune dysregulation. This inter-action between psychological and immunological systems begins with the activation of the sympathetic nervous system during emotional arousal. During a stress response, the hypothalamus secretes a series of hormones that eventually initiate the release of cortisol from the adrenal cortex. The primary physiological role of cortisol is stimulation of glucogenesis, protein degradation, and lipolysis, all of which generate energy supplies to support resistance to the perceived stressor *(1)*. A secondary effect of cortisol release is temporary immune dysregulation from the binding of cortisol to receptors located on immune cells *(2, 3, 4)*.

The biological interaction between the central nervous system and the immune system provides a theoretical basis for the notion that psychological factors can in some cases directly impact disease onset or progression. One example is evident among patients infected with human

From: *Current Clinical Neurology*
Myasthenia Gravis and Related Disorders
Edited by: H.J. Kaminski, DOI 10.1007/978-1-59745-156-7_17, © Humana Press, New York, NY

immunodeficiency virus (HIV). For example, using a longitudinal design Ironson et al. *(5)* observed that ratings of depression and hopelessness at baseline predicted change in immune system health and viral load of the disease, even after accounting for medication regimen and adherence to medications. Similarly, there has been some question as to whether psychological factors affect either the onset or the course of MG. As described below, to date there is no empirical evidence to support this claim.

A few early case reports and case series described a relationship between psychological illness and the precipitation of MG *(6, 7, 8, 9)*, but no controlled studies have been conducted to examine this association. Furthermore, studies that have examined premorbid personality characteristics of MG patients have not identified meaningful relationships between personality type and onset of MG. As an example, in 1969 MacKenzie et al. *(10)* reported that 8 of 25 MG patients "came from markedly abnormal family backgrounds", yet the subjective nature of this conclusion (e.g., some of the examples would not be considered unusual today) and the use of clinically limited terms such as "abnormal" limit the utility of the data. One interesting note from the study is the report of "abnormal" profiles on the Minnesota Multiphasic Personality Inventory (MMPI) among nearly half of the study cohort. Scores on the MMPI were not reported, but elevations were noted on three subscales that measure depression, concerns about disease and illness, and specific somatic symptoms of physical disease. Obviously, these scales are sensitive to symptoms of bona fide medical illnesses *(11)*, and therefore it is not surprising that MG patients obtained higher scores on these scales. Unfortunately, actual scores were not provided, making it impossible to determine whether the elevated scores were clinically meaningful. Furthermore, elevations were not observed on scales more sensitive to premorbid personality characteristics (e.g., obsessive–compulsive behaviors or psychotic symptoms) arguing against the notion of a premorbid or reactive "myasthenic personality".

Nearly identical results were reported in a second study conducted by Schwartz and Cahill *(12)*. In that study, 17 MG patients who volunteered for a psychotherapeutic intervention study were administered the MMPI and elevations were again observed on scales measuring depression, concerns about illness, and somatic symptoms of disease. The authors interpreted these results as a reflection of general reactive distress typical of patients with chronic disease. An elevated score was also obtained on a scale that measures psychotic behavioral characteristics, which was interpreted to reflect the tendency of MG patients to experience embarrassment and social isolation rather than reflecting "overt psychosis". Though there may be some validity to these findings, it is important to recognize the selection bias present in this study. The participants were volunteers in a psychological intervention study, and MG patients seeking mental health treatment reportedly initiated the investigation. The fact that the participants in the study were self-selected for psychological treatment raises the possibility that this group does not adequately represent the larger population of MG patients. This point is particularly relevant since most patients with MG do not experience clinically severe depression, and individuals with MG who are experiencing sufficient emotional distress to seek counseling may not accurately reflect the larger population.

A few reports suggest psychological health can affect the clinical course of MG by inducing exacerbations. These effects are based on very dated and descriptive studies *(6, 10, 13)*. No controlled studies have demonstrated that psychological distress influences acetylcholine receptor antibody levels or alters receptor properties in the presence of antibodies. In the only prospective study, Magni et al. *(14)* found no association between the type or number of significant life events and disease severity in a year of observation. In this study, 51 patients with a range of disease severity reported positive (e.g., marriage, promotion, birth of child) and negative (death of loved one, loss of job, etc.) life events during the 12 months immediately preceding entry into the study. Disease status and the number and type of life events were examined again after 12 months. At follow-up assessment, individuals who exhibited improvement in their symptoms reported no difference in the number or type of life events during the study period compared to individuals who exhibited either no change or more severe disease symptoms. These results do not address the possibility of acute changes in the disease status associated with stressful experiences, but the findings indicate that overall clinical status is not affected by psychological events.

While there is limited evidence that disease is directly affected by psychological health in MG, it is important to recognize that a patient may experience greater symptom severity during times of emotional distress, despite no change in disease activity. This point represents the critical distinction between a biomedical condition and how patients interpret and experience symptoms of the condition. Previous researchers have referred to the former as the "disease" and the latter as the "illness" *(15)*. These two dimensions of health covary, but they do not perfectly overlap. A full range of illness severity is possible at a given level of disease, and psychological health is an important factor that determines the point on the continuum that illness severity is expressed. This effect emanates from a patient's cognitive framework (i.e., schemata) that works to interpret life events and assign psychological valence to the experiences. During times of stress or depressed mood, individuals may perceive experiencing much greater symptoms of a chronic disease. Patients may be prone to identify these episodes as exacerbations of the "disease", though they actually reflect exacerbations of the "illness". This relationship highlights the importance of focusing on variables that affect illness behavior in chronic disease. One obvious variable is the manner in which a disease directly challenges the mental health of patients.

3. THE IMPACT OF MG ON THE PSYCHOLOGICAL AND SOCIAL HEALTH OF PATIENTS

Advanced medical technologies have improved diagnostic accuracy across many diseases and introduced effective pharmacotherapy, yet outright cures remain elusive. As a result, the number of people living with chronic disease has increased, leading to an appreciation of its impact on psychological well-being. This evolution has clearly taken place in the MG community, as life expectancy was significantly shorter only 30 years ago. Now, many MG patients are living longer secondary to sound medical care and the implementation of effective treatments. However, in the absence of a cure, living longer means managing the practical matters associated with a chronic medical condition.

Remarkably, little is known about how MG affects the psychological health of patients or the manner in which patients respond to these challenges. This is unfortunate because the fatigue associated with MG has the potential to significantly challenge psychological well-being. MG patients report that the fatigue associated with their disease produces moderate interference in their ability to complete physical activities and mild to moderate interference in their ability to participate in social functions *(16)*. This is most acute for patients whose speech degenerates into incomprehensibility during a conversation, as well as for those patients who describe embarrassment in social situations when muscles of the face fatigue and alter their ability to produce facial expressions, or others who experience personal discomfort in social environments when eating becomes difficult *(17)*. Clearly, the potential impact of MG on the quality of patients' lives highlights the importance of examining mental health.

3.1. Psychological Health in MG

We pointed out at the start of the chapter that little research has been directed toward understanding the mental health aspects of MG and this situation has not changed since the first publication of this chapter. This situation is likely a direct result of the lower prevalence of MG compared to other chronic conditions and consequently a comparatively lower interest in the scientific field to study aspects of the disease. Based on published papers, it appears that research into the mental health aspects of MG has been sparse as only a few studies have specifically examined the frequency of mood disturbance in MG. Two published studies focused on the frequency of anxiety disorders among MG patients *(18, 19)*, while two other studies focused on depression in this population *(20, 21)*. Each studied a small cohort of patients and they did not consistently include healthy control subjects. Thus, while these studies provide some important insights regarding the psychological health of MG patients, significant gaps remain in understanding the incidence and prevalence of psychiatric comorbidity in this population.

Paradis et al. *(18)* studied 20 individuals with MG and 15 patients with polymyositis/ dermatomyositis (PMD). Forty-three percent of the entire sample was characterized as severely symptomatic and 24% of the MG patients required respiratory-assisted breathing during the course of their illness. Results of the study revealed a generally high frequency of anxiety disorders using self-report and clinical methods of assessment. However, normative comparisons were not included and the results were not subdivided by neuromuscular group on some of the measures. For example, 43% of the patients had an anxiety disorder, but it is impossible to know if one of the two neuromuscular groups contributed more than the other to the results or if the percentage was equally distributed across the two samples.

Paradis et al. reported that significantly more MG patients than PMD patients were diagnosed with panic disorder and both groups endorsed "extensive" symptoms of depression on the Beck Depression Inventory (BDI), though the mean scores for both groups on the BDI were within the normal range (MG = 6.75, PMD = 7.33). When the prevalence of psychiatric illness was compared to a sample of patients with Parkinson's disease, significantly more MG patients met criteria for anxiety disorders. However, the most striking difference between the two groups was in regard to the prevalence of depression, which was greater than three times more common in the Parkinson's group.

In contrast, the frequency of panic disorder in MG was not elevated in another study. Magni et al. *(14)* examined 74 patients using a semistructured interview performed by a psychiatrist. Fifty-one percent of the patients met criteria for a psychiatric disorder (Axis I or II), with adjustment disorders representing the most common diagnosis (22%). Personality (18%) and affective disorders (14%) were also observed, while anxiety and somatoform disorders (5.5%) were infrequently identified. Individuals with more severe disease were more likely to meet criteria for psychopathology than were individuals with milder disease severity, but no associations were evident with use of medications or thymectomy.

The results of the studies described above contrast sharply. Paradis et al. identified an elevated prevalence of panic disorder, while Magni et al. found a low frequency of anxiety disorders, but a high prevalence of adjustment disorders. The discrepancy may reflect methodological differences. Paradis et al. examined a small cohort of MG patients, many of whom had significant disease severity, including several with respiratory distress, which is particularly anxiety provoking. By contrast, Magni et al. included a much larger group of patients with a greater range of disease severity. Furthermore, Magni et al. conducted psychiatric interviews to determine psychopathology. The larger sample size and use of clinical assessments represent important strengths, and therefore it is tempting to conclude that an elevated frequency of adjustment disorders best characterizes the psychological status of MG patients. Nevertheless, the absence of quantified psychological distress among the patients in these studies limits the clinical utility of the findings.

Our research group performed two investigations involving quantification of psychiatric symptom severity in MG *(20, 21)*. In the first, the frequency of depressive symptoms in MG was examined using a self-report measure that independently examined mood symptoms (e.g., feeling sad) and vegetative symptoms (e.g., feeling tired) of depression *(22)*. The importance of assessing these symptoms separately is predicated on the assumption that vegetative symptoms of depression (e.g., feeling tired) overlap with symptom complexes of neuroimmune diseases, such as fatigue *(21)*. We administered the measure of depression to a sample of 29 MG patients and 34 healthy control subjects with similar basic demographics. MG patients were recruited from local support groups and regional chapters of the National Myasthenia Gravis Foundation. The two groups did not differ significantly on the mood subscale (MG mean = 20.9, control mean = 18.6). By contrast, there was a significant group difference on the vegetative subscale (MG mean = 35.6, control mean = 24.0). The percentage of individuals who exceeded cutoffs for clinical significance differed by group only on the vegetative subscale (MG = 41%, control = 8%), reflecting the prominent expression of physical symptoms among the MG patients. Furthermore, median daily dose of prednisone was not significantly associated with the score on the mood subscale but it was significantly associated with the score on the vegetative subscale. The latter observation is noteworthy since use of prednisone is known to interfere with affect regulation, including alterations in depression, mania, and psychosis *(23)*, and there has been concern that MG patients are at risk for these mood disturbances in the context of steroid treatment. However, as noted above we did not observe any correlation between prednisone use and mood symptoms, and when present these effects of prednisone are likely dose dependent and less common among MG patients taking relatively low daily doses.

The results of that study were consistently observed in a second investigation, which examined quality of life and well-being in MG patients *(21)* in the same cohort of patients. Most of the patients from the previous study *(20)* were administered the Medical Outcomes Study Short-Form General Health Survey (SF-36). This self-report measure taps both physical and emotional health and the impact of disease on both these factors. Eight scales are derived from the measure including Physical Function, Social Functioning, Role Disruption-Physical, Role Disruption-Emotional, Mental Health, Vitality, Bodily Pain, and General Health. A particular strength of this measure is the assessment of how illness *interferes* with completion of activities and individual roles.

Table 1 lists the scores on the SF-36. The MG group had lower scores compared to the healthy control subjects (higher scores reflect better function) on all but one scale (Mental Health). On this subscale, the scores for the two groups were nearly identical. When a clinical criterion of 1.5 standard deviations below the mean of control subjects was used to determine the significance of scores between the two groups, only the score on Role Disruption-Physical exceeded the cutoff. Importantly, the overall composite score for quality of life/well-being did not differ between the two groups. In this study we also compared results on the SF-36 to those obtained from individuals with arthritis, hypertension, or congestive heart failure. Again, the greatest differences

Table 1
Quality of Life Ratings on the SF-36 for MG Patients and Health Control Subjects: Mean (Standard Deviation). Data Reported Previously by Paul et al. *(21)*

SF-36 scale	MG patients	Normative data
Physical Functioning	52.7 (24.9)	84.5 (22.9)
Social Functioning	72.2 (22.2)	83.6 (22.4)
Role Disruption-Physical	25.9 (35.0)	81.2 (33.8)
Role Disruption-Emotional	70.3 (39.5)	81.3 (33.0)
Mental Health	74.2 (14.5)	74.8 (18.0)
Vitality	45.1 (22.8)	61.1 (20.9)
Bodily Pain	70.5 (27.9)	75.5 (23.6)
General Health	56.4 (20.4)	72.2 (20.2)
Overall quality of life	58.4 (25.9)	73.0 (24.3)

Range = 0–100; higher scores reflect better function. Note the large standard deviations for both groups and the similarity in scores between the two groups on the Mental Health scale.

between these groups were in the domains of Physical Function and Role Disruption-Physical, with greater impairment evident for the MG group compared to the other patient groups. Overall ratings of quality of life and well-being did not differ between the MG patients and the other patient groups.

More recent studies on quality of life among MG patients have been conducted by Padua and colleagues and this work, to our knowledge, represents the only evolution of the field in this clinical population. Both the studies published by Padua et al. *(24, 25)* represent very important extensions of previous work from our lab. In the first study *(24)*, the authors examined the relationships between patients' reports of quality of life, Osserman classification, anti-acetylcholine receptor antibody levels, and measures of muscle disease (i.e., clinical exam, repetitive nerve stimulation, and single-fiber electromyography). These clinical measures of disease severity were not obtained in our original studies of quality of life in MG. Results of the Padua et al. study revealed significant correlations between the Physical Composite Score from the SF-36 scale and both actual and nadir Osserman classification and RNS amplitude. Interestingly, there were no relationships between perceived quality of life and either facial or ocular impairments or levels of anti-acetylcholine receptor antibody levels. Further, the outcomes did not differ according to thymectomy status or medication treatment regimen.

Perhaps most relevant to the current chapter is that the Mental Composite Score was low among patients regardless of Osserman classification. This differed significantly from the Physical Composite Score, which was fairly linearly related to Osserman status. Further, the Mental Composite Score did not correlate strongly with the other clinical disease measures. As reported by the authors, these findings suggest that mental health in MG may not be dependent on the severity of disease for any given patient (i.e., even patients with mild disease may have significant reductions in mental health). An important aspect of this study is that the participants were recruited from a neurology service, whereas in our series of studies the patients had been recruited from regional support groups. It is likely that the impact of disease status on mental health differs significantly between these cohorts.

The second, and most recent, study conducted by Padua and colleagues examined the relationship between regional disease severity (general, bulbar, and ocular) and health-related quality of life *(25)*. Forty-five Swedish patients recruited from the Department of Neurology at Uppsala Hospital were included in the study and all participants completed a measure of disease severity that separately assessed general, bulbar, and ocular weakness. Results of this study revealed that of the three types of fatigability, only bulbar involvement significantly correlated with reduced mental health ratings among patients. In addition, difficulties with ptosis correlated with ratings of Vitality and Social Function, but diplopia did not relate to any measure of the SF-36. These findings suggest that bulbar difficulties and ptosis are important physical determinants of mental health in MG. Collectively, these two recent studies indicate that among patients treated in a hospital setting, quality of life may be significantly affected primarily due to bulbar involvement and ptosis, and mental health may be significantly impacted even among patients with relatively mild disease severity.

4. COGNITION AND MENTAL FATIGUE IN MG

Involvement of the central nervous system (CNS) in MG has been a subject of debate for more than a decade. Acetylcholine receptors are located in both the CNS and the periphery, and a number of investigators have suggested that antibodies to the peripheral receptors cross-react with central receptors (for review, see *(26)*). There is, however, essentially no evidence that central acetylcholine receptors are affected in MG. Whiting et al. *(27)* reported that MG antibodies do not bind to acetylcholine receptors in the brain. Though the tissue sample used for analysis was obtained from a patient with AD, the authors found no evidence that this comorbid disease affected receptor binding. The fact that MG antibodies do not bind to central receptors should not be surprising since the antibodies also do not bind to receptors in the autonomic ganglia. It is important to note that even if the antibodies do bind to central receptors, the concentration of antibodies in the CNS may not be sufficient to interfere with central function *(28)*.

Despite the lack of evidence for direct central involvement by MG, approximately 60% of patients describe experiencing memory loss *(29)*. This percentage is much larger than that reported by age- and education-matched healthy control subjects, suggesting that memory may be affected in MG. We recently reviewed the literature pertaining to cognition and MG and found that more than half of the studies reported significantly poorer performance of MG patients compared to control subjects on at least one measure of cognitive function *(26)*. However, important methodological limitations were present in every one of these studies. The methodological concerns included no adequate control for differences in disease type (e.g., generalized vs. ocular), comorbid mood disturbance, medications, or levels of fatigue. Furthermore, discrepancies between patients and control samples on demographic factors, such as age and education, were not addressed in some studies, while others failed to exclude participants with visuomotor deficits. The effects of these factors may have artificially increased group differences on cognitive measures in some studies, while minimizing differences in other studies *(26)*.

4.1. Neuropsychological Function in MG

We attempted to address some of the methodological limitations of previous investigations using many of the subjects who participated in our previous studies *(30)*. All subjects were screened for severe ophthalmoplegia and symptoms of diplopia. Depression was assessed with a measure that only examined the mood symptoms of depression (e.g., feeling sad), thus avoiding the differentiation of fatigue and other vegetative symptoms of depression.

Neuropsychological measures were administered to examine five cognitive domains including attention, response fluency (generating words that begin with specific letters and words that represent exemplars of a category), information processing speed (orally substituting numbers for symbols), verbal learning and retention, and visual learning and retention. When compared to a sample of healthy control subjects similar in age and education, there were no significant differences on the measures of attention, retention of learned verbal information and retention of learned visual information. By contrast, the MG patients exhibited significantly poorer performance on the measures of response fluency, information processing speed, and learning of verbal and visual information. Regression analyses revealed no associations between cognitive performance and either mood disturbance or daily dose of medications.

The study provided some support for patients' concerns that memory function is affected in this disease. It is important to note, however, that retention of information did not differ between MG patients and healthy control subjects. This pattern of performance is markedly different than what is observed in degenerative diseases affecting the central cholinergic system (i.e., Alzheimer's disease). This latter point supports the hypothesis that cognition is not affected by central receptor dysfunction in MG. Alternative etiologies of cognitive dysfunction have been suggested, including neuronal dysfunction associated with cytokine activity in the brain, sleep apnea associated with respiratory distress, and fatigue *(26)*. To date, only fatigue has emerged as a likely candidate.

4.2. Fatigue in Myasthenia Gravis

The cardinal symptom of MG is fatigability of striated muscles. Physiologically, this manifests as decreased ability to sustain muscle activity, but behaviorally patients describe this experience as increased feelings of physical fatigue *(17, 31)*. Importantly, patients also report experiencing feelings of mental fatigue in addition to the symptoms of physical fatigue. This is not unusual considering that perceived levels of physical and mental fatigue covary, with increases in one dimension accompanied by increases in the other dimension. Furthermore, studies of healthy control subjects and neurological patients have revealed that mental and physical effort can independently produce increased levels of mental and physical fatigue *(32, 33)*. We recently demonstrated this phenomenon in patients with MG *(34)*.

4.3. The Relationship Between Fatigue and Cognition in Myasthenia Gravis

Considering that perceived levels of mental and physical fatigue are elevated in MG and the two dimensions of fatigue covary, we were interested if subjective fatigue was associated with cognitive difficulties in MG. To examine this relationship, we obtained ratings of mental and physical fatigue before and after completion of a cognitive battery *(34)*. The battery included demanding measures of attention, memory, visuospatial function, and information processing speed. Fatigue was assessed using the Multicomponent Fatigue Inventory (MFI; *30*), a self-report measure that separately assesses mental and physical components of fatigue. The MFI was designed to measure *change* in cognitive and physical fatigue (e.g., pre- and post-testing). Examples of items from the MFI include: "Is your attention span less than usual right now?" and "Do you currently feel weak?" Prior to completing the cognitive battery, subjects rated their level of fatigue using a scale that ranged from 1 ("not at all") to 5 ("a great deal"). The subjects were then administered the cognitive battery, with rest breaks taken throughout testing. After working on the cognitive battery for approximately 1.5 hours, subjects rated the change in their level of perceived fatigue compared to their previous rating.

Baseline ratings of mental and physical fatigue were elevated compared to healthy control subjects. In addition, the post-test ratings of fatigue revealed significant increases in both mental and physical fatigue for the MG patients but not for healthy control subjects. As such, the patients reported feeling much more tired after completing the cognitive test session, whereas the healthy control subjects reported no change in fatigue levels after completing the test session. When correlations were computed between performances on cognitive measures and fatigue ratings, we found significant relationships between change in mental fatigue and performance on tests of response fluency ($r = -0.50$), information processing speed ($r = 0.51$), verbal learning ($r = -0.39$), and visual retention ($r = -0.46$). In each of these cases, cognitive performance declined as the severity of perceived fatigue increased. There were no significant relationships between ratings of physical fatigue and cognitive performance.

The results provide strong evidence that factors associated with fatigue influence cognitive performance. Obviously, causal relationships cannot be inferred from the correlational analyses. However, the fact that change in mental fatigue strongly correlated with the cognitive measures that best discriminates MG patients from healthy controls suggested that the observed relationships are more than coincidental. It will be important to determine in future studies if cognitive performance actually declines following increased severity of fatigue in MG. This has recently been demonstrated in a sample of patients with multiple sclerosis *(35)*. Since fatigue among multiple sclerosis results from involvement of the CNS, it will be interesting to determine if fatigue resulting from peripheral sites (as in the case of MG) also produces deleterious effects on cognition *(36)*.

Overall, the status of cognitive function in MG remains unresolved. It appears that some individuals experience difficulty in demanding cognitive measures that require rapid processing of information. Our recent studies suggest that obvious demographic or treatment factors (e.g., medications) do not account for these performances. It also seems unlikely that direct involvement of the CNS by the autoimmune disease is responsible for the effects on cognition. At least based on the currently available data it appears likely that the feeling of mental fatigue that develops following cognitive effort is a significant determinant of patients' abilities to perform optimally on cognitive demanding tasks.

5. FACTORS THAT SUPPORT PSYCHOLOGICAL HEALTH IN MYASTHENIA GRAVIS

We have focused on the frequency of psychological disorders in MG and the controversy regarding cognition in this disease. These sections describe the psychological challenges that are associated with MG. In the next section, we focus on the variables that facilitate or mediate effective psychological adjustment. With rare exception, studies have not specifically addressed this topic in MG, but extensive information exists in other patient populations, and there is good reason to believe that many of these findings extend to MG. Four factors that are consistently identified as important constructs that predict psychological adjustment to adult chronic disease include perceived control, illness uncertainty, illness intrusiveness, and adequate social support. In the last section, we review the impact of these variables, with a particular focus on their relationship to MG.

5.1. Perceived Control

An individual's level of perceived control significantly influences how that person responds to threatening situations. Individuals seek a sense of control in their environment, and the absence of perceived control results in significant distress that is manifested both biologically and psychologically *(37)*. Diseases challenge one's sense of control or mastery. In the case of acute illness, control may be of less importance given the temporary nature of the circumstances. In chronic disease, however, the illness experience becomes integrated into the core features of personal identity, and issues surrounding control become paramount.

For many MG patients, control is an important consideration from the very beginning of their illness experience. The initial recognition of disease symptoms is frequently associated with acute distress regarding the etiology of the difficulties and the possible concern that the worst may be "unfolding". This process is amplified by the fact that the initial symptoms of many neuroimmune diseases such as MG, multiple sclerosis, and systemic lupus erythematosus are nebulous and fluctuating signs that are difficult for patients to describe, and even more difficult to associate with a specific disease. It is not surprising, therefore, that many patients are evaluated more than once before receiving a diagnosis of MG *(38)*. The length of time until the diagnosis may be long for some patients (3 years or more in certain cases *(38)*), and approximately one-third of patients are initially misdiagnosed with a psychiatric condition *(16, 38)*. Most patients do not experience these difficulties, but for those who do, feelings of inadequate control and emotional distress can be significant. It should not be surprising that some patients experience a reduction in anxiety and fear after receiving their diagnosis *(7)*.

The unpredictable and fluctuating course of MG represents another area where a person's sense of control can be challenged. Symptoms of many neuroimmune diseases fluctuate, with episodes of remission and exacerbation. This pattern of disease produces a sense of uncertainty regarding one's health and well-being. It is difficult for individuals to feel in control of their lives when they have limited ability to predict their general health. Planning of family events, work responsibilities, and personal activities becomes complicated by the uncertainty of one's physical state. For some patients, the possibility of acute respiratory distress represents a source of concern. This may be especially true for patients who travel frequently or for individuals who do not reside in close proximity to medical care. We know of one patient who took the necessary steps to ensure that her local ambulance service was educated about the medical aspects of MG, including the possibility of respiratory crisis.

Finally, the lack of control that patients have in the treatment of their disease is difficult to manage from a mental health perspective. The fact that treatment is administered and managed by an individual other than the patient is of course necessary but also potentially distressing. Even when the patient recognizes the expertise of the treating physician, there is an inherent desire to participate in the decision-making process, and for this reason many individuals greatly appreciate receiving information about treatment options throughout the course of their disease. Since MG is not a common disease, it is also possible that nonspecialist nurses and other critical care support staff will be less familiar with MG compared to clinicians working in neuromuscular clinics and this can serve as a source of concern for some patients.

5.2. Uncertainty in Illness

A hallmark characteristic of chronic illness is the cognitive experience of uncertainty. The unpredictable, variable nature of many chronic illnesses, in conjunction with complex and often intrusive and painful treatment regimens, combine to create such an appraisal context. As a construct, illness uncertainty has been defined as a cognitive experience elicited in situations in which the meaning of illness-related events are unclear and outcomes are unpredictable *(39)*. Mishel *(39)* conceptualized the overarching construct of illness uncertainty as being comprised of four distinct subcomponents, including ambiguity regarding the cues and the state of the illness, unpredictability of illness course and outcomes, complexity regarding the treatment and the health care system, and lack (or inconsistency) of information regarding the illness or treatments.

An extensive literature now exists that consistently demonstrates a robust association between perceptions of uncertainty and psychological distress during diagnosis, treatment, and stabilization periods of an illness (e.g., *40, 41*) across a wide array of adult illness groups, including myocardial infarction *(42)*, multiple sclerosis *(43)*, and cancer *(44, 45)*, among others. Collectively, these findings suggest that perceptions of uncertainty may be a critical cognitive factor that place individuals with chronic illness at increased risk for experiencing poor adaptation to their illness or experiencing clinically significant psychological distress. Importantly, uncertainty also represents a potential target for intervention. Psychosocial interventions that target uncertainty among adults with cancer have demonstrated improvements in adaptation to the illness post-intervention *(46, 47, 48, 49, 50)*. It would certainly appear that this construct is quite relevant to individuals with MG; however, to date, no research has been conducted that has addressed perceptions of uncertainty, or attempts to cope with uncertainty, in this population.

5.3. Illness Intrusiveness

As mentioned previously, quality of life for the individual with MG is potentially compromised by fatigue and the associated restrictions on activity that subsequently occur. Devins et al. *(51)* has referred to such illness-induced lifestyle disruptions as illness intrusiveness. Specifically, illness intrusiveness is defined as both the objective and perceived intrusiveness of an illness resulting from "illness-induced barriers" which may restrict engagement in activities and interests of value to the individual. Hypothetically, intrusiveness influences quality of life and adjustment outcomes in two ways. First, there are reduced opportunities for obtaining reinforcing outcomes as a function of being unable to engage in desired, meaningful activities. Second, reductions in personal control occur as the individual is limited in their ability to obtain positive outcomes or avoid negative outcomes *(52)*. Further, it is hypothesized that disease and treatment factors exert their influence on quality of life by indirectly influencing illness intrusiveness, which in turn exerts a direct influence on adjustment outcomes. From a conceptual framework, illness intrusiveness and perceived control are conceptually related but distinct constructs, which has been borne out in studies of patients with end-stage renal disease *(51)*.

A large body of research across a wide range of both adult and pediatric chronic health conditions provides substantial support for the illness intrusiveness model. Both disease (e.g., disease severity, fatigue) and treatment variables (e.g., type of treatment and time required for treatment) are strongly associated with illness intrusiveness *(53, 54)*. Further, increased levels of illness intrusiveness have been associated with a variety of specific and global adjustment outcomes (e.g., depressive symptoms, self-esteem, marital satisfaction) across multiple chronic health conditions and diseases *(55)*. A number of factors also appear to moderate the illness intrusiveness–adjustment outcome relationship. Such variables include the age or stage in the life cycle of the individual *(56)* as the stigma of the illness itself *(57)*.

Thus, the construct of illness intrusiveness would appear quite relevant to the experience of the individual with MG. To the extent that valued activities are restricted by fatigue or embarrassment, these individuals may be at higher risk for a variety of adjustment problems and reduced quality of life.

5.4. Social Support

The fourth major factor that influences psychological health in chronic disease is social support. Social networks consisting of family, friends, and colleagues provide an important buffer to the psychological challenges associated with any stressful experience *(58, 59, 60, 61)*. A series of studies have demonstrated that individuals with greater levels of social support experience better coping and adjustment in the face of stressors. This effect is due to the influence of social support on coping styles. Adequate social support contributes to "adaptive" coping strategies that involve enacting a plan to reduce personal distress. By contrast, the absence of social support is associated with reliance on avoidant coping strategies, which represents a less-effective coping response *(62)*.

Studies have shown that social support influences patients' sense of personal control and overall well-being. Rheumatoid arthritis patients who report inadequate support experience feelings of limited control and competence in coping with their condition. In turn, these patients report greater levels of depression and poorer life satisfaction *(63)*. Blixen and Kippes *(64)* have shown that satisfaction with social support is strongly associated with better overall quality of life in patients with osteoarthritis despite elevated symptoms of depression, pain, and discomfort. These results suggest that social support can moderate the severity of aversive symptoms, which in turn serves to preserve life satisfaction.

Like many chronic diseases, the sources of social support for MG patients predominately exist through family and friends, but additional support is available through services provided by national organizations. Regional chapters of these organizations sponsor events throughout the year, including regular meetings that provide a forum for patients to discuss concerns and share insight about experiences and effective compensatory strategies. Additional resources include newsletters that provide patients the opportunity to keep abreast of research, clinical care issues, and regional social functions. A third opportunity for social support is available online. A MG Internet "chat room" provides patients who are located thousands of miles away the opportunity to communicate and share experiences. The obvious advantages of the Internet are that patients have immediate access to other individuals with similar disease experiences and contact can be established without travel. As such, the social relationships that are available through the MG chat room may serve especially important functions for individuals with more severe disease and disability.

6. SUMMARY

In this chapter we reviewed the psychological challenges associated with MG as well as the factors that support effective adjustment to the condition. We concluded that there is no empirical evidence to support the hypothesis that psychological distress affects the onset or the course of MG. We also found that despite 40 years of research, our understanding of the prevalence of psychiatric comorbidity in MG is underdeveloped. Current data indicate that most MG patients adjust well to the disease, with generally good levels of mental health and preserved quality of life. However, these conclusions are based on a limited number of patients who volunteered for nonmedical research studies. It is possible that greater psychological distress exists among clinic patients who do not participate in these types of studies. Clearly, comprehensive studies are needed to better characterize the psychological health of individuals with MG.

We reviewed the controversial issue of cognitive impairment in MG. Slightly more than one-half of MG patients complain of memory loss that they believe is associated with their disease; yet, there is no empirical evidence that MG affects the CNS. The vast majority of studies include too many methodological problems to sufficiently address the issue of cognition in MG. Nevertheless, some preliminary data suggest that a subset of patients experience subtle difficulties with information processing and learning of information. Mood, medications, and visual disturbances do not readily account for these effects, but the impact of mental fatigue appears important. Future studies are needed to better clarify the effect of fatigue on cognitive performance in this population.

Finally, we examined the factors that are believed to foster good psychological adjustment to MG. Literature on other chronic diseases reveals that feelings of control over one's physical condition, managing the uncertainty and intrusiveness inherent in the illness, and having adequate social support networks engender psychological health in patients with chronic disease. For MG patients, concerns surrounding perceived control may begin with the initial recognition of symptoms and may continue if they experience delay before diagnosis and the initiation of treatment. Control continues to be important throughout the course of the disease, given the fluctuating nature of MG and the necessary dependence on the medical community. To the extent that patients can learn to exert control over circumstances that are indeed controllable, and cope with those situations that are not, quality of life may be enhanced. Further, learning to actively manage perceptions of uncertainty may also result in greater quality of life. In addition to perceived control, social support is an important determinant of psychological health. Adequate social support fosters adaptive coping methods, which in turn promotes psychological well-being and preserved quality of life.

Clinicians are in a unique position to positively influence psychological health in MG. Collaborative relationships that include review of diagnostic methods, treatment options, and advances in the scientific understanding of MG can provide patients with an important sense of involvement in their medical care. Clinicians are also in an ideal position to further develop the social support network of patients. This can be done informally by posting fliers in the waiting room that announce patient-based meetings and events or more formally by serving as the link between the patient and the resources available through the charitable organizations involved with MG. Such efforts are effective adjuncts to patient care, with the primary focus of maintaining the psychological health of patients. Preliminary evidence that many MG patients cope well with their disease and experience good quality of life signifies the importance of these efforts. Extending these studies to clinic-based populations represent an important direction of future research.

REFERENCES

1. Phillips LH. The epidemiology of myasthenia gravis. Ann NY Acad Sci 2003;998, 407–412.
2. Sherwood L. Human Physiology: From Cells to Systems. Second Edition. West Publishing Company: New York, 1993.
3. Blalock JE, Smith EM. A complete regulatory loop between the immune and neuroendocrine systems. Fed Proc 1985;44:108–111.
4. Maes M, Bosmans E, Meltzer HY. Immunoendocrine aspects of major depression. Relationships between plasma interleukin-6 and soluble interleukin-2 receptor, prolactin and cortisol. Eur Arch Psychiatry Clin Neurosci 1995;245:172–178.
5. Psychosocial factors predict CD and viral load change in men and women with human immunodeficiency virus in the era of highly active antiretroviral treatment. Ironson G, O'Cleirigh C, Fletcher MA, Laurenceau JP, Balbin E, Klimas N, Schneiderman N, Solomon G. Psychosom Med 2005;1013–1021.
6. Meyer E. Psychological disturbances in myasthenia gravis: A predictive study. Ann NY Acad Sci 1966;135:417–423.

7. Martin RD, Flegenheimer WV. Psychiatric aspects of the management of the myasthenic patient. Mt Sinai J Med 1971;38:594–601.

8. Shinkai K, Ohmori O, Ueda N, Nakamura J, Amano T, Tsuji S. A case of myasthenia gravis preceded by major depression. JNeuropsychiatry Clin Neurosci 2001;13:116–117.

9. Chafetz ME. Psychological disturbances in myasthenia gravis. Ann NY Acad Sci 1966;135:424–427.

10. MacKenzie KR, Martin MH, Howard FM. Myasthenia gravis: psychiatric concomitants. Canad Med Assn J 1969;100:988–991.

11. Graham JR. MMPI-2. Assessing Personality and Psychopathology. Second Edition. Oxford University Press: New York, 1993.

12. Schwartz ML, Cahill R. Psychopathology associated with myasthenia gravis and its treatment by psychotherapeutically oriented group counseling. JChron Dis 1971;24:543–552.

13. Santy PA. Underdiagnosed myasthenia gravis in emergency psychiatric referrals. Ann Emerg Med 1983;12:397–398.

14. Magni G, Micaglio G, Ceccato MB, Lalli R, Bejato L, Angelini C. The role of life events in the myasthenia gravis outcome: a one-year longitudinal study. Acta Neurol Scand 1989;79:288–291.

15. Kleinman A. The Illness Narratives. Basic Books: New York, NY, 1988.

16. Paul RH, Cohen RA, Gilchrist J, Goldstein J. Fatigue and its impact on patients with myasthenia gravis. Muscle Nerve 2000;23:1402–1406.

17. Sneddon J. Myasthenia gravis: a study of social, medical, and emotional problems in 26 patients. Lancet 1980;1:526–528.

18. Paradis CM, Friedman S, Lazar RM, Kula RW. Anxiety disorders in a neuromuscular clinic. Am J Psychiatry 1993;150:1102–1104.

19. Magni G, Micaglio GF, Lalli R, Bejato L, Candeago MR, Merskey H, Angelini C. Psychiatric disturbances associated with myasthenia gravis. Acta Psychiatr Scand 1988;77:443–445.

20. Paul RH, Cohen RA, Goldstein J, Gilchrist J. Severity of mood, self-evaluative and vegetative symptoms in myasthenia gravis. JNeuropsych Clin Neurosci 2000;12:499–501.

21. Paul RH, Nash JM, Cohen RA, Gilchrist JM, Goldstein JM. Quality of life and well-being of patients with myasthenia gravis. Muscle Nerve 2001;24:512–516.

22. Nyenhuis DL, Rao SM, Zajecka JM, Luchetta T, Bernardin L, Garron DC. Mood disturbance versus other symptoms of depression in multiple sclerosis. JINS 1995;1:291–296.

23. Brown ES, Suppes T. Mood symptoms during corticosteroid therapy: a review. Harv Rev Psychiatry 1998;5:239–246.

24. Padua L, Evoli A, Aprile I, Caliandro P, Mazza S, Padua R, Tonali P. Health-related quality of life in patients with myasthenia gravis and the relationship between patient-oriented assessment and conventional measurements. Neurol Sci 2001;22:363–369.

25. Rostedt A, Padua L, Stalberg EV. Correlation between regional myasthenic weakness and mental aspects of quality of life. Europ J Neurol 2006;13:191–193.

26. Paul RH, Cohen RA, Zawacki T, Gilchrist JM, Aloia MS. What have we learned about cognition in myasthenia gravis? A review of methods and results. Neurosci Bio Behavior Rev 2001;25:75–81.

27. Whiting J, Cooper J, Lindstrom JM. Antibodies in sera from patients with myasthenia gravis do not bind to nicotinic acetylcholine receptors from human brain. JNeuroimmunol 1987;16:205–213.

28. Keesey JC. Does myasthenia affect the brain? J Neurol Sci 1999;170: 77–89.

29. Ochs C, Bradley J, Katholi C et al. Symptoms of patients with myasthenia gravis receiving treatment. J Medicine 1988;29:1–12.

30. Paul RH, Cohen RA, Gilchrist J, Aloia MS, Goldstein JM. Cognitive dysfunction in individuals with myasthenia gravis. JNeurol Sci 2000;179:59–64.

31. Grohar-Murray ME, Becker A, Reilly S, Ricci M. Self-care actions to manage fatigue among myasthenia gravis patients. JNeurosci Nurs 1998;30:191–199.

32. Steptoe A, Bolton J. The short-term influence of high and low intensity physical exercise on mood. Psychol Health 1988; 2:91–106.

33. Fiske AD, Schneider W. Controlled and automatic processing during tasks requiring sustained attention: a new approach to vigilance. Human Factors 1981;23:737–750.

34. Paul RH, Cohen RA, Gilchrist J. Ratings of subjective mental fatigue relates to cognitive performance in patients with myasthenia gravis. JClin Neurosci 2002;9:243–246..

35. Paul R, Beatty WW, Schneider R, Blanco CR, Hames KA. Cognitive and physical fatigue in multiple sclerosis: relations between self-report and objective performance. Applied Neuropsychology 1998;5:143–148.

36. Krupp LB, Elkins LE. Fatigue and declines in cognitive functioning in multiple sclerosis. Neurology 2000;10:934–939.

37. House JS, Landis KR, Umberson D. Social relationships and health. Science 1988;241:540–544.

38. Nicholson GA, Wilby J, Tennant C. Myasthenia gravis: the problem of a "psychiatric" misdiagnosis. Med J Australia 1986;144:632–638.

39. Mishel MH. Reconceptualization of the uncertainty in illness theory. JNurs Sch, 1990;22:256–262.

40. Mishel MH. Perceived uncertainty and stress in illness. Res Nurs Health 1984;7:163–171.

41. Mishel MH, Braden CJ. Uncertainty: a mediator between support and adjustment. West J Nurs Res 1987;9:43–57.

42. Bennett SJ. Relationships among selected antecedent variables and coping effectiveness in postmyocardial infarction patients. Res Nurs Health 1993;16:131–139.

43. Mullins LL, Cote MP, Fuemmeler BF, Jean VM, Beatty WW, Paul RH. Illness intrusiveness, uncertainty, and distress in individuals with multiple sclerosis. Rehabil Psychol 2001;46:139–153.

44. Mishel MH, Sorenson DS. Uncertainty in gynecological cancer: a test of the mediating functions of mastery and coping. Nurs Res 1991;40:167–171.

45. Mast ME. Adult uncertainty in illness: a critical review of research. Sch Inq Nurs Pract: Int J 1995;9:3–23.

46. Braden CJ, Mishel MH, Longman AJ, Burns LR. Self-help intervention project: women receiving treatment for breast cancer. Cancer Pract 1998;6:87–98.

47. Badger TA, Braden C, Mishel MH. Depression burden, self-help interventions, and side effect experience in women receiving treatment for breast cancer. Oncol Forum 2001;28:567–574.

48. Mishel M, Belyea M, Germino BB, Stewart JL, Bailey DE, Robertson, Mohler JLC. Helping patients with localized prostate cancer manage uncertainty and treatment side effects: nurse delivered psycho-educational intervention via telephone. Cancer 2002;94(6): 1854–1866.

49. Mishel MH, Germino BB, Belyea M, Stewart JL, Bailey DE, Mohler J, Robertson C. Moderators of outcomes from an uncertainty management intervention for men with localized prostate cancer. Nurs Res 2003;52(2): 89–97.

50. Mishel MH, Germino BB, Gil KM, Belyea M, LaNey IC, Stewart J, Porter L, Clayton M. Benefits from an uncertainty management intervention for African-American and Caucasian older long-term breast cancer survivors. Psychooncology 2005;14:962–978.

51. Devins GM, Binik YM, Hutchinson TA, Hollomby DJ, Barre PE, Guttmann RD. The emotional impact of end-stage renal disease: importance of patients' perceptions of intrusiveness and control. Int J Psychiatry Med 1983–1984;13:327–343.

52. Devins GM. Illness intrusiveness and the psychosocial impact of lifestyle disruptions in chronic life-threatening disease. Adv Ren Replace Ther 1994;1:251–263.

53. Devins GM, Seland TP, Klein GM, Edworthy SM, Saary MJ. Stability and determinants of psychosocial well-being in multiple sclerosis. Rehabil Psycho 1993;38:11–26.

54. Devins GM. Illness intrusiveness and the psychosocial impact of end-stage renal disease. In MA Hadry, J Kiernan, AH Kutscher, L Cahill, AI Bevenitsky eds. Psychosocial Aspects of End-Stage Renal Disease: Issues of Our Times. Haworth Press: New York, 1991, 83–102

55. Devins GM, Shnek ZM. Multiple sclerosis. In: RG Frank, TR Elliott eds. Handbook of Rehabilitation Psychology. American Psychological Association: Washington DC, 2000, 163–184.

56. Devins GM, Beanlands H, Mandin H, Paul LC. Psychological impact of illness intrusiveness moderated by self-concept and age in end-stage renal disease. Health Psycho 1997;16:529–538.

57. Devins GM, Stam HJ, Koopmans JP. Psychosocial impact of laryngectomy mediated by perceived stigma and illness intrusiveness. Can J Psychiatry 1994;39:608–616.

58. Hough ES, Brumitt GA, Templin TN. Social support, demands of illness, and depression in chronically ill urban women. Health Care Women Int 2000;20:349–362.

59. Cott CA, Gignac MAM, Badley EM. Determinants of self-rated health for Canadians with chronic disease and disability. J Epidemiol Community Health 1999;53:731–736.

60. Brown GW, Andrews B, Harris T, Adler Z, Bridge L. Social support, self esteem and depression. Psychol Med 1986;6:238–247.

61. Holahan CK, Holahan CJ. Self-efficacy, social support, and depression in aging: a longitudinal analysis. J Gerontol: Psychol Sci 1987;42:65–68.

62. Schreurs KMG, de Ridder DT. Integration of coping and social support perspectives: implications for the study of adaptation to chronic diseases. Clin Psychol Rev 1997;17:89–112.

63. Smith CA, Dobbins CJ, Wallston KA. The mediational role of perceived competence in psychological adjustment to rheumatoid arthritis. J Appl Soc Psychol 1991;21:1218–1247.

64. Blixen CE, Kippes C. Depression, social support, and quality of life in older adults with osteoarthritis. Image J Nurs Sch 1999;31:221–226.

18

Myasthenia Gravis: Classification and Outcome Measurements

Gil I. Wolfe and Richard J. Barohn

1. INTRODUCTION

Myasthenia gravis is the best characterized and understood autoimmune disease of the nervous system *(1)*. It is the most common neuromuscular transmission disorder followed in neurology clinics. Nevertheless, standardized classification and grading systems and outcome measures for MG were not developed until the last decade and continue to be refined. In 1997 the Medical Scientific Advisory Board (MSAB) of the Myasthenia Gravis Foundation of America (MGFA) formed a task force to design classification and outcome measures for the disease *(2)*. The task force's primary charge was to create a system of uniformity in recording and reporting clinical data and outcomes research. Chairing the project was Dr. Alfred Jaretzki III, a cardiothoracic surgeon who had published extensively with colleagues at Columbia Presbyterian Medical Center on thymectomy for MG *(3, 4)*. He expressed concern that published outcomes following thymectomy could not be reliably compared across studies due to the lack of universally accepted measures. It was soon recognized that essentially all MG clinical research suffered from this same dilemma.

After 3 years of regular meetings, literature reviews, and input from MG experts, a consensus document on recommendations for MG clinical research standards was published in 2000 *(5)*. The essential elements of the task force's recommendations are summarized in this chapter, supplemented by other clinical measures for MG that have been developed and utilized in recent studies.

2. CLINICAL CLASSIFICATION OF MG

Clinicians have struggled to devise an adequate classification system for MG for a half-century. In 1958, Osserman and colleagues *(6, 7)* proposed placing patients in five groups: I, localized (ocular); II, generalized (mild or moderate); III, acute fulminating; IV, late severe; and V, muscle atrophy. A separate category for neonatal and juvenile forms was created. Osserman and Genkins *(8)* later divided group II into A (mild) and B (moderate) subclassifications and dropped the muscle atrophy group. Various modified Osserman criteria were suggested over the years and some new schemes were developed *(4, 9, 10, 11, 12, 13)*. The most widely used modification of Osserman's original classification is found in Table 1. However, the Osserman approach has been criticized over the years for several shortcomings. These include vague descriptive terminology, the potential for subjects to satisfy descriptions for more than one class at the same point of time, and the absence of a category for asymptomatic patients.

It became clear early on to the MGFA task force that they would have to tackle the difficult topic of MG classification, even though it did not bear directly on research outcome measures. The term "class" is used for the five broad disease subdivisions, and the descriptors "mild, moderate, and severe" borrowed from prior classifications with full realization of the subjective nature of such terminology (Table 2) *(5)*. For instance, two experienced MG clinicians might

From: *Current Clinical Neurology*
Myasthenia Gravis and Related Disorders
Edited by: H.J. Kaminski, DOI 10.1007/978-1-59745-156-7_18, © Humana Press, New York, NY

Table 1
Modified Osserman Classification for MG

Group 1: ocular
Group 2: mild generalized
Group 3: moderate to severe generalized
Group 4: acute, severe, developing over weeks to months
Group 5: late, severe with marked bulbar involvement

Table 2
MGFA Clinical Classification *(5)*

Class I	Any ocular muscle weakness; may have weakness of eye closure. All other muscle strength is normal.
Class II	Mild weakness affecting muscles other than ocular muscles; may also have ocular muscle weakness of any severity.
	IIa. Predominantly affecting limb, axial muscles, or both; may also have lesser involvement of oropharyngeal muscles.
	IIb. Predominantly affecting oropharyngeal, respiratory muscles, or both; may also have lesser or equal involvement of limb, axial muscles, or both.
Class III	Moderate weakness affecting muscles other than ocular muscles; may also have ocular muscle weakness of any severity.
	IIIa. Predominantly affecting limb, axial muscles, or both; may also have lesser involvement of oropharyngeal muscles.
	IIIb. Predominantly affecting oropharyngeal, respiratory muscles, or both; may also have lesser or equal involvement of limb, axial muscles, or both.
Class IV	Severe weakness affecting muscles other than ocular muscles; may also have ocular muscle weakness of any severity.
	IVa. Predominantly affecting limb, axial muscles, or both; may also have lesser involvement of oropharyngeal muscles.
	IVb. Predominantly affecting oropharyngeal, respiratory muscles, or both; may also have lesser or equal involvement of limb, axial muscles, or both.
Class V	Defined as intubation, with or without mechanical ventilation, except when employed during routine postoperative management. The use of a feeding tube without intubation places the patient in class IVb.

not always agree on whether a patient has mild or moderate generalized disease (class II or III). A brief summary of the MGFA clinical classification follows. A pure ocular MG patient is denoted as class I; there can be orbicularis oculi weakness, but any additional axial, limb, or oropharyngeal weakness placed the patient in classes II–V. The task force appreciated that some patients with MG have selective bulbar muscular weakness; therefore, separate distinctions for classes II through IV are designated "a" if weakness is predominantly limb/axial or "b" if weakness is predominantly oropharyngeal/respiratory. A patient requiring intubation is class V. Use of a feeding tube without intubation places the patient in class IVb. The most severely affected muscle groups should guide class assignment, and a "maximum severity designation" denotes the most severe pretreatment classification, which can be used as a historical point of reference. It is not intended for the clinical classification to be used as an outcome measure. Its primary purpose is to identify subgroups of patients who share similar clinical features. The classification is being used in several ongoing clinical trials including a federally funded study of thymectomy in non-thymomatous MG *(14)*, and was utilized in a 19-center Japanese survey of disease outcome *(15)*.

3. QUANTITATIVE MG SCORE (QMG)

The quantitative MG score (QMG) is the best studied objective outcome measure in MG. The current QMG is an expansion and modification of the scale first developed by Bessinger and colleagues in the early 1980s *(10, 16)*. The original QMG consisted of eight items, each graded

Table 3
Quantitative MG Score (QMG)

Test items Weakness (score)	None (0)	Mild (1)	Moderate (2)	Severe *(4)*	Item score (0,1,2, or 3)
1. Double vision on lateral gaze right or left (**circle one**), seconds	61	11–60	1–10	Spontaneous	
2. Ptosis (upward gaze), seconds	61	11–60	1–10	Spontaneous	
3. Facial muscles	Normal lid closure	Complete, weak, some resistance	Complete, without resistance	Incomplete	
4. Swallowing 4 oz./ 120 ml water	Normal	Minimal coughing or throat clearing	Severe coughing/ choking or nasal regurgitation	Cannot swallow (test not attempted)	
5. Speech following counting aloud from 1–50 (onset of dysarthria)	None at #50	Dysarthria at #30–49	Dysarthria at #10–29	Dysarthria at #9	
6. Right arm outstretched (90° sitting), seconds	240	90–239	10–89	0–9	
7. Left arm outstretched (90° sitting), seconds	240	90–239	10–89	0–9	
8. Vital capacity (% predicted) mouthpiece or facemask (**circle one; best of 3**)	≥80%	65–79%	50–64%	< 50%	
9. Right hand grip: (**best of 2**)					
Male	≥45	15–44	5–14	0–4	
(kg W) Female	≥30	10–29	5–9	0–4	
10. Left hand grip: (**best of 2**)					
Male	≥35	15–34	5–14	0–4	
(kg W) Female	≥25	10–24	5–9	0–4	
11. Head, lifted (45° supine), seconds	120	30–119	1–29	0	
12. Right leg outstretched (45° supine), seconds	100	31–99	1–30	0	
13. Left leg outstretched (45° supine), seconds	100	31–99	1–30	0	
			Total QMG score (range 0–39)		

0–3, with a score of 3 being the most severe. Tindall et al. expanded the scale to 13 items and used this version of the QMG as the primary efficacy measurement in two trials that determined that cyclosporine was effective therapy for MG *(17, 18)*. Barohn et al. *(19)* replaced three items in Tindall's scale that were subjective in nature – facial muscles, chewing, and swallowing – so that each item on the QMG was scored objectively (Table 3). Interrater reliability testing was performed on this scale in advance of a randomized, placebo-controlled study of intravenous gammaglobulin in MG that utilized the modified QMG as a primary outcome measure *(20)*. At a 95% confidence level, QMG scores did not differ by more than 2.63 units, translating to a required sample size of 17 patients per treatment arm in a placebo-controlled trial to detect a significant difference at a power of 0.80 *(19)*.

Tindall's version of the QMG was found to have concurrent validity, meaning that in a single visit, the QMG, manual muscle testing, functional scores, and patient self-evaluation show good agreement *(21)*. The latest version of the QMG was recently tested for responsiveness, the QMG's sensitivity to clinical change versus "noise" that can be expected even in clinically unchanged subjects, as well as longitudinal construct validity, meaning how changes in the score correspond to clinically relevant change *(22)*. The calculated index of responsiveness of 1.45 is considered excellent for clinical outcome measures. The direction of QMG changes validly reflected gestalt impressions of patient status (improved, unchanged, or worse) and correlated tightly with changes in a manual muscle testing score ($p<0.0001$).

The MGFA task force recommended that the QMG be used in all prospective studies of therapy for MG *(5)*. In addition, it encouraged proposals to improve the QMG, including "weighting" of subscores derived from the total scale to reflect regional impairment seen in many MG patients, a point emphasized by other authors *(22)*. The QMG has been used in recent studies of mycophenolate mofetil, tacrolimus, and etanercept *(23, 24, 25, 26)*. An instructional training video and manual for the QMG is available from the MGFA *(27)*. The QMG can be completed in 20–30 min. The only specialized equipments required are a spirometer and a hand-held dynamometer.

4. MYASTHENIA GRAVIS MANUAL MUSCLE TEST (MG-MMT)

Investigators at Duke University Medical Center have developed a disease-specific manual muscle test that can be performed at the bedside without specialized equipment (Table 4) *(28)*. Thirty muscle groups usually affected by MG (6 cranial nerve/24 axial limb) are measured on a 0–4 scale (0 = normal; 1 = 25% weak/mild impairment; 2 = 50% weak/moderate impairment; 3 = 75% weak/severe impairment; 4 = paralyzed/unable to do). The MG-MMT demonstrates good interrater reliability with a mean difference between scores of 1.3 ± 1.8 points. It correlates well with the QMG *(22, 28)*. However, there is wide scatter of MG-MMT values within general disease classifications, an issue also observed with the QMG (Wolfe et al., unpublished data). Advantages of the MG-MMT over the QMG are that it can be performed by the physician as part of a routine clinic visit, takes less time, and requires no specialized equipment. Unlike the QMG, the MG-MMT does not directly assess swallowing, speech, or respiratory function. The MG-MMT has been used in open-label or retrospective studies of mycophenolate mofetil and etanercept in MG *(23, 26, 29)* and awaits further use in prospective clinical trials.

5. MYASTHENIC MUSCLE SCORE (MMS)

This scoring system has been used by French investigators and was developed by Gajdos and colleagues in 1983 *(30)*. The MMS summates nine independent functions that encompass cranial, neck, truncal, and limb strength (Table 5). The total score ranges between 0 and 100. Unlike the scales described earlier, a higher score on the MMS connotes better strength and function.

Table 4
Myasthenia Gravis Manual Muscle Test

	Right	Left	Sum
Lid ptosis	—	—	—
Diplopia	—	—	—
Eye closure			—
Cheek puff			—
Tongue protrusion			—
Jaw closure			—
Neck flexion			—
Neck extension			—
Shoulder abduction (deltoid)	—	—	—
Elbow flexion (biceps)	—	—	—
Elbow extension (triceps)	—	—	—
Wrist extension	—	—	—
Grip	—	—	—
Hip flexion (iliopsoas)	—	—	—
Knee extension (quadriceps)	—	—	—
Knee flexion (hamstrings)	—	—	—
Ankle dorsiflexion	—	—	—
Ankle plantar flexion	—	—	—
Total score			═

Score each function as follows: 0, normal; 1, 25% weak/mild impairment; 2, 50% weak/ moderate impairment; 3, 75% weak/severe impairment; 4, paralyzed/unable to do. In addition, record any condition other than MG causing weakness in any of these muscles.

Pulmonary function is not assessed in the MMS. The MMS has high interrater reliability and showed slightly higher agreement between observers than the QMG in one study *(31)*. This study also determined that the high interobserver agreement was not dependent on a significant degree on any single item, suggesting a high reliability for individual constituents of both the MMS and the QMG. Both the MMS and the QMG correlated well with a five-grade functional disease classification. Less robust was the correlation with a patient self-evaluation tool *(21, 31)*.

6. MG ACTIVITY OF DAILY LIVING PROFILE (MG-ADL)

The MG-ADL is a simple eight-point questionnaire which focuses on common symptoms reported by MG patients (Table 6) *(32)*. It was developed at the University of Texas Southwestern Medical Center to complement the QMG. Each item of the profile is graded 0 (normal) to 3 (most severe). A nurse, technician, or physician asks the patient the eight questions and records the responses. The MG-ADL correlates well with the QMG and can serve as a secondary efficacy measurement in clinical trials *(32)*. No specialized training is required, and it can be administered in less than 10 min.

The MG-ADL has been applied widely in retrospective and prospective clinical studies over the last few years *(14, 20, 33, 34)*. It is worthy to note that the MG-ADL is not a quality of life (QOL) scale, and the MGFA task force recommended that a disease-specific QOL instrument be developed for MG *(5)*.

7. FATIGUE TESTING

Quantification of static fatigue during isometric strength testing has been studied in diseases such as multiple sclerosis and amyotrophic lateral sclerosis, but surprisingly has received little attention in MG. We recently performed static fatigue testing (SFT) in 77

Table 5
Myasthenic Muscle Score

Task	Score
Maintain upper limbs horizontally outstretched: 1 point per 10 s	0–15
Maintain lower limbs above bed plane while lying on back: 1 point per 5 s	0–15
Raise head above bed plane while lying on back	
Against resistance	10
Without resistance	5
Impossible	0
Sit up from lying position	
Without help of hands	10
Impossible	0
Extrinsic ocular musculature	
Normal	10
Ptosis	5
Double vision	0
Eyelid occlusion	
Complete	10
Mild weakness	7
Incomplete with corneal covering	5
Incomplete without corneal covering	0
Chewing	
Normal	10
Weak	5
Impossible	0
Swallowing	
Normal	10
Impaired without aspiration	5
Impaired with aspiration	0
Speech	
Normal	10
Nasal	5
Slurred	0
	Total score

MG patients by comparing the maximum force generated in the initial 2–7 s time epoch with the 25–30 s epoch *(35)*. Our initial results suggest SFT correlates better with decremental responses on repetitive stimulation than either the QMG or the MG-ADL. However, further study of this tool is necessary before it can be recommended for use in clinical trials.

8. MGFA THERAPY STATUS

The MGFA task force developed a system to describe treatment regimens for MG patients enrolled in clinical studies (Table 7). The various abbreviations may be used individually or in combination to summarize management. In addition, the duration of treatment, current and prior doses of pertinent medications, and schedule of plasma exchange or intravenous gammaglobulin infusions should be recorded.

Table 6
MG Activities of Daily Living (MG-ADL) Profile

Grade	0	1	2	3	Score (0,1,2, or 3)
1. Talking	Normal	Intermittent slurring or nasal speech	Constant slurring or nasal, but can be understood	Difficult to understand speech	
2. Chewing	Normal	Fatigue with solid food	Fatigue with soft food	Gastric tube	
3. Swallowing	Normal	Rare episode of choking	Frequent choking necessitating changes in diet	Gastric tube	
4. Breathing	Normal	Shortness of breath with exertion	Shortness of breath at rest	Ventilator dependence	
5. Impairment of ability to brush teeth or comb hair	None	Extra effort, but no rest periods needed	Rest periods needed	Cannot do one of these functions	
6. Impairment of ability to arise from a chair	None	Mild, sometimes uses arms	Moderate, always uses arms	Severe, requires assistance	
7. Double vision	None	Occurs, but not daily	Daily, but not constant	Constant	
8. Eyelid droop	None	Occurs, but not daily	Daily, but not constant	Constant	
				MG-ADL score total (items 1–8)	=

Table 7
MGFA MG Therapy Status

NT	No therapy
SPT	Status post-thymectomy (record type of resection)
CH	Cholinesterase inhibitors
PR	Prednisone
IM	Immunosuppresive therapy other than prednisone (define)
PE(a)	Plasma exchange therapy, acute (for exacerbations or preoperatively)
PE(c)	Plasma exchange therapy, chronic (used on a regular basis)
IG(a)	IVIg therapy, acute (for exacerbations or preoperatively)
IG(c)	IVIg therapy, chronic (used on a regular basis)
OT	Other forms of therapy (define)

9. MGFA POSTINTERVENTION STATUS (PIS)

Assessment of postintervention status (PIS) designates the clinical state of MG patients at any time after initiation of treatment *(5)*. It can be used for routine follow-up or in formal clinical trials (Table 8). PIS should be determined by a clinician skilled in the evaluation of MG patients as it is based on careful clinical evaluation with both symptoms and signs taken into account. Of note, the MGFA task force concluded that isolated weakness of eyelid closure does not necessarily represent a sign of active disease and, therefore, this finding would not exclude patients from

Table 8
MGFA Postintervention Status

Complete stable remission (CSR)	The patient has had no symptoms or signs of MG for at least 1 year and has received no therapy for MG during that time. There is no weakness of any muscle on careful examination by someone skilled in the evaluation of neuromuscular disease. Isolated weakness of eye closure is accepted.
Pharmacologic remission (PR)	The same criteria as for CSR except that the patient continues to take some form of therapy for MG. Patients taking cholinesterase inhibitors are excluded from this category because their use suggests the presence of weakness.
Minimal manifestation (MM)	The patient has no symptoms or functional limitations from MG but has some weakness on examination of some muscles. This class recognizes that some patients who otherwise meet the definition of CSR or PR do have weakness that is only detectable by careful examination.

MM-0: The patient has received no MG treatment for at least 1 year.
MM-1: The patient continues to receive some form of immunosuppression but no cholinesterase inhibitors or other symptomatic therapy.
MM-2: The patient has received only low dose of cholinesterase inhibitors (< 120 mg pyridostigmine per day) for at least 1 year.
MM-3: The patient has received cholinesterase inhibitors or other symptomatic therapy and some form of immunosuppression during the past year.

Change in status

Improved (I)	A substantial decrease in pretreatment clinical manifestations or a sustained substantial reduction in MG medications as defined in the protocol. In prospective studies, this should be defined as a specific decrease in QMG score.
Unchanged (U)	No substantial change in pretreatment clinical manifestations or reduction in MG medications as defined in the protocol. In prospective studies, this should be defined in terms of a maximum change in QMG score.
Worse (W)	A substantial increase in pretreatment clinical manifestations or a substantial increase in MG medications as defined in the protocol. In prospective studies, this should be defined as a specific increase in QMG score.
Exacerbation (E)	Patients who have fulfilled criteria for CSR, PR, or MM but subsequently developed clinical findings greater than permitted by these criteria.
Died of MG (D of MG)	Patients who died of MG, of complications of MG therapy, or within 30 days after thymectomy. List the cause (see Morbidity and Mortality table)

qualifying for complete stable remission or pharmacologic remission. Patients receiving cholinesterase inhibitors were excluded from the pharmacologic remission category, however, since these agents may mask myasthenic symptoms and signs.

For prospective studies, it was recommended that PIS change of status categories (improved, unchanged, or worse) be linked to predetermined changes in the QMG score. In certain studies, the PIS may be the most important element in outcome determination. PIS classification is being used to guide corticosteroid dosing adjustments in prospective trials of mycophenolate mofetil and transsternal thymectomy in MG *(20)* and was included in long-term outcome assessment in open-label studies of tacrolimus *(25, 36)*.

REFERENCES

1. Vincent A, Palace J, Hilton-Jones D. Myasthenia gravis. Lancet 2001;357:2122–2128.
2. Barohn RJ. Standards of measurement in myasthenia gravis. Ann N Y Acad Sci 2003;998:432–439.
3. Jaretzki III A. Thymectomy for myasthenia gravis: analysis of the controversies regarding technique and result. Neurology 1997;48(suppl 5):S52–S63.

4. Jaretzki III A, Penn AS, Younger DS, Wolff M, Olarte MR, Lovelace RE, et al. "Maximal" thymectomy for myasthenia gravis. Results. J Thorac Cardiovasc Surg 1988;95:747–757.

5. Jaretzki III A, Barohn RJ, Ernstoff RM, Kaminski HJ, Keesey JC, Penn AS, et al. Myasthenia gravis: recommendations for clinical research standards. Neurology 2000;55:16–23.

6. Osserman KE. Clinical aspects. In: Osserman KE, editor. Myasthenia Gravis. New York: Grune & Stratton; 1958. pp. 79–80.

7. Osserman KE, Kornfeld P, Cohen E, Genkins G. Studies in myasthenia gravis: review of two hundred eighty-two cases at the Mount Sinai Hospital, New York City. Arch Int Med 1958;102:72–81.

8. Osserman KE, Genkins G. Studies in myasthenia gravis: review of a twenty-year experience in over 1200 patients. Mt Sinai J Med 1971;38:497–537.

9. Barohn RJ, Jackson CE. New classification system for myasthenia gravis (abstract). J Child Neurol 1994;9:205.

10. Besinger UA, Toyka KV, Heininger K, et al. Long-term correlation of clinical course and acetylcholine receptor antibody in patients with myasthenia gravis. Ann NY Acad Sci 1981;377:812–815.

11. Olanow CW, Wechsler AS, Sirotkin-Roses M, Stajich J, Roses AD. Thymectomy as primary therapy in myasthenia gravis. Ann NY Acad Sci 1987;505:595–606.

12. Wolfe GI, Barohn RJ. Neuromuscular junction disorders of childhood. In: Swaiman KF, Ashwal S, Ferriero DM, editors. Pediatric Neurology: Principles and Practice. 4th ed. Philadelphia: Mosby Elsevier; 2006. pp. 1941–1968.

13. Oosterhuis HJ. Observations of the natural history of myasthenia gravis and the effect of thymectomy. Ann N Y Acad Sci 1981;377:678–690.

14. Wolfe GI, Kaminski HJ, Jaretzki III A, Swan A, Newsom-Davis J. Development of a thymectomy trial in non-thymomatous myasthenia gravis patients receiving immunosuppressive therapy. Ann NY Acad Sci 2003;998:473–480.

15. Kawaguchi N, Kuwabara S, Nemoto Y, Fukutake T, Satomura Y, Arimura K, et al. Treatment and outcome of myasthenia gravis: retrospective multi-center analysis of 470 Japanese patients, 1999–2000. J Neurol Sci 2004;224:43–47.

16. Besinger UA, Toyka KV, Homberg M, Heininger K, Hohlfeld R, Fateh-Moghadam A. Myasthenia gravis: long-term correlation of binding and bungarotoxin blocking antibodies against acetylcholine receptors with changes in disease severity. Neurology 1983;33:1316–1321.

17. Tindall RSA, Phillips JT, Rollins JA, Wells L, Hall K. A clinical therapeutic trial of cyclosporine in myasthenia gravis. Ann NY Acad Sci 1993;681:539–551.

18. Tindall RSA, Rollins JA, Phillips JT, Greenlee RG, Wells L, Belendiuk G. Preliminary results of a double-blind, randomized, placebo-controlled trial of cyclosporine in myasthenia gravis. N Engl J Med 1987;316:719–724.

19. Barohn RJ, McIntire D, Herbelin L, Wolfe GI, Nations S, Bryan WW. Reliability testing of the Quantitative Myasthenia Gravis Score. Ann NY Acad Sci 1998;841:769–772.

20. Wolfe GI, Barohn RJ, Foster BM, Jackson CE, Kissel JT, Day JW, et al. Randomized, controlled trial of intravenous immunoglobulin in myasthenia gravis. Muscle & Nerve 2002;26:549–552.

21. Sharshar T, Chevret S, Mazighi M, Chillet P, Huberfeld G, Berreotta C, et al. Validity and reliability of two muscle strength scores commonly used as endpoints in assessing treatment of myasthenia gravis. J Neurol 2000;246:286–289.

22. Bedlack RS, Simel DL, Bosworth H, Samsa G, Tucker-Lipscomb B, Sanders DB. Quantitative myasthenia gravis score: assessment of responsiveness and longitudinal validity. Neurology 2005;64:1968–1970.

23. Meriggioli MN, Ciafaloni E, Al-Hayk KA, Rowin J, Tucker-Lipscomb B, Massey JM, et al. Mycophenolate mofetil for myasthenia gravis: an analysis of efficacy, safety, and toleratibility. Neurology 2003;61:1438–40.

24. Nagane Y, Utsugisawa K, Obara D, Kondoh R, Terayama Y. Eur Neurol 2005;53:146–150.

25. Ponseti JM, Azem J, Fort JM, Codina A, Montoro JB, Armengol M. Benefits of FK506 (tacrolimus) for residual, cyclosporin- and prednisone-resistant myasthenia gravis: one-year follow-up of an open-label study. Clin Neurol Neurosurg 2005;107:187–190.

26. Rowin J, Meriggioli MN, Tüzün E, Leurgans S, Christadoss P. Etanercept treatment in corticosteroid-dependent myasthenia gravis. Neurology 2004;63:2390–2392.

27. Barohn RJ. Video: how to administer the quantitative myasthenia test. Myasthenia Gravis Foundation of America, Inc. 1821 University Ave.W. Suite S256,St. Paul, MN 55104; 1996.

28. Sanders DB, Tucker-Lipscomb B, Massey JM. A simple manual muscle test for myasthenia gravis. Ann N Y Acad Sci 2003;998:440–444.

29. Ciafaloni E, Massey JM, Tucker-Lipscomb B, Sanders DB. Mycophenolate mofetil for myasthenia gravis: an open-label pilot study. Neurology 2001;56:97–99.

30. Gajdos P, Simon N, de Rohan Chabot P, Goulon M. Effets a long terme des echanges plasmatiques au cours de la myasthenie. Resultats d'une etude randomisee. Presse Med 1983;12:939–942.

31. Gajdos P, Sharshar T, Chevret S. Standards of measurement in myasthenia gravis. Ann N Y Acad Sci 2003;998:445–452.

32. Wolfe GI, Herbelin L, Nations SP, Foster B, Bryan WW, Barohn RJ. Myasthenia gravis activities of daily living profile. Neurology 1999;52:1487–1489.

33. Takamori M, Motomura M, Kawaguchi N, Nemoto Y, Hattori T, Yoshikawa Y, et al. Anti-ryanodine receptor antibodies and FK506 in myasthenia gravis. Neurology 2004;62:1894–1896.

34. Kawaguchi N, Yoshiyama Y, Nemoto Y, Munakata S, Fukutake T, Hattori T. Low-dose tacrolimus treatment in thymectomised and steroid-dependent myasthenia gravis. Curr Med Res Opinions 2004;20:1269–1273.
35. Herbelin LL, Nations SP, Wolfe GI, Foster BM, Bryan WW, Barohn RJ. The correlation between static fatigue testing and the quantitative myasthenia gravis score and activities of daily living profile (abstract). Neurology 2001;56:A63.
36. Ponseti JM, Azem J, Fort JM, López-Cano M, R V, Buera M, et al. Long-term results of tacrolimus in cyclosporine- and prednisone-dependent myasthenia gravis. Neurology 2005;64:1641–1643.

Index

Printed in the United States of America

DATE DUE

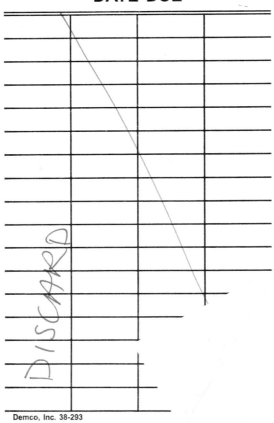

DISCARD